A TREATISE ON VITICULTURE

This edition is for
Pierre and Marietjie

and

To the Memory of

ABRAHAM IZAK PEROLD

20 October 1880 – 11 December 1941

A TREATISE ON VITICULTURE

BY

A. I. PEROLD, B.A., DR. PHIL.

PROFESSOR OF VITICULTURE AT THE UNIVERSITY OF STELLENBOSCH

First published in 1927

This paperback edition first published
in Great Britain in January 2012
by Inform & Enlighten
Copyright © 2012 Inform and Enlighten Ltd

ISBN: 978-0-9561523-2-9

**

Transcribed and edited by Peter F May

INFORM & ENLIGHTEN LTD
47 FONTMELL CLOSE
ST ALBANS AL3 5HU

PREFACE

THIS work is addressed particularly to readers in California, Australia and South Africa, for although there are several excellent works in French, German and Italian dealing with the science and practice of viticulture, there is no work that I am aware of in English that embraces the whole subject; and in preparing this work for the press I have endeavoured, however inadequately, to fill that gap.

This book is intended to serve both the student and the practical grape-grower. There are in it technical passages that will appeal more to the student, e.g. the chapters dealing with the biology of the vine, its external and internal morphology, the theory of grafting. My remarks on the practice of viticulture, such as those dealing with the propagation, manuring and pruning of the vine, the production of table grapes for export, will, it is hoped, assist the practical grape-grower as well as the student.

I should like to thank the following publishers who have kindly allowed me to reproduce certain illustrations: Messrs. Paul Parey, Berlin; Masson et Cie, Paris; Hachette et Cie, Paris; H. D. Tjeenk Willink en Zoon, Haarlem. And I wish to record here my thanks to Prof. Dr. K. Kroemer, of Geisenheim-a.-Rh., under whose kind and able guidance I studied the anatomy of the vine in 1907-1908 and made my drawings of the anatomical sections reproduced in this book.

A. I. PEROLD.

STELLENBOSCH,
    SOUTH AFRICA.

PREFACE TO THE 2012 EDITION

A TREATISE ON VITICULTURE has been out of print, since its first edition appeared in 1927, as a result of its printing plates being destroyed during bombing in the Second World War. The book is still listed on university reading lists, second hand copies are very expensive and remaining copies are being stolen from libraries. This edition makes Perold's Treatise available to all.

Abraham Izak Perold was one of the world's foremost experts in viticulture and winemaking and wrote more than eighty publications, none of which are still in print, with the exception of one pamphlet reprinted in 2011 as *A Year in Paarl*.[1]

This is a new reprint, not a photocopy, but the text has been aligned to match the original pagination. This was done so that references made elsewhere to a page in the Treatise will be valid in this edition.

Original photographs have been reproduced on the same page as in the original, with the exception of two colour plates, Fig 49 and Fig 78 which are shown on pages xii and xiii in this edition in monochrome. Tables have been reconstructed where possible and reproduced where not.

Original spellings and formatting, bolding and italics have been used, with the exception of decimal points which in the original are shown raised to mid-line and here as full-stops (periods), and spaces before punctuation such as colons, question and exclamation marks being removed.

Any reader who thinks there is a mistake in transcription is requested to contact me for clarification.

PETER F. MAY
peter@pinotage.org

ST. ALBANS
ENGLAND
JANUARY 2012

---

[1] *A Year in Paarl with A I Perold: Vine and Wine Experiments 1916* by Peter F May & A I Perold, Inform & Enlighten Ltd, 2011, ISBN: 978-0956152312

CONTENTS

CHAPTER I

GENERAL INTRODUCTION

PAGE
1-25

1. Origin of modern viticulture. 2. Geographical distribution of the vine: *(a)* general, *(b)* South Africa, (c) United States, *(d)* Australia. 3. Influence of climate: *(a)* altitude and latitude, *(b)* great mountain ranges, (c) special site, *(d)* large masses of water, *(e)* winds. 4. Influence of soil: soil requirements, the physical state of the soil, the soil's temperature, the soil's colour and its temperature, chemical composition of the soil. 5. Geological origin of the soil.

CHAPTER II

THE EXTERNAL AND INTERNAL   26-104
MORPHOLOGY OF THE VINE

A. External morphology of the vine: 1. the root; 2. the stem— *(a)* the stem of the seedling, *(b))* the stem of the cultivated plant: the permanent stem, the bark, the buds and eyes, position of wood- and fruit-eyes on the cane, formation of fruit-eyes, adventitious buds (eyes), the bourgeonnement or very young shoots, the young shoots, the cane, position of inflorescences and tendrils on the shoot, origin and nature of the tendrils; 3. the leaf; 4. the flower; 5. the fruit: the bunch, its shape, the peduncle or stalk, the pedicel, the berry (skin, pulp, seed), shape of the berry. B. Internal Morphology (Anatomy): 1. the structure of the cell; 2. the different kinds of cells and tissues —(i.) parenchymatous cells and tissues, (ii.) boundary cells and tissues, (iii.) the mechanical tissue system (collenchyma and sclerenchyma), (iv.) the conducting tissues (xylem and phloem); 3. grouping of tissues in the various organs of the vine—*(a)* structure of the root (primary and secondary), *(b)* structure of the stem (primary and secondary, comparative anatomy of the stem), (c) structure of the leaf, *(d)* structure of the flower, *(e)* structure of the fruit.

CHAPTER III

CHAPTER IV

CHAPTER V

CHAPTER VI

CHAPTER VII

CHAPTER VIII

CHAPTER XIII

A. The production of table grapes for export. I. Conditions to be satisfied by table grapes for export. II. The factors and operations that play an important part in the production of table grapes for export: climate, soil, site, varieties of grapes, stocks, preparation of soil, planting and trellising of vines, manuring, soil cultivation, pruning, limiting the crop, trimming and thinning the bunches, removal of leaves and tendrils, control of diseases.
B. The selling of table grapes for export. 1. Picking the grapes, 2. final trimming of bunches, 3. the grading of the grapes, 4. packing the grapes, 5. transportation from farm to market, 6. the sale of export grapes overseas, 7. the economics of the growing of table grapes for export.

CHAPTER XIV

Vinegar: the acetic fermentation, manufacture and uses of vinegar, etc. *Grape Syrup:* 1. preparation of the grape juice (extraction, clarification, de-acidification of the juice), 2. the preservation of the juice, 3. the desired concentration and how it is determined, 4. concentrating the juice, 5. treatment of the syrup, 6. determination of the degree of concentration of the syrup. *Unfermented Grape Juice:* 1. the grapes, 2. the extraction of the juice, 3. clarification of the juice, 4. first pasteurisation of the juice, 5. the storage of the juice, 6. the bottling and re-pasteurisation of the juice for the trade. *Raisins:* A. Muscatels—Málaga raisins and Muscat of Alexandria lye raisins; B. Sultanas; C. Currants.

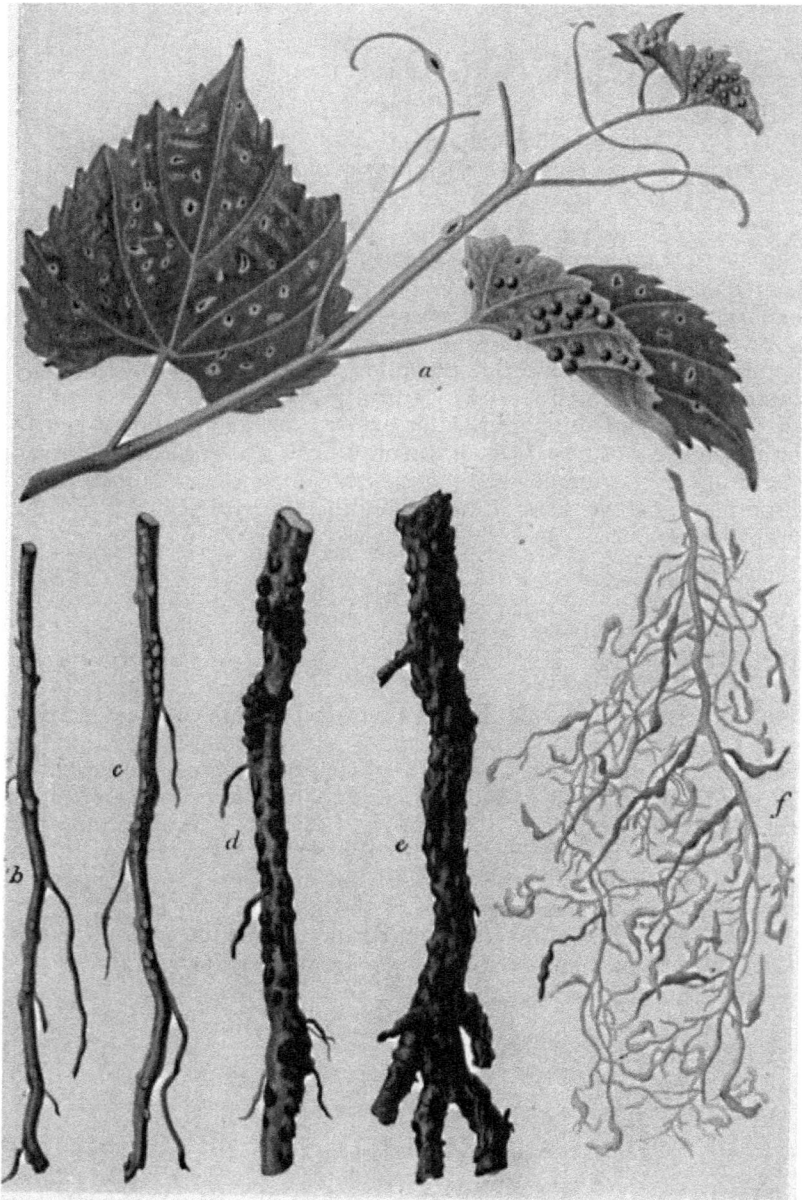

Fig. 49. Effects of Phylloxera on the Vine, after Viala

Editor's Note, 2012 Edition: Fig. 49 is a colour plate that appears between pages 188-199 in the original.

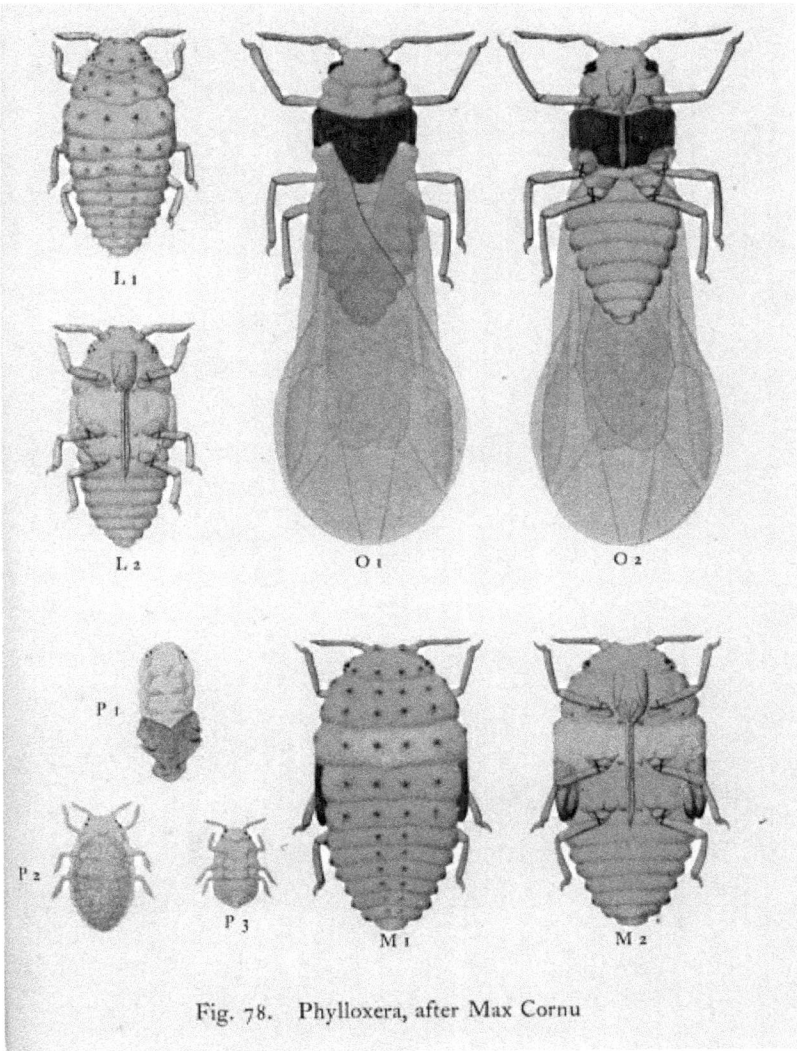

Fig. 78. Phylloxera, after Max Cornu

Editor's Note, 2012 Edition: Fig. 78 is a colour plate that appears between pages 488-499 in the original.

CHAPTER I

GENERAL INTRODUCTION

1. ORIGIN OF MODERN VITICULTURE

FOR a long time it was generally believed that the vine (*Vitis vinifera*, Linné) originated in the regions south of the Caspian Sea, *e.g.* Northern Persia,[1] and from there was introduced into Western and Central Europe and Northern Africa by Semitic nations, the Greeks and the Romans. The discovery of wild indigenous vines in North America undermined this theory. And the subsequent finding of grape seeds in pre-historic lake-dwellings in Europe, and of fossilised vine leaves and bunches of grapes in the brown coal strata of the tertiary geological period, made this theory no longer tenable.[2] We know from evidence of fossil vine leaves and grape seeds that in the tertiary period the vine flourished in Germany, France, England, Iceland, Greenland, North America, and Japan. Later, during the Ice Age, in the northern hemisphere the vine undoubtedly suffered and was forced southwards, but afterwards regained its own and spread over a large area.

The fossil grape-vines of prehistoric time [*Vitis teutonica* (found in Germany), *Vitis islandica* (found in Iceland), etc.] resemble in outward appearance the American *Vitis cordifolia* far more than the *Vitis vinifera*, Linné. The latter, however, occurs in the younger strata together with fossil plants of pre-historic age, *e.g.* near Montpellier in the south of France and in Italy.[3] Grape seeds have also been found in prehistoric tombs in Greece. During the Bronze Age,

---

[1] This theory is still upheld by Tamaro, *Uve da tavola*, 1915, p. 147.

[2] Compare Bassermann-Jordan, *Geechichte des Weinbaues*, 1907, vol. i.p.4.

[3] A. de Candolle, *L'Origine des plantes cultivées*, 1886, pp. 152-153.

therefore, man ate the grapes of the vines that climbed the forest trees. Also, wild vines (not vines that have become wild!) are found in Germany and elsewhere even to-day.

The vine and wine have always gone together, even in prehistoric periods. According to A. de Candolle, *l.c.* p. 154, the documents concerning viticulture and wine-making in Egypt go back some 5000-6000 years. In Palestine viticulture was practised very early, and the Phoenicians no doubt spread it to Northern Africa and Spain. Into Greece it was probably introduced from Asia. We know from Homer that viticulture flourished in Greece in his time, and the art was known probably much earlier. It is fairly certain that the Romans gained their knowledge of grape-growing and wine-making from the Greeks. For a long time they fancied only the Grecian wines, and it was only during the first century of the Christian era that the Italian wines began to acquire a good reputation in their own country.

Gallic or French viticulture took its origin in Marseilles, where Ionian Greeks from Phokaia in Asia Minor founded a Grecian colony in 600 B.C. under the name Massalia. They probably began to cultivate and make wine from the indigenous grape-vines near Marseilles. Massalian wine culture was certainly independent of that of Italy. From Marseilles the culture of the vine gradually spread up the Rhone valley. The conquest of the whole of Gaul by Julius Caesar (58-51 B.C.) helped very considerably to spread the culture of the vine in these parts. Indeed, Pliny tells us that Gallic wine had such a good reputation in Rome that it was imported in large quantities.

It was at the latest in the second century, during the peaceful reign of the great emperors Trajanus, Hadrianus, Antonius Pius, and Marcus Aurelius, that viticulture spread along the left bank of the Rhine in Alsace, the Rhenish. Palatinate, Rhenish Hessia, and the Moselle valley. This viticulture was Gallic in origin.

In the fifteenth century viticulture had spread to Madeira (1421) and the Canary Islands, and later to South Africa, Australia, North and South America.

2. GEOGRAPHICAL DISTRIBUTION OF THE VINE

(a) General. — In the northern hemisphere the viticultural industry is carried on between 20° and 51° N., and in the southern hemisphere between 20° and 40° S.. North Germany, France, Spain, Portugal, Italy (including Sicily and Sardinia), Tunis, Algeria, Switzerland, Austria and the Succession States, Hungary, Bulgaria, Roumania, Russia, Greece, Turkey, Palestine, Japan, the United States of America (particularly California), Madeira, and the Canary Islands. In the southern hemisphere viticulture is practised in Chili, Argentina, Peru, South Africa, and Australia. We find the most northern viticultural area in Germany, as far as 51° N.

The successful production of good table grapes in the open is much more limited than that of wine grapes. Owing to the higher revenue to be derived from the growing of table grapes, we find that these are grown under glass and in hot-houses in England, South-West Holland, Belgium, and Northern France, where they could not be grown in the open. Here the existence and the distribution of the grape industry depends largely upon the price of the necessary fuel.

(b) South Africa — In South Africa viticulture as an important industry is limited to the south-western districts (winter rainfall area), namely: Paarl, Worcester, Stellenbosch, the Cape, Caledon, Malmesbury, Tulbagh, Ceres, Robertson, Montagu. Of these, the first three are the most important as regards the quantities produced, the fourth as regards the quality of the light dry red wines produced, whilst the last two together with Worcester produce the bulk of our raisins, particularly sultanas. The bulk of our table grapes for export and for local markets is grown in Paarl, Constantia, and the Hex River valley, whilst Stellenbosch (Banhoek and the environs of the town of Stellenbosch) grows a fair quantity of table grapes for export. In the near future the area producing table grapes for export will no doubt be largely increased.

In the town of Graaff-Reinet vines are grown on a fairly large scale, mainly for the production of table grapes for local markets.

Along the Orange River from Upington to Kakamas vines are now being grown more and more on the rich irrigated alluvial soils. The principal varieties grown are Hanepoot or Muscat of Alexandria for table grapes and raisins, and Sultana (Thompson's Seedless) for the production of sultana raisins. This area has great possibilities in this direction and may yet become an important grape-growing centre.

In the areas of limited rainfall, such as Worcester, Robertson, Montagu, Graaff-Reinet, and the grape-belt along the Orange River, vines are grown under irrigation. This is done also in parts of Ceres and Tulbagh, particularly where the soil is somewhat shallow. In the remaining and greater part of our viticultural area, irrigation was frequently practised prior to the advent of phylloxera, but has been replaced almost completely in our grafted vineyards by the system of dry land farming, our average annual rainfall in this area being 25-30 inches.

In 1655, three years after the establishment of a Dutch settlement at the Cape, the Commander, Jan van Riebeeck, brought the first vines into South Africa. Godée Molsbergen[1] states that Van Riebeeck imported Hanepoot (Muscat of Alexandria), Muscadel, and Stein, which are to-day still amongst our favourite and most valuable varieties. The arrival in 1688 of the French Huguenots, many of whom came from the south of France, gave a great impetus to the development of the wine industry at the Cape. The famous Governor, Simon van der Stel (1679-99), laid out and developed the farm "Groot Constantia" which in the past produced the famous "Constantia Wine". Within fifty years of Van Riebeeck's first importation of vines into South Africa, viticulture was definitely established as an important industry at the Cape.

(c) **United States**.—In the United States the viticultural industry is of the greatest importance to the state of California. "Outliers of this main region of the Pacific slope run north into Oregon, Washington, Idaho, and even into British Columbia. . . ."[2] Here mainly Vinifera varieties are

---

[1] Godée Molsbergen, *De Stickier van Hollandsch Zuid-Afrika* Jan van Biebeeck, 1912, pp. 146-148

[2] 2 U. P. Hedrick, *Manual of American Grape Growing*. The Macmillan Co., 1919.

grown. Hedrick[1] divides the grape regions of America as follows :

The *Pacific slope*, of which California is the centre. The northern boundary of the grape growing area has just been given. Eastwards it runs into Nevada, Arizona, New Mexico, and even Utah and Colorado, though in most of these districts it is still insignificant according to Hedrick. Southwards it extends into Mexico, where Spanish priests introduced vines at the beginning of the seventeenth century, but the industry is of no great importance. California is by far the most important viticultural state of America.

The *Chautauqua grape-belt*, lying along the north-eastern shore of Lake Erie in New York, Pennsylvania, and Ohio, is the second most important grape region in America. In Eastern America, owing to the climate and consequent diseases and pests, Vinifera varieties cannot be grown successfully as a paying proposition. Hence indigenous varieties and their natural or artificial crosses, mostly belonging to the Labrusca species, are the ones grown here. In the Chautauqua grape-belt, according to Hedrick, *l.c.* p. 19, Concord occupies 90 per cent of the acreage under vines, Niagara 3 per cent, Worden 2 per cent, and a dozen or more varieties, including Moore Early and Delaware, occupy the remaining 5 per cent. The grapes are sold as table grapes (65,000 tons out of a crop of 100,000 tons) and for the manufacture of grape-juice.

The *Niagara region* is Canada's chief grape-producing area. It lies on the southern shore of Lake Ontario, across the end of Lake Erie and across the narrow isthmus of Niagara. The varieties of grapes grown, methods of pruning, etc., are roughly the same as in the Chautauqua belt. The Niagara grape originated here, and is grown more extensively here than elsewhere.

The *Central Lakes Region* of New York.— "In the central part of western New York are several remarkable bodies of water known as the Central Lakes. Three of these are large and deep enough to give ideal climatic conditions for grapes, and about these lakes are grouped several important areas of vineyards, making this the third most important grape region in America."[2] Here Catawba and Delaware take the place of Concord and Niagara.

---

[1] Hedrick, *l.c.* pp. 17-22.
[2] Hedrick, *l.c.* pp. 20-21

Minor Grape Regions.—These include several islands in Lake Erie, south-western Michigan, etc.

(*d*) **Australia.**—In Australia the viticultural industry is found in the south-east of the continent and mainly in the states of Victoria and South Australia. According to Portes and Ruyssen,[1] Australian viticulture was started in 1813 or 1814 by Gregory. James Bushy also grew and distributed vines before 1830. He toured France and Spain and imported 574 grape varieties about the year 1831 into New South Wales. According to Thudichum and Dupré,[2] 550 of these varieties grew in an experimental garden adjoining the Government garden in Sydney in 1832. Portes and Ruyssen further tell us that Bushy had planted five acres of vines at Camden in 1833. Thus New South Wales was the first viticultural state in Australia, but it was soon overtaken by Victoria and South Australia.

In New South Wales[3] "the Hunter River district produces the finest dry wines of the Hock, Chablis, Claret, and Burgundy types. It is practically the only area remaining in Australia to-day which produces the finest dry quality wines." Around Albury and Corowa, along the Murray River, more full-bodied and sweet wines are made.

In the Murrumbidgee irrigation areas vines are now grown fairly extensively both for wine (medium quality) and more particularly for Sultanas, Zante Currants, and raisins (Gordo blanco or Muscat of Alexandria or Hanepoot, and to a small extent Waltham Cross). In the Camden and Cumberland districts, practically at the back-door of Sydney, table grapes are grown.

In Victoria vines are largely grown in different parts of the state. In the Mildura settlement on the Murray River they are grown under irrigation, whilst in the Rutherglen district along the same river they are grown without irrigation.[4] In the Lily dale and western districts good light dry wines, white

---

[1] Portes et Ruyssen. *Traité de la vigne et de ses produits*, vol. i. p. 261; Paris, 1886.

[2] Thudichum and Dupré, *A Treatise on the Origin, Nature and Varieties of Wine being a complete manual of Viticulture and Oenology*, pp. 738-739; Macmillan & Co London and New York, 1872.

[3] According to a letter, dated 29th December 1924, which I received from Mr. G. Valder, Under-Secretary and Director of Agriculture for N.S.W.

[4] According to T. C. Angove, "The Wine Industry"', in the *Journal of the Dept. of Agric. of South Australia*, vol. xxvi. p. 624 (1923).

and red, are produced, whilst in Rutherglen we have heavy types of sweet full-bodied wines.

In South Australia, according to a letter received from Mr. Arthur J. Perkins, Director of Agriculture in South Australia, dated Adelaide, 15th December 1924, the main centres of the wine-making industry are: the neighbourhood of Adelaide, around Angaston, Tanunda, Calire, and Auburn, McLaren Vale, Lyndoch. On the River Murray irrigation areas vines are grown chiefly for their dried fruit, and are represented by the Gordo blanco Muscatel, Sultana vine, and the Zante Currant. Within recent years some grapes have been grown specially for the manufacture of strong spirits, the variety used being the Spanish Doradillo [Jaen] (according to Arthur J. Perkins).

In Queensland and Western Australia, too, viticulture is practised to a small extent.

3. INFLUENCE OF CLIMATE

This is one of the most important factors influencing the production of grapes. Under the climate of any place we include: temperature and its variation, rainfall, and particularly the amount of moisture present in the air and soil during the course of the year, frost, hail, the hours and distribution of sunshine during the year, winds, etc. As a remedy for drought we can sometimes use irrigation with success. Frost can be combated in various ways, as we shall see later. From strong winds we can protect the vines to some extent by planting wind-breaks. On the whole, we cannot change an unfavourable climate into a favourable one. Therefore a favourable climate is a *sine qua non* if the growing of grapes is to be a commercial success.

The summer and autumn must be long and warm enough to enable the grapes and the vines to be perfectly matured before the leaves fall and the vines become dormant. It is of the greatest importance that little rain should fall from the time the grapes turn colour ("veraison") till after the vintage. Abundant rains during this period will promote the development of disease, will cause cracking and rotting of the berries, and give watery grapes which will travel badly and be of inferior quality. Hence we need not be surprised to find that in the grape-growing regions of the Argentine (Mendoza and San Juan), California, Australia,

South Africa, etc., winter rains predominate and warm bright weather is the rule when the grapes are ripening. In the European and North African wine districts the quantity, and particularly the quality, of the vintage depend on favourable weather (i.e. little rain and plenty of sunshine) during this critical period

When the vines have entered their new period of growth in spring, mild weather is most desirable. Frost at this stage can cause tremendous damage to the coming crop and to the vine itself. Where severe frosts set in early in autumn the grapes may be damaged on the vines and the leaves will fall prematurely, thus preventing the vine from ripening its wood properly. During the dormant period in winter, the air and soil should not become so cold as to freeze the vine.

Let us now briefly consider the influence of altitude and latitude, of great mountain ranges, of a special site, of large masses of water, and of winds on the climate of an area or locality, and therefore upon its suitability for grape-growing.

(a) *Altitude and Latitude*—The temperature of the air sinks 1° C. for a rise of 160-200 metres, or 1° F. for a rise of 292-364 ft. Hence a locality at a high altitude may some times be more suitable for viticulture than one at a low altitude that is farther away from the equator. The nearer we get to the equator, the higher the "grape-line" rises, just as the snow-line does. Generally speaking, the upper limit for grape-growing lies 2000-2500 metres or 6560-8200 ft below the snow-line. The direction of the mountain range and the distance from the sea exert an influence, upon this limit. Along the Rhine at 60° N. lat. the upper limit for grape-growing is 260 metres or 853 ft. above sea-level; in South Germany it is 400 metres or 1312 ft.; along the Andes mountains in Peru the vine is still successfully grown at an altitude of 1200 metres or 3936 ft., and in South Africa it is successfully grown on a commercial scale near Johannesburg (Lombardy Estate) at an altitude of nearly 6000 ft. At such high altitudes it is usually almost impossible to grow table varieties with success in the open.

In section 2 it was stated between which degrees of latitude the viticultural industry exists. It will be noticed that the belt between 20° N. lat. and 20° S. lat. was excluded. In this belt the vine actually grows during the whole or greater part of the year, and therefore does not get its much-needed winter rest. The result is an inferior and worthless crop.

The *warm temperate zone* which lies between 34° and 45° N., and between 34° and 45° 8., is the most suitable for grape-growing. It is interesting to note that the viticultural areas in South Africa. Australia, and the Argentine lie respectively between 33° and 34°, 32° and 38°, and 31° and 34° S. lat. This correspondence is remarkably close.

(*b*) *Great Mountain Ranges.*—Where high mountain ranges run for hundreds of miles in one direction, they can exert a tremendous influence on the climate of the regions they traverse. The Alps, running from France through Switzerland into Austria, cut on the cold northerly winds and thus help to give Northern Italy its mild climate. Where such ranges run from east to west, their southern flanks are warmer than their northern in the northern hemisphere and vice versa in the southern hemisphere. The example of Northern Italy just mentioned is a case in point. The effect of high mountain ranges on rainfall is sufficiently well known not to need any elaboration here.

(*c*) *Special Site.*—This is a matter of the greatest importance. In every country some parts are known to be more subject to frost, hail-storms, cold and strong winds, etc., than other parts. We have already seen how, in the southern hemisphere, a northerly slope is the warmest, because the mid-day sun here stands in the north; the next warmest is a north-easterly, then north-westerly, then easterly, then westerly site, and the coldest is a southerly site. For the northern hemisphere we must simply alter "north" to "south" and *vice versa.* In countries where the climate is not very favourable for grape-growing, a slight improvement in the lie of the land often makes a very great difference to the quality of the grapes produced and the wine made from them. In the hot wine countries, level and low-lying lands also provide good sites for vineyards if the soil is suitable.

Usually a warm site is the best for early varieties and for varieties that are very late in ripening. A moderately warm, dry site is best for varieties that are picked very late (*e.g.* Chasselas Fontainebleau), not because they ripen very late, but because they keep well on the vines after having reached maturity. On the whole, a warm site will suit most varieties of grapes best. The best quality of table grapes and wine is produced along slopes or on fairly high land with a good natural drainage. Moist, low-lying land will usually produce heavy crops of more or less soft and watery grapes, which might give a medium quality wine, but will give inferior table grapes that will not carry well. When we remember that table grapes must be able to keep and stand long transportation, whilst the crop must usually be reduced considerably to ensure good quality, we see that the low site with its greater production possesses only disadvantages without any advantage to the grower of good table grapes or the producer of high-class wines. The most favourable slope is one facing the north (in the southern hemisphere), so that the rows of vines can run from east to west. In this way we get a warm site without unduly exposing our grapes to sunburn. Where frost is to be feared, a low site, particularly along the banks of a river will be the most dangerous.

(d) Large Masses of Water.—It is a generally known fact that the sea, large lakes, and large rivers exert a moderating influence upon the climate in their neighbourhood. This is due to the high specific heat of water, which means that water takes up a relatively large quantity of heat for every degree it rises in temperature, and gives off a relatively large amount on cooling. The water thus acts as a regulating reservoir of heat far more than soil or air does. Therefore the water in a large dam or river is cooler during the day than the soil in the neighbourhood, and vice versa at night. The Cape Peninsula illustrates these influences very well. Because of the maritime climate, grapes ripen five to six weeks later in Constantia than in Paarl. The Constantia area is near to False Bay, where the water is much warmer than in Table Bay. Also it is fairly well protected by Table Mountain against the cool south-westerly winds which blow from the Atlantic Ocean. The result is that, whilst grapes ripen well in Constantia, they do not do so in Camps Bay, Sea Point, and along the west coast of the peninsula generally.

Madeira and the Canary Islands enjoy a much milder climate than places on the corresponding degrees of latitude on the African continent, simply on account of the moderating influence exerted by the Atlantic Ocean. Thus they are more suitable for the production of good wines than the corresponding regions on the African continent.

On account of the Atlantic Ocean and the Baltic Sea, Western Europe possesses a distinctly maritime climate which depresses the northern limit of its viticultural area. Thus the vineyards in the Champagne near Reims (about 49° 22′ N. lat.) are the most northern in France, whereas the famous vineyards along the Rhine in the Rheingau and along the lower Moselle lie considerably farther north (over 50° N. lat.). This is largely due to the lesser influence of the sea, which is, however, still strong enough to prevent the vines from being frozen in winter. This would no doubt take place much farther inland on the same degree of latitude (*e.g.* at Breslau, Krakau, Kiew, etc., with their continental climate and severe winters which make viticulture impossible in those parts).

In parts of England and Belgium, with the same mean temperature for the year as the Rheingau, viticulture cannot be practised because the influence of the sea keeps the summer and autumn too cool to enable the grapes to ripen properly in the open. In the Rheingau the vineyards are to be found almost exclusively on the southerly, *i.e.* warmest, slopes along the northern bank of the Rhine. On a northerly site grapes ripen badly in the Rheingau. We further find that most vineyards are considerably higher than the river. The vineyards on the low-lying land near the river do not produce such sweet grapes or good wines as those situated on a higher level, while grapes grown at too great a height do not ripen so well through insufficient warmth.

In the Medoc (this is the famous red wine district of Bordeaux along the left bank of the Gironde to the Gulf of Biscay) the influence of the sea becomes more noticeable the nearer one approaches the coast, with the result that the viticultural area comes to a stop long before the coast is reached, whilst the last vineyards produce only a poor very light ordinary wine, and table grapes cannot be grown in the open at all. The river here is really an estuary several miles wide.

In the United States we find the viticultural industry in the eastern coastal state of New York, the western coastal state of California, and in the states bordering on the great lakes of the interior. In the states far in the interior, with a severe continental climate, we do not find any grape-growing of importance.

In our own viticultural area we find the same state of affairs. Constantia, on the Cape Peninsula, with its sea climate produces our finest dry red table wines. Stellenbosch too, being near enough to the sea to be influenced by it, produces fine dry red table wines. The Paarl district is much warmer and produces mostly heavy wines. Its red table wines are on the whole not so fine as those of the afore-mentioned districts. Its white wines are, however, quite as good. In Paarl the grapes ripen two to three weeks earlier than in Stellenbosch, and at Stellenbosch several weeks earlier than in Constantia. In the Warm Bokkeveld (Ceres District) the grapes ripen about the same time as in Constantia, owing to the influence, not of the sea, but of the altitude, which is 1500-1600 ft. above sea-level.

(e) Winds—The influence of the winds on the rainfall of a certain area is well known. But I would like to remind the reader that it is due to the north-western winds blowing mainly in winter, and to the configuration of the mountains, that the viticultural area of the Western Province (Cape) has a sufficient rainfall (25-30 in. per annum) to grow vines without irrigation. In the districts of Worcester, Robertson, and Montagu, where the annual rainfall is frequently less than half this amount, most vineyards are irrigated.

Winds, however, influence the climate in many other important ways. Where they blow across high mountains covered with snow and ice or, generally, across the ice-fields, they can suddenly bring about a spell of cold weather. This is frequently the cause of severe frost in spring, which may cause great damage to the vines in the south-east of France, in Germany, etc., if the vines have already commenced their new growth.

The sirocco blowing across the scorching hot Sahara desert sometimes causes very considerable damage to the grape crop in Tunis and Algeria. Any hot wind will greatly increase the loss of water from the vines and from the soil by evaporation, and may cause the grapes to be sunburnt or prevent the vines from getting sufficient water for the grapes to ripen properly. For the grower of table grapes a hot wind is a very serious matter. The berries cannot become large owing to insufficient moisture in the soil, and the grower must suffer.

When the south-east wind blows in the Western Province of the Cape during the summer months, it dries out the soils. Should it subside during the forenoon, the afternoon is usually very hot and the grapes easily get sunburnt.

Strong winds can cause great damage to a vineyard in several ways. They may blow off the young shoots during their early growth when they are still tender, and thus remove a considerable portion not only of the coming crop, but also of the wood that was to serve as bearers for the next crop. It the canes are already sufficiently lignified to withstand the pressure of the wind, the developing buds may be damaged, and leaves may be torn. The bloom of table grapes may be rubbed off and sand, blown against the ripening grapes, may render them unfit for export. Hence it is best to avoid planting table grapes where they would be exposed to strong winds during the growing period, and more particularly during the last month before the grapes are picked. Should it be unavoidable it is necessary to provide wind-breaks and adopt a suitable method of trellising and training.

Where the dust nuisance exists next to a public road, a dense hedge of, say, Australian Myrtle or *Eucalyptus cornuta* should be grown in order to keep the dust off the grapes as far as possible.

It is during the blooming and setting period that cold winds can cause most damage, as the fertilisation is thereby hindered and the berries in consequence set badly. Later on, continuous cold winds will delay the ripening of the grapes.

4. INFLUENCE OF SOIL

Although the climate of a certain area really decides whether it is suitable for viticulture, it is a fact that the soil and its exposition are of the utmost importance for success in viticulture. Soil and climate must guide us in the choice of the varieties of grapes we are to grow. Where phylloxera exists, we ought further to consider which stocks will answer in the soil and whether the varieties we propose growing will thrive well on them.

Soil Requirements.—The first and foremost requirement is that the soil should be fairly deep, cool (i.e. always possessing sufficient soil moisture), and well drained.

At the Cape the best export grapes are grown on soils that need not be irrigated. This does not exclude the production of good export grapes under judicious irrigation. The great thing every grape-grower should aim at, is always to have enough moisture in the soil—not too much and not too little. In this connection it must be pointed out that high-class table grapes, where very big berries are aimed at, require more water in the soil than wine grapes, where the size of the berries is not so material provided that their juice is sweet enough.

Good drainage is necessary to keep the soil warm enough and well ventilated, whilst at the same time preventing it from becoming too wet. For grape-growing we give preference to soils that are naturally well drained. For this reason heavy clay soils should preferably not be used for vineyards. the best soil for the growing of table grapes is a gravelly coarse decomposed granite which is at least 2-3 ft. deep *e.g.* along the Paarl Mountain in the Cape. A coarse sandy loam, 15-24 in. deep, above a layer of gravel which in turn rests on clay, also provides good soil for producing good table grapes the clay in the subsoil prevents the soil losing too much water and plant food. The gravel should be permeable for the roots of the vines.

It is the physical state and structure of the soil more than anything else which determine its suitability for the growing of grapes, and more particularly of high-class table grapes. Hence we find suitable soils on various geological formations. Very fertile soil is undesirable for the best quality. A soil of medium fertility is best. Fairly moist soils, rich in humus, are unsuitable, as they contain too much nitrogen which gives rise to rank growth and irregular crops, whilst the vines and grapes will be particularly susceptible to pests and diseases. The table grapes produced on such soils will be too soft for distant transportation, whilst the wine grapes will yield wines rich in nitrogenous compounds which are on this account very susceptible to bacterial diseases and take a long time to get bright, if they ever reach that stage. Again, these soils are usually low-lying and therefore not suitably situated for producing grapes of good quality.

The presence of fairly large quantities of potash and phosphoric oxide in a soil is very useful, but it is not so important as a good physical state.

The presence of fairly large quantities of lime is good, but not absolutely necessary. Too much lime may cause chlorosis. Hence calcareous soils demand special lime resistant stocks.

The Physical State of the Soil.—Soil consists mainly of stones, gravel, sand (coarse and fine), silt, clay, and humus. According to the relative proportions in which we find these constituents in a soil, we speak of a stony, gravelly, sandy soil, loam, silt, clay, or muck soil. Under a "broken" soil I understand something between a clay and a sandy soil which remains loose fairly easily, consists largely of sand, including coarse sand, with sufficient clay to make it decidedly heavier than a sandy soil.

Sandy soils are loose and easy to till, and make the creation and preservation of a good soil mulch an easy matter. Drainage, if not good, can easily be arranged. Where the subsoil contains enough clay or is sufficiently impervious, sandy soils will retain moisture well. Where such is not the case, they may lose too much water and become too dry. A coarse sandy soil with a fair quantity of fine sand, clay, and humus is usually a good vineyard soil. It is usually some-what poorly provided with plant food, but responds well to suitable manuring, and can thus yield heavy crops. The grapes are usually of good quality and ripen fairly early, unless the soil is decidedly moist. In some sandy soils the European or Vinifera vine is fairly immune to phylloxera, and can therefore be grown on its own roots without appreciably suffering from attacks by phylloxera.

Clay soils are difficult to till, usually badly ventilated, frequently badly drained, and difficult to drain properly. They are frequently fertile and can then yield good and even heavy crops, which are usually, however, of poor quality. Whoever aims at high quality should therefore not plant vines on a clay soil.

Broken soils are looser and easier to cultivate than clay soils, and they usually are the best vineyard soils. The decomposed granite soils of the Western Province of the Cape, which have been formed under a humid climate and are mostly still in contact with the mother rock, belong to this type of soil, and are on the whole eminently suitable for producing wine and table grapes of very high quality.

The Soil's Temperature—Stones help to keep the soil loose and improve its natural drainage. They also tend to cause a higher soil temperature and thus may promote the growth of the vine. Where the soil moisture is rather limited stones and pebbles, when present in large quantities, may cause a shortage of moisture in the soil which may result in the grapes not growing out properly and the berries remaining too small. In parts where more warmth is desirable, stones will be an advantage in a soil. In France we find the most famous vineyards on stony soil. The following analysis of a vineyard soil of the Graves in the Bordeaux district exemplifies this:[1]

	Stones.	Gravel.	Fine Soil.	Total.
Top Soil	54.6	16.4	29.0	100.0
Subsoil	57.2	17.7	25.1	100.0

In Germany, along the Rhine and Moselle, the vineyard soil is frequently covered with a layer of dark-coloured slag or broken shale in order to promote warmth and thus facilitate the proper ripening of the grapes.

We have already seen how the exposure of the soil influences its suitability for grape-growing.

The state of cultivation of a soil exerts a great influence upon its capacity for absorbing and losing heat. Prof L Ravaz[2] of Montpellier has shown that the temperature of the air above uncultivated vineyard soil is generally higher (up to 4 C.) than above cultivated soil. This explains why a vineyard that is exposed to frost in spring is more damaged by frost where the soil has been cultivated. This danger disappears as soon as a good rain has fallen or a surface irrigation has been given. In such cases it is therefore best to retard cultivation until the dangerous period is over In summer it is good for the same reason to keep the top layer of the soil loose for a couple of inches, as this helps to keep the air above the soil cool and thus diminishes the loss of soil moisture by evaporation. This is due to two causes.

---

[1] J M Guillam, *Etude générale de la vigne* p. 333; Paris, 1905.

[2] L. Ravaz, *Recherches sur la culture de la vigne*, p. 9; Montpellier, 1909.

The loose soil contains much more air than the firm, uncultivated soil, and therefore gets heated much more slowly during the day, stationary air being a very bad conductor of heat. At night the loose soil, with its much greater exposed surface, loses considerably more heat by radiation than the uncultivated soil.

Prof. Ravaz, in the previously cited publication, also showed that it is dangerous to draw furrows in the middle of the rows and plough the soil towards the vines during the

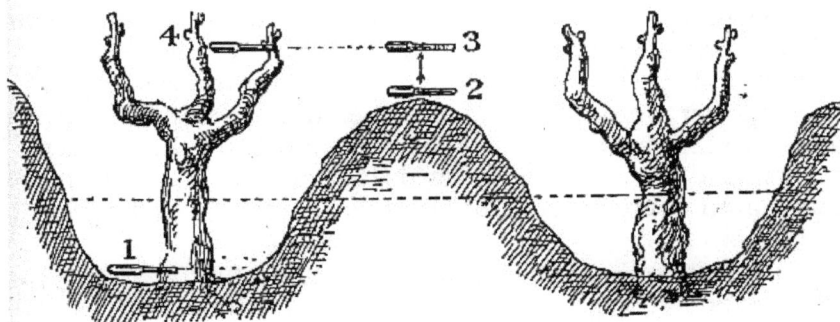

Fig 1. Measuring temperature of air. From Ravaz, *Recherches sur la culture de vigne*, 1909, Coulet et fils, Montpellier

dangerous period. He placed thermometers in the positions 1, 2, 3, 4, which registered the lowest temperatures during the night. The experiment was conducted at Montpellier; in whiter. The following are some of his results:

Dates in 1907.	Minimum Temperature in degrees C. in position.				Remarks.
	1.	2.	3.	4.	
Jan. 12	−3·0	−4·2	−4·0	−3·5	White frost.
,, 19	−4·4	−6·3	−5·4	−5·2	Very severe white frost.
,, 20	−5·2	−7·4	−7·1	−6·3	,, ,,
,, 21	−6·0	−7·7	−7·5	−6·9	,, ,,
Feb. 1	−5·0	−6·2	−6·3	−6·0	Very strong wind.
,, 2	−6·1	−7·0	−6·9	−6·4	,, ,,
,, 7	−7·0	−9·0	−8·8	−8·2	White frost.
,, 10	−4·5	−5·2	−5·3	−5·0	,,

From these data we see that it was least cold in the hollow (position 1). Next comes position 4 just above 1, whilst positions 2 and 3 (i.e. on top of the ridge) were the coldest. There was usually only a slight difference in temperature between positions 2 and 3, although 2 was usually colder than 3. The explanation is that the loss of heat by radiation at position 1 takes place upwards and in a slanting direction past the tops of the adjacent ridges, whilst at the top of the ridge (position 2) the reverse is true, and the angle of radiation there is 360° minus that at position 1. Hence the soil and air cool more rapidly at 2 than at 1. At position 4 the radiated heat from the two adjacent sides of the ridges cross and maintain the temperature of the air in this position somewhat higher than in position 3, which was at the same height.

Had the soil been ploughed towards the row of vines, then the air would have been colder near the vines' bearers and bursting buds or young shoots, and thus the danger of frost would have been increased. It should be borne in mind that half a degree more frost frequently results in damage to the vines.

The Soil's Colour and its Temperature.—Where the climate is not very warm, the vineyards are by preference established on dark-coloured soils, since they become warmer than the white or light-coloured soils, and therefore cause the vines to grow more vigorously and to ripen their grapes better. Particular attention is paid to this in the northern viticultural areas, *e.g.* along the Rhine (in the Rheingau and neighbourhood) and Moselle, where the vineyard soil is often covered with a layer of dark shale or slag in order to increase the heat-absorbing capacity of the soil and thus to favour the better development of the vine. In Constantia, where the mean temperature is considerably lower than in Paarl, the black varieties of grapes are preferably grown on the reddish soils and the white varieties on the light-coloured soils, as the former require extra heat to ripen properly and form sufficient colouring matter in their husks.

Experiments conducted by Prof. Ravaz, as related in his previously quoted publication, have shown that, at a depth of 4 in., soil which is red or black on the surface gets much warmer than where its surface is white. He further showed that the higher soil temperature increased the vine's vigour. In this connection he carried out an experiment where the soil was covered with a layer of concrete excepting for 6 in. round the stem of every vine.

He subdivided this plot into three equal plots and coloured these respectively white, red, and black. The vines grew most vigorously where the concrete was coloured black. Next came the red plot, whilst those on the white plot showed least vigour. The weights of the canes produced on the white, red and black plots we as 1 : 1.33 : 1.47. Hence the red and black plots yielded respectively 33 per cent and 47 per cent more canes by weight than did the white plot on the same soil. Here we have an example of how the practical wine farmer realised what influence the soil's colour will have on the vine before science had yet proved and explained it.

Chemical Composition of the Soil.—This important factor will be discussed later under "Manuring".

5. GEOLOGICAL ORIGIN OF THE SOIL

Through the action of air, cold and heat, water (including ice) and plants, the rocks forming the earth's crust are continuously changed and broken up on the earth's surface, and thus what we know as soil is gradually formed out of them. This process we call *weathering*. We may therefore call soil "weathered rocks". Gradually plants and animals begin to live in this product and to influence it. Their remains are mixed with the soil and give rise to the formation of that valuable soil ingredient *humus*, which is the dark, more or less decomposed vegetable matter in the soil which helps to render the soil fit for the production of agricultural crops.

Although our agricultural soils are therefore more than merely weathered rocks, it will yet be clear that the character and suitability of soil will in a large measure depend upon the mother rock or rocks out of which it has been formed, and also upon the conditions under which this formation took place. Soils can rest upon their mother rocks (residual or *in situ* soils), or they may have been removed therefrom by wind (*aeolian, loess* soils) or water (*alluvial* soils) to accumulate in the places where we now find them. While it is a simple matter to indicate the mother rock in the first case, it requires a considerable amount of study to do so in the subsequent cases, yet it is possible. In order to determine the value and suitability of an agricultural soil in a certain area, it is very useful to find out from what rocks it was formed.

For the further discussion of this matter it will be necessary to say a few words about these rocks and their formation.

We distinguish between *plutonic* and *sedimentary* rocks. The first class includes (*a*) granite, which, at a considerable depth beneath the earth's surface, slowly cooled down from a hot intrusive liquid mass to the hard granite as we know it to-day, and (*b*) basalt, which has flown as a molten mass from the crater of a volcano or out of a fissure in the earth's crust and cooled down in the open air to the hard dark rock which, after being exposed to the agencies of weathering for some time, usually presents a reddish outer surface.

The sedimentary rocks we find in layers called strata, and their nature readily shows us that they were originally formed out of a mass of loose parts which were deposited in layers, generally in water. Plants and animals which then lived and were interred in this material are now found there as *fossils*. It is mainly their presence which enables us to determine the order in which these strata have been deposited, and to identify them wherever they may occur

Some strata were formed before the earth was inhabited by living organisms. This is proved by the fact that no fossils have yet been found in them nor in the underlying strata. In South Africa the Malmesbury shales and Table Mountain sandstone are examples of such non-fossiliferous strata or formations. Of these two the former is the older, as it underlies the latter—the bottom stratum is always the older as it must have been deposited first, provided the strata still remain the way they were deposited. The Bokkeveld beds are younger than the Table Mountain sandstone, and they contain many fossils—they are our oldest fossiliferous strata. The Karroo shales and mudstones again are much younger than the Bokkeveld beds.

We can now proceed to consider briefly what kinds of soil are formed out of the various rocks and how far they are suitable for vine-growing.

In arid regions *granite* gives rise to the formation of a coarse sand, whereas in humid regions, *e.g.* the Western Province of the Cape, it gives rise to good agricultural soils with a fairly coarse, gravelly texture where the soil still rests in contact with the mother rock.

Such soils are termed decomposed granite soils. Along the slopes of Paarl Mountain (and Groenberg, which are now simply large masses of granite mostly covered with its decomposition products, we find excellent examples of such granite soils. The shales and sandstones into which this granite originally intruded have long ago been decomposed by weathering and have been washed away, thereby exposing the masses of granite which now form these mountains. In the Western Province of the Cape we find many instances where the intruded granite has been exposed in kloofs and along mountain slopes, although the main mass of granite is still covered by the overlying Malmesbury shale and Table Mountain sandstone. In these cases, *e.g.* in Constantia, the Stellenbosch, Paarl, Malmesbury districts, soils have been formed out of granite and one or both of the other rocks mentioned. Sometimes limited areas of almost pure decomposed granite are met with, but these soon pass over into other soils according as more of the products of decomposition of the sandstone and (or) shale are mixed with those of the granite. Near the granite the soil is usually of a fairly coarse texture. Where the products of the weathering of granite have been carried for some distance we usually find a heavy dark clay or pot-clay soil, as the fine particles of clay remain longer in suspension and are therefore transported farther by water than the coarser particles.

Granite consists mainly of the three minerals: quartz, felspar, and mica. The latter can be the white Muscovite or the black Biotite. Most of the granites in the South African viticultural area contain the latter. Mica shines and easily splits into thin blades. It is the first to be decomposed by weathering, and is the origin of the iron compounds which give the decomposing rock and the granitic soil a yellow or reddish-yellow or reddish-brown colour. It gives the clay soils from decomposed granite their bluish-black colour. When burnt into bricks the iron compounds are turned into ferric oxide which gives these burnt bricks their red colour.

The next to weather are the felspars. These are mainly the potash felspars, orthoclase and microcline; these are gradually decomposed into clay and potash salts which are absorbed by the resulting soil. The quartz grains are of a grey colour and are the most highly resistant against weathering.

They are therefore seen as sand grains in the decomposed granite. Thus it is the mica, and more so the felspars, which determine the fertility of the decomposed granite soil. As the felspars in granite contain hardly any of the plagioclases (soda-lime felspars), the resulting granitic soil is *fairly rich in potash but poor in lime*. These soils will therefore be much improved by a periodic dressing of crushed limestone for crop production.

The granitic soils are very suitable for most American stocks, and therefore offer no great difficulty in reconstituting a vineyard on American stocks. They are eminently suitable for vine-growing, and give good crops of wine and high-class table grapes where the climate, site, etc., are favourable.

Basalt and other dark or basic volcanic rocks usually weather fairly rapidly. They do not contain free quartz and are therefore termed basic in contrast with granite and other rocks which contain free quartz and are therefore called acid rocks. They, however, do contain iron, calcium, sodium, potassium, magnesium, alumina, silica, etc., and give rise to the formation of red, brown, and sometimes dark-coloured soils which, being fairly rich in lime, are sweet soils. Along volcanoes like Vesuvius and Etna we find these rocks, together with the soils that were formed from them. They are very suitable for vine-growing. Along the slopes of Vesuvius the famous Lacrima Christi wine is produced on such a soil. Round about Catania much wine is made along the south-eastern slopes of the Etna. Such soils are not met with in the Cape viticultural area.

The Malmesbury shales are the predominating geological formation in the viticultural area from Cape Town to Tulbagh and from there along the Breede River Valley to near Worcester, wherever it is not interrupted by intruded granite or covered by Table Mountain sandstone. They give rise to the formation of a clayey, cloddy soil with a grey to reddish colour. This soil retains its moisture well, and it is easy to cultivate it and keep a fine loose mulch on the surface where it occurs by itself. It is fairly well supplied with potash, but very poor in phosphoric oxide and lime. For vine-growing it is very suitable, and, under favourable local conditions and with good manuring, can produce table and wine grapes of good quality and in sufficient quantity.

On the Stellenbosch University Farm a large portion of our wine and table grapes are grown with great success on fairly pure decomposed Malmesbury shales. The soil has a reddish colour. Generally the soils from these shales are mixed with decomposition products of Table Mountain sandstone or granite or both.

Table Mountain sandstone covers the tops of all the high mountains in the Cape viticultural area. As it consists almost entirely of quartz, it obviously represents the extreme type of an acid rock. Hence its residual soil also is sour. Where it does not yet contain a considerable amount of humus, it is a pure sandy soil. Its physical state is good. It is easily drained, aerated, and kept in good tilth. Where the subsoil is clay or is sufficiently impermeable, it retains water well and loses it by evaporation slowly. Such a sandy soil is naturally poor, but may be made highly productive by proper and sufficient manuring. Where it contains a fair amount of humus, as is shown by its darker colour, it retains plant food much better than where humus is wanting. It is better where mixed with some decomposed granite or shale or both.

Such a sandy soil is suitable for vine-growing if it contains neither too much nor too little moisture. Grapes grown on such a soil usually ripen early and become very sweet. Somewhat moist, heavily manured sandy soils give abundant grape crops, but the quality will not be very good. For the production of good table grapes on sandy soils, they must evidently contain a fair amount of clay and not too much soil moisture, whilst the crops have to be limited to a reasonable amount and the manuring must be regulated accordingly.

The *Bokkeveld beds* occur in the Cape viticultural area in the following parts: Warm Bokkeveld, Hex River Valley from de Dooms upwards, Nuy and other parts of Worcester, Robertson, Montagu, Ladismith, Oudtshoorn, Swellendam, Riversdale, etc. Most of the vines in these parts are grown on alluvial soils along the rivers. These soils are mainly formed out of the decomposed Bokkeveld shales, are exceedingly fertile, and generally require irrigation. They are mostly grey to reddish loams with a considerable amount of clay.

Where they are mixed with sand from the decomposed Table Mountain sandstone covering the Hex River Mountains, Langeberg, etc., they are of a lighter nature.

These soils are well provided with lime and are therefore sweet. They are most highly productive and give very abundant crops, 17 tons of grapes per acre or 250 hl. wine per ha. being not at all uncommon. Practically the whole of our Sultana crop is produced on these soils, where the Sultana vine produces much heavier and more regular crops than in the area between Cape Town and Tulbagh, where the annual rainfall (25-30 in.) is double that of the Sultana area. In the latter the Sultana vine does not suffer from anthracnose, whilst in the Cape Town-Tulbagh area its whole crop is often almost completely destroyed by this disease.

These sweet soils also produce our finest and sweetest Muscadel wines. Here White French (Palomino) can give very good light wines, good sherries, and very sweet white jeripico. Hermitage and other black varieties have thus far produced only a very inferior dry red wine on these soils. Hanepoot (Muscat of Alexandria) produces good raisins on them, though no better than what it produces on less sweet or even somewhat sour soils. The table grapes produced on this sweet soil are usually somewhat soft for long transport, although very big berries can be obtained. With very judicious irrigation it may be possible to produce here as well table grapes that will stand transportation, especially if very hardy varieties such as Barlinka are grown. These soils constitute the irrigated portion of our vineyards, while those on the previously described formations are worked almost exclusively on the dry land system.

The *Karroo formation* gives very fertile soils, but these everywhere require to be irrigated. They lie outside our real viticultural area, and are of no importance to our wine industry. Vines grow and bear well on them where climatic conditions are favourable. The same applies to the silt soils along the Orange River and on its islands. As these soils mostly lie within our summer rainfall belt, viticulture is on the whole an uncertain undertaking in these parts.

Limestone generally gives rise to a stony, dry soil. Where the top soil contains a fair amount of clay and humus, it may retain sufficient moisture to make vine-growing a success without irrigation, if the annual rainfall and its distribution are favourable, although the subsoil may be mainly limestone. Examples of this we find in Champagne, Upper Burgundy, Cognac, Jerez de la Frontera, etc., which produce world famous wines, Cognac producing brandy. It is an advantage if the soil is fairly rich in lime. But too much lime in the soil may cause lime chlorosis, as we shall see later on.

From the foregoing it will have been noticed that the vine thrives in the most diversified soil types. Where the vine, under suitable climatic conditions, will not thrive, few, if any, agricultural crop plants will succeed. Hence it is frequently difficult to replace the vine by another crop plant which will give anything like the same return. Where first-class table grapes are to be produced, the soil and climate must be more suitable than where grapes are grown for another purpose, as we shall see more clearly in the chapter dealing specially with the production of table grapes.

CHAPTER II

THE EXTERNAL AND INTERNAL MORPHOLOGY OF THE VINE

IN order to free ourselves from a mere mechanical application of certain rules of thumb in viticultural practice and thus to prepare the way for progress, we must strive to get an insight into the life of the vine itself. This requires a good deal of study, but the reward compensates us for the trouble taken. He who knows why a certain operation is required, does it much better than the person who has to act merely according to instructions. This knowledge further enables him eventually to improve upon the current practice.

In commencing our study of the vine, we notice, in the first instance, that it lives partly in the air and partly in the soil. Both science and practice have taught us that there exists an intimate relation between the development of the parts of the vine above the ground and those in the ground, and that, as we shall see later on, they mutually influence each other.

In this study of the vine we shall first consider its external morphology, then its internal morphology, *e.g.* anatomy, and leave the discussion of its biology for the next chapter.

A. EXTERNAL MORPHOLOGY OF THE VINE

Here we shall study the different parts of the vine, their form and grouping, their origin and development. The branch of botany which deals with this subject is called morphology. The vine consists of the following parts : root, stem, leaf, flower, and fruit. On account of the practical as well as the scientific aims of this work, I shall, during the discussion of the various parts of the vine, from time to time take the liberty of indicating the function of the various parts.

26

1. THE ROOT

According to their origin, we distinguish between *germination roots and adventitious roots.* The former are produced when grape seed germinates, as can be seen in Fig. 2, whereas the latter arise out of the stem. When grape seed germinates a clear tap-root is formed. Soon roots are developed around its neck and later on lower down as well. On these lateral roots of the first degree develop lateral roots of the second degree; on these latter, roots the third degree, and so on. Except for the production of new varieties out of seed, the germination roots are of no importance to the grape-grower, as in actual practice the vine is always propagated vegetatively, i.e. by planting out cuttings or by grafting.

Of all the more importance, therefore, are the *adventitious* roots which, as has already been stated, are formed on parts of the stem. Usually they are to be found only underground. Where the vine, however, grows in a very moist atmosphere and in the shade, as for example in green-houses under glass, *aerial roots* are sometimes formed on those parts of the stem which are several years old. These, as a rule, do not develop lateral roots, but dry up again. Where they are, however, formed near the ground, as I have already noticed on young vigorously growing scions of grafted vines, they cease to be air roots as soon as they enter the soil and may become

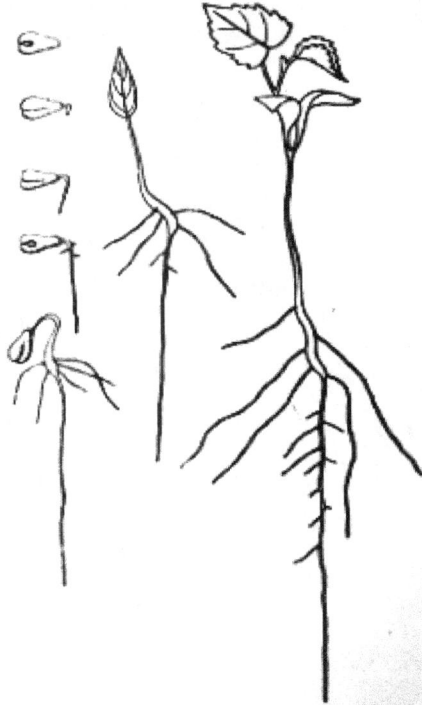

Fig. 2.—Development of seedling out of the grape-seed (after R. Goethe). From Babo u. Mach, *Handbuch des Weinbaues*, 1923. Paul Parey, Berlin.

27

strong roots. As we shall see later on, such roots have to be cut off early to prevent the scion from freeing itself from the stock and returning to the ungrafted state.

The Development of Adventitious Roots—Where a young growing shoot is more or less covered with moist soil, it readily forms adventitious roots. This also happens when layering growing or dormant canes of vine. When cuttings of dormant canes are planted out under suitable soil conditions, they will generally form adventitious roots quite readily. Even a single bud with a bit of wood attached to it will suffice to form a new vine through the development of adventitious roots out of the wood (See Fig. 3.) The various varieties of grapes do not, however, strike root with equal facility from cuttings. Whilst most species varieties strike root very or fairly easily it is difficult matter to get cuttings of *Vitis Berlandieri* to strike root. Jacquez cuttings are not quite so bad in this respect, but they still strike root with some difficulty.

Fig. 3.—Formation of adventitious roots on a one bud vine cutting. From Babo u. Mach, *Handbuch des Weinbaues*, 1923. Paul Parey, Berlin.

Most of the roots— and the most vigorous ones—are formed at the nodes. On the internodes far fewer roots are developed, and these usually remain much weaker than those formed at the nodes. Some varieties, for example Jacquez, form nearly all their roots only at the bottom node, whilst others like Riparia Gloire, Aramon × Rupe Ganzin Nos. 1 and 2, etc, develop numerous roots higher up the cutting also.

The *primary* roots thus formed develop *secondary* roots or lateral roots of the first degree, and on these are developed *tertiary* roots. The latter are sometimes called *hair roots*, because they are very fine and thin in comparison with the other (primary and secondary) roots. They absorb nearly all the water and plant food which the vine takes up from the soil. They are formed every year on the other roots of the vine as soon as its dormant period of winter rest is over and the vine awakens to new life.

Kroemer[1] distinguishes the following four parts in every individual root: the root-point, the stretching zone, the absorption zone, the transportation zone. The *root-point* comprises the youngest part of the root and is 2.5 mm. or $2/25$-$1/5$ in. long. It serves exclusively for the formation of new cell tissue and for the elongation of the root. The external layers of its tissue are called the *root-cap*, because they cover the inner layers of the root-point's tissue (*i.e.* the cone of meristematic cells) like a cap.

The *stretching zone* follows immediately upon the root-point and is only a few millimetres in length. Here the new cell tissues formed in the root-point grow out to their full length and undergo their full development.

The *absorption zone*, which follows immediately upon the previous zone, is always several centimetres (hence 1 in. or more) long, and can be recognised by its yellowish colour and the presence of the fine *root-hairs* on it. Here the soil solution is absorbed. Hence the name "absorption" zone.

The *transportation zone* extends over the rest of the root, and can easily be recognised by its brown colour. Its external layers of cells are suberised (corky), and it serves for trans-porting the absorbed soil solution to the aerial parts of the vine, and for transporting the manufactured food back from there to the roots, and particularly to the growing points.

At the termination of the vine's growing period, *i.e.* late in autumn or near the beginning of winter, the whole of the absorption zone then existing is converted into a transportation zone, as the whole root now turns brown through suberisation of its external cell layers. The roots start growing again as soon as the soil, after winter, becomes warm enough.

---

[1] Babo und Mach, *Handbuch des Weinbaues* (4th ed. 2 vols., 1923-24), vol. i. p.263.

Where a strong root starts sprouting at its end, the young white point is considerably thicker than the old root; where a thin rootlet sprouts at its end, the new part is just as thick as the old.

As the vine in a few years will reach its full aerial development (that is up to the point of development desired by its owner), and is kept at about the same stage of development by the annual winter pruning, it follows naturally that the root-system also is hereby limited in its development.

As Müller-Thurgau[1] first pointed out, a competition takes place between the different roots of the vine for the manufactured food that flows from the leaves to the roots as the vine grows older. The result is that only the strong roots can continue their growth, whilst the weaker ones perish through exhaustion. Hence old vines show only a small number of roots that emanate directly from the root-stump. We frequently find only two or three such roots. As long as the vine is still young, some new roots are developed on the root-stump. Later on this hardly ever takes place. On the main roots the lateral roots also decrease in number for the same reason, although new lateral roots may be formed on them too. When a root is broken off when cultivating the vineyard, one or more roots are formed at its end (i.e. at the wound). If it was a thin root (1 mm. or $\frac{1}{25}$ in. diameter), usually only one root is formed, according to Kroemer,[2] but if the roots are fairly thick (5.15 mm. or $\frac{1}{5}$-$\frac{3}{5}$ in. diameter) a whole bunch of new roots is usually formed. This phenomenon is well known to every attentive grape-grower.

The different species of the genus *Vitis* differ considerably in the development of their root-systems. Thus we find that *Vitis riparia's* root-system consists of a large number of thin roots, whilst that of *Vitis cordifolia* consists of a small number of thick fleshy roots. The consequence of this is that the former does well in moist soils, whilst the latter thrives also in very dry soils. From this we see that the development and nature of a vine's root-system is of the greatest importance, particularly when we wish to use it as a stock for grafting other varieties (or species) on, as is now so commonly done in order to protect the European vines against phylloxera. Later on we shall see the variety grafted

---

[1] According to Kroemer in Babo u. Mach, *l.c.* vol. i. p. 261.
[2] In Babo u. Mach, *l.c.* vol. i. p. 267.

on a stock can exercise a great influence on the latter's development, and can thus enhance or diminish its usefulness in different soils.

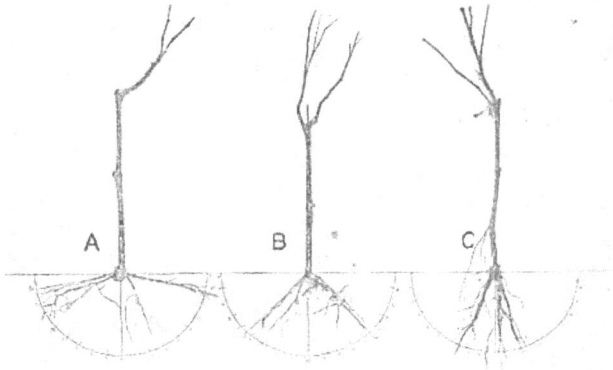

Fig. 4.—Measuring the angle of geotropism. A Riparia gloire ; B. Rip. × Rup. 3309 ; C. Rup. du Lot. From Guillon, *Étude générale de la vigne*, 1905. Masson et Cie., Paris.

Direction of Roots in Soil.—When we grow different varieties and species in the same soil under identical conditions, we find that the roots of some of these grow in an almost horizontal direction or relatively near the surface, whilst those of others show a steeper downward growth. This phenomenon falls under the general conception of geo-tropism, whereby is expressed the property of the various parts of a plant to grow in a definite direction with reference to the direction of the force of gravity.

The directive influence of the sun (*heliotropism*) does not affect the roots, as these grow underground, and is therefore negative.

According to Guillon,[1] the directive influence of the soil moisture (*hydrotropism*) is usually stronger than that of geotropism. The roots are inclined to follow the moisture. Here we must certainly only take heed of the main roots. When we deal with a species or variety whose roots are inclined to stay near the surface of the soil, we may expect that it will sooner suffer from drought than where the opposite is the case. The latter varieties, as, for instance,

---

[1] J.-M. Guillon, Etude generals de (a vigne (Paris, 1905), p. 181.

Rupestri du Lot (see Fig. 4), will no doubt resist drought better in deep soil, but will perish sooner in wet soil.

Under the *angle of geotropism*, Guillon understands the angle made with the vertical by the main roots originating in the root-stump, *i.e.* their deviation from the plumb-line. In order to determine this angle, he allows cuttings to develop roots in a nutrient medium such as that of Knop, where nothing hinders the direction of growth of the roots, or he grows them in deep, moist, loose soil, and observes for some years the directions in which their roots grow. Only the directions of the main roots are here taken into account. The following table contains some of the results obtained by Guillon[1] in his experiments conducted at the Viticulture Experiment Station of Cognac:

VALUE OF THE ANGLE OF GEOTROPISM OF THE BOOTS OF SOME OF THE PRINCIPAL AMERICAN STOCKS

Rupestris du Lot	20°
Berlandieri Resseguier No. 2	30°
Aramon × Rupestris Ganzin No. 1	40°
Aramon × Rupestris Ganzin No. 2	35°
Riparia × Rupestris 3309	45°
Chasselas × Berlandieri 41 B	45°
Riparia × Rupestris 101-14	60°
Mourvedre × Rupestris 1202	60°
Berlandieri × Riparia 420A	65°
Berlandieri × Riparia 34E.M.	65°
Riparia × Cordifolia-Rupestris 106-8	70°
Riparia Gloire de Montpellier	80°

2. THE STEM

That part of the vine to which the leaves are attached is called the *stem*. It unites the leaves with the roots. The parts of the stem to which the leaves are attached are called nodes, and the part between two *nodes* is called an *internode*.

(a) The Stem of the Seedling

---

[1] J.-M. Guillon, *l.c.* p. 194

In the first year the little stem is weak. It grows straight up and often forms only between six and ten leaves above the two cotyledons, and these leaves are grouped in a spiral around the little stem. On the Stellenbosch University Farm I have observed seedlings out of hybrid seeds growing in their first summer to a height of over 3 ft. and forming over 30 leaves. After about the 12th leaf the spiral grouping of the leaves on the stem ceases and the leaves are grouped normally as on the cultivated plant. The tip of the seedling's stem stands straight up, whereas that of growing canes on ordinary vines is always curved. In every leaf-axil, also in that of the cotyledons, there is a bud out of which a shoot can be formed during the next growing season with its leaves right and left of its axis. It is usually only after five or six years that this seedling commences to form flowers and develop fruit.

(b) The Stem of the Cultivated Plant

The Permanent Stem.—As in viticultural practice the vine is always propagated by means of cuttings, the stem at first is part of a cane of which the smaller portion is above ground and the rest underground. On this stem the roots and new canes are formed. In the first winter these canes are cut back and some are cut away clean against the young stem. This is repeated every year with the result that, in course of time, the permanent stem or *trunk* becomes divided into *arms*, on which are situated the pruned back canes or bearers. The underground portion of the originally planted cutting becomes the *root-stump*. It usually grows thicker from the bottom upwards. This is easily understood when we bear in mind that higher up the root-stump some roots still emanate.

The trunk gets thicker than the root-stump, but the transition is gradual. Only with grafted vines do we find the curious phenomenon of a striking thickening of the stem or trunk just at or above the joint of stock and scion. This thickening is caused after grafting by the cutting and the subsequent disturbance of the tissues for conducting the dissolved plant food, which in turn, to a greater or lesser extent, causes an accumulation of the descending elaborated plant food, in consequence of which an extra amount of growth and the thickening of the stem takes place at and just above the joint. Where scion and stock are of the same variety, the amount of the disturbance is the least, and hence in this case we see hardly any thickening of the stem near the joint. But where they are of different varieties, and worse still, where they belong to different species (as always happens when European vines are grafted on American stocks), the thickening can be very considerable, although in such cases it also varies greatly. We shall return to this point when discussing the influence of grafting.

The Bark— The bark of the growing shoots can be smooth or ribbed, glabrous or more or less covered with short straight hairs (pubescent) or with woolly hairs (cobwebby). Its colour may be green with or without reddish stripes, yellowish green, reddish or almost red, or greenish white. It may be covered with *bloom*, a waxy layer, or not. The ripe canes may also be smooth or ribbed, slightly pubescent (seldom) or glabrous, covered with bloom or not; their colour may be light, dark, reddish or yellowish brown, or grey; the bark may show brown or almost black spots and may not. In any case the bark adheres firmly to the cane in the first year.

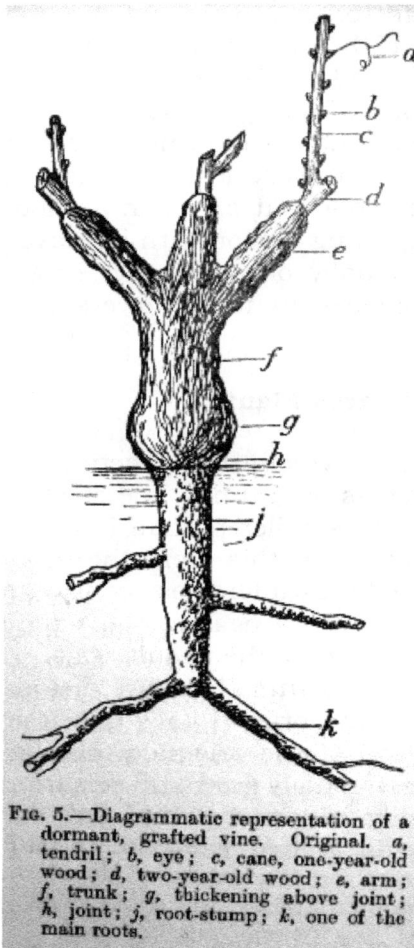

Fig. 5.—Diagrammatic representation of a dormant, grafted vine. Original. *a*, tendril; *b*, eye; *c*, cane, one-year-old wood; *d*, two-year-old wood; *e*, arm; *f*, trunk; *g*, thickening above joint; *h*, joint; *j*, root-stump; *k*, one of the main roots.

In the second year the bark becomes a little looser, whilst from the old wood of the trunk it peels off in flakes or strips. This gives the old trunk its peculiar appearance. At the same time this fairly thick layer of dead bark protects the permanent stem against the hot rays of the sun and prevents it from cracking. On the other hand, it offers shelter for the

34

insect pests and fungoid diseases for hibernating, whence they emerge to attack the vine during its growing season.

Where this holds good tor the European varieties, I must point out that the American vines we use as stocks form a smoother and thinner bark on the old wood than the former. Hence we must expose as little as possible of the American stem above ground, as otherwise it might easily crack during hot dry weather, and thus the whole vine's life would be threatened.

The Buds and Eyes.—In view of their great importance in grape-growing, and of the fact that the shoot (cane) is only the product of the development of a bud, we shall discuss it before studying the other parts of the stem. Superficially considered, we may say that in every leaf-axil a new bud is formed on the young shoot. As the leaves are grouped alternately to the right and left of the stem, the same naturally holds good for the buds. On closer examination it soon becomes apparent that as a rule two buds are formed in every leaf-axil.

The smaller of the two generally develops during the growing season in which it was formed, and then gives rise to the formation of a *sucker* or *lateral*. On an unpinched and vertically growing shoot this sucker usually remains short, its wood does not ripen, and it drops off in winter. Where the shoot, however, grows horizontally or is pinched at an early stage of its development, the sucker may develop into a strong lateral. By pinching a shoot that grows horizontally and removing the greater part of the suckers whilst they are still small (say 1-2 in. long), the remaining ones will develop more vigorously, so that they can be used as bearers when pruning during the next winter. This is illustrated in Fig. 96. Under summer pruning I shall revert to this point. Frequently these laterals bear fruit the same season they are formed. Such grapes are known as *second crop*.

The larger bud consists in reality of three buds, as is seen in Fig. 6, and hence such a compound bud is called an eye to distinguish it from a single bud. Unfortunately the term "bud" is frequently used for "eye". It is very desirable that in describing the vine we should strictly adhere to Hedrick's[1] definition of a *bud* as "an undeveloped shoot" and an eye as "a compound bud". The larger bud which we see

---

[1] U. P. Hedrick, Manual of American Grape-growing, 1919, p. 304.

with the naked eye on a vine shoot or cane is really an eye, and therefore I shall henceforth call such buds "eyes" according to the more correct terminology. Unfortunately Hedrick himself does not adhere strictly to his own definition of a bud and an eye on the very page (304) where he defines them.

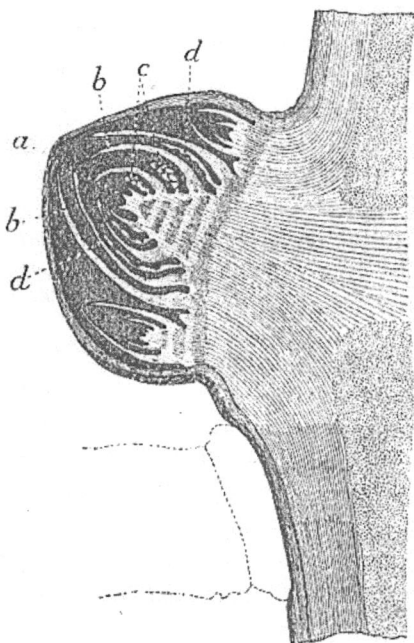

FIG. 6.—Eye in longitudinal section. Strongly magnified. *a*, main shoot; *b*, rudimentary leaf; *c*, rudimentary inflorescence; *d*, rudimentary tendril. From Babo u. Mach, *Handbuch des Weinbaues*, 1923. Paul Parey, Berlin.

In contrast with most fruit trees on which we can distinguish fruit-buds from leaf-buds we cannot do this with the vine. Here we have to do with leaf-buds and with leaf- and fruit-buds in one. Thus the larger eye, which remains dormant till the next spring, usually has three buds, namely, the central or *main bud* and the *secondary buds* on either side of the main bud. This is well illustrated in Fig. 6. Of these the secondary buds usually do not give fruit-bearing shoots, whilst the main bud frequently develops into a fruit-bearing shoot. In Fig. 6 we can see the rudimentary inflorescence in the central bud. As we see the eye and not the buds it contains when looking at a vine shoot or cane, it would be more appropriate to speak of *fruit-eyes* and *wood-eyes* rather than fruit-buds and wood-buds, it being understood that a fruit-eye is one containing a bud which will develop into a fruit-bearing shoot, and a wood-eye being one which does not contain any bud that can develop into a fruit-bearing shoot.

According to Ottavio-Marescalchi[1] the wood-eyes are usually larger and thicker. It does not follow, however, that

---

[1] Ottavio-Marescalchi, *I principii della viticoltura*, 1909, p. 200.

36

all the thick eyes contain buds that will develop into fruit-bearing shoots. It is only by examining an appropriate section of such an eye under the microscope that we can tell for certain whether it contained a bud that could have developed into a fruit-bearing shoot.

The main bud usually contains two rudimentary inflorescences, which may be as many as four. On Alicante Bouschet, Seibel 127 and several other self-bearers, shoots with four inflorescences frequently occur. Usually in these cases the lower two develop into normal bunches whilst the upper ones form only small bunches. Where the shoot that developed out of the main bud is killed by frost or hail, or is broken off, one of the secondary buds usually develops into a fresh shoot, which seldom bears fruit.

On carefully examining such a dormant eye, particularly when it starts swelling in spring and is on the point of bursting, we notice that it is covered by two brown scales. Just below them we further notice a number of woolly threads which are a bad conductor of heat and thus help to protect the enclosed buds against damage by cold in winter. Hence we find that practically all opening buds are at first covered by a mass of brown woolly threads which remain visible for some time at the basis of the young shoot.

Position of Wood- and Fruit-eyes on the Cane.—As the microscopic investigation of the dormant eyes shows, and as will be visible on the shoots developing out of them in spring, the eyes near the basis of the cane (*i.e.* the place where it comes out of the two-year-old wood) frequently do not contain a bud which can give rise to a fruit-bearing shoot, whilst those farther away generally contain buds which can develop into fruit-bearing shoots. Thus we usually find the fruit-eyes some distance away from the basis of the cane. The different grape varieties show considerable differences in this respect, and this has to be borne in mind when pruning them.

As Hermitage (Cinsaut) and Muscat of Alexandria (Hanepoot) usually have fruit-eyes near the basis of the cane, they are pruned short, only two well-developed eyes being left. Where we know that the lower eyes of Sultana (Thompson's Seedless), Ohanez (Almeria grape), and Cabernet Sauvignon are usually only wood-eyes, we prune these varieties long, leaving at least 10-12 good eyes on the long bearers or rods that are to produce fruit.

Formation of Fruit-eyes.—According to Müller-Thurgau and Behrens[1] the formation of the rudimentary inflorescences in the fruit-eyes begins very early, when the young eyes are still soft and just show the first strong swelling. As a rule it commences with the lower, *i.e.* oldest eyes which are strong enough, and then proceeds upwards to the higher eyes as they develop on the growing shoot. In Europe it commences in the lowest eyes for the first bunch about the middle of June, for the second bunch at the beginning of July, and about the beginning of August all the rudimentary inflorescences that can develop into bunches during the next year have been formed. The further development of these inflorescences proceeds till October, when the eye enters upon its winter rest, which lasts till March of the following year. Now the eye begins to develop further until it bursts and a new shoot with fruit develops out of it; but after the winter rest not a single new inflorescence is formed in any resting eye; only such in florescences as are then in existence develop further.

In the Western Province of the Cape the corresponding dates will be respectively about the middle of November, beginning of December, beginning of January, March, and August.

From this we clearly see that the possible size of the vintage is already determined during the preceding growing season. In the northern hemisphere the possible size of the 1925 vintage is determined in 1924, whilst in the southern hemisphere it is determined during the 1923-24 growing season. In the lowest fruit-eyes the primitive development of the inflorescences is complete before the winter rest begins. With eyes high up the cane such is usually not the case, and hence in their later development they frequently produce something which is half inflorescence and half tendril.

It will now be clear that the development of the buds is of the greatest importance to the grape-grower. It is therefore his business to do all he can from a few weeks before the time of bloom and for the next two months to promote a good development of the eyes which he will have on his bearers during the following season, as upon their development the possible size of the resulting vintage will in the first instance depend. He can do this, for instance, by

---

[1] Quoted from Babo u. Mach, *l.c.* i. pp. 278-279

topping the young shoots and removing the superfluous ones, as we shall see later on. Again, it is undoubtedly necessary to ensure, through sufficient and suitable manuring and cultivation, a vigorous but not excessive growth in the vines, if they are to give good crops. Heat and sunshine during the first four months of the vine's growing season (September till the end of the year at the Cape) will undoubtedly also exercise a favourable influence upon the formation and development of the fruit-eyes.

When the first shoots are eaten off by animals or are badly damaged by hail or frost so that their inflorescences are gone, we cut the young shoots close to the bearers in order to allow the other eyes to develop, which sometimes form some fruit-bearing shoots.

Adventitious Buds (Eyes).—Out of the old wood, buds sometimes develop which previously were not to be seen. They are called *adventitious buds*. The shoots developing out of them are called *water sprouts*, and these usually bear no fruit. Where they are not radically removed early in summer or during the winter pruning, some of the remaining buds may subsequently develop fruit-bearing shoots. They serve an excellent purpose when they are required to form a new arm on an old trunk, for they soon give normal fruit-bearing canes. When an old trunk is cut off, the numerous young shoots issuing from it reveal the unsuspected presence of a large number of such adventitious buds. To a large extent, at any rate, these buds seem to me to be simply the buds at the base of the canes which do not develop in spring and remain sleeping on the vine from year to year. In this way the trunk will in course of time have a large number of such buds of which some will develop when a part of the permanent stem or its whole upper portion is suddenly removed, and thus reveal their presence once more.

The "Bourgeonnement" or Very Young Shoot.— Under "Bourgeonnement" Pulliat[1] understands the early stage of development of the young shoot from the time the eye bursts until the young shoot is so far developed that the first inflorescence just becomes distinctly visible. For the description of a vine an accurate knowledge of the young shoot at this stage is of considerable value. The very early stage, when the bud-scales have just burst and the

---

[1] V. Pulliat, *Mille varietes de vignes*, 3rd ed., 1888, p. xiii.

developing bud is just visible, is of little value for this purpose, as the bud in nearly every instance is then more or less densely covered with brown woolly hairs. From the stage when the separate young leaves show themselves distinctly until the first inflorescence becomes distinctly visible, this young shoot offers valuable characteristics for distinguishing between the different varieties of the same species. The *hairiness* or *pubescence* can then still be strong, moderate, or already very slight on both sides of the little unfolding leaves or only on their under (then really outer) side, and on the little leaves as a whole or only on their nerves. The *colour* of this young shoot is of great value in describing a variety. It can be green, yellowish green, garnet red, reddish violet, etc. The colour is most strongly developed along the edges of the young leaves, but sometimes it is spread over the whole leaf-blade. Sometimes, again, it is particularly developed upon the felty covering of hairs on the under side of the young leaves of certain varieties, especially towards the end of the stage of development we are here concerned with.

The Young Shoot.—From now onwards the young shoot develops rapidly. Opposite the third, fourth, or fifth leaf we see the first inflorescence according to variety. Frequently we notice a second inflorescence opposite the next leaf. The shoot bears up to four inflorescences. The number depends amongst other things upon the variety. Where the inflorescences cease, we find tendrils opposite the leaves. This can be seen on Fig. 7, c.

The Cane.—When the shoot becomes lignified and ripens, and particularly when it has shed its leaves, we call it a cane. When we examine such a cane in winter we notice that it is thickened in some places, and also that it is here that we find the eyes and tendrils, and where formerly the leaves were borne. These particular places can be considered as joints in the stem (cane) and are called *nodes*. The part between two nodes is called an *internode*. The different species and varieties differ considerably in the amount of thickening at the nodes and in the length of the internodes.

Also the internodes are shorter near the base of the cane and towards its end than in between. The actual dimensions of the cane of one and the same variety depend largely upon the conditions under which they have grown. Shoots that grow very vigorously will give thicker canes with longer internodes than those growing less vigorously.

A   B   C

FIG. 7.—A. Branching shoot of Sultana with lateral in leaf axle below branching. B. Vigorous shoot of Sultana with a branch that has developed into a tendril carrying leaves. C. Shoot of Ferdinand de Lesseps, showing the relative positions of the leaves, inflorescences, and tendrils on the shoot. Original.

On a longitudinal section of a cane, such as in Fig. 8, B, we distinguish the bark, wood, and pith in the internode. The relative thicknesses of these three may vary considerably with different species and varieties. Sultana has a thicker pith than most other European or Vinifera varieties. The pith is a dead, brown, soft tissue. We notice that the pith is interrupted at the nodes by a layer of woody tissue called the *diaphragm*. We find this with all Vitaceae except *Vitis rotundifolia*, in which the pith runs uninterruptedly through the cane.

Hedrick[1] points out that Prof. Millardet of Bordeaux was the first to draw attention to the value of the diaphragm in distinguishing between the different species. It is the presence or absence of a diaphragm, its thickness and its shape, which are here of value. Thus the diaphragm of *Vitis riparia* is very thin, that of *Vitis rupestris* is somewhat thicker, whilst that of *Vitis cordifolia, Vitis aestivalis,* and *Vitis labrusca* is thick. This can best be studied in canes one year old.

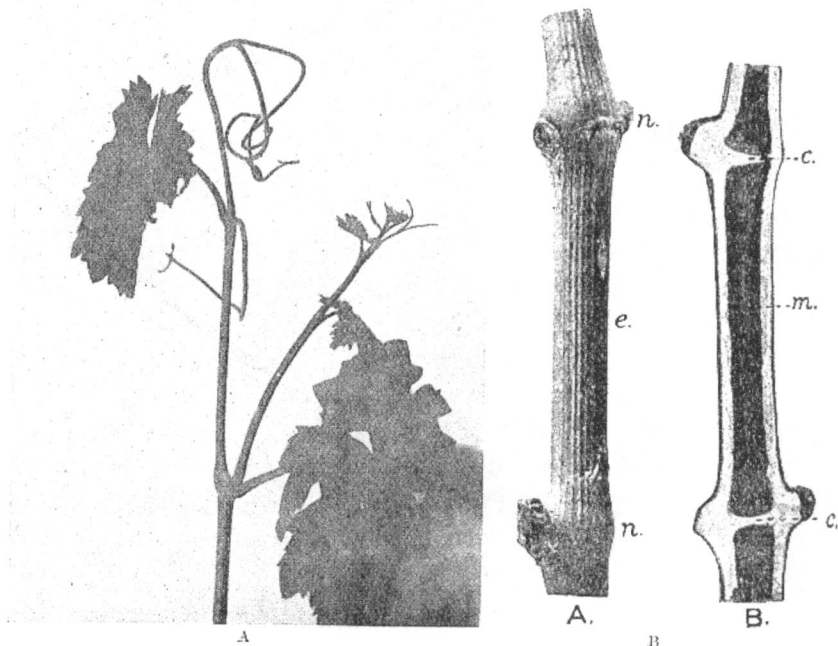

Fig. 8.—A. Shoot of Blauer Portugieser that has developed into a tendril, with a lateral to continue the shoot's further growth in length. Original.
B. Part of cane in winter. (A) External view: *n*, node; *e*, internode. (B) In longitudinal section: *c*, diaphragm; *m*, pith. From Guillon, *Étude générale de la vigne*, 1905. Masson et Cie., Paris.

Position of Inflorescences and Tendrils on the Shoot.— As was previously stated, we find the first inflorescence opposite the third, fourth, or fifth leaf. The actual position varies with different varieties. On Muscat of Alexandria the first inflorescence is lower down on the shoot, whilst it is fairly high up on Raisin blanc.

---

[1] U. P. Hedrick, *l.c.* p. 102

In some seasons the inflorescences are on the average higher than usual, and farmers regard this as indicating a poor vintage. They are quite right, for in such a season we find many shoots with only one inflorescence where we are accustomed to find two. Where the inflorescences cease, the tendrils begin. If we number the nodes on the shoot 1, 2, 3, etc. (starting at the basis), and the first inflorescence occurs at 3, the next inflorescence or tendril will be at 4 but on the other side of the shoot, the next at 6 and on the same side of the shoot as that at 4, the next at 7 on the same side of the shoot as that at 3, the next ones at 9 and 10, 12 and 13, etc. After every two inflorescences or tendrils comes one node without either inflorescence or tendril. This *discontinuity* of the tendrils we meet with in all the species of the genus *Vitis* except *Vitis labrusca*, which has an inflorescence or a tendril opposite nearly every leaf, so that its tendrils are mostly continuous. The above gives the general rule for the grouping of the tendrils on the shoot, but slight deviations from this rule occur from time to time.

Origin and Nature of the Tendrils.—The tendrils of the vine are stem tendrils or *metamorphosed shoots*.[1] The inflorescence is a panicle taking the place of a tendril; intermediate forms between inflorescences and tendrils are of frequent occurrence (Strasburger, *l.c.* p. 658). There are various views concerning the morphological nature of the tendrils (see Viala-Vermore[2] in vol. i. pp. 284-298). Guillon[3] regards the tendril as the product of a normal bud on the stem (cane), a sort of sucker which developed rapidly whilst remaining united ("soudé") with the stem till it reaches the next node. Kroemer in Babo u. Mach[4] regards the tendrils of the vine as thread-like stems which have arisen through a transformation of fertile terminal buds which have been displaced to one side by the development of the axillary shoot. As a proof of this origin of the tendril he mentions the fact that wherever it divides into two there is always a bracteal leaf underneath the point of division. The latter represents a transformed leaf. This is proved by the fact that it some times grows out to a normal green leaf, as can be

---

[1] See Strasburger's *Text-book of Botany*, 1921, p. 182.

[2] Viala-Vermorel, Traité général de viticulture, ampélographie, 7 vols., 1901-10.

[3] J.-M. Guillon, *l.c.* p. 236.

[4] Babo u. Mach, *l.c.* vol. i. p. 279.

well seen on Fig. 7, B, which represents a vigorously growing Sultana shoot. The tendril on the left is rather like a shoot in appearance and has a normal green leaf at its point of division, but no bud has developed in this leaf-axil.

Fig. 7, A, represents a vigorously growing shoot of Sultana and gives us an example of *dichotomous branching* in the vine. Although this phenomenon is not of frequent occurrence, I have seen it several times on very vigorous shoots. Further, it is interesting to note on Fig. 7, A, that, whereas the leaves on the undivided shoot are to the right and left of it, those on the branch shoots or daughter axes are in front of and behind these. This means that the position of the leaves on the stem above the point of branching showed a difference of 90° compared with the leaves below the branching.

Fig. 8, A, represents a shoot of Blauer Portugieser or Portugais bleu; here we have the rare case of a shoot developing into and ending in a tendril. Its uppermost leaf had no bud in its axil. The lateral in the bottom leaf-axil at the last normal node had already started a fairly strong growth, and it would have continued the axis if the shoot had not been cut for taking the photograph. Here we have a direct proof of the tendril being a metamorphosed shoot. This shoot grew on a vine grafted on Jacquez in my Ampelographic collection on the Stellenbosch University Farm.

That inflorescence and tendril possess a common origin is proved by the following facts: they occupy the same relative positions on the stem (shoot); both show bracteal leaves underneath every bifurcation; the stalk of the bunch of grapes develops sometimes a secondary bunch and sometimes a tendril at its node; lastly, intermediate forms between inflorescence and tendril are frequently observed.

The *function* of the tendrils is to tie the shoots to something, and thus to expose the leaves more to the sunlight. In its native state the vine is a climbing plant, and therefore it needs the tendrils. They are irritable to contact and encircle shoots, leaves, grape bunches, wires, branches of trees, and anything with which they come in contact. The young leaves when encircled are greatly hindered in their development and subsequent functions for the vine. Tendrils that encircle and penetrate bunches of table grapes should be removed whilst young, as later on this operation will damage the bunch.

3. THE LEAF

In many respects the leaf is one of the most important organs of the vine. Later on we shall see that it plays a most important part in the nutrition of the vine, as it is the seat of the process of photosynthesis. Leaves through evaporation of water help to keep the vine cool and assist in its other life functions. They also protect the grapes against sunburn by keeping them in the shade.

The full-grown foliage leaf of the vine consists of the *leaf-stalk* (petiole) and the *leaf-blade* (lamina). The latter is generally traversed by five main nerves or veins which arise together out of the leaf-stalk where it is attached to the blade. Lateral veins spring from all the main nerves, and from these first lateral veins others spring. On holding up the green leaf-blade against the light we see that it is traversed by a fine network of weaker veins which are interconnected (see Figs. 9 and 37). The vine-leaf is therefore *netted veined*. In between the network of veins which forms the *leaf-skeleton* lies the *mesophyll*. In this are embedded the finer veins.

The *leaf-stalk* serves to connect the leaf-blade with the stem and to place it suitably with respect to the light. It therefore also serves for transporting the solution of absorbed mineral food or raw sap into the leaf, and the elaborated sap from the leaf to the other parts of the vine.

The leaf-stalk varies from long to short and from stout to slender according to the species and variety of vine, and to the conditions of growth. Its cross-section is usually round with a slight depression on the upper side (sometimes very pronounced). Its colour is green, sometimes reddish (particularly on the sunny side), and more or less striped. Some-times, however, it has very little colour. Its surface can be glabrous or more or less pubescent. The hairs may be woolly or may be short, stiff, and straight (well illustrated on Rip. × Rup. 3306).

The leaf-stalk gives the leaf a certain amount of mobility which helps to protect it against damage by wind, and further enables the leaf-blade to take up the most favourable position for the vine relative to the direction of the rays of the sun.

The nerves serve to transport the aqueous solution of plant food (both raw and elaborated) in the leaf and to keep it spread out for proper exposure to the sunlight. For a careful description of a vine and for purposes of identification, use is made of the relative lengths of the main nerves, the size of the angles between them, and of the angles between some lateral nerves and some main nerves.

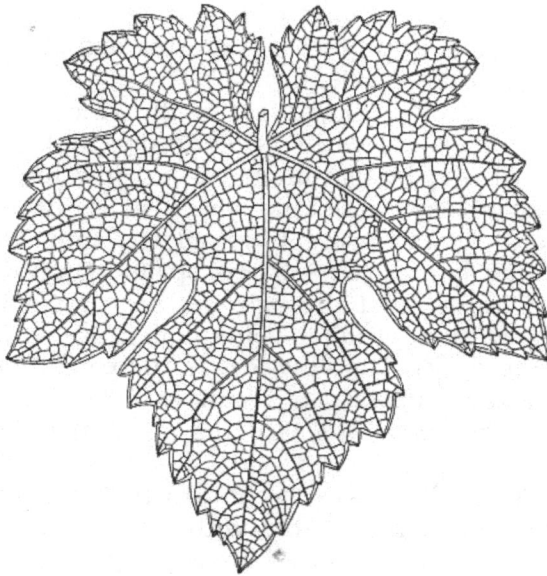

Few varieties (of vines) possess entire leaves. Usually they are *lobed*. Fig. 9 shows us a 5-lobed leaf. The recess 1 or bay between two lobes is called a sinus. The sinus at the attachment of the petiole to the leaf is known as the *petiolar* sinus. The next two are the basal sinuses, and the two sinuses towards the apex of the blade are the lateral sinuses (see Hedrick, *l.c.* p. 307). The development of these sinuses is of considerable value in describing any vine. Vine leaves can be entire, or 3- or 5-lobed (see Fig. 10). We seldom meet with 7-lobed leaves (*e.g.* Laubscher's Gem). On the Parsley Grape (Chasselas Ciotat) the divisions are so deep that the lobes terminate on the five main nerves before they meet and these lobes are again further divided, as can be seen on Fig. 10.

The margin of the vine leaf is always more or less divided. These divisions are known as *teeth*. The teeth may be rounded or pointed, short or long, narrow or broad, of equal size or varying fairly regularly in size.

The surface of the leaf-blade may be rough or smooth, uneven or even [in some instances (*e.g.* Alicante Bouschet) the marginal zone is curved downwards].

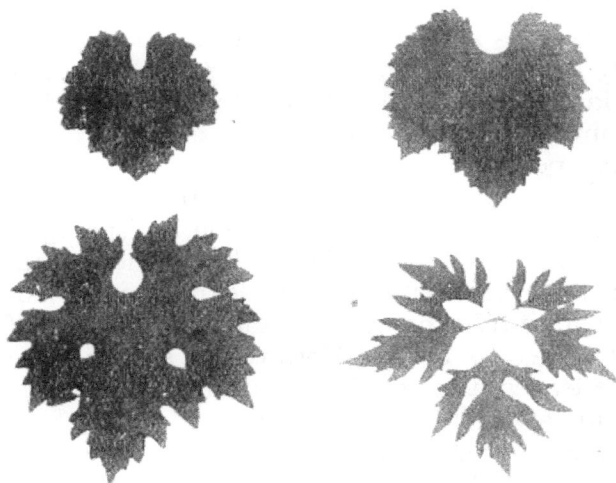

Fig. 10.—Leaf forms: entire, 3-lobed, 5-lobed (Cabernet Sauvignon), Parsley leaf shaped (Chasselas ciotat). Original.

The *pubescence* of the leaf is a very important means of identification, and is used in the classification, of vines. The very young leaves of nearly all vines are more or less pubescent. As they grow older the pubescence usually decreases. It may consist of short straight hairs, loose woolly threads (tomentum), cobwebby threads, or a dense felt-like tomentum. These hairs can be isolated or dense, can occur only on the petiole and nerves or also on the mesophyll. They can occur on both sides of the leaf or, as is generally the case, only on the under side. The full-grown leaves of some varieties are glabrous on both sides, *e.g.* Rosaki.

The *colour* of the full-grown leaf may be dark green, green, light green, yellowish green, or reddish.

It is always more intense on the upper than on the under surface. In the young leaves green is less prominent, whilst reddish violet, garnet red, and yellow are frequent, particularly on the margins of the leaves. And we frequently find that, where the young leaves show a dense felt-like tomentum on the under surface, this is coloured reddish violet (*e.g.* Ferdinand de Lesseps, Jacquez, etc.), especially in the marginal zone.

With some varieties, when the grapes turn colour, the leaves begin to develop red spots (*e.g.* Gros Colman), or the whole leaf gets fairly coloured (*e.g.* Pontac). Sometimes the leaf now develops a large number of red spots (*e.g.* Alicante Bouschet). When the grapes are ripe they develop more colour. Many leaves of varieties with red juice, like Pontac and Alicante Bouschet, are almost entirely red when the grapes are ripe.

Most varieties do not develop the *autumn colours* in their leaves till they are near the point of being shed. We then notice that the leaves of the white varieties turn yellow and later light brown, whilst those of the red and black varieties turn red. In the latter instance, only parts of the leaf are coloured red as a rule. However, it sometimes happens that whole leaves turn red. The autumn colours are valuable in describing a variety.

Finally, I wish to point out that some varieties have thick leaves (*e.g.* Gros Golman) and others thin leaves (*e.g.* Rosaki); that some leaves are brittle and others tough; that some are dull and others shiny. The nerves of some varieties are sunk well in the leaf; of others they are prominent, particularly on the under surface of the leaf. The nerves are generally coloured green, but on some varieties (*e.g.* Prune de Cazouls and Laubscher's Gem) they have a red or reddish colour. Not a single leaf is precisely like any other leaf on the same shoot or vine, but the principal characters remain sufficiently constant to provide valuable data for determining the identity of a species or variety. It is also remarkable that no two halves of a leaf are precisely alike. The vine leaf is more or less asymmetric.

4. THE FLOWER

On the vine (*Vitaceae*) the individual flowers are grouped in a *racemose inflorescence* which on the vine is a *panicle*. We have already seen that the inflorescence is present in the bud in a fruit-eye. When still very young, the inflorescence may be glabrous or pubescent. Its colour may be light green, green, and sometimes red (*e.g.* Rupestris du Lot), or it may be green with a reddish tip (*e.g.* Sultana). These characters can be made use of in describing a vine.

With some varieties (*e.g.* Muscat of Alexandria, Gros Colman, etc.) the inflorescences are fairly big and with others small (*e.g.* Hermitage, Barlinka, etc.), whilst the full grown bunches may not differ so much in size. At this stage of their development the former therefore give us the impression that they are going to give us a much heavier crop than the latter.

As we have previously seen, the inflorescence is closely related to the tendril (it takes its place on the stem). On the *peduncle* or stalk of the flower cluster (inflorescence) there is a node or joint at which we frequently find a side cluster developed or a tendril bearing some flowers and, later on, berries. Inversely, we sometimes find a tendril bearing some berries.

About two and a half months after the buds have burst, the flowers start blooming. Now for the first time we see the individual flowers. The various species of the genus *Vitis* are all *polygamous-dioecious*, *i.e.* some varieties of the same species bear only male flowers, whilst others bear *hermaphrodite flowers* [the former have only stamens whilst the latter stamens and pistils]. According to Guillon, *l.c.* p. 239, this seems to be the rule for all species without exception.

Most European vines (*Vitis vinifera*) have hermaphrodite flowers. On Fig. 11 we see such a flower just before opening, opening, and when fully open. The flower is attached to a short stalk, the *pedicel*, which is thickened at the top and a low rim around the base of the flower, which can still be recognised by its fine low teeth as the remainder of a *calyx* with 5 sepals. Inside the calyx is the pentamerous *corolla* with 5 green petals united by their tips, which become loosened at the base when the flower opens. The corolla is shed as a little cap (see Fig. 11, *bk*) with the bottom ends of the petals curled outwards and upwards.

The opened flower shows the male and female genital organs. The former are called *stamens* (generally 5 in number); and consist of the *filament* and the anther. Each anther is formed of two *thecae* or pairs of pollen-sacs joined by the continuation of the filament, the *connective*. Each pollen-sac is filled with *pollen*. When the flower is in full bloom, the pollen-sacs in each theca are in free communication with each other (after the septum has broken down), and both sacs open by a common split in the wall and pour out the pollen. The pollen consists of small, oval, yellowish grains which are enclosed in a cell wall with three thin walled bands running from pole to pole, in the middle of which a *germinal pore* for the subsequently developing pollen-tube is to be seen. Between the 5 stamens and at their base we find 5 *nectaries* which give off the pleasant smell of blooming grape flowers and from which honey can be made. They serve to attract insects and in this way promote pollination, and particularly cross-pollination.

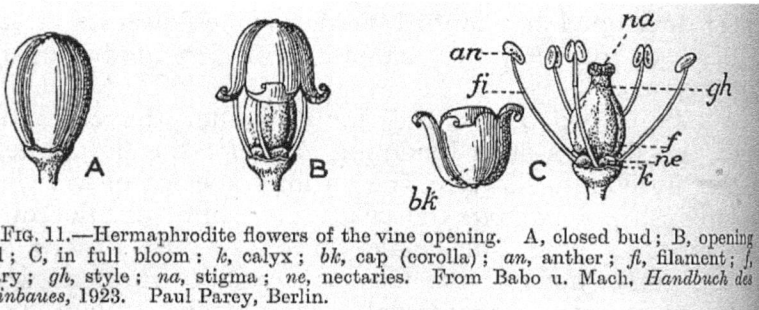

Fig. 11.—Hermaphrodite flowers of the vine opening. A, closed bud; B, opening bud; C, in full bloom: *k*, calyx; *bk*, cap (corolla); *an*, anther; *fi*, filament; *f*, ovary; *gh*, style; *na*, stigma; *ne*, nectaries. From Babo u. Mach, *Handbuch des Weinbaues*, 1923. Paul Parey, Berlin.

The *number of stamens* is usually 5, but can be 4-8, according to Viala-Vermorel, *l.c.* vol. i. p. 126. They maintain that 4 stamens are very rare and that they occur only on female flowers with rudimentary stamens. I have, however, frequently observed hermaphrodite flowers with 4 stamens on Pinot Chardonnay and Sabalkanskoi, both of which, in my collection, are very subject to coulure and millerandage. The 4 stamens did not differ at all in appearance from those of flowers with 5 stamens. Hence it would seem that the opinion of Viala-Vermorel, quoted above, does not apply in every case.

On inflorescences of Muscat of Alexandria, Ferdinand de Lesseps, Raisin blanc (Servan blanc), and many other varieties I have found many flowers with 6 stamens next to ones with 5. They are almost too numerous to be regarded as exceptions. It would be more correct to say that the flowers have mostly 5 stamens with numerous instances of 6 stamens. On Ferdinand de Lesseps and Red Hanepoot I have found flowers with 7 stamens and 7 petals. Also, I have always found that the number of stamens is equal to that of the petals. Hence Viala-Vermorel, *l.c.* i. p. 126, are wrong in asserting that the number of the petals remains 5, although the flower may have 6 or more stamens. On counting the number of petals in 70 caps that dropped on to a sheet of paper which I held under an inflorescence of Red Hanepoot in full bloom, I found 30 (*i.e.* 43 per cent) with 5 petals, 37 (i.e. 43 per cent) with 6 petals, and 3 (*i.e.* 4 per cent) with 7 petals. As previously stated, the number of stamens was the same as that of the petals.

The stamens surround the female part of the flower called the *pistil* or *gynaecium*, which usually has the shape of a champagne bottle. It consists of two and very rarely of three united *carpels*. Each carpel, and in the present instance of a syncarpous gynaecium, each pistil consists of the *ovary*, *style*, and *stigma* (respectively *f, gh,* and *na* on Fig. 11, C). The ovary is considerably swollen, the style generally very short, and the stigma somewhat broader than the style. This is well illustrated in Figs. 11 and 12. The ovary here, therefore, is usually double, and each ovary contains two anatropous *ovules*.

The most important part of the ovule is the *nucellus*. This contains one particularly large cell in which the process of fertilisation takes place, and which is called *embryo-sac.* The nucellus is invested by an internal and external *integument*, which have a small passage, the *micropyle,* through which the pollen-tube must enter. The base of the nucellus where the integuments arise is called the *chalaza.* The *funiculus* is a short stalk which attaches the nucellus to the carpel. The *style channel (canalis stilinus)* is the opening in the style through which the pollen-tube enters the micropyle.

The structure of female flowers is similar to that of the hermaphrodite flowers. Their stamens, however, are different, being very short and bent downwards. See Fig. 12, B.

According to Rathay, quoted from Babo u. Mach, *l.c.* vol. i. p. 289, their pollen-grains cannot germinate and are therefore unable to fertilise the ovules. These pollen-grains are spherical and have no germinal pores. We find them on some cultivated varieties of the European grape (*e.g.* Bakator, Madeleine Angevine, White Malvasia), also on certain American varieties (*e.g.* Clinton and Solonis), and on some wild vines.

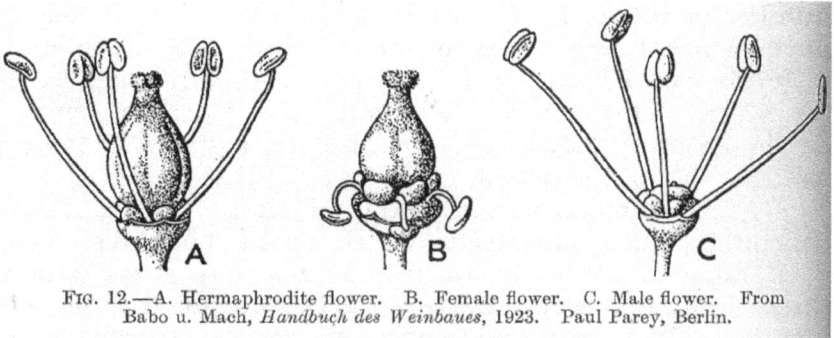

FIG. 12.—A. Hermaphrodite flower. B. Female flower. C. Male flower. From Babo u. Mach, *Handbuch des Weinbaues*, 1923. Paul Parey, Berlin.

Male flowers (see Fig. 12, C) we find mainly on wild and American vines, but they occur also on individuals of European varieties. According to Mach, quoted from Babo u. Mach, *l.c.* vol. i. p. 290, male as well as hermaphrodite flowers occur on the inflorescences of Cabemet franc, Blaue Kadarka, and Harslevelü. Sometimes, however, we find inflorescences with only male or only hermaphrodite flowers. The male flowers differ from the hermaphrodite flowers only in the build of the pistil, which is undeveloped, and has no stigma and style, but only a small hemispherical ovary which cannot be fertilised. Such a flower, therefore, cannot develop into a fruit, although its pollen-grains germinate quite easily and can fertilise flowers with a normal pistil.

According to Rathay's researches it follows that the wild varieties generally form only male and female flowers, which always occur on separate vines. On the cultivated varieties of the European grape we find as a rule only hermaphrodite and female flowers, which are likewise on separate vines. Rathay assumes that the hermaphrodite flowers have developed out of male flowers (see Babo u. Mach, *l.c.* vol. i. p. 291).

5. THE FRUIT

During full bloom pollen-grains come on to the stigma. Here they germinate under favourable conditions; the pollen-tube penetrates the tissue of the stigma and passes through the style channel and the micropyle into the nucellus, which is then fertilised. This results in the ovary developing into the fruit. The latter is a berry fruit and is our well-known *grape-berry*. The ovules become the grape seeds and the whole inflorescence becomes the grape *bunch* or *cluster*, which consists of *peduncle* or *stalk, pedicels*, and *berries*.

During its further development the bunch becomes longer and broader till it reaches its full development.

The Bunch. — Before entering on the discussion of its different parts, I wish to say something about the shape of the bunch.

Shape of the Bunch.—Some varieties have bunches of bout the same thickness from top to bottom. These we say are *cylindrical* in shape. Others again produce bunches of the most common shape, *i.e.* they are broad at the top and become narrower as they reach the point. These are *pyramidal* or *conical* in shape. Some varieties again form bunches of which the upper portion is cylindrical and the lower portion conical. These we describe as *cylindro-conical* in shape. Finally we find that some varieties generally have *round* or *globular* bunches.

With one and the same variety the shape of the bunches may sometimes vary considerably. While the bunches hanging free are long, those resting on the ground will often be fairly round. This, however, does not alter the fact that the prevalent shape of the bunch is a useful character when describing any variety.

The p*eduncle* or *stalk* branches repeatedly until we meet with the final branches to which a variable number of pedicels are attached. Each pedicel bears a berry. Below every branching of the stalk we find on the outside a bracteal leaf on which is small, almost colourless, narrow, curved, folded, without nerves, and soon shrivels up and disappears. They are particularly striking and well developed on White Prince or Centennial. At first the stalk is green, but it changes colour as the bunch, develops.

With some varieties (*e.g.* Bonnet de retord) the whole stalk becomes yellowish when the grapes are ripe, whilst with most varieties it is brown from where it comes out of the cane to where it starts branching (or sometimes before this), and the rest of it remains green. In the first instance the stalk remains soft and can easily be broken off with one's fingers; in the latter instance it is lignified and so tough that it must be cut. At the joint or node the stalk breaks off much more easily than anywhere else, particularly when it is not yet properly lignified.

The *pedicel* can be short or long, thick or thin, brittle or tough. For the production of high - class table grapes the pedicel should not be too short, as the ripe berries can then be farther apart and still nicely fill the bunch. Also, it is much easier to thin a variety with brittle pedicels (*e.g.* Prune de Cazouls) than one with tough pedicels. For the latter the thinning scissors must always be kept very sharp. Where thinning is done by the finger nails, the advantage of brittle pedicels is still greater. The pedicel gradually becomes thicker as it approaches the berry, and near where it enters the berry it shows a peculiar thickening which the French call "bourrelet". On some varieties it is more developed than on others. Its surface is generally rough and warty. The end of the pedicel projecting into the berry is called the *brush*.

The bunches of some varieties have a short peduncle, if we consider its unbranched portion only, whilst other varieties have long peduncles. Where the shoot grows more or less horizontally and the bunch hangs freely, it is inclined to develop a somewhat longer peduncle than where it grows straight up or obliquely. It, however, remains a fact that some varieties (*e.g.* Pontac or Teinturier, Stein, etc.) have very short peduncles, whilst others (*e.g.* Gros Colman, St. Jeannet tardif, etc.) have fairly long or even very long peduncles. In describing varieties of grapes, this should be taken into account.

The Berry. — Here we deal with the fruit of the vine. This constitutes its most valuable part. As the vine, in common with all other living organisms, wants to propagate itself, we have to regard particularly the ripe seed as its fruit and its most important product from the point of view of the vine.

From our own point of view, we regard the whole berry, and particularly its pulp, as the most important part, because we can eat it or drink it in the shape of grape-juice or wine. The berry consists of skin, pulp, and seeds (see Fig. 13, A).

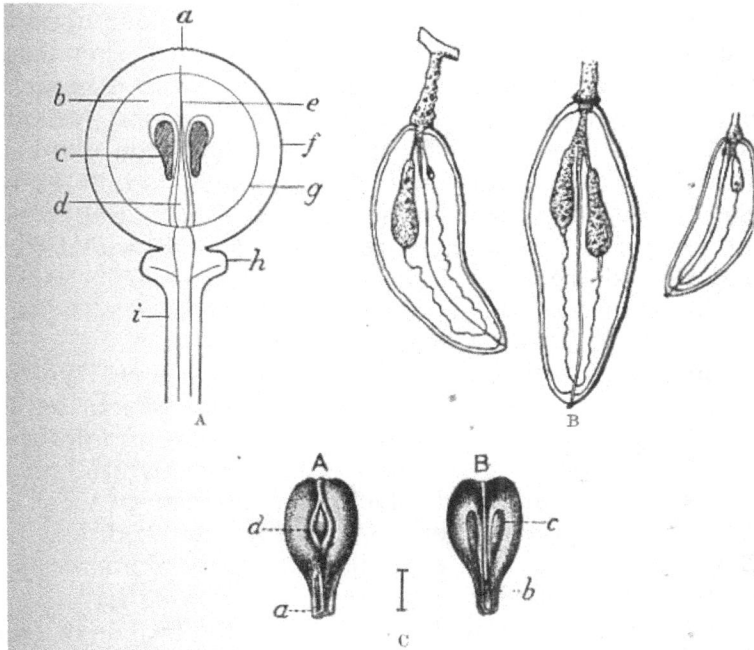

FIG. 13.—A. Structure of grape berry: *a*, mark of stigma; *b*, pulp; *c*, grape-seed; *d*, brush; *e*, central vascular bundles; *f*, skin; *g*, superficial vascular bundles; *h*, bourrelet (thickened part of pedicel); *i*, pedicel. From Guillon, *Étude générale de la Vigne*, 1905. Masson et Cie., Paris.

B. Berries of the White Acorngrape in longitudinal section (after Müller-Thurgau). From Babo u. Mach, *Handbuch des Weinbaues*, 1923. Paul Parey, Berlin.

C. Grape-seeds. A. Dorsal or back view. B. Ventral or front view. *a*, beak; *b*, raphe; *c*, seed-folds; *d*, chalaza. From Babo u. Mach, *Handbuch des Weinbaues*, 1923. Paul Parey, Berlin.

The **skin** has different colours in the various stages of its development. When the berry is young, the skin usually is of a green colour. This is why the unripe fruit is called green. A few varieties (*e.g.* Chasselas rose), however, develop a red colour in the skin at a very early stage; others do it a little later (*e.g.* Pontac). Bonnet de retord very early shows faint red stripes where the bloom will show in white stripes on the ripe black berries; White French shows a whitish skin when the berries have grown to about half their full size.

55

In its development the berry reaches a point where it becomes somewhat soft and changes colour. This is called by the French *véraison*. The skin of white varieties now becomes greenish white, yellowish white, light whitish green, whilst that of red varieties becomes pink and that of black varieties red or reddish black. From now the change of colour usually increases until the grapes are ripe. The skins of the white varieties then show a colour which varies from greenish, greenish yellow, whitish, light yellowish, golden yellow, brownish yellow, to reddish yellow or yellowish with a reddish tint. On the sunny side the colour is usually more strongly developed than in the shade. Red varieties at complete maturity have a light red or pink, light violet-red, full to dark red, reddish - violet or dark reddish - violet colour. Black varieties at complete maturity have a slightly reddish-black, violet-black, or jet-black colour.

The various varieties of grapes have their typical colours. But we meet with some varieties which agree precisely in all their characteristics with the exception of the colour of their ripe grapes. Such are: Pinot (white, red,- black), Mondeuse (white, red, black), Hanepoot or Muscat of Alexandria (white, red), Greengrape (white, red), Sultana or Thompson's Seedless (white, red), etc. This phenomenon occurs with old varieties that have been cultivated for a very long time. Sometimes, as with Greengrape, we find shoots on the same vine bearing respectively white and red grapes. Sometimes a white and a red bunch occur on the same shoot; sometimes white and red berries occur in the same bunch, and sometimes even individual berries are coloured partly white and partly red.

Some years ago I saw red next to white Sultana bunches on the same vine near Robertson. There was no difference between the white and red Sultana grapes except the colour. Mas and Pulliat[1] are of opinion that in such cases the black variety gave rise to a red variety out of which there later arose a white variety. Their reason for this belief is that we do not know of a white and black variety (type) where there does not also exist a red variety. According to this view we should expect to find black, red, and white, or black and red, or red and white, but not only black and white varieties or types of the same variety.

---

[1] Mas et Pulliat, *Le Vignoble* (3 vols.), vol. ii. pp. 119-120 ; Paris, 1874-79

The skin owes its colour to grains of pigment that are usually deposited in its three or four outer layers of cells. This applies to the bulk of the red and black varieties, the pulp in such cases is white or greenish but not red, and the fresh juice pressed out of such grapes is similarly not red. It is therefore possible to make white grape-juice and wine out of such grapes. This is actually done in the Champagne district in France, where white Champagne is made out of the black Pinot grapes.

With some varieties the pigment is deposited in the inner layers of cells of the skin (*e.g.* Teroldigo, Isabella, Mourvedre × Rupestris 1202, Jacquez, etc.), hence they can yield only red coloured products. Finally, there are some varieties, *e.g.* Pontac, Alicante-Bouschet, Petit Bouschet, direct producers like Seibel 117, etc., the pulp of which is coloured red when ripe. They give very dark red wines or grape-juice which can be used for colouring similar products.

According to Pacottet[1] the inner cells of the skin which come into contact with the pulp contain the *aromatic substances* which give the wine its characteristic flavour. They are volatile essential oils and appear in the berry at the same time as the pigment. It is, however, especially during the last eight days when the grapes get fully ripe that they develop their full characteristic taste and flavour. The aromatic substances then penetrate the pulp and the skin, through which they partly diffuse into the air. Hence we must allow the grapes to get completely ripe if we want as much as possible of their particular flavour in the product we wish to make out of them. If, on the other hand, we desire to make a more neutral product with as little as possible of the flavour of the grapes in it, we must harvest the grapes as soon as they are ripe enough for our purpose. The foregoing applies to such varieties as Muscat of Alexandria, Muscadel, Ferdinand de Lesseps, Isabella, Concord, etc. Many American grapes, and their hybrids with European grapes that are used as direct producers, have a foxy taste, which is most strongly developed in the fully ripe grapes. If such grapes are harvested before they are fully ripe, the resulting product will have much less of the foxy taste than if this operation is delayed until the grapes have reached full maturity.

---

[1] P. Pacottet, *Vinification*, 1904, p. 19.

The colour of the skin, and hence of the grape, is to a greater or lesser extent affected by the *bloom* on it. This is a waxy coating on the berries which becomes visible when they are about the size of buck-shot, and continues developing until the grapes are ripe. The bloom may appear white to bluish in colour, and it may be thick or only slight. It may be spread more or less uniformly or in stripes, as on the berries of Bonnet de retord. While we find very little bloom on the berries of Rosaki, those of Barlinka and Gros Colman are covered by a dense bluish bloom. On Prune de Cazouls and Barbarossa the bloom is only of medium thickness and whitish in colour.

Grapes with a well-developed and undamaged bloom look much fresher and prettier than grapes with hardly any bloom or with damaged bloom. Hence every grower of high-class table grapes does his utmost to land his grapes on the market with as much bloom on them as possible. Whoever has succeeded in packing his grapes without appreciably damaging their bloom will at the same time have prevented any damage being done to them, and will therefore stand a good chance of landing them on the market in a tip-top condition. We shall see later on how this can be done.

The function of the bloom is not merely to improve the outward appearance of the grapes, or to collect the yeast cells and other fungi from the soil and air and facilitate their growth, or to assist in the formation and preservation of the characteristic aromatic substances of every variety of grape. All this it does, but I regard the main function of the bloom as a protection against sunburning, other diseases, and mechanical damage to the skin. It is well known that table grapes that have been thinned late in the season, when the bloom will unavoidably have been considerably damaged (rubbed off), are very liable to sunburn if hot weather follows shortly after thinning.

The skin can be thick or thin, soft or tough. When we grow table grapes which have to endure a fairly long transport to distant markets, we should give preference to varieties with tough and thick skins, *e.g.* Ohanez or White Almeria, Barlinka. Thin but tough skins are not entirely unsuitable for long transport, but a thin and soft-skinned variety usually is.

It is furthermore of the utmost importance that the berries should adhere very well to the pedicels. If the berry gets detached because the pedicel has dried out and broken it is not so serious, but if the skin of the berry gets torn where the pedicel enters it, or the berry gets pulled off completely from the pedicel and brush, decay nearly always sets in very soon, and the whole bunch and sometimes all the grapes in the box may perish completely.

The **pulp** may be tough (*e.g.* Concord, Ferdinand de Blesseps, etc.), or firm and crisp (*e.g.* Flame-coloured Tokay, Ohanez or Almeria, Molinera Gorda, etc.), or soft (*e.g.* Green-Grape, Muscat Hamburgh, etc.). Varieties with tough pulp are met with especially in those of the Labrusca species. Their skins slipoff easily from the pulp hence they are known as slip-skin varieties.

The colour of the pulp at maturity is light green, yellowish, whitish and sometimes (*e.g.* Pontac) red. Its aroma and flavour may be strawberry-like (*e.g.* Isabella, Ferdinand de Lesseps etc.), or be of a muscat character (*e.g.* Muscat of Alexandria, Muscadel, and other Muscat varieties), or be less pronounced and sometimes fairly neutral (most varieties).

Its taste at maturity may be very sweet or only slightly so, mild and only slightly acid (White French or Palomino, Molinera Gorda, etc.), or pleasantly acid (*e.g.* Stein, Barlinka), or acid (Janesville), or somewhat astringent and characteristic (*e.g.* Cabernet Sauvignon with an herbaceous taste), or foxy (*e.g.* Isabella, Concord, Jacquez, etc.). Generally speaking, the taste may on the whole be pleasant, more or less neutral, or even unpleasant.

The **seeds** are 0-4 in number, usually 2-3. they can be large to small, light coloured, light to dark brown, and sometimes reddish at maturity. Sometimes they are fairly flat, sometimes plump, sometimes short, and sometimes long. Fig 13, C, shows us the back and front of a grape seed. On the back we see more or less in the centre the *chalaza*, which is the place where the seed-coats and kernel are connected. On the front we notice the *raphe*, which is the ridge connecting the *hilum* (*i.e.* the scar left where the seed was attached to the seed-stalk or funicle and situated near the beak) and chalazal region, and the two depressions which Hamböck[1] termed "Kernfalten" (i.e. seed-folds). The narrower prolonged base of the seed is the *beak*.

The characters of the seeds are accounted of much more value in determining species than for differentiating between different varieties of the same species, although they may sometimes be of great value here also. The size and weight of the seeds vary very much in different species, as they do in varieties of the same species. "The shape and colour of seed offer distinguishing marks, while the size, shape, and position of the raphe and chalaza furnish very certain marks of distinction in some species."[2]

The wild vines (V. *riparia*, V. *rupestris*, V. *Berlandieri*, etc.) usually have small berries with a fairly large number of seeds. The pulp in these is very little developed, so that the berries consist largely of skin and seeds. This need not surprise us when we remember that the function of the berry, in the first instance, is to store the seeds and allow them to develop properly, and that the main object of the vine is to propagate itself by means of its seeds. We find the same thing happening with many other wild fruits.

By his cultural methods man has in many instances improved the wild fruits to what they have now become under cultivation.

Grape varieties with very large berries, *e.g.* White Prince, Gros Colman, etc., usually have fairly large seeds, but in the same variety the size of the seed varies very much less than the size of the individual berries. Also, the ratio of the weight of the seeds to that of the rest of the berry is much smaller with big berries than with small berries.

Fig. 13, B, clearly shows that the berry develops unsymmetrically if seed is formed on the one side only. Where seed was formed normally on both sides, we get the normally shaped berry. Where properly developed seed is not formed, or where the berry remains seedless, it will remain much smaller than those of the same variety containing normal seeds. It is common knowledge that small round berries often occur on the bunches of Muscat of Alexandria in between the large oval berries. The former are always seedless, and, when raisins are made from such grapes, they

---

[1] From Babo u. Mach, *l.c.* p. 297.
[2] U. P. Hedrick, *l.c.* p. 308.

are separated from the large raisins which contain seeds, because the fact that they are seedless increases their commercial value. The seedless berries of other varieties, the berries of which normally have seeds, are always perfectly round. This I observed on ring-barked shoots of Bicane, Olivette Blanche, Ohanez, Gros noir des Beni-Abès, and on Rosaki, Btc. If the berries are normally seedless, they can be oval, etc. Sultana and Black Monukka.

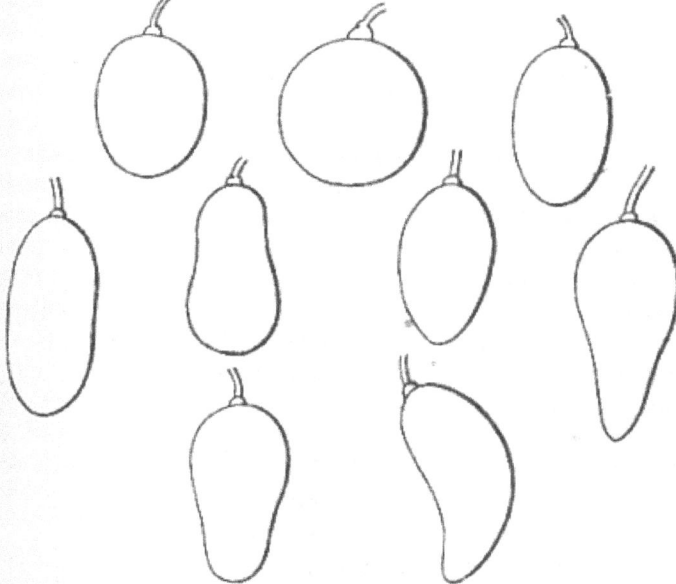

FIG 14 Shapes of grape berries (after R. Goethe). From Babo u. Mach, *Handbuch des Weinbaues*, 1923. Paul Parey, Berlin.

A small number of varieties, *e.g.* Sultana, Currants, Black Monukka, produce only seedless berries. The Cape Currant (white and black) is simply a seedless type of our Muscadel. Sometimes one finds a few large berries on the bunch of the Cape Currant, which then grow to the size of the Muscadel berry and contain seed.

Shape of the Berry.—As we shall see later on, the shape of the berry is made use of in classifying the European grape varieties. Where the berries hang loose in the bunch and can develop freely, we find the following shapes: *spherical* (*e.g.* White Crystal, Palomino, Gros Colman, Barbarossa or Gros Guillaume, etc.), *short oval* or *subovoid* (Molinera Gorda, Bariinka, etc.), *oval* or *ovoid* (Hermitage or Cinsaut, Muscat of Alexandria or Hanepoot, Bicane, etc.), *long oval* or *elongated* (Rosaki, Acorngrape, White Prince, etc.), *irregular* (Ohanez, White Cornichon or Lady's Finger, etc.). (See Fig. 14.) In one and the same variety we sometimes find deviations from the normal shape. Thus Molinera Gorda and Barlinka sometimes develop nearly spherical berries where their normal shape is a short oval; Muscat of Alexandria forms short oval next to the normal oval berries. The conditions of growth can exercise a certain amount of influence in this respect.

B. INTERNAL MORPHOLOGY (ANATOMY)

1. THE STRUCTURE OF THE CELL

Hitherto we have studied the different organs or parts of the vine in so far as it was possible to do so with the naked eye. If, however, we take very thin sections of any part of the vine and examine these under the microscope, we notice that all the parts of the vine—like those of any other plant— are built up out of a large number of very small cells, which look and are grouped differently in different parts of the vine. In this study of the internal morphology or the anatomy of the vine, we shall first of all study the cell as such, then the different kinds of cells and tissues, and finally their grouping in the various organs of the vine.

Every cell is formed by cell division out of an already present cell. Every young cell is enclosed by a delicate membrane of cellulose, called the cell wall, and is completely filled by a turbid, granular mass called the protoplast which is present in all living cells.

"Apart from the protoplast and cell wall, we can distinguish yet other parts in all cells. Sometimes these appear as particular organs of the protoplast, *e.g.* the nuclei and chlorophyll grains, then again as deposits of definite chemical substances, *e.g.* the crystals and the starch grains. These latter objects all agree with each other in this respect that they are always formed in the protoplast, and that they can be formed nowhere without its co-operation. Therefore, like the cell wall, they must be regarded as products of the activity of the protoplasts. . . . *Out of this it follows that the protoplasts are the seat of the vital phenomena, the bearers of life; all other parts of the plant are only secondary in relation to them."* [1]

The protoplast consists of cytoplasm, nucleus, and chromatophores, which are the living constituents of the cell.

The *Cytoplasm* is a viscous, colourless mass. Besides water and mineral substances, it always contains nitrogen compounds, and may in addition contain carbohydrates and fats. As the most important vital processes (*e.g.* nutrition and growth) occur in the cytoplasm, it is very important to know that it is killed by heat, when heated above 50° C. Or 122° F., and by a number of chemical substances such as alcohol, ether, chloroform, strong acids, salts of mercury, carbolic acid and other phenols, picric acid, formaline, etc. Also, cytoplasm can only live and function properly in the presence of oxygen or air which contains plenty of oxygen.

Whilst the protoplast completely fills the very young cell, we soon notice during its further growth that more transparent areas are formed in the protoplast. They are called *vacuoles* and consist of an aqueous solution, the *cell sap*. These vacuoles are always inside the cytoplasm. During the cell's further growth the different vacuoles usually unite to form a single vacuole, which usually fills the greater part of oldish cells, whilst the cytoplasm in a thicker or thinner layer covers the cell wall. Sometimes, as in Fig. 15, we notice strands of protoplasm running through the vacuole. This always happens when the nucleus is in the middle of the cell, but is not limited to this case only.

The *Nucleus* occurs, without exception, in every protoplast in the living tissue cells of the higher plants. In the young cells it has as a rule a spherical form, and is large in relation to the cell. In the older cells, on the other hand, it usually has a lenticular form and is relatively small, since it has meanwhile not increased in size. It usually contains a sharply outlined, highly refractive body called the *nucleolus*. Sometimes it contains several of these nucleoli. The consistence of the nucleus seems to be that of a tenacious fluid. "Some insight into the finer structure of the nucleus is obtained from properly fixed and stained preparations. In these a deeply staining reticulum of *chromatin*, which appears to consist mainly of proteids containing phosphorus,

---

[1] Oudemans en de Vries, *Leerboek der Plantenkunde*, 4° druk, 1906, vol. i. p. 3.

is evident. . . . The nucleoli stain deeply but differently from the chromatin." [1]

In its chemical composition the nucleus agrees very closely with the cytoplasm. Both consist mainly of proteids.

The great difference lies in the substance composing the chromatin granules, which is peculiar to the nucleus. This substance is composed of some representative or other of the group of the nucleids. On cell division, the contents of

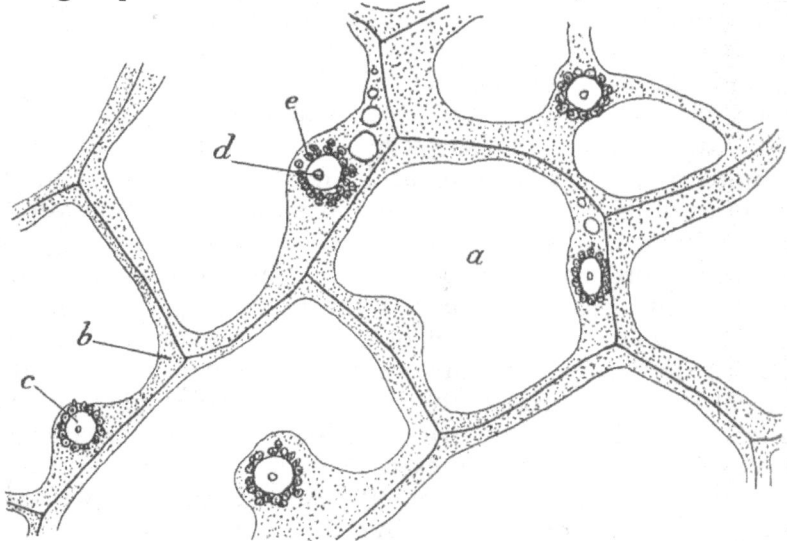

FIG. 15.—Cells of pearl-gland of the vine. *a*, vacuole filled with cell sap; *b*, cytoplasm; *c*, nucleus; *d*, nucleolus; *e*, crystals of calcium oxalate. ×431. Original.

the nucleus is divided in a somewhat complicated manner (*Karyokinesis*) in equal and in every respect precisely similar halves between the two daughter cells. Similar processes of division take place when two protoplasts are merged into one during fertilisation. From this we can conclude that the nucleus is the main bearer of the hereditary characters.

The *Chromatophores* of the young cells consist of spherical or lenticular bodies of plasmatic material, which usually lie in fair numbers near or around the nucleus, but can be carried along by the cytoplasm in its movements. Both nucleus and chromatophores lie in the granular cytoplasm. (See Fig. 15.)

The latter can multiply themselves by direct division into equal halves. In older cells they develop into

[1] Strasburger's *Text-book of Botany*, 1921, p. 17.

chloroplasts, leucoplasts, or chromoplasts. "In parts of plants which are exposed to the light the chromatophores usually develop into chlorophyll bodies or chloroplasts. These are generally green granules of a somewhat flattened ellipsoidal shape, and are scattered, in numbers, in the parietal cytoplasm of the cells ...   The green pigment, chlorophyll, is essential for the decomposition of carbon dioxide in the chloroplasts."[1]

In parts of the plant where the light does not penetrate, the rudimentary chromatophores develop into leucoplasts. These are sometimes also called *amyloplasts*, because it is in them that sugar is transformed into grains of starch. When exposed to light, the leucoplasts can be transformed into chloroplasts. This explains why potatoes get a green colour when they show above ground and are thus exposed to sunlight. Out of the leucoplasts, and also out of the chloroplasts, can be formed the chromoplasts which give the yellow and red colour to many parts of the plant, especially to the flowers and fruit.

Wherever the vine grows, there are always young cells dividing themselves into two daughter cells. During this cell division the nucleus first divides mitotically into two identical daughter nuclei, between which there is then formed a new cell wall in the middle of the mother or dividing cell and at right angles to its cell walls. This new cell wall at the same time divides the rest of the protoplast (chromatophores and cytoplasm) equally between the two daughter cells. These latter grow until they have reached the size of the original mother cell, and then in their turn proceed to divide each into two equal halves. Through the continuation of this process a group of similar cells is formed which constitute a cell tissue.

In their very early stage the tissues of plants still consist of small, thin-walled, and similar cells which are rich in cytoplasm and multiply themselves by cell division. Such a cell tissue is known as *meristem*, and the youthful cells are hence called *meristematic* cells. The cambium is such a meristem. With increasing age the meristematic cells and tissues develop into other cells and tissues.

---

[1] Strasburger, *l.c.* pp. 17-18.

At first the young cells stop dividing into two and grow much bigger. It is at this stage that the previously discussed vacuoles make the appearance in the cell. Now also an internal pressure arises between the cell wall and its content, which is called *turgor* or *sap pressure* (*osmotic pressure*) in living cells. This turgor plays a very important role in the life of the plant. The more elastic the cell wall is, the more the cell can grow in size under the influence of the turgor. Cells whose cell walls have been

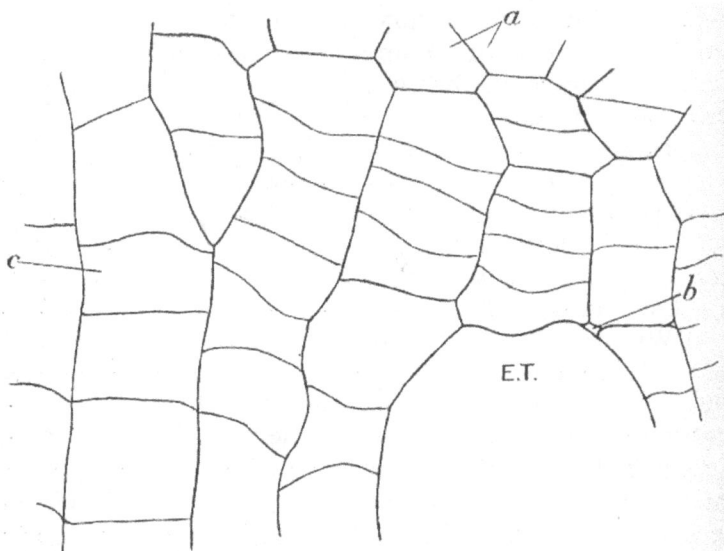

FIG 16 —Cambium in the cross-section of a young vine shoot, a, cells of the phloem ; b, top of cell cut off and not an intercellular space; c, interfascicular cambium; E.T., embryonic trachea, x 1088. Original.

distended by the turgor are said to be *turgescent*; through loss of water they become flaccid—this is what happens when plants wilt.

It is under the influence of this pressure that the cytoplasm is pressed close against the cell wall. Whereas the cell wall retained its thickness, *i.e. grew in surface*, whilst the cell grew bigger, the protoplast now that the cell is full-grown deposits material on the existing cell wall in the form of new layers or *lamellae*. The oldest lamella which was originally formed in the meristematic stage is known as the *primary* or *middle lamella*, whilst those formed subsequently are called *secondary* or *thickening lamellae*.

(See Fig. 17, where the thin places in the cell walls show the primary lamella the rest of the cell walls having been uniformly thickened as just described.) The unthickened parts are known as *pits*. These can be circular, elliptical, or spindle-shaped areas. The two first named we find in the cells of the medullary rays on Fig. 17, and the last named on the sclerenchymatous fibres in Fig. 26. On a cross-section through the

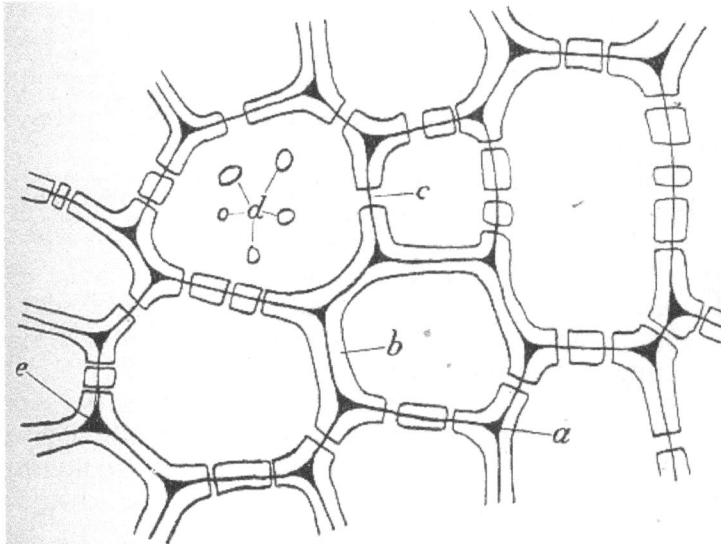

FIG. 17.—Tangential section of a medullary ray of one-year-old wood of Muskat gutedel. a, intercellular space (characteristic of these cells) ; b, thickened cell wall; c, thickened cell wall (pit) ; d, unthickened areas in cell wall (pit surface), x 1125. Original

thickened cell wall, these pits look like *c* in Fig. 17. Where the thickening of the cell wall has been very considerable, the pits develop into fine tubular canals (*pit canals*). These we see on Fig. 26, c, in the strongly thickened sclerenchymatous fibres. It will be noticed that the pit canals correspond exactly, as the middle lamella thickens symmetrically on both sides.

On closer examination it appears that the unthickened middle lamella (pit membrane) which closes the pit canals is not completely closed, but contains a certain number of minute pores. Through these pores extremely fine filaments of cytoplasm (known as *plasmodesms*) pass.

They serve to connect the protoplasts in adjoining cells, i.e. throughout the whole plant body. Although they are mostly confined to the pit membrane, they may also penetrate the whole thickness of the cell wall. "The existence of these connecting filaments of living substance between the protoplasts confers an organic unity on the whole body of the plant, serving for the conduction both of substances and of stimuli."[1] In this way the really living part of the vine consists of one large body of protoplasm. Through the pit membranes and the plasmodesms liquids can pass much more readily than would have been possible through the thickened and frequently hardened cell wall. The thickening of the cell walls increases the mechanical rigidity of the cells, and hence of the tissues and of the plant.

Sometimes the cell walls on the whole remain unthickened whilst only certain parts are thickened. If the pits are elliptical and close together, we get a *reticulate thickening* of the cell wall. On some kinds of cells we find a *spiral thickening* of the cell wall. These two ways of thickening make the cell walls more rigid whilst hardly diminishing their permeability.

The *bordered pits* are thus called because they appear as two concentric rings. They are formed by the secondary lamellae closing the pit more and more as the thickening of the cell wall proceeds.

During their growth in size and during the thickening of the cell walls, most cells round themselves off to a certain: extent. During this process the middle lamella splits at the angles of the cells, and in this way narrow *intercellular spaces* filled with air are formed. These intercommunicate and form a connected system of ventilation which communicates also with the outside air. It is of the utmost importance to the life of the vine that this system of ventilation should remain intact, as the gases required for respiration and those emitted by the various tissues must be introduced and removed almost exclusively by this system.

---

[1] Strasburger, *l.c.* p. 44

2. THE DIFFERENT KINDS OF CELLS AND TISSUES

When the young meristematic cells develop further, they form different kinds of cells and tissues to meet the different requirements of the vine. According to Kroemer in Babo u. Mach, *l.c.* p. 302, we can summarise these as follows: nutrition and metabolism, protection and physiological separation of the cell masses, rigidity of the organs, transportation of the plant sap. The kinds of cells and tissues serving these purposes are the *parenchymatous* cells and tissues, the *boundary* cells and tissues, the *mechanical* cells and tissues, and the *conducting* cells and tissues.

(1) Parenchymatous Cells and Tissues

The parenchyma cells can assume various shapes. The *parenchyma* is the tissue occupying the space between the other organs. It differs from the other tissues in possessing a ventilating system of *intercellular spaces* (see Fig. 17). Their cell walls always consist of cellulose and are very seldom lignified. They are only moderately thickened and provided with round or elliptical pits, thus facilitating the movement of water and dissolved food in its tissues. Their protoplast is strongly developed and always encloses one or several large vacuoles. Their chromatophores are either leucoplasts or chloroplasts. In the leaves the parenchyma, by means of its chlorophyll granules, brings about the formation of organic substance out of carbon-dioxide (photosynthesis). In the stem and roots the leucoplasts (amyloplasts) in the parenchyma cells serve in the formation and storage of reserve food materials (mainly starch) for subsequent use in metabolism. Here they also participate in the transportation of food in the plant. Very important is the fact that the parenchyma cells can at any time again form meristematic cells by cell division. The most important vital processes of the plant take place in the parenchyma.

Closely related to the parenchyma cells are the *calcium oxalate* cells of the vine, which occur in all its organs. They resemble the parenchyma cells in nearly every respect. They differ from these, however, in containing calcium oxalate crystals. These crystals occur in the vacuoles of

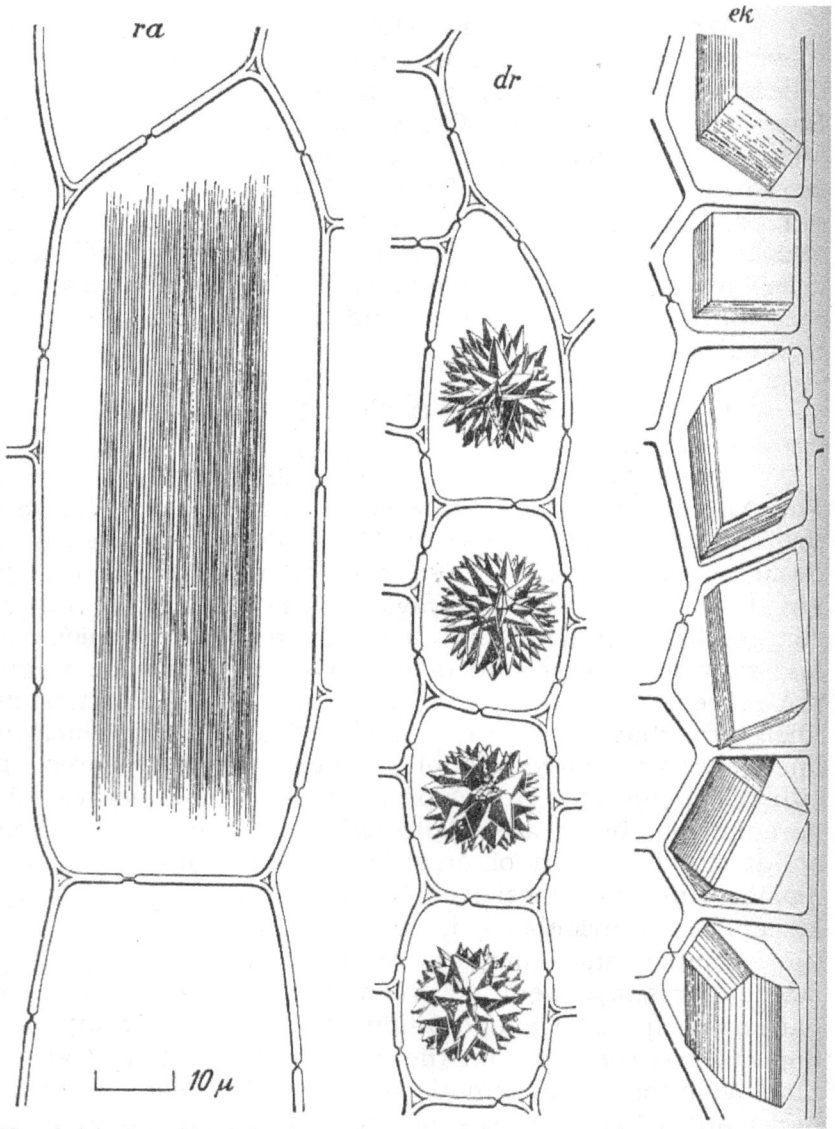

FIG.—Calcium oxalate cells of the vine. *ra*, a cell from the bast parenchyma of the root filled with a bundle of raphides : *dr*, stellate crystal aggregate of calcium oxalate in a row of cells of the leaf-stalk ; *ek*, large solitary crystals in the marginal cells of a medullary ray. x 1000. From Babo u. Mach, Handbuch des Weinbaues, 1923. Paul Parey, Berlin.

these cells, but are originally formed in the cytoplasm. They can assume the shape of individual crystals, star-like clusters of crystals, or a bundle of needle-shaped crystals (raphides), as is beautifully illustrated in Fig. 18.

(2) Boundary Cells and Tissues

To this group of cells, in the first instance, belong the epidermal cells which constitute the *epidermis*. This tissue invests the plant body as a protecting coat and consists of a angle layer of closely united cells without intercellular spaces between them. The stomata, which occur only above ground, constitute the only exception to this rule.

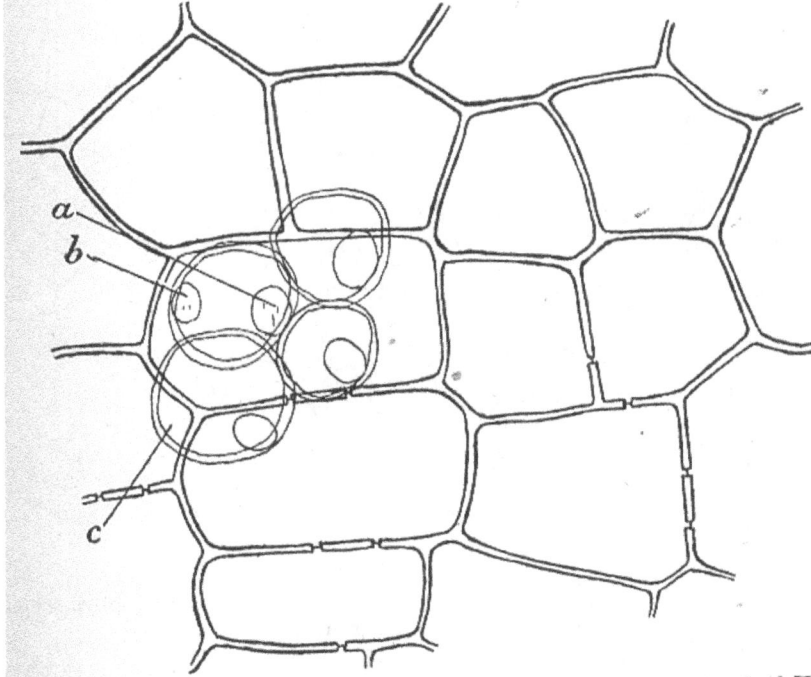

FIG. 19.—Epidermis on upper side of a young, undeveloped leaf of Weisser Calabreser. x 1088. Original.

The outer walls of the subaerial epidermal cells are strongly thickened and the external lamellae are cutinised and impervious to water. The outermost lamella is called the cuticle. It runs as a continuous cutinised film from cell to cell. Sometimes it is covered by a layer of wax which constitutes the bloom on the berries and shoots. All these devices serve to reduce the evaporation of water through these epidermal cells to an absolute minimum. The outer walls of the epidermal cells on the young roots are not thickened, and fey remain permeable to water and dissolved salts. The lateral and inner walls of the epidermal cells are usually thin and permeable to cell sap. Their cell content closely corresponds with that of the parenchymatous cells. Their chromatophores are leucoplasts or chloroplasts.

The epidermis serves to protect the internal parts of the plant against mechanical injury, and to regulate the absorption and evaporation of water, as well as the exchange of gases through its openings, and in this way to bring the plant into contact with its environment.

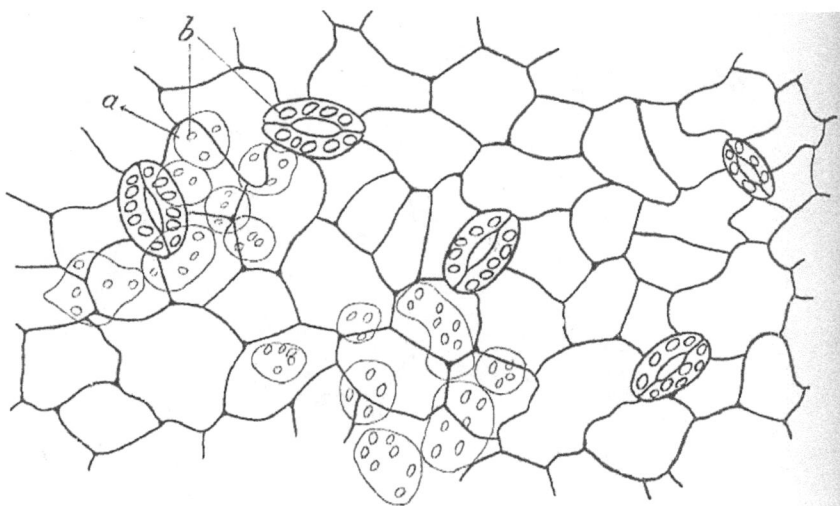

FIG. 20.—Epidermis on lower side of vine leaf with stomata. *a*, mesophyll seen with deep focussing; *b*, chlorophyll grains, x 495. Original.

In the leaf these openings are known as *stomata*. They occur mostly on the lower surface of the leaf. On the upper surface we find a few here and there on the nerves. Every stoma is an intercellular passage bounded by a pair of half-moon-shaped cells called *guard cells*. On Fig. 21 we see the two guard cells with their strongly thickened cell walls. Immediately adjoining these are two large subsidiary cells which facilitate the opening and closing of the air passage by the guard cells.   Underneath the guard cells and the subsidiary cells we see the large air-filled intercellular space which is in direct communication with the other intercellular spaces of the parenchyma.

When the vine contains plenty of water and the pressure inside the guard cells is high, these, in view of the peculiar thickening of their cell walls, assume more of a round shape, with the result that the cell walls at the opening of the stoma move backwards or farther apart, thus opening the passage. This in turn promotes a more rapid transpiration and gaseous exchange. Inversely, if the vine's supply of water becomes low the pressure inside the guard cells diminishes, which ultimately leads to their closing the passage or pore. When the passage is fully opened transpiration is at its maximum, and it sinks to zero when the passage is completely closed.

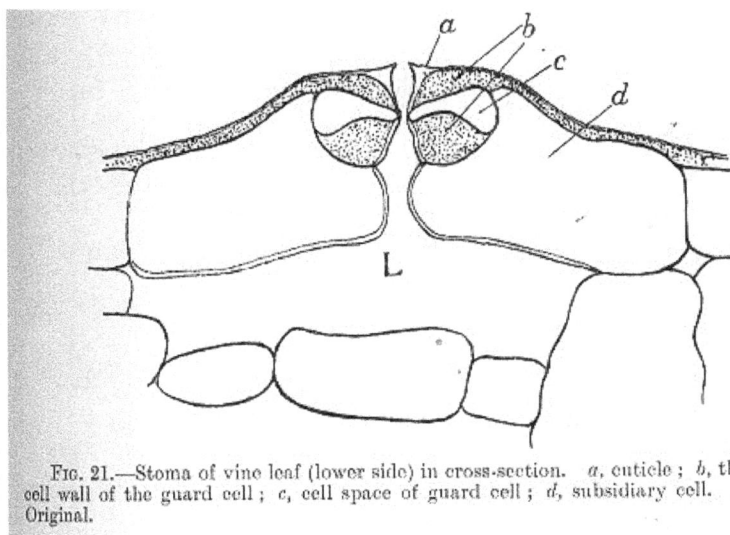

Fig. 21.—Stoma of vine leaf (lower side) in cross-section. a, cuticle ; b, thickened cell wall of the guard cell ; c, cell space of guard cell ; d, subsidiary cell. × 1088. Original.

The gaseous exchange (i.e. taking in of carbon dioxide and air and giving off of water vapour, carbon dioxide, and oxygen), and thus the ventilation of the plant, takes place mainly through the opened stomata. For the purposes of respiration oxygen can usually penetrate into the plant in sufficient quantity through the cuticle and epidermal cells.[1]

Sometimes the epidermal cells grow out in the shape of *hairs* which, on the organs growing above the ground, can be straight hairs or woolly threads, while on the young roots

---

[1] See Strasburger, *l.c.* p. 52.

they are known as *root-hairs*. (See Fig. 30.) These latter are of the greatest importance to the vine, as they absorb the main portion of the dilute soil solution containing the nutrient salts which the vine requires. Their growth in length gives them a much larger area of cell wall through which the soil solution can enter than that possessed by the outer wall of an ordinary epidermal cell.   They grow between the soil particles and come into immediate contact with the film of water surrounding these particles.   Hence the formation of root-hairs is a device of the plant to ensure a sufficient absorption of water and nutrient substances from the soil.

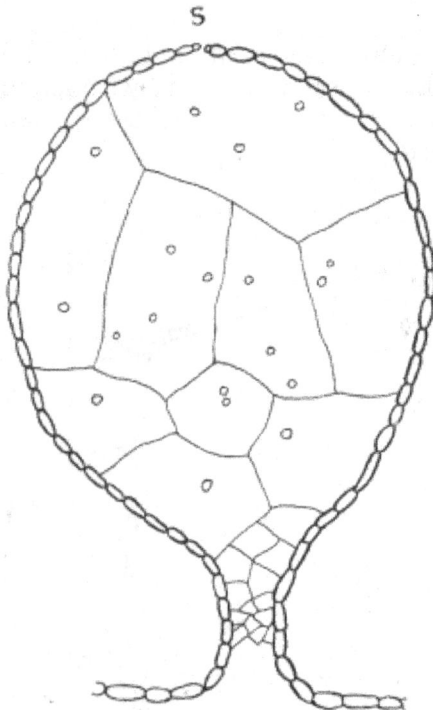

FIG. 22.—Pearl-gland of the vine. s, stoma . x 57. Original.

We find the hairs on the following parts of the vine growing above ground: shoots, leaf-stalks nerves; sometimes also on the finer nerves, but hardly ever on the parenchyma.

They are more numerous on the young shoots and leaves than on the older ones.

In the early stage of vigorous growth we notice small, roundish, translucent, white little pearl-like glands on the young parts.They also are an outgrowth of the epidermis, and are called *pearl-glands*. (see Fig. 22). They disappear after a while, and at any rate have nothing to do with any disease. They also contain stomata (see Fig. 22). To the boundary cells also belong the endodermal cells, the exodermal cells, and the cork cells which at first are not boundary cells, but separate some tissues from each other, and afterwards actually do become boundary cells.

The endodermal cells form the *endodermis* of the root. They are usually more or less elongated. The young cells have thin, unsuberised (not corky) walls which allow water to pass through them. Their radial cell walls are thickened and cutinised on a narrow strip, which appear as dark dots (the *caspary* dots) in a cross-section (see Fig. 23). The older cells form a thin lamella of suberin all over the wall, and on this again a soft lamella of cellulose is deposited. In this state the walls of the endodermal cells are completely suberised and are therefore almost completely impermeable to cell sap. As these cells contain living protoplasts, their walls must be permeable to water at least to a certain extent. As we see in Fig. 23, the older endodermal cells are rich in tannin and have a dark colour.

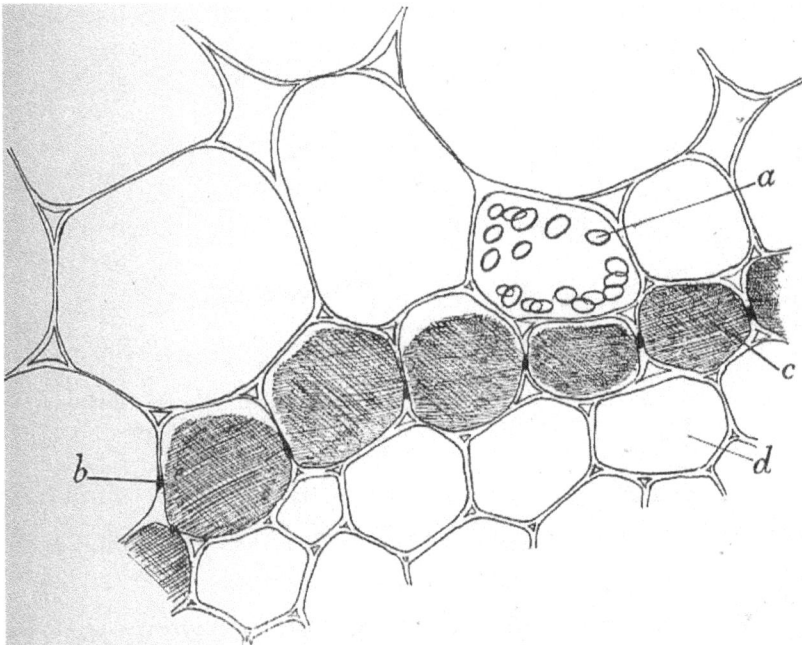

FIG 23 —Endodermal cells: cross-section of a young vine-root, *a*, tannin m parenchyma cell; *b*, Caspary dot; c, endodermis rich in tannin; *d*, pericycle. x 1125). Original

The endodermis consists of a single layer of closely united cells which separates the primary cortex from the central cylinder (see Fig. 30), and is intended to regulate the flow of sap between these two tissues.

Closely related to the endodermal cells are the exodermal cells, which form the *exodermis* of the young root (see Fig. 30). This consists of one to two layers of closely united cells which form a sheath (exodermis) between the epidermis and the cortex, and prevent the transfusion of nutrient substances from the cortex.

The *cork* cells usually have a tabular shape. Their cell walls are completely suberised. In contrast with the endodermal cells, they are dead. The complete isolation brought about by the suberisation of their walls soon causes their protoplasts to die, when the cell space is filled by air. These cork cells in several layers form a dense cork tissue which, on old stems and roots, protects and limits the living cortex on its outside. When discussing the powers of resistance of

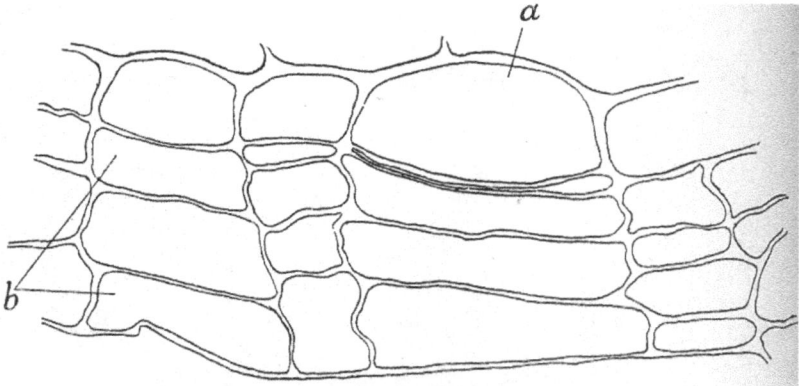

FIG. 24.—Cork cells : cross-section through one-year-old cane of Muskat gutedel. *a*, phelloderm; **b**, cork layer, x 1088. Original.

certain vines against the attacks of phylloxera, we shall meet with an important example of protection given by this cork tissue.

(3) The Mechanical Tissue System

Certain cells and tissues have a purely mechanical function. These tissues are the *collenchyma* and *sclerenchyma*, and are formed of *collenchyma cells* and of *sclerenchyma cells* (stone cells) and *sclerenchyma fibres* ("bast fibres") respectively.

The collenchyma (see Fig. 25) consists of long, living fibres which are usually somewhat pointed at the ends, and are subdivided by cross walls into several cells. The walls are strongly thickened, particularly at the angles. They consist of cellulose and pectic substances, highly refringent and not lignified. Their cell walls are soft, possess a high water content, and appear faint bluish white under the microscope. Their cell content consists of protoplast and cell sap, sometimes with chlorophyll grains. They resemble the parenchyma cells except for the unequal thickening of their walls, and occur in the outer parts of the cortex of the young shoots and of the leaf-stalks. In the leaves we find them over the vascular bundles of the nerves. "They occur exclusively in the young parts of the plant that are still capable of extending, and serve in the first place to give rigidity to the cell groupings, but at the same time also participate in the plant's functions of nutrition."[1]

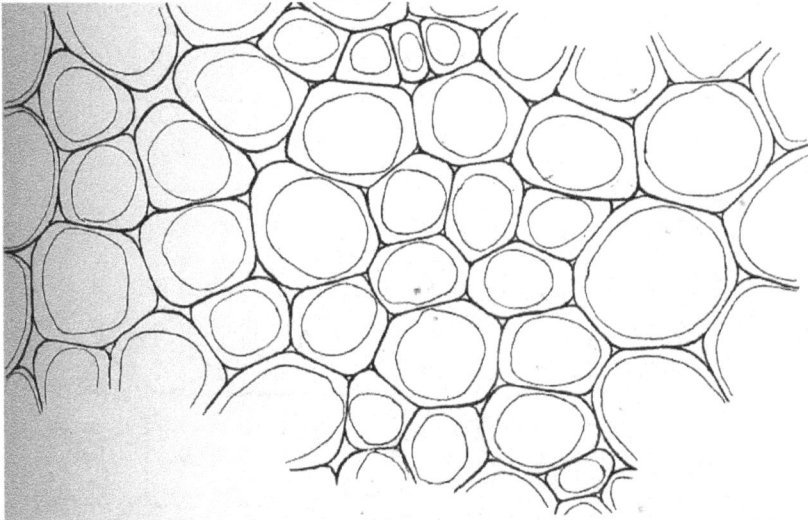

FIG. 25.—Collenchyma of a young vine shoot in cross-section, x 1125.
Original.

The *sclerenchyma fibres* can readily be recognised in a cross-section by their uniform thick cell walls, which are usually lignified. Hence they are usually coloured red by phloroglucin and strong hydrochloric acid. Also, their cell walls are much more highly refringent than those of the surrounding tissues. The fibres are pointed sharp at both ends and are generally very long. The one in Fig. 26, A, has

---

[1] Babo u. Mach, *l.c.* vol. i. p. 311.

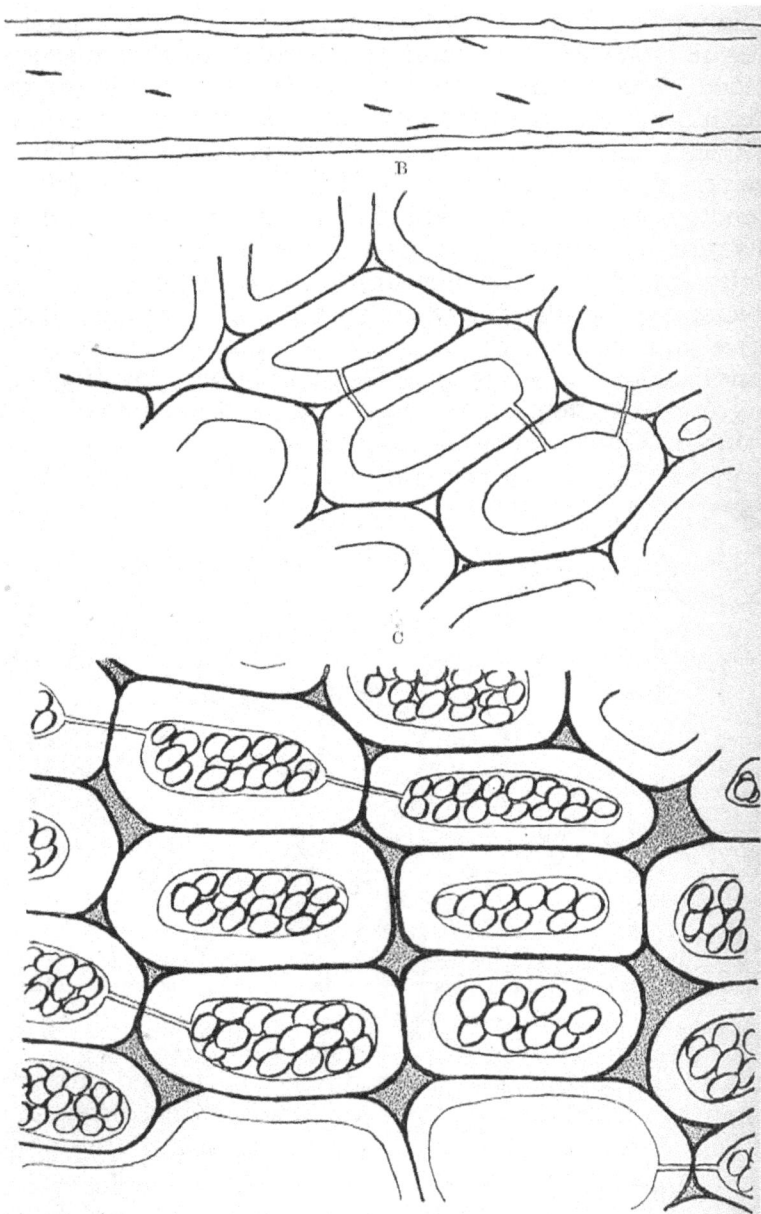

FIG. 26.—A. Sclerenchyma fibre from the primary cortex of the vine. × 43. Original.
B. Part of the same sclerenchyma fibre, × 430. Original.
C. Sclerenchyma: cross-section through the bast of a one-year-old vine shoot. × 1088. Original.
D. Sclerenchyma (bast fibres) of a vine root in secondary growth, living and filled with starch grains. Cross-section x 1088. Original.

a length of nearly 3 mm. (In Flax they are 20-40 mm., in Stinging Nettle 70 mm. long.) Sometimes they are sub-divided by thin cross-walls.

They possess obliquely placed, narrow, elliptical pits (see Fig. 26, B), which appear as pit canals on cross-sections (see Fig. 26, c and r). Their protoplasts usually remain alive only until the cell wall has reached its full thickness, when they soon die and the small cell space is filled with air or an aqueous liquid. Sometimes, however, they remain alive, and can then serve both as a mechanical tissue and as a place for storing reserve food like starch (see Fig. 26, D). These fibres are usually grouped in strands.

Fully developed collenchyma and sclerenchyma as a rule do not occur in the same organs, at least not simultaneously. "At a later stage the collenchyma frequently changes into sclerenchyma; in other cases the latter is directly formed as such."[1]

In the vine *sclerenchyma cells* (stone cells) occur exclusively in the seeds. Here they form the dense, hard seed-coat which encloses the softer pearls of the seed to protect them against desiccation and mechanical injury. As with the sclerenchyma fibres, their walls are much thickened, but to a greater extent. Their thickened walls contain numerous pit canals (see Fig. 27) and are lignified. These stone cells are fairly short and not pointed as the bast fibres are.

It is due to the collenchyma and the sclerenchyma fibres that the young shoots of the vine do not break off very easily. When we ring-bark the shoot against coulure, we cut these mechanical tissues, and then the shoot breaks off very easily.

(4) The Conducting Tissues

As the vine develops it requires more rapid conduction the nutrient substances from one organ to another (i.e. from roots to leaves and conversely) than can be brought about by diffusion through the walls of the living parenchyma cells even though they are pitted. For meeting this demand the vine possesses, in common with most plants, special conducting tissues called *vascular bundles* which lie in the main direction of conduction. Here each

---

[1] Oudemans en de Vries, *l.c.* i. p. 90.

vascular bundle consists of two separate strands or a multiple thereof, one of which is called *xylem* and the other *phloem*. The xylem consists of *tracheae* and *tracheides*, and serves for conducting the water

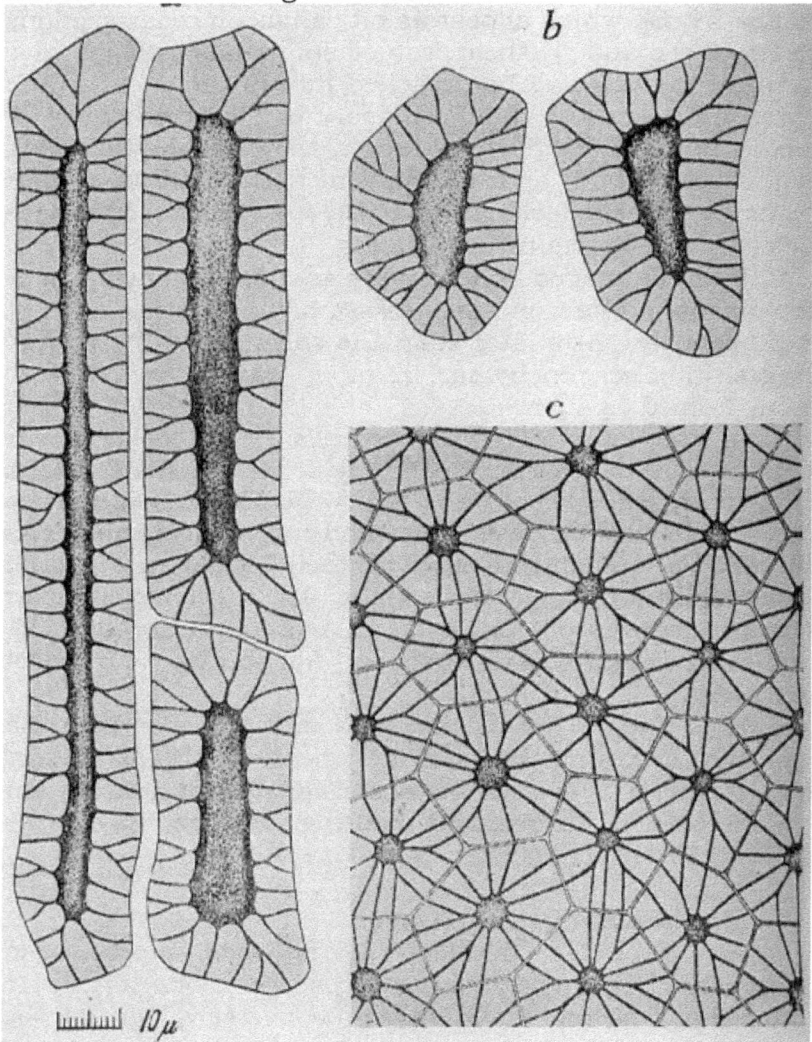

FIG. 27.—Sclerenchyma cells (stone cells) of the vine. *a* and *b*, isolated stone cells from the stone cell layer of the grape-seed; *c*, hard layer of the grape-seed in a surface section, × 720. From Babo u. Mach, *Handbuch des Weinbaues*, 1923. Paul Parey, Berlin.

and dissolved substances absorbed by the roots (*i.e.* the raw sap or mineral food) to the points of growth, and particularly to the leaves. The phloem consists of *sieve-tubes* with *companion cells* and parenchyma cells, and serves to conduct the elaborated sap or organic food (proteids and carbohydrates). Although they are generally near each other, the conduction the xylem and phloem usually takes place in opposite directions.

The tracheae and sieve-tubes represent cell tubes which have been formed through the fusion of meristematic cells situated in rows above each other. They are therefore not ordinary cells but cell-tubes, each of which arose out of a number of cells, and they form continuous channels for rapidly conducting the various nutrient materials. In the vascular bundles the cells and cell-tubes are closely united, hence we do not find any intercellular spaces in these bundles.

The sieve-tubes consist of long rows of elongated cells which are in open communication with each other by means of open pores which are formed in the transverse walls during be above-mentioned fusion of these cells. These perforated transverse walls are called *sieve-plates*, whence the name sieve-tubes. With the vine these sieve-plates are placed obliquely in the sieve-tube, thus allowing them a greater surface than if they had been placed transversely. This greater surface means a larger number of pores and hence more rapid conduction. On a tangential longitudinal section, as in Fig. 28, A, the sieve-plates are represented by very oblique lines with a series of knobs.

The lateral walls of the sieve-tubes are also provided with sieve-plates, are always thin, and never lignified. Each of the elements of a sieve-tube contains a thin living layer of cytoplasm lining the wall, one nucleus, a few leucoplasts with starch grains, and a large vacuole filled with a coagulable cell sap rich in albuminous substances and frequently in carbohydrates and inorganic salts (phosphates). The layers of cytoplasm on the two sides of each sieve-plate are in direct communication with each other by means of strands of cytoplasm that pass through the pores in the plates.

After one or a few vegetative periods they cease functioning because their sieve-plates become closed by *callus-plates*, which prevent any further conduction. These callus-plates consist of *callose* which is coloured dark blue by aniline blue aqueous solution, and is highly refractive. At the approach of the dormant period, the same blockage through the formation of callose takes place in the young sieve-tubes, but, in their case, the callose is redissolved in the following spring, again to allow the conduction of the organic food through the sieve-tube. On Fig. 28, B, we see such a callus-plate as it looks during winter.

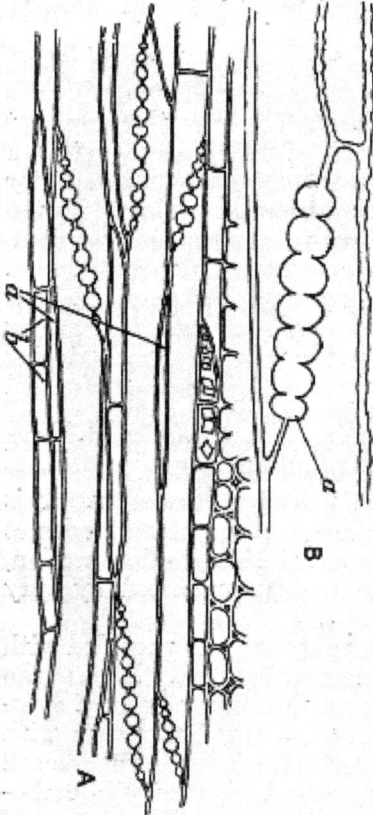

Directly associated with the sieve-tubes we find companion cells and parenchyma cells. In the bast these parenchyma cells, according to Hugo de Vries,[1] serve for conducting and temporarily storing sugar and starch.

The *tracheae* also arise out of longitudinal rows of cells, the transverse walls of which disappear during their development, thereby giving rise to these tube-like vessels. They are characterised by the absence of a protoplast (they are therefore dead) and by the presence of lignin in their walls. Hence the walls are coloured red by phloroglucin and strong hydrochloric acid, as does not happen to the sieve-tubes. We find them exclusively in the true and in the xylem of the vascular bundles. They are sometimes filled with air, but usually with air and water, and sometimes with water only. During rapid evaporation of water through the leaves they contain more air than water; otherwise the converse holds good.

FIG. 28.—Phloem of Muscat gutedel in a one-year-old shoot. Longitudinal tangential section. A*a*, companion cells; A*b*, parenchyma cells of the phloem. × 225. Original.
B. Sieve-plate with winter knobs of callose. × 495. Original.

---

[1] See Oudemans en de Vries, *l.c.* i. p. 117

82

The walls of the fully developed tracheae contain a large number of elliptically bordered pits and show annular, spiral, netted (reticulate) thickening (see Fig. 29). These vessels are so large in the old wood of the roots that they can easily be observed with the naked eye, especially in a thin section (See Fig.- 32).

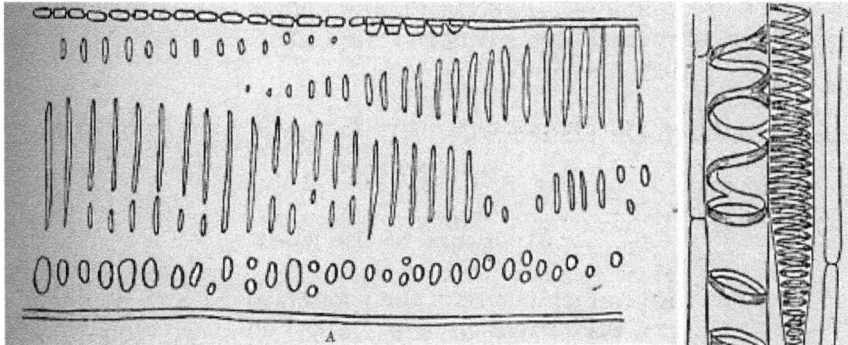

Fig. 29.—A. Trachea in secondary wood of a shoot of Muskat gutedel (macerated after Schultze). × 500. Original. B. Ring and spiral vessels. From Oudemans en de Vries, *Leerboek der Plantenkunde*, 1906. H. D. Tjeenk Willink en Zoon, Haarlem.

The later vessels been formed, generally the wider they will be. The *tracheides* were each of them formed out of a single meristematic cell. Their ends are fairly pointed. Their cell wall is built similarly to that of the vessels. Like the tracheae their function is to conduct water and the dissolved mineral food. They   are usually placed among the tracheae. Their cell walls may be completely, partially, or not at all lignified.

According to the grouping of the strands of phloem and xylem, we can distinguish between different vascular bundles. In the vine we find only radial and collateral vascular bundles. The *radial vascular bundle* (see Fig. 31) lies in the middle of the young root of the vine and consists of several strands of xylem and an equal number of strands of phloem, which, however, are placed on different radii. The *collateral vascular bundle* (see Fig. 33) consists of a strand of xylem and a strand of phloem lying one behind the other on the same radius. In the stem the strand of xylem is directed inwards and that of the phloem outwards (see Fig. 33). In the leaves the vascular bundles are collateral, the xylem being nearer the upper and the phloem nearer the lower surface of the leaf.

3. GROUPING OF TISSUES IN THE VARIOUS ORGANS OF THE VINE

(a) **Structure of the Root**

The internal structure of the root changes with its age.

Primary Structure of the Root.—This we find in the absorption zone of the young root, of which Fig. 30 gives us a cross-section. First of all we find the epidermis. Its thin walled cells readily absorb the soil solution, as they are free from lignin. Stomata and a cuticle are not met with on the epidermis of the root. In order to facilitate the absorption of the soil solution, some of these epidermal cells grow out into root-hairs, as can be seen on Fig. 30. In the young root the *cortex* consists of a girdle of parenchyma cells which is bounded on the outside by the *intercutis* and on the inside by the *endodermis*. The thickness of the parenchyma varies with that of the root. It consists of colourless, elongated parenchyma cells with well-developed intercellular spaces. They contain some large cells with raphides of calcium oxalate

The radial vascular bundle (central cylinder) of the finest rootlets contains two strands of xylem and phloem respectively (see Fig. 30). In the moderately thick long roots we usually find five of each. The outermost layer of cells of the central cylinder, lying just inside the endodermis, is called the *pericycle*. When the root develops lateral branches, these originate from the pericycle. The first and smallest vessels are formed against the pericycle. More towards the centre they become bigger and younger. Their development is therefore centripetal whilst that of the sieve-tubes is centrifugal.

Secondary Structure of the Root. — The secondary structure of the root commences where the absorption zone passes over into the transportation zone. Here the young root loses its epidermal cells, which now die and become separated from the root. Now the intercutis becomes the boundary and protective tissue. The walls of its cells now

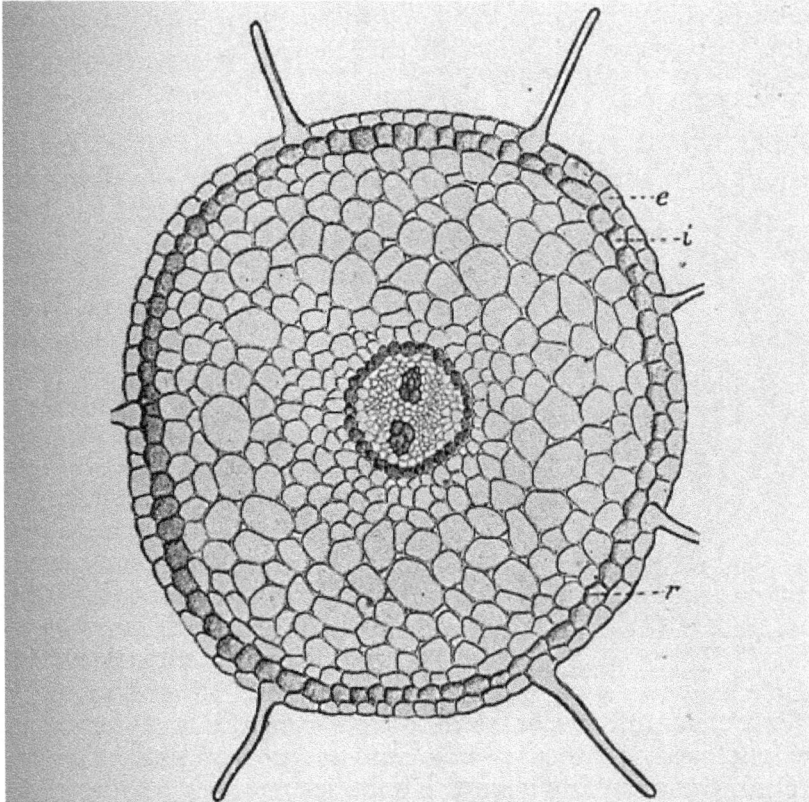

FIG. 30.—Primary structure of the root. Cross-section through the root of the vine in the absorption zone. *e*, epidermis; *i*, intercutis; *r*, cortex. In the centre is the vascular bundle with two strands of xylem and phloem respectively, enclosed by the endodermis. × 75. From Babo u Mach, *Handbuch des Weinbaues*, 1923. Paul Parey, Berlin.

become completely suberised and it assumes a brown colour. It then encloses the parenchyma as a closed thin layer of cork. As the root grows thicker we notice new tissues being formed in the central cylinder. These are the two characteristic tissues: *cambium* and *phellogen*. In the central cylinder we notice some parenchymatous cells of the pericycle just on the outside of the xylem strands divided in two, mainly by tangential transverse walls. This also takes place with the parenchymatous cells just inside the phloem strands and in those in the intervening parenchyma. In this way a wavy ring of young meristematic cells is formed which is called *cambium* (see Fig. 31).   By the activity of the cambium this ring soon becomes more even. On Fig. 16 the young transverse cell walls in the cambium are seen to be thinner than the older walls.

The newly formed cells retain their character only in the middle of the meristematic ring or cylinder. On the inside and opposite the primary phloem strands they develop into secondary wood, and towards the outside they form *bast* (see Fig. 31).

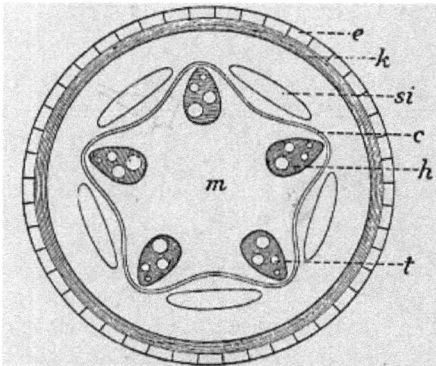

Fig. 31.—Beginning of secondary growth in the root. Formation of root cambium (diagrammatic). *e*, endodermis; *k*, cork; *si*, phloem; *h*, xylem; *c*, cambium ring; *k*, vessels. From Babo u. Mach, *Handbuch des Weinbaues*, 1923. Paul Parey, Berlin.

Simultaneously with the formation of the cambium ring, the cell walls of the endodermis are suberised, thus cutting off all nutrition of the primary cortex from the tissues of the central cylinder. The cortex therefore dies.

Owing to the root's growth in thickness as a result of the activity of the cambium, the primary cortex bursts and becomes detached from the root, which is now enclosed by the endodermis as its boundary tissue. Through radial cell division the endodermis for a certain time keeps pace with the root's growth in thickness.

Before the primary cortex is completely lost, we notice new tangential cell divisions taking place in the cells that lie immediately inside the endodermis. This leads to the formation of a new tissue called *cork cambium* or *phellogen*, Through continuous cell division it forms cork on the outside and *phelloderm* on the inside. The cork and phelloderm with the phellogen in between form the *periderm*.

The layer of cork now cuts off the root completely from its environment. During the root's progressive growth in thickness, new layers of cork are continually formed deeper in the bast, each of which in turn cut off the outer parts from all nutrition. They therefore die, burst because the root becomes thicker, and gradually get separated from the root, on which they remain as loose bits of dead bast.

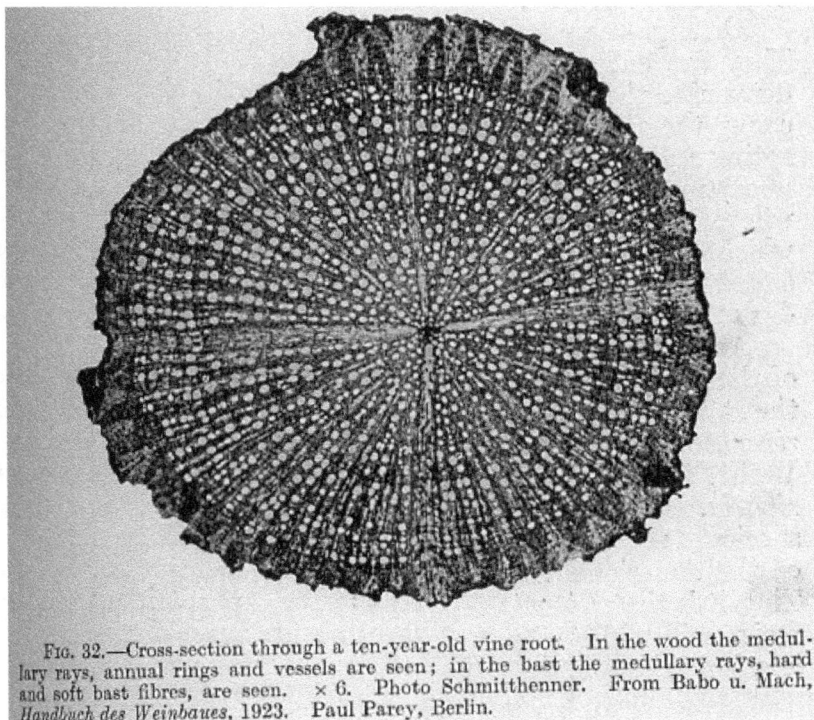

FIG. 32.—Cross-section through a ten-year-old vine root. In the wood the medullary rays, annual rings and vessels are seen; in the bast the medullary rays, hard and soft bast fibres, are seen. × 6. Photo Schmitthenner. From Babo u. Mach, *Handbuch des Weinbaues*, 1923. Paul Parey, Berlin.

The root now shows its secondary structure and consists of the layer of cork with loose bits of dead bast, the bast, cambium, secondary wood, primary wood, and pith.

Once the root has grown thick, it consists mainly of secondary wood. With some species the bast is fairly thick and fleshy, whilst with others it is thin. On Fig. 32 the portion with the white patches represents the secondary wood, at the centre of which we see the black primary wood like a cross. The white patches represent the sections through the vessels.

The whitish stripes which radiate from the centre outwards, are the *medullary rays,* which consist of living parenchymatous cells (see Fig. 17) filled with starch grains. The medullary rays pass between the vascular bundles. Those that penetrate to the primary xylem strands are called *great* or *primary medullary rays* and are developed from the fascicular cambium. These medullary rays traverse the wood and the bast. See further under "Secondary Structure of Stem".

On Fig. 32 we see also black concentric rings in the wood. They are called *annual rings* of growth, and enable us to determine the age of the root. On parts of Fig. 32 we can count the rings on this section of a ten-year-old root. In spring and early summer larger vessels and wider cells (*early wood*) are formed than later on during the vegetative period, when, towards its end, flat thick-walled cells are formed which compose the dense and frequently darker coloured late wood that can be recognised with the naked eye as a dark ring.

According to Guillon[1] the width of the vessels, the breadth of the medullary rays, the proportion of bast, and the size of the starch grains vary with the different species. With *Vitis riparia* the vessels are wide and the medullary rays narrow. With *Vitis rupestris* the vessels are narrower than with *Vitis riparia.* "The anatomical structure of the roots of the species of the vine explains many points about the adaptation of the American vines to the soil" (Guillon, *l.c.* p. 189). Thus it is that *Riparia* with its wide vessels demands a fertile moist soil, whilst *Rupestris* with its narrower vessels exacts less from the soil.

(6) Structure of the Stem

As the eyes contain the growing points, they are the first to grow when the vine commences its new vegetative period, and give rise to the young shoots. With the opening of the bud, the rudimentary parts of the future shoots, that were present in the bud, develop further.

Primary Structure of the Stem. — In the youngest internodes of the shoot we find the primary structure of the stem, which now consists of epidermis, cortex, and central

---

[1] J.-M. Guillon, *Étude générale de la vigne* (1905), pp. 188-189.

cylinder. The epidermis is covered by the cuticle and possesses stomata, though these are much fewer than in the leaves, and hairs. The (primary) cortex is thin and is composed of parenchyma and collenchyma. The strands of collenchyma form the ribs of the young shoot. They are not equally developed in different varieties and species.

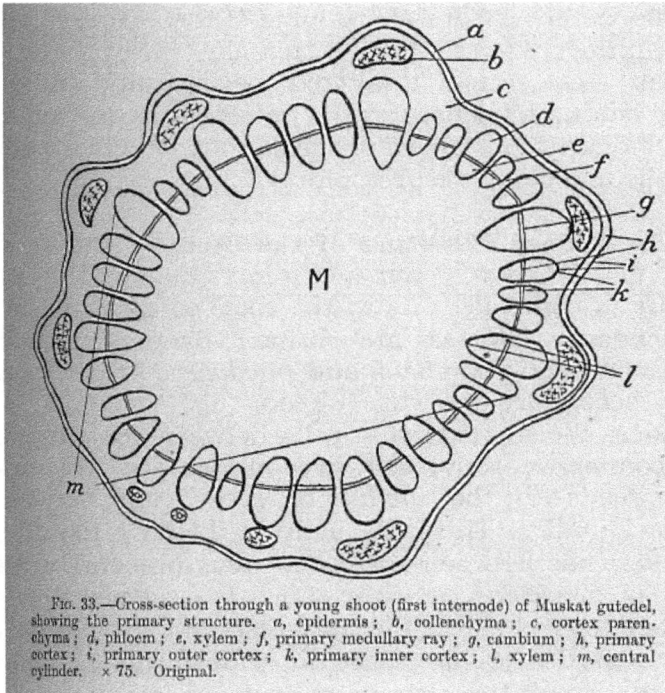

FIG. 33.—Cross-section through a young shoot (first internode) of Muskat gutedel, showing the primary structure. *a*, epidermis; *b*, collenchyma; *c*, cortex parenchyma; *d*, phloem; *e*, xylem; *f*, primary medullary ray; *g*, cambium; *h*, primary cortex; *i*, primary outer cortex; *k*, primary inner cortex; *l*, xylem; *m*, central cylinder. × 75. Original.

We find them strongly developed in the young shoots of Gros Maroc, Cabernet × Berlandieri 333, etc., whilst those of Barlinka, Riparia × Rupestris 101-14, etc., are fairly smooth. The last (innermost) layer of cells of the (primary) cortex can be recognised by their regular shape and by the numerous, fairly large starch grains which they contain. They show the characteristics of a *starch-sheath*. The peripheral layers of parenchyma cells in the cortex of the young green shoot contain chlorophyll, while those farther in are colourless.

In contrast with what we saw on the young root, we find that here the cortex is relatively much less developed than the *central cylinder,* which consists of the vascular bundles, medullary rays, and pith (see Fig. 33). The vascular bundles are grouped around the edge of the central cylinder in such a way that the phloem is next to the cortex and the xylem opposite the phloem, as in Fig. 33. Here, therefore, we have the open or collateral grouping of the vascular bundles. Their number varies, but is usually fairly large. Their cambium arose out of unaltered meristematic cells of the growing point, and separates the phloem from the xylem. The pith is large. It consists of short parenchymatous cells with thin walls and complete protoplasts.

Secondary Structure of the Stem.— The secondary growth of the stem commences immediately the primary structure is completed. As in the root, so also in the stem, the secondary tissues are formed by two cell-forming secondary tissues: *cambium* and *phellogen.* At the beginning of the secondary growth, already in the fairly young internodes, the meristematic cells between the phloem and xylem commence active cell division (see Fig. 16), and this soon spreads to the medullary rays lying between the vascular bundles. This new actively dividing tissue is the *cambium.* That between the xylem and phloem bundles is known as *fascicular cambium,* and that formed across the medullary rays is called the *inter-fascicular cambium,* and it connects the fascicular cambium, thus forming a complete cambium ring. The cambium retains its meristematic character only in the middle of the ring. On the inside it forms *wood* and on the outside bast. The tissues formed by the activity of the fascicular cambium resemble the xylem and phloem of the primary vascular bundle. By the activity of the interfascicular cambium the primary medullary rays are continued through the wood and bast. In its further growth, during the first year the external appearance of the cross-section of the cane alters only slightly at first. The medullary rays remain about the same width, so that the vascular bundles merely seem to get wider as the shoot grows in thickness.

Gradually, however, this changes. The sclerenchymatous fibres on the outside of the phloem bundles thicken their walls, and in the bast new groups of sclerenchymatous or bast fibres are formed. As with the older roots, the bast of the stem consists of medullary rays and *bast rays, i.e.* the parts of the bast between the medullary rays. Every bast ray of a one-year-old shoot that has ripened properly consists of 2 to 3, and in isolated cases up to 6,[1] tangential layers of phloem tissue or soft bast which alternate with the strands of bast fibres or hard bast.

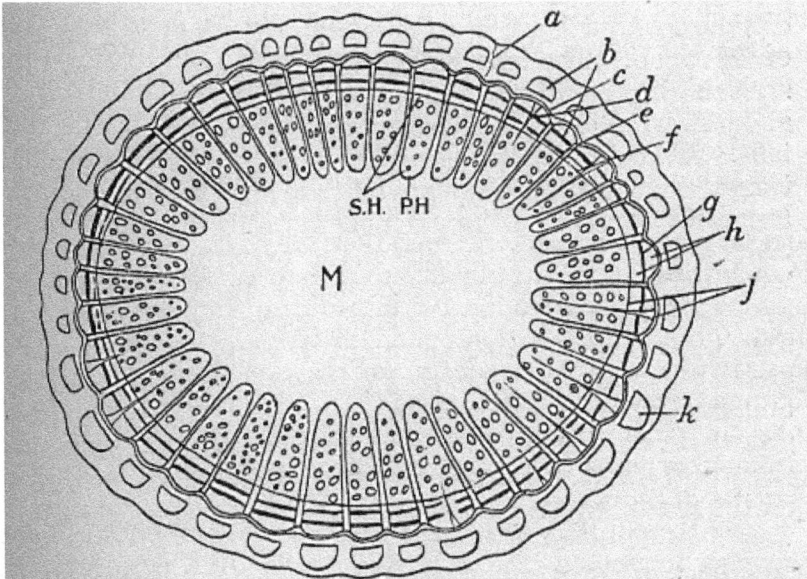

FIG. 34.—Cross-section through a one-year-old shoot of Muskat gutedel. *a*, dead bark; *b*, sclerenchyma; *c*, cork; *d*, secondary medullary ray; *e*, primary medullary ray; *f*, cambium; *g*, hard bast (sclerenchyma); *h*, soft bast (phloem tissue); *j*, bast ray; *k*, sclerenchyma fibre strands (primary hard bast); P.H., primary wood; S.H., secondary wood; M., pith. × about 14. Original.

On Fig. 34 the hard bast is represented by the fairly broad, black tangential lines inside the wavy layer of cork. Here we notice 1-2 layers of hard bast and 2-3 layers of soft, bast in every bast ray. Where a secondary medullary ray (Fig. 34 d) has been developed in a bast ray, we find double this number of layers of hard and soft bast.

The hard bast is composed of bast fibres with pit canals which can be dead or alive and filled with starch (see Fig. 26,D).The soft bast, as in the root, is composed of sieve-tubes, companion cells, and parenchymatous cells. The

---

[1] According to Babo u. Mach, *l.c.* vol. i. p. 323.

phloem bundles and the parenchyma are never lacking in the bast as they are required for the conduction and temporary storing of the assimilated (organic) food—proteids in the phloem, and sugar and starch in the parenchyma.

The wood is composed of tracheae, tracheides, sclerenchyma fibres, replacement fibres ("Ersatzfasern"), and parenchyma cells.[1] The tracheae are narrower than those of the root, but large in comparison with the other elements of the wood. The tracheides have the shape of fibres, the replacement fibres consist of elongated cells which are pointed at both ends, and are characterised by their living cell content. Sometimes they are subdivided by strongly pitted transverse walls. "They have been called replacement fibres because they replace the parenchyma cells in the real wood. They are the living elements of the wood rays (*i.e.* parts of wood between the medullary rays, A. I. Perold), and as such, besides giving rigidity to the stem, serve in the first instance for storing foodstuffs. In autumn they are filled with starch grains which disappear in spring when the vine requires new building material for the opening buds and the resumed growth in thickness."[2] They constitute the main mass of the wood. The replacement fibres and the parenchyma cells of the medullary rays are the only living elements of the wood.

The medullary rays pass through both wood and bast. "The parenchyma of the medullary rays thus connects the parenchyma of the bast with that of the wood, and unites all the living tissue of the stem and root into a single system. Assimilated material moving downwards in the bast can thus pass radially into the wood and be carried in this for some distance upwards or downwards, to be stored as starch in the living parenchymatous cells."[3]

Towards the end of the vegetative period during which the cambium ring was formed, a thin layer of a new meristem, the *phellogen,* is formed in the outer layer of the bast. This gives rise to the formation of a continuous layer of cork, the periderm, which consists of a few rows of thin-walled cork cells and a feebly developed *phelloderm.*[4] This cork tissue or periderm is seen as a wavy line on Fig. 34. It

---

[1] According to Babo u. Mach, *l.c.* i. p. 326

[2] Babo u. Mach, *l.c.* i. p. 328.

[3] Strasburger, *l.c.* p. 156.

[4] See Guillon, *l.c.* p. 201.

completely cuts off the food supply of the tissues outside of it. They die and adhere to the ripe one-year-old cane (stem) as brown dead bark. The *bark* includes the whole of the dead tissue outside of the phellogen. The cork now serves as the boundary tissue of the cane, and cuts it (hence also its medullary rays) off from its environment.

During the continued growth in thickness in the second and subsequent years, new layers of cork are continually formed deeper in the bast, which every time results in the death of the tissues external to it. In this way plenty of bark is formed in course of time on the old stem, which peels off in flakes and strips and thus gives old vine stems their peculiar appearance, and also protects them against loss of water and against overheating.

Annual rings also occur in the wood of the stem, but are not always directly visible. By staining cross-sections of the wood with phloroglucin and hydrochloric acid or with aniline hydrochloride, we can easily demonstrate their presence. The thickening of the rings, in stems growing horizontally, is less above than underneath. If the stem is trained along a wall, the rings are thicker on the side towards the light than against the wall. According to Rathay[1] the thickness of the annual rings on *Vitis vinifera* varies from 0.06 mm. to 3.39 mm., and averages 0.63 to 1.30 mm.

The *pith* of the stem dies during its secondary growth and the cells assume a brown colour, after which they only play a passive role.

Thus far we have discussed the anatomy of the internodes. It will now be necessary briefly to discuss that of the nodes. Here we notice a thickening of the stem on the side of the eye and also on that of the tendril or cluster where present. In the young shoots the collenchyma is much more strongly developed in the nodes than in the internodes. In its secondary structure the medullary rays are broader and the vascular bundles smaller than in the internodes. The pith is interrupted in the node by a new organ, the *diaphragm*, so that the pith of every internode is separated from that of every other internode. This we find with all the *Euvites* or true vines.   Only with the *Muscadinia* (*Vitis rotundifolia*) is the diaphragm wanting.

---

[1] See Babo u. Mach, *l.c* vol. i. p. 328.

The thickness of the diaphragm varies with different species and varieties. In every instance the diaphragm is a living bridge in the hollow wood cylinder of the one-year-old canes. Both on its upper and lower surface it is separated from the dead pith by a layer of cork. During the first year the diaphragm is composed almost exclusively of living parenchymatous cells.

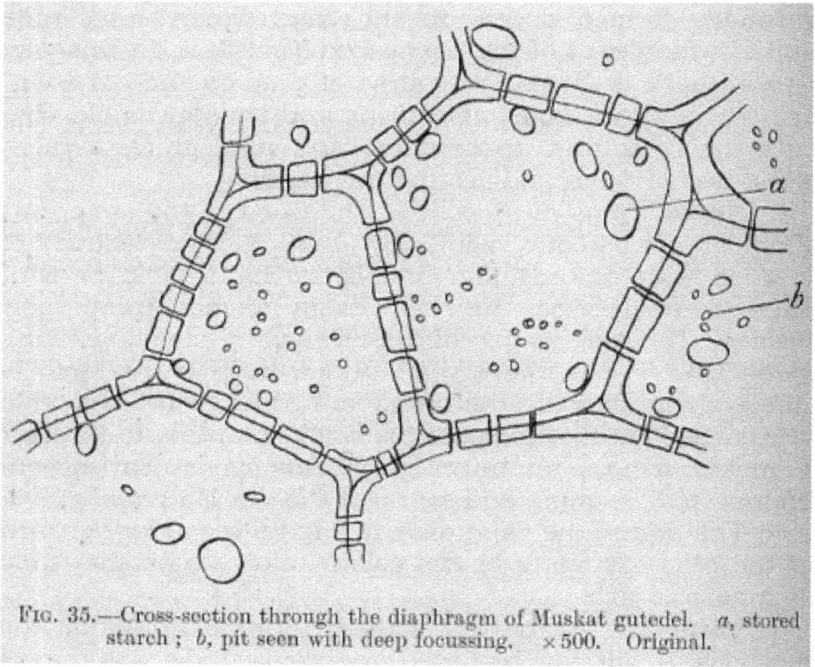

FIG. 35.—Cross-section through the diaphragm of Muskat gutedel. *a*, stored starch ; *b*, pit seen with deep focussing. × 500. Original.

Their walls are fairly thick, lignified, and possess numerous pits (see Fig. 35). In the ripe wood the node possesses numerous grains of stored starch because it contains many cells that belong to the medullary rays, which, as we have seen above, are broader here than in the internodes. Also in the cells of the diaphragm starch is found, and sometimes tannin too. The large quantity of stored starch in the internodes is necessary to provide food rapidly for the developing bud. This also explains why a cutting that is planted out develops most of its roots at the nodes.

The distinct *green colour* of the wood of the cane is one of its striking characteristics. We notice it not merely in the green shoots, but also in the ripe one-year-old canes during winter. It is caused by the chlorophyll content of the chromatophores in the living cells of the wood (*i.e.* the replacement fibres and the parenchymatous cells of the medullary rays). They can assimilate a little carbon dioxide where it reaches them, but their main function is to transform the sugar coming from the leaves into starch.

Comparative Anatomy of the Stem.— According to Guillon, *l.c.* p. 205, the differences in structure in the different species concerns particularly the shape of the epidermal cells, the development of the collenchyma, the shape of its elements, the width of the sieve-tubes, the relative proportions of soft and hard bast, the size of the starch grains, etc. We have already referred to the different development of the collenchyma in different species and varieties, and to the absence of the diaphragm in *Vitis rotundifolia*. In this last-named species the cork is not formed deep in the bast, but in its first layer, and beneath every stoma on the shoot a *lenticel* is formed which is very striking from the outside.

The lenticels bring the ventilating system of the deeper lying tissues into open communication with the air. They are natural openings in the otherwise uninterrupted layer of cork. In the lenticels we find, in contrast with the rest of the layer of cork, intercellular air spaces and round, loosely connected elements with living protoplasm and aqueous cell sap.[1] At the commencement of winter they are covered by a thin layer of cork, but the opening is restored during the following spring.

On the canes of *V. rotundifolia* the lenticels can be seen with the naked eye as prominent warty growths. Its pith does not die (as happens with the Euvites, all of which have a diaphragm) and contains lots of starch.[2]

"*Vitis riparia* possesses many and wide sieve-tubes, very wide tracheae, and small starch grains. With *Vitis rupestris* the sieve-tubes are narrower and less numerous, the starch grains bigger, and the pith narrower. With *Vitis vinifera* the vessels are normal and the starch grains big."[3]

---

[1] According to Oudemans en de Vries, *l.c.* vol. i. p. 104
[2] According to Guillon, *l.c.* p. 206
[3] Guillon, *l.c.* p. 205

(c) **Structure of the Leaf**

The structure of the leaf-stalk closely resembles that of the stem in its primary growth. Fig. 36 clearly shows this. It is composed of epidermis, cortex, and central cylinder. In the cortex occur broad strands of collenchyma and parenchyma tissue. The central cylinder contains open vascular bundles, pith, and medullary rays. Here also a continuous cambium ring separates the xylem (inside) from the phloem (outside) strands. The two uppermost vascular bundles are much bigger than the others. Above them we find two smaller vascular bundles for additional rigidity to keep the leaf in its proper position. The phloem bundles are bordered on their outside by thin strands of sclerenchyma (see Fig. 36).

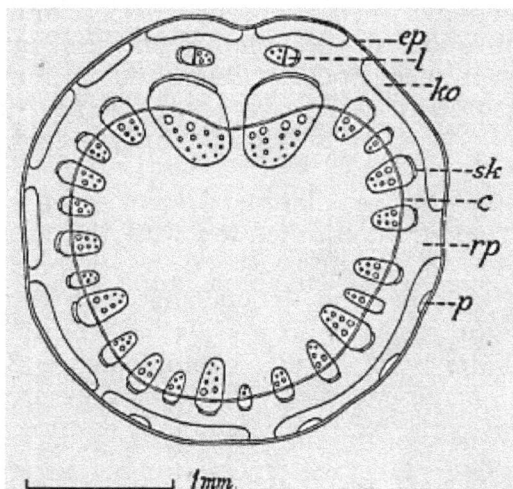

Fig. 36. — Cross-section through the petiole of the vine (diagrammatic). *ep*, epidermis; *ko*, collenchyma; *p*, parenchyma; *rp*, cortex parenchyma, *c*, cambium; *l*, vascular bundle; *sk*, sclerenchyma strand. × 20. From Babo u. Mach, *Handbuch des Weinbaues*, 1923. Paul Parey, Berlin.

The strands of collenchyma contain very little chlorophyll and colourless cell sap. They form the whitish stripes in the leaf-stalk. The parenchymatous cells, on the contrary, are rich in chlorophyll and sometimes contain red cell sap. This seldom occurs with the white grape varieties, and hence their leaf-stalks usually show whitish and green stripes. In red and black varieties the parenchymatous cells contain so much red sap that it covers the green colour, as a result of which we notice red and whitish stripes on the leaf-stalk, especially on the sunny side. The parenchymatous cells of the pith and medullary rays all contain chlorophyll, with here and there raphides as well.

At its base the vascular bundles of the leaf-stalk enter the parenchyma of the bast and soon unite with the conducting tissue of the stem (shoot). In autumn the leaf drops after suberisation has taken place at its base. Sometimes the leaf-blade drops from the leaf-stalk before the latter drops from the stem.

The leaf-blade is composed of a tissue of parenchyma called *mesophyll* which is traversed by the *nerves*, and is bounded on both sides by an *epidermis*.

The nerves are the continuation of the leaf-stalk, which they resemble very closely indeed in their structure. They also contain several vascular bundles, which lie next to each other with the xylem above (*i.e.* nearer the upper surface) and the phloem underneath the xylem. Every vascular bundle is surrounded by a sheath of parenchymatous cells, even in the finest branches (see Fig. 38). These cells are usually elongated and are closely united without intercellular spaces.

The vascular bundles lying on the sides of the main nerves gradually branch out into the side or secondary nerves. This ramification is repeated several times. Whilst Fig. 9, showed us the network of nerves in the leaf of the vine as seen with the naked eye, Fig. 37 gives us a magnified picture of the finer nerves in the leaf of the vine to where they end blindly in the mesophyll. On Fig. 38 we see a higher magnification of the termination of a vascular bundle in the leaf of the vine. In it we notice the sheath consisting of one row of elongated parenchymatous cells which completely encloses the vascular bundles, and possesses intercellular spaces only on its outside. Where the chlorophyll grains have been drawn, the intercellular spaces appear as blank white patches. In the rest of my drawing the intercellular spaces are those showing sharp corners, as those of the cells are always rounded off in nature. The illustration shows a surface section of the leaf and only of the upper portion of the nerve, so that we see here the xylem part of the vascular bundle with the spiral thickening of its walls.

On the phloem side of the strong nerves, i.e. their lower side, they are strengthened by sclerenchymatous strands on the outside of the phloem, as we saw happening in the leaf-stalk. This is wanting in the finer nerves, where it is replaced by strands of collenchyma on both sides of the nerve. The finest nerve endings possess no such strengthening tissue.

In the *mesophyll*, lying between the upper and lower

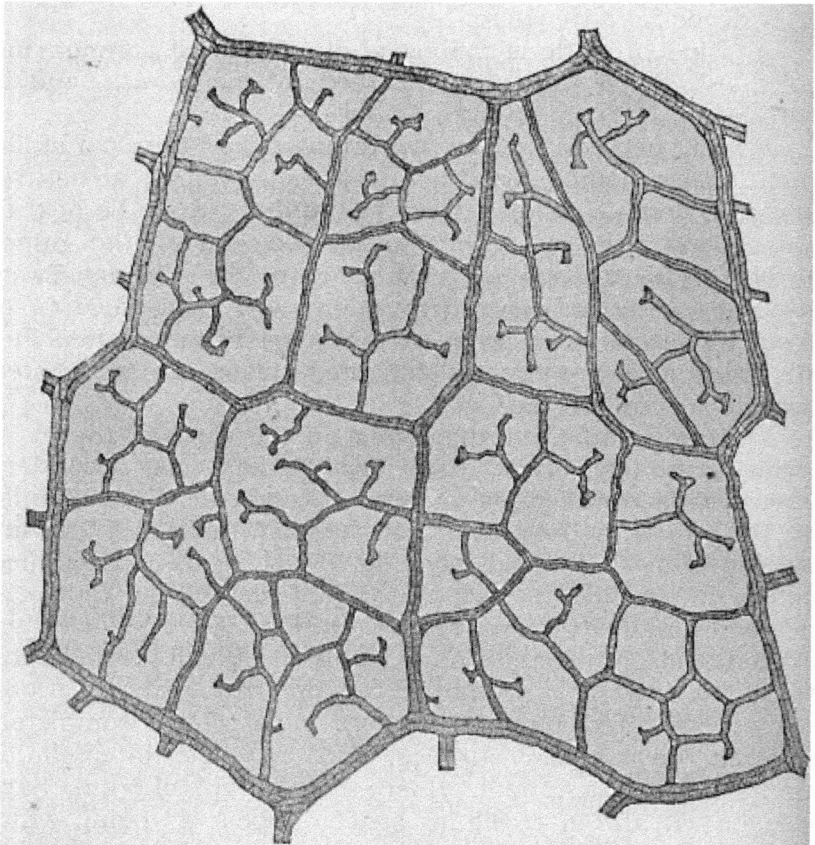

FIG. 37.—Network of nerves in the leaf of the vine (Riesling). × 45. From Babo u. Mach, *Handbuch des Weinbaues*, 1923. Paul Parey, Berlin.

epidermis and the nerves, we can clearly distinguish two layers. The upper layer consists of one row of elongated cells standing at right angles to the upper surface of the leaf, and known as palisade cells. They are partly separated from each other by intercellular passages. This can well be seen on a cross-section of the leaf, as in Fig. 39. They possess

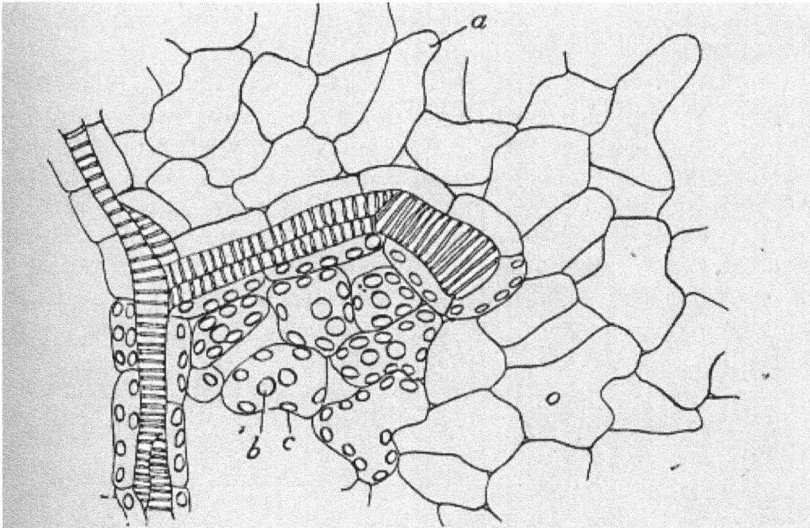

FIG. 38.—Termination of vascular bundle (nerve) in the leaf of Weisser Calabreser seen from above, hence only xylem visible. × 500. Original.
a, mesophyll parenchyma; b and c, chlorophyll grains.

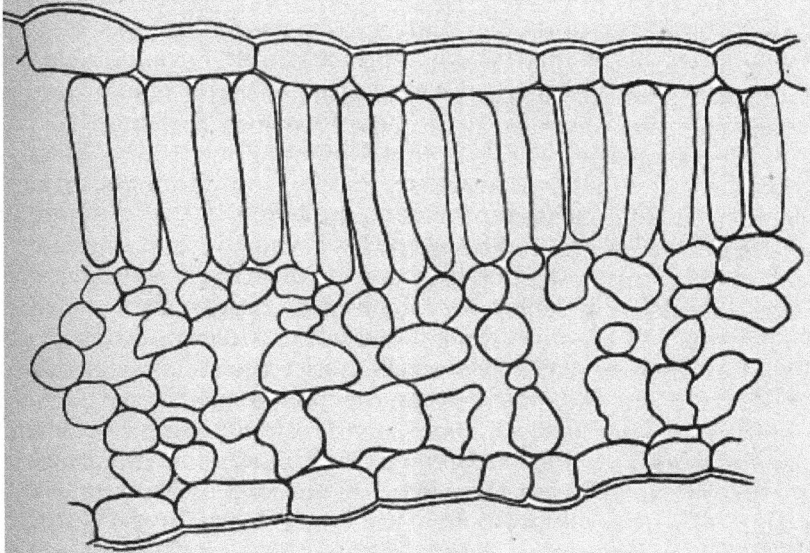

FIG. 39.—Cross-section through a vine leaf. × 500. Original.

large numbers of chlorophyll grains, and are pre-eminently the seat of carbon assimilation (photosynthesis).

The lower layer of the mesophyll consists of a mass of irregularly shaped cells with large intercellular spaces between them. Hence this part of the mesophyll is known as the *spongy parenchyma*. These cells contain far less chlorophyll grains than the palisade cells. This, together with the air spaces between them, causes the lighter colour of the lower surface of the leaf. Here the carbon assimilation is only slight. The main function of the spongy parenchyma, in conjunction with the stomata in the epidermis on the lower surface of the leaf, is to provide for the ventilation and transpiration of the leaf.

As in the other parts of the vine, we find also in the leaf raphides, crystal stars and individual crystals of calcium oxalate.

For the comparative anatomy of the leaf I refer the reader to Guillon, *l.c.* p. 224.

(d) Structure of the Flower

The leaves of the corolla are composed internally of a few layers of parenchymatous cells, and externally have a single row of thin-walled epidermal cells containing a few stomata, and in the middle of the leaf runs one vascular bundle. On the epidermis occurs an ornamentally striped cuticle.[1]

The filaments possess an equally simple structure. In the middle they have a vascular bundle, on the outside an epidermis, and inside a normal parenchyma.

The anther, in its imperfectly developed state, clearly shows two pairs of pollen-sacs in its two halves (thecae) which are connected by the connective, which is a continuation of the filament. The connective possesses a structure similar to that of the filament. In the pollen-sacs the pollen-grains are formed. The walls of the pollen-sac possess a feebly developed epidermis together with a strongly developed fibrous layer or endothecium (see Fig. 40, *f*). During the bloom the thecae open by one split in the wall where the two pollen-sacs have united (see Fig. 40).

The *pollen-grains* (according to Babo u. Macn, *l.c.* i. p. 339) average 25 μ in length and 15 μ in width (*i.e.* 0.025 × 0.015 mm. or 0.001 × 0.0006 in.), and are therefore very small. They seem to contain two highly plasmatic cells without a transverse wall but enclosed by a common membrane consisting of a

---

[1] According to Kroemer in Babo u. Mach, *l.c.* i. p. 338.

colourless internal layer, the *intine*, and a pale yellow external layer, the *exine*. On the pollen-grains of male and hermaphrodite flowers we notice three thin-walled bands running from pole to pole. In the middle of these bands the exine is perforated by a round pore. When the pollen-grain gets into water or sugar water, it swells and assumes a spherical shape, when a pollen-tube protrudes

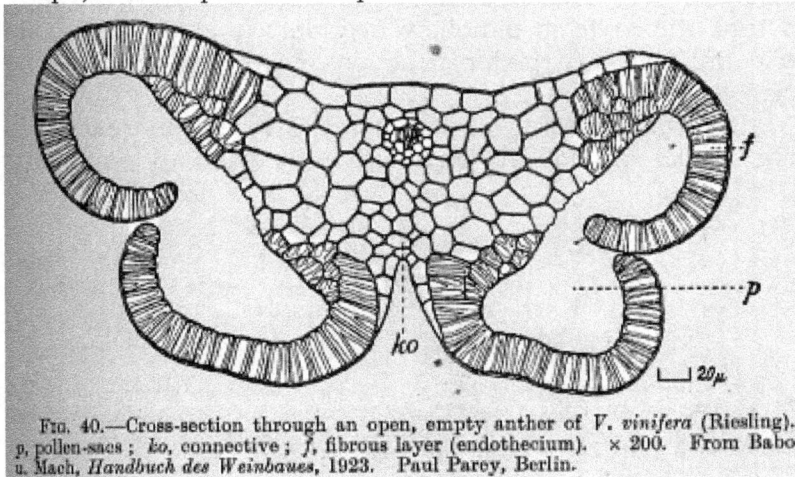

FIG. 40.—Cross-section through an open, empty anther of *V. vinifera* (Riesling). *p*, pollen-sacs ; *ko*, connective ; *f*, fibrous layer (endothecium). × 200. From Babo u. Mach, *Handbuch des Weinbaues*, 1923. Paul Parey, Berlin.

through a pore (see Fig. 41, *d, e, f*). The pollen-tube is formed by the intine protruding. As the pollen-tube grows in length, the protoplasm of the pollen-grain passes into it. In the same way the pollen-grain germinates when it gets on to the stigma under favourable conditions. Here it finds a sticky liquid which causes it to swell and the pores of the exine to be opened, through which the pollen-tube emerges and then penetrates the tissue of the stigma into the style canal and through the micropyle to bring about the fertilisation of the ovules. While this happens, the pollen-grain remains lying on the stigma, but is left by its protoplast.

The stigma is covered with papillae (conically protruding surface cells) which secrete a sticky liquid as soon as the pistil is ripe for being fertilised.

This liquid retains the pollen grains on the stigma and enables them to germinate.

The short style (according to Babo u. Mach, *l.c.* i. p. 341) consists of a ground mass of embryonic parenchyma tissue of small cells with numerous intercellular spaces. In these cells the large quantity of raphides is striking. On their outside they are bound by an epidermis. We have already seen that the style is a hollow cylinder. It is through this style canal that the pollen-tube passes on its way to the ovary.

For the anatomy of the ovary I refer the reader to special works on botany. Here I merely wish to point out that,

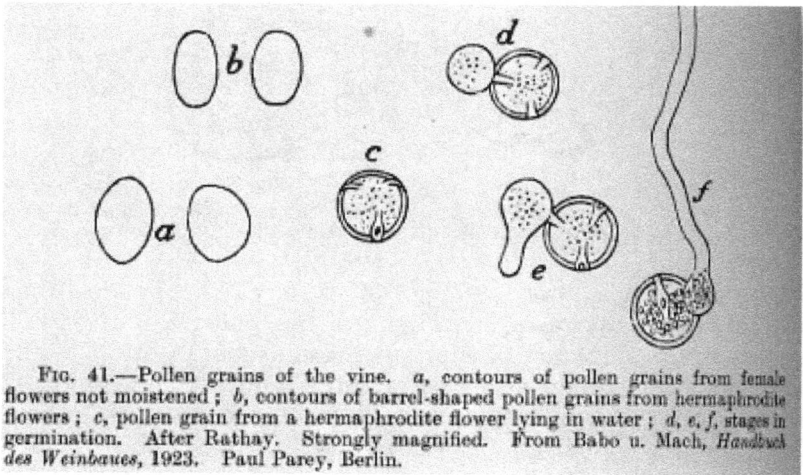

FIG. 41.—Pollen grains of the vine. *a*, contours of pollen grains from female flowers not moistened ; *b*, contours of barrel-shaped pollen grains from hermaphrodite flowers ; *c*, pollen grain from a hermaphrodite flower lying in water ; *d, e, f,* stages in germination. After Rathay. Strongly magnified. From Babo u. Mach, *Handbuch des Weinbaues,* 1923. Paul Parey, Berlin.

according to Portele,[1] 15-18 vascular bundles generally pass from the pedicel into the ovary and provide for the conduction of its required supply of water and food.

(e) Structure of the Fruit

The structure of the stalk and of the pedicels closely corresponds with the primary structure of the stem. Their epidermis also contains a few stomata. The pedicel merely possesses fewer vascular bundles than the stalk. Also its pith is smaller in relation to its xylem strands than happens in the stalk.

---

[1] See Babo u. Mach, *l.c.* i. p. 341.

The *berry* develops only from the cells that were formed in the ovary. Thus no new cells are formed during its development; the existing cells merely grow (see Babo u. Mach, *l.c.* i. p. 342). The cells in the *skin* are small and tangentially stretched. As long as the grapes are green, the berries contain chlorophyll. The external layer of cells forms the epidermis. The outer wall of the epidermal cells is thickened, and is covered by a thin cuticle which is covered with a thin coating of wax, the *bloom*. The network of vascular bundles which enter the berry like a brush, spread through the pulp of the berry. As was previously pointed out, these come out of the pedicel. Next to elongated cells and sugar - conducting elements, they are mainly composed of narrow spiral vessels. The *pulp* consists of large parenchymatous cells with big vacuoles. These cells usually show a radial elongation and possess a diameter of 300-400 μ. (*i.e.* 0.3—0.4 mm. or 0.012—0.016 in.), which is fully twelve times the length of the pollen-grains.

The *seed* is the entire product developed out of the ovule. It possesses a fairly complicated structure. Fig. 42 shows us two sections through a grape seed which

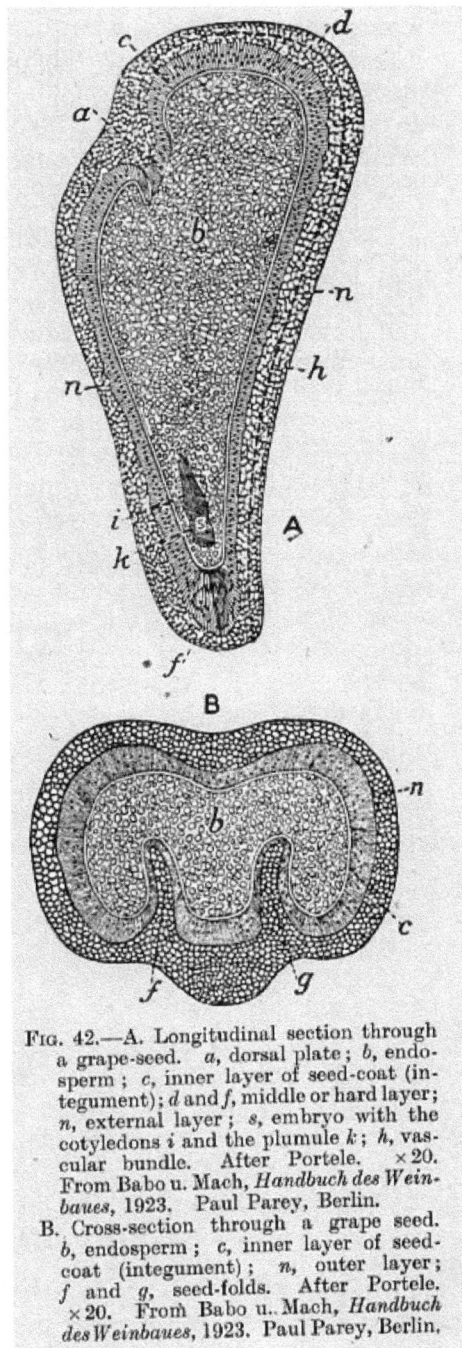

Fig. 42.—A. Longitudinal section through a grape-seed. *a*, dorsal plate; *b*, endosperm; *c*, inner layer of seed-coat (integument); *d* and *f*, middle or hard layer; *n*, external layer; *s*, embryo with the cotyledons *i* and the plumule *k*; *h*, vascular bundle. After Portele. ×20. From Babo u. Mach, *Handbuch des Weinbaues*, 1923. Paul Parey, Berlin.
B. Cross-section through a grape seed. *b*, endosperm; *c*, inner layer of seed-coat (integument); *n*, outer layer; *f* and *g*, seed-folds. After Portele. ×20. From Babo u. Mach, *Handbuch des Weinbaues*, 1923. Paul Parey, Berlin.

consists of *seed-coat, endosperm,* and *embryo.* The *seed-coat* contains an external layer consisting of parenchymatous cells rich in tannin[1] which dries out completely when the seed is ripe, but again swells out easily when the seed comes into contact with water. The middle layer of the seed-coat consists of hard stone cells (see Fig. 27) which are lignified and radially elongated. The inner layer of the seed-coat consists of small parenchyma cells.

The *endosperm* is composed of an embryonic tissue of small cells filled with proteid (aleurone grains), fat, and oil.

The *embryo* lies near the beak of the seed and is composed of the two *cotyledons* or *seed-leaves* and the rudiment of the apical bud (the real young plant), the *plumule,* which consists of *cauliculus* and *radicle.*

---

[1] It is largely this tannin which we find in red wine.

CHAPTER III
GENERAL INTRODUCTION

THE BIOLOGY OF THE VINE

THE nutrition of the vine will be discussed in a later chapter by itself. Here we shall briefly study the different stages in the biological cycle of the vine in the order in which they occur. First of all comes the germination of the grape seed which gives rise to a new vine. Once this vine is taken into cultivation, we notice it bleeding when pruned; later on its buds start opening and the young shoots develop until the flowers begin to open and the vine enters its blooming period. At this point we shall study its blooming and fertilisation, then the development of the grapes until they are ripe. Finally, the ripening of the wood and shedding of leaves complete the period of growth, and the vine enters upon its winter rest to re-commence the same cycle towards the beginning of spring.

1. THE GERMINATION OF GRAPE SEED

As was previously pointed out, the grape seed has a very hard seed-coat. For a favourable germination, the seed should not dry out more than can be avoided. Hence it is good practice to let the seed remain in the berry if it can turn into a natural raisin on the vine, and to take out the seed when it is to be planted. The seed should in any case be properly ripe, which means that the grapes containing these seeds should be allowed to get fully ripe on the vine before being picked. The ripe seeds can at once be taken out of the berries, freed from pulp and washed rapidly in clean water, when they are allowed to get air-dry in an ordinary room, and are stored in a corked bottle in a cool place. This clean dry seed will not develop mould.

Before planting out, the seed is soaked in cold water for three or four days, the water being renewed every twenty-four hours. As the seed requires moisture, oxygen (air), and a certain degree of temperature for its germination, it is planted in loose, moist garden soil that has received a liberal dressing of well-rotted farmyard manure. It is important that the seed should be planted at least 1-1½ inches deep, so that the soil can retain the empty seed-coat during germination. Seed that is planted too shallow, will, on germinating, give rise to a young seedling having its cotyledons still in the seed-coat when they are above ground. The seed-coat now dries out, gets hard, and can hardly be separated from the developing seedling without seriously injuring it.

As soon as the seed has absorbed enough water and the temperature is high enough (22-30° C. or 71.6-86° F. is the most favourable or optimum temperature for germination), the seed will start germinating if sufficient air is present. During the process of germination an active metabolism takes place in the albumen of the seed, which has now become soft through absorption of water. Out of the stored starch sugar is formed, and out of the proteids asparagine among other things. The cells of the endosperm expand and their walls grow thinner. This produces a great internal pressure which causes the seed-coat to burst from the seed-beak along the raphe. Meanwhile the embryo also develops. Gradually the radicle emerges (see Fig. 2) from the seed. If the seed has been planted deep enough, the seedling will in course of time show its little stem and the cotyledons above the ground whilst the seed-coat will remain in the ground.

Seeds that have ripened well will germinate best. According to Hamböck[1] it is inadvisable to keep the seed for more than three years. It is always best to use seeds of the previous crop. He asserts that an alcohol concentration of 7 per cent (by weight ?) acting for forty-eight hours at 16° R. or 20° C. almost completely destroys the seed's germinating power. Foëx[2] maintains that seeds that remained in contact with the must during fermentation, germinate just as

---

[1] C. Hamböck, quoted from Babo u. Mach, *l.c.* i. pp. 175-176.
[2] G. Foëx, *Cours complet de viticulture*, 4th ed., 1895, p. 283.

well as those that were removed directly from the ripe berries.

Fresh seed, under favourable conditions, usually completes its germination within one month. After that time only very few seeds will germinate. The further development of the seedling has already been discussed.

2. THE BLEEDING OF THE VINE

If we cut the cane of a vine towards the end of winter or the beginning of spring, we notice a watery liquid emerging from the wound. The vine is then said to be "bleeding". We notice that the vine starts bleeding when pruned only after the soil and air have risen somewhat in temperature and the vine has entered upon its new period of growth by forming young roots, but before new shoots and leaves have appeared. On examining the bleeding wound with a strong magnifying glass, we shall see that the liquid flows only out of the tracheae and the tracheides, hence out of the wood or xylem bundles.

The bleeding of the vine is ascribed to two causes, namely, the absorption of water by the newly formed young roots and the expansion of the gas bubbles (air and carbon dioxide) in the sap of the roots. Tamaro[1] is a supporter of the latter view. He says that the rise in soil temperature causes these gas bubbles to expand, thus producing a rise in the internal pressure. This again causes the sap to flow from a fresh wound. According to the researches of Houdaille and Gullion,[2] we must ascribe the bleeding of the vine almost exclusively to the absorption of soil moisture by the young rootlets. The gas pressure theory is insufficient to explain the bleeding of the vine, and can at most play only a secondary role in this connection.

Houdaille and Guillon have found that the vines start bleeding when the temperature of the soil, at a depth of 25 cm. or 10 in., has reached 10.2° C. (50-36° F.) with Aramon grafted on Riparia, 11° C. (51.8° F.) with Aramon grafted on Jacquez, and 14° C. (57.2° F.) with Aramon grafted on Beriandieri. It would seem that every variety of vine

---

[1] Prof. Dott. D. Tamaro, *Uve da tavola*, 4th ed., 1915, p. 157.
[2] See Guillon, *l.c.* pp. 255-273.

(mainly the stock with grafted vines) requires a definite soil temperature before it can bleed. This bleeding comes on suddenly, rapidly increases in intensity, then decreases gradually.

The *sap pressure*, when the vine bleeds, has been observed already by Hales[1] in 1725. He attached a vertical glass tube to a freshly decapitated vine in April (in France), and soon noticed that the sap had risen to a height of about 7 metres (22 ft. 11½ in.) in the tube. According to later observers, *e.g.* Wieler,[2] the highest sap pressure in the vine during the period of bleeding amounts to a pressure of 900-1100 mm. mercury or 12.24—14.96 metres (i.e. 40' 2"—49' 1") water or 1.18—1.45 atmospheres. Thus the sap pressure in this case can reach nearly 1½ atmospheres.

Measurement of Sap Pressure.—Houdaille and Guillon measured the sap pressure by fixing a capillary mercury manometer with an enclosed air column to a squarely cut cane by means of a piece of thick-walled rubber tubing, which they tied to the manometer tube and to the cane. Before attaching the manometer, they measured the length, L, of the air column, the atmospheric pressure, and the temperature of the air in the shade—the manometer was also kept in the shade. From the pressure and temperature of the air they calculated the length, L_0, which the enclosed air column would have occupied at 0° C. and under a pressure of 760 mm. mercury (= 1 atmosphere). They subsequently, from time to time, measured the length of the air column and the temperature of the air, both in the shade. In every case they calculated what the length of the air column would have been at 0° C. We will suppose it was L'_0 at a certain moment. If the sap pressure was x atmospheres, we get $\frac{x}{l} = \frac{L_0}{L'_0}$, as the length of the air column at the same temperature varies inversely with the pressure (here we assume that the bore of the capillary tube is perfectly uniform). Hence the sap pressure, x, is equal, to $\frac{L_0}{L'_0}$ atmospheres.

They also measured the *quantity of sap* lost, by attaching a capillary tube (2 mm. internal diam.) to a freshly cut cane

---

[1] See Guillon, *l.c.* p. 256.
[2] See Babo u. Mach, *l.c.* i. p. 408.

and collecting in a corked glass tube the liquid that flowed out. The maximum quantity they measured on a Berlandieri vine, which lost 5545 c.c. (nearly 1¼ gallons) sap. Their researches led to the following results:

1. The period of the maximum loss of sap corresponds with the period of maximum sap pressure.

2. The rapidity with which the sap flows out and the sap pressure do not stand in any direct relation to the external weather conditions.

3. The quantities of sap lost by the different arms on the same vine or by different vines of the same variety vary greatly.

4. During the period of bleeding the changes in the sap pressure are not directly related to the changes in the temperature of the air and the soil. The pressure varies from 1.05 to 1.85 atmospheres with an air temperature of 8.5—10° C. and a soil temperature of 9.7—11.3° C.

5. Observations made on an Aramon vine grafted on Riparia show that a direct relation exists just as little between the quantity of sap lost and the temperature of the air and soil. The sap flows at a fairly constant speed during great variations in the temperature.

6. A certain periodicity in the loss of sap is only noticed when the buds on the cut cane begin to open. Then it is at its maximum from midnight till 6 A.M., and at its minimum from 12 noon till 6 P.M. This is easily understood, as the phenomenon is now complicated by the loss of water through the transpiration of the young developing shoots. This loss is least between midnight and 6 A.M., and greatest between 12 noon and 6 P.M. Later on, when the first leaves have opened, the bleeding stops during the day, because of the increased loss of water by transpiration, to start again during the night. Still later, this bleeding too comes to a stop.

If we cut a shoot in summer and attach a manometer to it, it will show a negative pressure or an absorption. At such a wound water can be sucked in, which will go to the other aerial parts but not to the roots, since no negative pressure obtains in these latter. The suction is caused by the transpiration going on in the other shoots (leaves) of the vine. About the middle of autumn, however, when the absorption of water by the roots diminishes and the leaves now lose less water by transpiration, we find a negative pressure or absorption from a different cause. The vine is now no longer filled with water internally, but sucks up water through the capillary action of its vessels. According to Houdaille and Guillon's experiments, previously quoted, one kilogram of canes and roots cut from the vine in autumn can absorb respectively up to 130 and 150 grams, of liquid.

It is only towards the end of winter or the beginning of spring, when the roots once more begin sprouting, that this negative pressure changes into a positive sap pressure which shows itself by the bleeding of canes that are then cut.

Canstein,[1] in his experiments on the bleeding of the vine in spring, found a total loss of sap of 0.76—20.15 lit., or nearly 4½ gallons per vine. A high soil temperature always resulted in great loss of sap by bleeding. He found that the young shoots on bearers that have bled particularly heavily in spring, were backward in their development compared with those on bearers that bled only very little. Thus heavy bleeding might have a detrimental influence upon the opening and subsequent development of the buds, although this will usually be hardly noticeable in good moist soil. He summed up the results of his investigations as follows: [2]

"1. One can never, in spring, foretell when bleeding will be most copious, and it is not immaterial whether the vine is pruned on a day when it will lose 50 c.c. or on another when it will lose 950 c.c. sap.

"2. One should commence pruning at an early date, as soon as the soil permits.

"3. When one is obliged to prune later, one should select the cold days for this operation and not the warm days when bleeding is stronger.

"4. In vineyards containing several varieties, one should start with the varieties having thick canes and a strongly developed root-system, and finish up with the others.

"5. If one prunes too late and the vine loses much sap, it is not merely weakened by a loss of nutriment, but is subject to a morbid activity which shows itself in summer and in autumn in diseases and in an irregular development; it

---

[1] Canstein, "Uber das Tranen oder Bluten der Weinstocke im Fruhjahre" *Annalen der Oenologie*, iv. p. 499. Quoted from Babo u. Mach, *l.c.* i. p 408

[2] Quoted from Guillon, *l.c.* pp. 271-272.

blooms later, the grapes ripen in an irregular fashion, and the wood does not ripen."

Composition of the Sap (Tears).—According to Guillon,[1] the solids in 1 lit. Tears amount to about 2 grams, of which two-thirds consist of organic substances and one-third of mineral substances. He and Houdaille analysed the tears from separate arms on the same vines and from different vines with the following result:

DETERMINATION OF ASH AND DRY EXTRACT

Origin of Liquid.		Dry Extract (organic and mineral substances) per litre Sap.	Ash per litre Sap.
Aramon on Riparia *A*	. .	1·816 gram	0·645 gram
Aramon on Riparia *B* { 1st arm		1·716 ,,	0·596 ,,
{ 2nd ,,		2·133 ,,	0·608 ,,
Jacquez { 1st portion	. . .	1·918 ,,	0·376 ,,
{ 2nd ,,	. . .	1·556 ,,	0·380 ,,
Berlandieri { beginning	. . .	2·025 ,,	0·662 ,,
{ middle	. . .	1·872 ,,	0·662 ,,
{ end	. . .	2·020 ,,	0·525 ,,

From these results it is evident that the extract and ash contents vary in different arms of the same vine ; the ash content remained fairly constant with the same variety during the whole period of bleeding, whilst the organic extractive substances showed a greater variation.

According to analyses made by Neubauer,[2] one litre of tears contains on an average 1.58 gr. organic and 0.74 gr. inorganic materials. The qualitative analysis showed the presence of a magnesium compound $C_6H_{14}MgO_8$, gum, sugar, Ca-tartrate, inosite, succinic acid, oxalic acid, and other extractive substances. The average ash content of 1 lit. sap was 665 mg., of which 69 mg. was potash, 19 mg. phosphoric acid, and 275 mg. lime partly as carbonate. The nitrogen content varies from 150 to 450 mg. per litre sap.

From the foregoing it is clear that the bleeding of the vine causes a loss, though not a great loss, of plant food. On account of this fact, and more still because of the other detrimental

---

[1] Guillon, *l.c.* pp. 269
[2] Neubauer, "Untersuchung des im Fruhjahr aus denfnsch geschmttenen Keben ausfliessenden Saftes, der sog. Rebtranen", *Annalen Ser Oenologie*, iv. p. 499.

effects of copious bleeding previously mentioned, we should prune the vines when they do not bleed very much. Where vines are grafted on the spot, this should be done when they will not bleed too much on being decapitated, as heavy bleeding hinders the formation of callus, and hence the success of the operation.

3. THE OPENING OF THE BUDS AND DEVELOPMENT OF THE SHOOTS UP TO THE BLOOM

(a) **Budding**.—In spring the eyes of the canes start swelling until the brown scales covering the buds burst and a woolly bud becomes visible. This bud now develops rapidly. We notice the first little leaves enclosing the growing bud, and before long we see these little leaves separately on their leaf-stalks and the inflorescences also become visible. As soon as the first leaves become visible we say that the vine is *budding, i.e.* the buds are opening, and from this stage till the moment the first inflorescence becomes visible we call the young shoot *bourgeonnement*. Later on we speak of a young shoot, which again becomes a cane when ripe and ready for the dormant period.

According to Guillon[1] it takes twenty to thirty days from the time the vine starts bleeding until the buds begin to open. We have previously seen that the vine contains a very considerable amount of stored food (particularly starch) in winter. Before the buds can open, a portion of this food must be dissolved and be available, in aqueous solution, for the opening bud. The newly formed young roots absorb the soil solution which contains the mineral food. This, together with the dissolved organic food (sugar, etc.), is conducted to the growing points. Plain cuttings that have just been planted out usually start budding before they have developed roots. This proves that budding is not dependent upon the formation of roots. Such cuttings can absorb water by imbibition through the wound at the bottom end. Established vines have usually formed new roots by the time their buds begin to open, as they can only absorb the soil solution through their young roots. The buds of scions on freshly grafted cuttings or established vines usually begin to open

---

[1] Guillon, *l.c.* p. 274.

before the union has been cemented by the formation of callus, and such opened buds can easily die unless a proper union of callus is formed in time, and in bench-grafted cuttings, roots as well. Where scions have two eyes, the upper one may develop after the necessary union between scion and stock has been effected and thus continue to grow and give us a grafted vine. Even the eyes of a piece of cane lying on the ground may open if sufficiently moist, but these will soon die for want of moisture. Hence soil in which cuttings of Jacquez and other varieties that strike root with difficulty have been planted, must be kept sufficiently moist.

The temperature of the soil is the factor that decides when budding will commence.

Kovessi, *l.c.* pp. 39-40, says that the two principal varieties of the South of France, Aramon and Carignan, begin budding as soon as the average daily temperature reaches 11° C. Over a period of twenty-five years this temperature has been reached between 26th and 27th March. The American vines there start budding seven to twelve days earlier (average, ten days). Every variety, according to Kovessi, has its definite temperature at which its buds begin to open. Thus *Vitis rupestris* and Pinot noir begin budding at an average temperature of 10° C., whereas Aramon and Carignan begin at 11° C.

The time when budding will begin is of great importance in parts where frost is to be feared in spring. Many factors can influence the date of budding. Of these I wish to mention the following: variety of grape, date of winter pruning, method of pruning, vigour of vine, nature of soil, site, climate, and treatment of vine after having been pruned.

It is well known that, under identical conditions, some varieties will start budding earlier than others. At my request, Mr. J. C. van Jaarsveld, manager of the Paarl Viticultural Experiment Station, noted the dates on which the 120 varieties I had planted there in 1911 began to bud, bloom, and turn colour. The following table gives the dates of budding for 25 varieties during 1914-17 (according to Mr. J. C. van Jaarsveld).

Variety	1914	1915	1916	1917
Red Hanepoot	20th Sept.	13th Sept.	11th Sept.	29th Aug.
White Hanepoot	12th	8th	8th	5th
Muscat d'Alexandrie (imported in 1910)	14th	8th	8th	3rd
Gros Colman	3rd	1st	20th Aug.	20th Aug.
Rosaki	2nd	1st	22nd	27th
Sultana	31st Aug.	1st	22nd	23rd
Barbarossa	16th Sept.	3rd	6th Sept.	29th
Barlinka	7th	3rd	6th	27th
Ohanez (Almeria grape)	6th	13th	4th	1st Sept.
Raisin blanc	16th	3rd	8th	29th Aug.
Hermitage	22nd Aug.	11th	11th	3rd Sept.
Cinsaut	17th Sept.	13th	8th	3rd
Cabernet Sauvignon	19th	13th	11th	1st
Pinot fin	8th	3rd	27th Aug.	28th Aug.
Red Muscadel	2nd	1st	24th	16th
White Muscadel	2nd	1st	20th	10th
White Greengrape	22nd	3rd	8th Sept.	27th
Red Greengrape	21st	11th	11th	29th
Stein	17th Aug.	1st	27th Aug.	20th
White French	22nd	11th	8th Sept.	27th
Henab Turki	10th Sept.	3rd	8th	25th
Madeleine Angevine	2nd	1st	3rd	5th
Ferdinand de Lesseps	-	6th	22nd Aug.	20th
Madeleine royale	31st Aug	1st	22nd	5th

From these data we conclude:

1. That the same variety does not commence budding at the same date in different years.

2. That the period of budding, in case of the varieties mentioned, lasted

In 1914 from 17th Aug. to 20th Sept., i.e. 35 days
In 1915 from 1st Sept. to 13th Sept., i.e. 13 days
In 1916 from 20th Aug. to 11th Sept. i.e. 23 days
In 1917 from 5th Aug. to 5th Sept. i.e. 32 days

3. That the shortest period of budding occurred where it started the latest (1915), and that the period of budding lasted two to five weeks, according to the year.

4. That the earliest budding in any year took place on the 5th August and the latest on the 20th September.

Generally speaking, the earlier a vine is pruned, the sooner it will bud. In order to get an appreciable difference in the date of budding, the vines must be pruned very early. Thus vines pruned in May in the Western Province of the Cape, will start budding a good deal earlier than those pruned in July.

Pruning between 15th June and 15th July makes little difference in the date of budding. Sometimes, according to weather conditions, earlier pruned vines may start budding after those pruned later. Vines pruned in August—which is very late for the Cape—will certainly start budding late.

The system of pruning adopted also influences the date of budding. The eyes on rods or log bearers will begin budding later than those on spurs or short bearers. Vigorous vines will bud earlier than weak ones.

Site and soil conditions (moisture content, colour, etc.) can cause the soil to get warm quicker and hence make budding earlier, or vice versa. Climate has the same influence, e.g. in Paarl vines bud much earlier than in Constantia or in the Warm Bokkeveld (high altitude, 1500 ft. above sea-level).

Vines that have been treated with ferrous sulphate and sulphuric acid or only with the latter after having been pruned in autumn, will bud six to eight and even twelve to fifteen days later in consequence.

As has just been stated, climate influences the date of budding. If, in a winter rainfall area, the rains stop in the latter half or the middle of winter and fairly warm weather prevails for several weeks continuously, the vines will start budding near the end of winter. If cold rainy weather now sets in and lasts for about a month right into spring, the budding will be bad and irregular. In this case budding is spread over a much longer period than usual. This happened at Paarl in 1914 and resulted in the longest budding period (five weeks as against two weeks in 1915). In such a year vines will not stir till very late, but will give the impression of being sick. Later on they will bud and develop fairly normal shoots, but with very few grapes on them.

A cold and rainy winter of considerable length, followed by a mild and sunny spring, results in a somewhat late but short budding period (see 1915 in table on p. 114). In this case the vines do not make a false start in resuming their growth after winter. In a normal season at the Cape, when the rains and cold weather are over by the middle of August and are followed by a few odd rainy days in the latter half of

August and the beginning of September, the vines bud from about the 20th August till about the 15th September and the budding season lasts about three weeks.

In parts where frost is to be feared in spring, the vines must be pruned very late, and such varieties as normally bud late should be grown in preference.

(b) **Development of Shoots up to the Bloom**.—This is the period of preparation for the bloom and fructification. The shoots grow in length and thickness, and their various parts (leaves, tendrils, blossoms, eyes) gradually develop. The transpiration of water through the leaves causes an upward current of sap. The soil solution is absorbed by the young roots and their root-hairs, and is conducted through the wood to the leaf in which it is distributed through its network of veins. On the way up a portion of the soil solution with the dissolved mineral food passes by osmosis from the wood into the surrounding living tissues. In the leaves the carbon assimilation takes place and the organic food is built up, as we shall see later. This assimilated food is conducted to the other growing parts of the vine through the phloem bundles, whence it reaches the surrounding living tissues by osmosis.

The vine must breathe, and for this purpose of respiration takes up oxygen from the air and oxidises part of its organic food to carbon dioxide and water. These products of respiration are excreted mainly through the leaves and the young roots. In the latter instance the excreted carbon dioxide has a solvent action on the soil particles and helps to enrich the soil solution in mineral food. As the carbon assimilation proceeds only in the light, *i.e.* not during the night, whilst respiration proceeds all the time, and as the former process is much more intense than the latter, plants seem to give off oxygen during the day and carbon dioxide during the night. As we shall see later, this is merely the combined effect of the two processes.

It is of the greatest importance to the farmer to protect the young developing shoots against damage by insect pests and fungoid diseases. The oldest leaves are usually the most valuable to the vine. Through proper manuring and soil cultivation we must see to it in good time that the vine will have sufficient food, air, and water at its disposal to be able to develop well, and later on to give a good crop. The further development of the vine will then mainly depend upon weather conditions. Much sunshine and warm weather will greatly promote it. After budding, the young shoots rapidly grow in length. At first this growth proceeds at an increasing speed up to a point where it becomes fairly constant, to decrease later on until it comes to a stop. Guillon[1] gives the following data which he obtained from growing shoots at the Cognac Viticultural Experiment Station in 1901:

OBSERVATIONS UPON THE PERIODS OF GROWTH IN LENGTH OF SHOOTS

VARIETY	Date of Budding	Growth in Length for every 15 days.								Total Growth in Length
		On 17th May	On 1st June	On 15th June	On 1st July	On 15th July	On 1st Aug	On 15th Aug	On 1st Sept	
		cm.	cm.	cm.	cm.	cm.	cm.	cm.	cm.	cm.
Folle Blanche	9th April	45	56	42	34	43	40	13	5	278
Cabernet Franc	9th April	34	54	42	39	48	47	16	2	282
Pinot	13th April	32	68	44	29	36	35	13	3	260
Aramon	9th April	21	72	54	32	43	38	7	1	268

From this table we see that the growth in length from the 17th May till the 1st June, with every variety, was greater than in any subsequent half-monthly period. From 1st June to 1st August the growth in length was on the whole fairly uniform; from 1st to 15th August it was much less; and from 15th August to 1st September still less than in the preceding period. After the 1st September the growth in length was about nil. The figures in the first column (17th May) give the total length of the shoots on that date, whilst those in the subsequent columns give the growth in length during each period.

It is during the period of rapid growth that the flowers open and fructification takes place. When the grapes turn colour (i.e. at the "veraison") the shoots do not grow any longer, or at most only very little. As was previously stated, the oldest leaves are the most valuable, and therefore should not be removed—at least not until the grapes and wood are properly ripe.

---

[1] Guillon, l.c. p. 280.

4. BLOOMING AND FRUCTIFICATION

Here we have to do with a very critical part of the vine's development. On its success will depend whether we can expect an abundant or a poor crop. During the preceding period the young inflorescences and their flowers too naturally developed and naturally grew bigger. At the bloom the green connected leaves of the corolla become detached at their base, curl up outwards, and drop like a little cap. As soon as this cap has been shed, the now liberated stamens move farther away from the stigma. The pollen is now usually fit for fertilisation, and the pollen-sacs open by a common split through which the pollen escapes in the direction of the stigma. Where the latter is just about as high as the anthers or somewhat lower, the pollen easily gets on to it and fertilisation takes place readily. Where the stigma, however, stands much higher than the anthers, as for instance with Ohanez (Almeria grape), the pollen no longer gets on to the stigma so easily. It is for this reason that the farmers in the Almeria district repeatedly pass over the blooming inflorescences with soft woollen brushes in order to ensure better pollination. This operation brings about the desired result. Insects (mainly bees) and wind naturally assist pollination. This, as will be readily understood, can easily bring about *cross-pollination*, *i.e.* the pollen fertilising the ovary comes from the flower of a different variety (species) than the one that is fertilised. In practice this makes no difference, as it does not affect the grapes in any way and as the seeds are not used for propagating the vine, except where cross-pollination was intentionally carried out for producing hybrids. We have already seen that the pollen of female flowers usually cannot germinate and hence is incapable of bringing about fertilisation. In such a case fertilisation can only be brought about by cross-pollination.

Before entering upon a further discussion of fertilisation, I wish briefly to consider the blooming dates of grapes.

Blooming Dates of Grapes.—Like budding, blooming depends upon the weather. Bright warm weather brings on blooming earlier than cool rainy weather, and reduces the time it lasts. The following table gives the dates when the different varieties mentioned began to bloom. For the years

118

1914-17 the observations were made by Mr. J. C. van Jaarsveld on the Paarl Viticultural Experiment Station, and for 1922-23 they were made by myself in the ampelographic collection on the Stellenbosch University Farm.

Variety	1914	1915	1916	1917	1922	1923
Madeleine angevine	Oct. 26	Oct. 29	Oct. 21	Oct. 19	Oct. 28	Oct. 28
Madeleine royale	28	23	26	19	28	21
Black Prince	Nov. 2	Nov. 1	26	25	31	27
White Muscadel	Oct. 30	Oct. 29	16	25.	Nov. 2	19
Ferdinand de Lesseps	—	29	21	21	Oct. 28	19
Pinot fin	Oct. 26	26	27	27	Nov. 1	17
Hermitage	Nov. 2	Nov. 1	28	Nov.8	7	31
Pontac	Oct. 30	1	28	3	1	27
Rosaki	26	1	26	1	4	29
White Hanepoot	Nov. 6	1	Nov. 6	8	4	Nov. 4
Red Hanepoot	6	9	4	8	9	2
White French	2	1	6	Oct. 27	3	Oct. 27
White Greengrape	Oct.26	1	6	Nov. 1	3	29
Red Greengrape	Nov. 2	1	4	1	3	31
Stein	Oct. 30	Oct. 29	Oct. 26	Oct. 25	3	29
Sultana	Nov. 2	Nov. 1	28	Nov. 1	4	27
Gros Colman	Oct. 31	1	28	1	2	23
Barlinka	Nov. 6	6	Nov. 4	1	7	Nov. 2
Henab Turki	2	1	4	1	12	11
Ohanez (White Almeria)	Oct. 30	1	6	8	7	4
Barbarossa	Nov. 2	1	Oct. 30	8	7	4

For the varieties mentioned in this table, blooming commenced

> In 1914 on October 26 to November 6, an interval of 11 days
> In 1915 on October 26 to November 9, an interval of 14 days
> In 1916 on October 16 to November 6, an interval of 21 days
> In 1917 on October 19 to November 8, an interval of 20 days
> In 1922 on October 25 to November 12, an interval of 18 days
> In 1923 on October 17 to November 11, an interval of 25 days

or an average blooming season of eighteen days.

The blooming dates of the different varieties vary with the year and the stock, but so far it has not been possible to discover any regularity in this variation. It must not for a moment be supposed that the blooming dates indicate precisely the order in which the grapes will ripen. Thus the earliest variety, Madeleine angevine, during 1914-17 and 1922-23 began blooming respectively 4, 3, 16, 20, 10, 12 days before the latest variety, Ohanez. On the whole, however, the early varieties bloom before the late ones, although the latest variety is not always last to bloom.

According to Guillon,[1] the higher the mean temperature is during the period of bloom, the shorter will be this period (*i.e.* from when the first variety begins to bloom till when the last begins). At Cognac he fixed the blooming period at eight days in 1902 with a mean temperature of 21-5° C., and in 1903 at fourteen days with a mean temperature of 13-5° C. Hedrick[2] gives the average length of the blooming season for grapes as twenty days, and, as previously stated, I found it to be eighteen days.

Millardet, according to Guillon,[1] has found that the opening of the flowers depends mainly upon the temperature. At 15° C. a few flowers open from time to time, at 17° C. they open normally, and at 20-25° C. they open fast. The sunlight as such is without influence on the opening of the flowers, except in so far as it may cause a rise in temperature. Millardet has observed that the flowers of Chasselas, in fine warm weather, start opening at 7 A.M. if the temperature of the air has then reached 15° C. For the first hour, and in any case until 17° C. has been reached, only a few flowers open. After 17° C. they open quicker, and about 9 A.M. they open from minute to minute, sometimes several simultaneously on the same cluster. Then the speed of opening rapidly diminishes. Between 10 and 11 A.M. it nearly ceases, and after that till evening only one or two more flowers will open on the cluster.

In cold rainy weather the flowers open badly and irregularly, and also not before the temperature of the air has reached 15-17° C. If this unfavourable weather lasts for two to three days in succession, the flowers do not open completely. The corolla-cap is lifted somewhat but does not drop. Under such circumstances both pollination and fertilisation are very bad and the berries set badly.

The Fertilisation.—Once the pollen-grains have got on to the stigma, they are kept there by the sticky fluid on its

---

[1] Guillon, *l.c.* p. 282.

[2] U. P. Hedrick, *Manual of American Grape-growing*, 1919, p. 288.

120

surface, and, under favourable conditions, germinate in it.

The ovary is now ready to be fertilised. If the weather is favourable, *i.e.* if the temperature of the air is not too low, the pollen germinates and the pollen-tube penetrates to the ovules which will become the future grape seeds after fertilisation and subsequent development. The process of fertilisation is complicated and will not be discussed here at greater length. For more information on this point the reader is referred to Babo u. Mach, *l.c* i. pp. 343-347, or any standard treatise on botany.

Cold weather is more harmful to fertilisation than rain without cold weather. The cold weather prevents or hinders the germination of the pollen-grains and the real, process of fertilisation. At the Paarl Experimental Station a cold south-west wind started blowing one afternoon whilst the block of Hanepoot (Muscat of Alexandria) vines was in full bloom and the berries had partly begun to set; the result was that that same evening I noticed the ground was almost covered with the green flowers that had been shed, and that season these vines had only scraggy bunches with few decent berries.

Non-setting, Millerandage, and Parthenocarpy

By *non-setting* we mean the dropping of the flowers or rather pistils that have not been fertilised. Where this has been bad, such bunches will be scraggy, showing bare stalks with berries here and there.

Parthenocarpy means the formation of fruit without fertilisation. The small seedless berries formed on the bunches of Hanepoot, Rosaki, etc., if rainy weather prevailed during the blooming season, have been formed parthenocarpically. They are always round, although, as in the example just quoted, the normal berries are oval. With the Sultana, Currant, and other seedless grapes, we have probably to do with a particular type of parthenocarpy. By covering up bunches of Sultana in paper bags just before the flowers began to open (during the 1924-25 season on the Stellenbosch University Farm), I have proved that cross-pollination is not required in this case, as the bunches formed inside the paper bags were just as well set and the berries just as big and good as those on the uncovered bunches. I am of opinion that the seedless varieties have only defective pistils or ovaries. It is not yet known how far some sort of irritation, caused by pollen-tubes, may be required to bring about the setting of the berries of seedless varieties. This question is again discussed in the chapter on vine diseases.

Bunches that have set badly and show a fairly large number of small seedless berries show the phenomenon known as *Millerandage* (see Fig. 73). It is often met with at the Cape in Madeleine angevine, Bicane, Pinot Chardonnay, Codega, Gros Noir des Beni-Abès, etc.

After a while the stamens dry up and drop. The real process of fertilisation continues, and we soon notice that the tiny berries begin to set. Those that are not setting drop, whilst those that set now grow fairly rapidly in size.

5. THE DEVELOPMENT OF THE GRAPES

We can divide this development into the three stages: when the grapes are green, ripe, and over-ripe.

(*a*) **The Green or Growing Stage**.—This stage lasts from the setting of the berries up to the véraison, when they turn colour. The berries and stalk are now green. The former rapidly grow in size, the latter less rapidly. During this period, which lasts quite two months, the berries grow from 2 mm. diameter at the beginning up to about 20 mm. diameter at the end. As they are still green and therefore contain chlorophyll, photosynthesis now takes place in them also.

As we have previously seen, the berry does not grow bigger through the formation of new cells, but merely through the extension of the existing cells. The chemical composition of the berry does not alter appreciably during this period, and differs very little from that of the other green parts of the vine. According to Guillon[1] it now contains 10 gr. sugar per litre sap, of which 8 gr. are dextrose and 2 gr. levulose. Its total acidity amounts to about 25 gr. per lit. (calculated as sulphuric acid) and 38-3 gr. per lit. (calculated as tartaric acid), and consists mainly of tartaric and malic acids and their acid salts. Macagno,[2] quoted from

---

[1] Guillon, *l.c.* p. 286.
[2] Macagno, "Recherches sur les fonctions des feuilles de la vigne", *Comptee rendus*, lxxxv., 1877, p. 810.

Guillon,[1] has analysed leaves, shoots, and berries during different stages of the vine's development, and obtained amongst others the following results per kilogram of green material analysed.

At the beginning of this period:

	Sugar.	Bitartrate of Potash.
Leaves from the end of fruit-bearing shoots	14.21 gr.	7.41 gr.
„ „ „ base .	10.81 „	5.12 „
„ „ „ non-fruit-bearing shoots	11.93 „	4.91 „

Towards the end of this period:

	Sugar.	Bitartrate of Potash.
Leaves from the end of fruit-bearing shoots	15.31 gr.	12.52 gr.
Berries „ „ „	10-0 „	. .

Here we do not notice much change during the period of growth. Neither do we find a great difference in composition between the berries and the leaves near the bunches, i.e. those at the base of the fruit-bearing shoots.

Lewis[2] in 1909-10 studied the development of the grape at Groot Constantia (Cape). In this study he found the well-known relations about the composition of the grape during its development. The sugar content changed little till near the end of the green stage. Only after the grapes had begun to turn colour (véraison) did the *sugar* content increase rapidly. This happened between 14th and 28th January, but nearer the latter date. The dextrose was appreciably more than the levulose until the grapes were ripe, when they were about equal.

The *total acidity* increased till the end of the green stage, after which it decreased continuously. The free tartaric acid continuously grew less from the 17th December, whilst the bitartrate of potash increased until the grapes were ripe, and at full maturity dropped something in some cases. The malic acid increased up to the 28th January when the grapes had, roughly speaking, turned colour, after which it grew less.

The following table gives some of the results obtained by Lewis (*l.c.* pp. 9-11):

---

[1] Guillon, *l.c.* p. 286.
[2] J. Lewis, M.A., "The Development of the Grape", *Agric. Journ.*, Dept. of Agric., Cape of Good Hope, Nov. 1910.

123

Variety.	Chemical Constituent in gr. per lit.	Dates on which Samples of Grapes were picked.						
		17.12.09.	31.12.09.	14.1.10.	28.1.10.	11.2.10.	25.2.10.	11.3.10.
Cabernet Sauvignon, planted in 1901, grafted on American stock and trellised.	Sugar	5·5	6·3	7·4	71·8	152·7	193·6	234·3
	Dextrose	5·3	..	7·1	51·7	93·7	112·4	115·6
	Levulose	0·0	..	0·1	18·3	58·2	80·0	117·6
	Total acidity (as tartaric acid)	30·9	30·9	37·0	34·6	15·2	10·5	6·8
	Pot. bitartrate	3·67	3·76	3·85	5·08	6·58	7·05	6·30
	Free tartaric acid	16·1	12·5	12·4	9·2	5·7	2·5	0·31
	Malic acid	12·0	15·2	20·6	20·8	6·1	4·6	3·6
White French, planted in 1901, grafted on American stock, trellised.	Sugar	6·4	7·6	10·8	56·5	114·3	168·3	192·8
	Dextrose	6·2	..	..	39·9	65·9	93·2	98·5
	Levulose	0·0	..	..	15·6	47·7	74·7	94·9
	Total acidity (as tartaric acid)	27·0	27·8	30·0	27·5	10·3	5·9	4·6
	Pot. bitartrate	3·57	3·39	3·95	4·61	5·26	6·02	6·02
	Free tartaric acid	14·3	11·7	11·2	8·9	4·2	1·8	0·29
	Malic acid	10·1	13·1	15·4	14·6	3·6	2·2	1·9
Hermitage, planted in 1904, grafted on American stock, not trellised (bush vines).	Sugar	3·6	5·9	7·7	32·7	124·5	168·9	217·9
	Dextrose	..	..	..	24·7	75·9	90·8	109·1
	Levulose	..	..	..	7·3	47·6	78·0	107·2
	Total acidity (as tartaric acid)	27·9	35·9	37·5	30·0	14·3	8·5	6·2
	Pot. bitartrate	3·57	3·57	3·95	4·32	6·20	5·83	5·92
	Free tartaric acid	14·8	12·3	11·7	9·7	3·5	1·4	0·43
	Malic acid	10·5	19·8	21·5	24·6	7·4	4·3	3·0
Stein, planted before 1885, ungrafted, untrellised (bush vines).	Sugar	5·8	7·8	12·8	69·3	157·7	213·9	236·4
	Dextrose	..	..	11·9	47·9	83·0	112·9	116·9
	Levulose	..	..	0·4	20·3	73·9	101·5	117·6
	Total acidity (as tartaric acid)	27·3	33·8	37·6	33·7	14·4	10·1	5·9
	Pot. bitartrate	3·48	3·67	38·5	4·70	5·08	6·11	5·92
	Free tartaric acid	14·9	11·5	11·0	7·7	4·5	1·2	0·00
	Malic acid	11·6	18·6	22·3	21·6	7·1	5·8	3·2

During the green stage the berries do not show much bloom. The young berries of Bonnet de retord now show reddish streaks running from pole to pole where later on the stripes of white bloom will run. The small berries of Chasselas rouge are red instead of green, as are its shoots and leaf-stalks and, to a great extent, the leaves.

(b) **The Ripening Stage.**—The ripening of the grape commences with the véraison, when the berries are almost full-grown. They become somewhat soft, and those of the white varieties more or less translucent. The green colour changes to greenish or yellowish white (with white varieties), pink (with red varieties), and red to black (with black varieties) at the véraison. As the ripening proceeds, the colour increases. When *fully ripe* (*i.e.* when the absolute quantity of sugar in the berries does not increase further) the colour is slightly greenish to yellowish, red or violet red, and violet black or black with white, red, and black varieties respectively.

From the commencement of the véraison the bloom is rapidly formed. The growth of the vine now almost stops, and the leaves lose turgor.

If rain now falls after a period of dry weather, the berries swell out considerably, and with some varieties, *e.g.* Gros Colman, can easily crack. This is due to the greatly increased pressure inside the berry caused by the absorption of a large quantity of water. The increased sugar content favours this absorption of water.

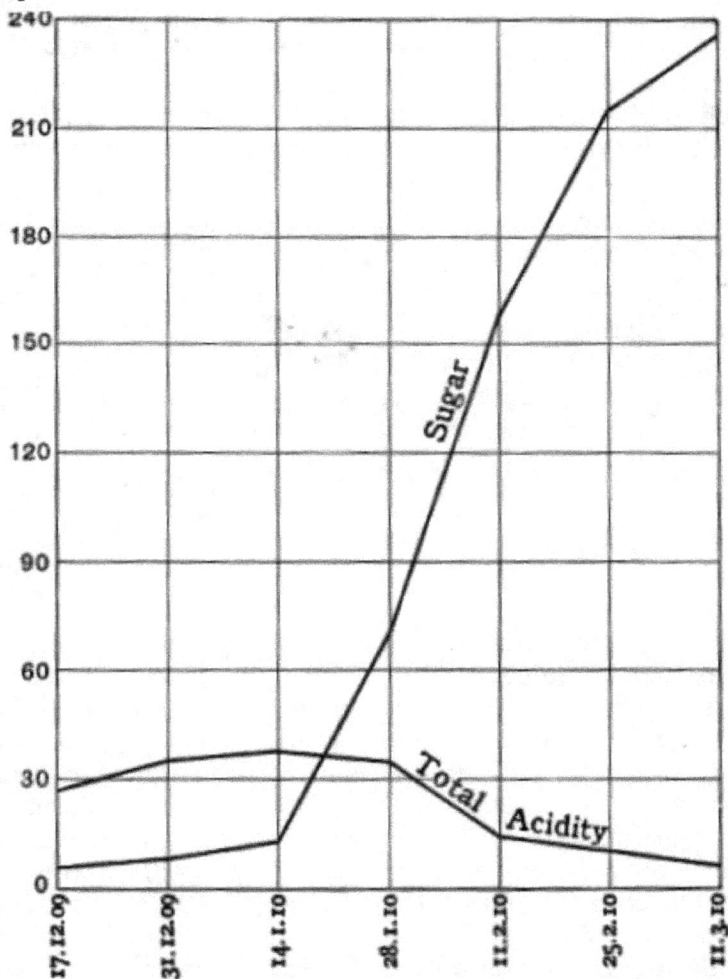

Diagram showing the changes in sugar and total acidity content of the must of ripening Stein grapes (according to Lewis's analyses).

Potash now begins to decrease in the shoots and to increase in the berries. The same happens with the phosphoric acid, only to a lesser extent. Lime and magnesia seem, as heretofore, to accumulate in the leaves.

The following table gives the dates when the varieties mentioned turned colour (véraison) at the Paarl Viticultural Experiment Station in 1914-17 (according to Mr. J. C. van Jaarsveld).

The first four are early varieties; the last three are late varieties; Hermitage and Pontac are early mid-season; Sultana, Rosaki, Gros Colman, mid-season; White and Red Hanepoot are mid-season to late mid-season varieties. The different varieties therefore ripen more or less in the order in which they turn colour, but they do not take exactly the same length of time from the véraison up to full maturity.

Variety	1914	1915	1916	1917
Madeleine angevine	Dec. 5	Dec. 6	Dec. 4	Nov. 26
Black Prince	30	25	18	Dec. 18
Pinot fin	23	23	28	22
Ferdinand de Lesseps	—	24	28	Jan. 12
Hermitage	Jan. 2	Jan. 8	Jan. 2	7
Pontac	Dec. 30	Dec. 23	Dec. 18	Dec. 22
Rosaki	Jan. 12	Jan. 8	Jan. 14	Jan. 12
White Hanepoot	20	13	18	21
Red Hanepoot	6	10	18	16
Sultana	10	8	4	12
Gros Colman	2	3	16	5
Barlinka	18	13	20	25
Henab Turki	15	10	20	16
Ohanez (Almeria)	17	18	20	10

The Total Acidity.—We have already seen that the acidity of the juice of the berries increases till near the veraison, when it again diminishes. The curve for the total acidity of the ripening Stein grape (according to Lewis's analyses) on p. 125 shows us that the highest point was reached on 14th January or shortly before or after this date. This was the date of the veraison for this grape. We notice further that the sugar curve rises rapidly from this same date. On 28th January the acidity stood at 33.7 and on 31st December it stood at 33.8, so that in the fortnight following 14th January it fell as much as it had risen in the preceding fortnight. At the beginning of the ripening stage the total acidity stands at its highest, and consists mainly of free tartaric and malic acids, potassium bitartrate, and acid salts of malic acid. As is shown by Lewis's results, the free tartaric acid gradually diminishes
whilst the potassium bitartrate increases, and the malic acid diminishes until full maturity is reached. The rise in the

potassium bitartrate represents only a small fraction of the loss of free tartaric acid. In the fully ripe grapes Lewis found that the must contained only 0.00-0.43 gr. free tartaric acid per litre—*i.e.* nothing or very little.

Grünhut[1] maintains that fully ripe grapes do not contain any free tartaric acid. This is fairly true in hot countries, but in countries with a temperate climate the must then still contains an appreciable quantity of free tartaric acid and much more free malic acid. According to analyses made by Girard and Lindet,[2] the must of ripe Folle blanche grapes in Cognac still contained 6.1‰ free acids in 1893 and 9.7‰ in 1894. Of the latter amount 2.4‰ consisted of free tartaric acid and 7.3‰ of free malic acid, etc. Pacottet[3] quotes the following analytical data for the vintage of the year 1900:

	Gamay. Gr. Per lit.	Pinot. Gr. Per lit.	Aramon. Gr. Per lit.	Folle Blanche. Gr. Per lit.
Free tartaric acid	4.2	0.0	2.3	5.5
Pot. bitartrate	5.0	6.5	5.2	3.2

In the must, about half of the total acidity consists of malic acid which is mostly free.

If the summer is very dry and the grapes must ripen and be harvested in the dry weather, they can contain abnormally high quantities of free acid. This happened during the 1922 summer in France. Analyses made by Fonzes Diacon[4] of French red wines of the 1922 vintage, show an abnormally high quantity of free tartaric acid which frequently reached 1.50-3.50 gr. per litre. He attributes it to the drought, which prevented sufficient potash being absorbed from the soil to neutralise partially the tartaric acid to the usual extent.

Muller-Thurgau[5] seeks the *origin of the acids* in the berry itself. He is of opinion that the acids are formed by oxidation (during the process of respiration in the berry) out

[1] Dr. L. Grünhut, "Die Chemie des Weines", in *Sammlung chem. u. chem.-tsch. Vorträge* herausg. von Prof. Ahrens, vol. ii. p. 73 (1897).

[2] Quoted from Viala-Vermorel, "Traité general de viticulture", *Ampélographie* (7 vols., 1901-10), vol. ii. p. 212.

[3] P. Pacottet, *Vinification*, 1904, p. 9.

[4] Fonzes Diacon, Revue de Viticulture, tome lix. p. 231

[5] See Grünhut, *l.c.* p. 74.

of the sugar that came from the leaves. A. Mayer[1] was the first to propound and substantiate this theory. It is now generally accepted. Out of sugar, tartaric and oxalic acids can be formed by oxidation, and malic acid can be formed from tartaric acid by reduction.

The *fall in the acidity* of ripening grapes we ascribe to: (a) Partial neutralisation by potassium and calcium compounds absorbed from the soil and entering the berry; when tartaric acid changes to pot. bitartrate, it loses half its acidity as it changes from a dibasic into a monobasic acid. (b) Respiration in the berries, which not only forms the acids out of sugar, but also oxidises these acids, particularly the malic acid, to carbon dioxide and water. This process is more intense in warm than in temperate climates, and hence the must of ripe grapes contains a lower acidity in warm than in temperate climates. Grapes hanging in the sun will contain as much sugar as grapes hanging in the shade, but their acidity will be lower on account of the enhanced respiration taking place in the sun. Müntz[2] gives the following illustration of this:

	Must from Grapes	
	In the Sun	In the Shade
Sugar per 100 c.c.	17.96 gr.	17.96 gr.
Acidity per litre (as H_2SO_4)	4.96 gr.	5.66 gr.

Some authors have maintained that part of the acids is changed into levulose, but this is merely a theory without an experimental basis, and it does not seem to me to be feasible. I am therefore of opinion that the whole fall in acidity must be ascribed to respiration and, in a lesser degree, to the partial neutralisation, possibly aided by conduction of part of the acids to the other parts of the vine. In this connection Czapeck[3] says: "The sweetening of fruit that ripens does not rest on a change of the organic acids into sugar, as was formerly assumed, but upon a consumption of the organic acids during the oxygen-respiration whilst the sugar content of the fruit continuously increases".

---

[1] See Babo u. Mach, *l.c.* i. p. 448

[2] Müntz, *l.c.* p. 562.

[3] F. Czapeck, "Atpiung der Pflanzen", in *Handworterbuch der Naturwissenschaften* (in 10 vols., 1912-15), vol. i. p. 716 (Jena, 1912).

The Sugar Content.—We have already seen that the sugar content of the sap in the berry is low during the green stage, and that it increases slowly towards the end of this period. At this stage the levulose is much less than the dextrose. After the véraison the sugar content of the must rises rapidly until the maximum is reached at full maturity, when dextrose and levulose are present in about equal quantities. In over-ripe grapes the sugar content of the must is usually higher, but the berries do not contain more sugar—their must has merely been concentrated by an evaporation of water.

Under identical conditions different varieties of grapes will, at full maturity, usually contain different amounts of sugar. The same applies to the total acidity. Again, under varying conditions the same grape will show a varying sugar content at maturity, *e.g.* White French in the Paarl and Stellenbosch districts of the Cape Province generally has only about 20 per cent sugar when ripe, whilst in the fertile sweet karroo soils and the alluvial soils of the Robertson district it reaches and even exceeds 25 per cent sugar.

The Origin of the Sugar.—It is established beyond doubt that all the sugar in the grapes comes out of the leaves. In the green berries too a small amount of sugar is formed, but this is only very small and is not even as much as is consumed in respiration (see Babo u. Mach, *l.c.* i. p. 449). In the berries we do not find starch, and this explains why grapes cannot ripen once they have been cut. In apples and pears we find stored up a great deal of starch which is changed into sugar during the ripening of these fruits after they have been picked.

At the carbon assimilation, which we shall discuss later, the dextrose and levulose we find in ripe grapes are formed in the leaves, whence they are conducted to the berries. Therefore, if we remove a large percentage of the leaves before the grapes are ripe, we should expect to find less sugar in the ripe bunches on such shoots than on those that have retained all their leaves.

The Effect of removing Leaves.—Müntz[1] conducted an

---

[1] Müntz, *l.c.* p. 555.

experiment with Malbec and Merlot in the Gironde in order to test this. On the 2nd October, when the grapes were not fully ripe, he took samples of grapes from his experimental vines and analysed the must in each sample. Then he removed 25 per cent of the leaves on each shoot (starting from below). The grapes were then hanging in the sun.

Next to every partly defoliated vine he kept a control with all its leaves intact. On the 13th October the grapes in the vineyard were ripe and pressing had begun. Then he again took samples of grapes from the experimental vines, and analysed the must with the following results:

	Malbec.			Merlot.		
	Oct 2	October 13		Oct 2	October 13	
	Before defoliation	Vines not defoliated	Vines defoliated 25%	Before defoliation	Vines not defoliated	Vines defoliated 25%
Spec. Grav. of must in degrees Baume	9.2	12.3	10.0	8.8	11.4	9.0
Sugar in gr. Per 100 c.c.	16.35	22.78	17.48	15.19	19.93	15.37
Acidity in gr. Per. Lit. (as H2So4	7.96	5.31	6.02	7.08	5.31	6.73
Colour of must	Slightly coloured	Colour fairly intense	Slightly coloured	Slightly coloured	Colour fairly intense	Slightly coloured

From these results we clearly see that the partial defoliation resulted in little sugar being formed during the eleven days that the experiment lasted, and much less than where no leaves had been removed, whilst the acidity fell less than where the vines had kept all their leaves. From the practical point of view, therefore, it was wrong to remove the leaves, as the grapes did not ripen as well as where the vines had kept all their leaves. During the eleven days of the experiment the weather remained bright. This experiment incidentally also proves that the lower or older leaves have more power than the higher or younger leaves to form sugar.

In the Sauternes district (France) it is customary to remove the lower leaves to above the highest bunch, but this is done only when the grapes have reached their full maturity, and serves firstly to prevent an undue development of the fungus causing the noble rot, and also to favour the concentration of the sap in the berries by thus directly exposing them to the rays of the sun.

The Composition of Ripe Grapes

When the grapes are fully ripe, they are usually fit for picking for wine-making. For the manufacture of non-alcoholic drinks the grapes are harvested somewhat earlier in order to get a higher acidity in the juice. For light wines the grapes are picked earlier than for heavy and sweet wines.

Kind of Determination	Cabernet Sauvignon	Riesling	White French	Stein	White Greengrape	Hermitage
Weight of bunch in grams	157.8	212.2	512.3	212.0	270.5	342.3
Average weight per stalk in grams	5.62	8.26	15.81	6.52	7.30	6.75
Weight of 100 berries in grams	133.8	142.0	211.2	148.8	260.4	355.6
Weight of seeds of 100 berries in grams	5.37	5.42	5.94	6.10	7.11	6.03
Average number of berries per bunch	113	153	228	141	103	96
Average number of seeds per 100 berries	156	149	185	168	191	175
Ratio of bunch : stalk .	28.1	25.5	32.5	32.4	37.0	56.0
Ratio of berry : seeds	24.9	26.2	36.1	24.2	36.6	59.0
Percentage of weight of berries in bunch	96.14:	96.00	96.91	96.92	97.30	98.03
Percentage of weight of stalk in bunch	3.56	3.91	3.09	3.08	2.70	1.97
Percentage of weight of seeds in berries	4.01	3.81	2.31	4.09	2.71	1.70
Composition of Must —						
Specific gravity at 15° C.. 15° C..	1.0984	1.0861	1.0809	1.1009	1.0975	1.0915
Total extract in grams per 100 c.c.	25.62	22.38	21.02	26.27	25.38	23.80
Ash in grams per 100 c.c.	0.327	0.286	0.283	0.332	0.270	0.281
Sugar in grams per 100 c.c.	23.43	20.28	19.28	23.64	22.93	21.79
Dextrose in grams per 100 c.c.	11.56	10.19	9.65	11.69	11.21	10.91
Levulose in grams per 100 c.c.	11.76	9.92	9.49	11.76	11.71	10.72
Total acidity (as tartaric acid)in grams per 100 c.c.	0.68	0.79	0.46	0.59	0.79	0.62
Pot. bitartrate in grams per 100 c.c.	0.630	0.639	0.602	0.592	0.611	0.592
Free tartaric acid in grams per 100 c.c.	0.031	0.046	0.029	0.000	0.053	0.043
Malic acid in grams per 100c.c.	0.36	0.44	0.19	0.32	0.37	0.30
Total dissolved N in must in grams per 100 c.c.	0.0228	0.0228	0.0157	0.0228	0.0420	0.0228
Dissolved proteids [(Dissolved N—ammoniacal N) x 6.25]in grams per 100 c.c.	0.1075	0.0988	0.0719	0.1119	0.1839	0.1031
Potash in grams per 100 c.c.	0.194	0.186	0.150	0.149	0.162	0.161
Lime in grams per 100 c.c.		0.0092	0.0095	0.0106	0.0097	0.0081
Magnesia in grams per 100 c.c.	0.0108	0.0102	0.0091	0.0105	0.0108	0.0096
Phosphoric oxide in grams per 100 c.c.	0.0141	0.0187	0.0164	0.0234	0.0251	0.0277

Table grapes are picked as soon as they are sweet enough, and generally before they are fully ripe. For raisin-making, grapes are harvested just when they are fully ripe.

In order to find out when the grapes are just fully ripe, we must cut thoroughly representative samples of, say, eight bunches of them on every other day, crush each sample by hand, and strain it through muslin or filter it through coarse filter paper. The filtered must is then poured into a glass cylinder and the sugar (or really the total extract) determined by means of a saccharometer. As soon as the sugar content becomes nearly constant, the grapes will be fully ripe. Through experience we can determine this with fair accuracy by merely walking through the vineyard and tasting the grapes. When the berries at the tips of the bunches are properly sweet (these are the last to ripen) and an occasional berry at the top of the bunches shows signs of beginning to shrink, the grapes will certainly be fully ripe. On the whole, a saccharometer is most reliable for this purpose.

The relative proportions of the weights of stalk, husks, pulp, and seeds and the composition of the must vary with the different varieties and are also influenced by the soil, climate, site, manuring, cultivation, magnitude of the crop, stock (with grafted vines), and the degree of ripeness at which the grapes are examined. On p. 131 are given some of the results obtained by Lewis in his study of grapes at Groot Constantia when these were fully ripe (11th March).

The following analyses are taken from Girard and Lindet[1] (see Table on opposite page).

The great difference between the figures for France and South Africa we find in the higher sugar contents and the higher specific gravities of the South African musts. Otherwise the figures agree fairly closely.

The **stalks**, in the case of ripe grapes, form from just under 2 per cent to 4 per cent (sometimes up to 5 per cent) of the weight of the bunch, according to the variety of grape. The water content and further composition largely depend upon the degree of ripeness at which they are analysed. The riper the grapes are, the lower is the water content of their stalks. According to analyses made in San Michele,[2] the water content of the stalks of various varieties ran from 34 to 78 per cent, and averaged 64 per cent. They seem to contain no sugar, but a little starch. They contain a fair quantity of tannin running from 1.27 to 3.17 per cent. The stalks of 1000 lbs. of ripe grapes contains on an average 1.1 lb. of tannin.

---

[1] Girard et Lindet, "Recherches sur la composition des raisins des principaux cepages de la France", *Revue de Viticulture*, iv. 317, 341, vi. 173, 201, 225, 249.

[2] See Babo u. Mach, *l.c.* i. p. 365

COMPOSITION OF THE PRINCIPAL GRAPE VARIETIES OF FRANCE

	Aramon Per.cent.	Cabernet Sauvignon Per.cent.	Pinot Noir Per.cent.	Folle Blanche Per.cent.
Composition of bunch: stalks	4.07	2.94	1.61	3.19
Composition of bunch: berries	95.93	97.06	98.39	96.81
Average weight per berry	3.69 gr.	1.22gr.	1.20 gr.	1.55 gr.
Composition of berry: pulp	88.81	87.44	88.51	87.22
Composition of berry: skin	9.45	8.72	6.61	9.92
Composition of berry: seeds	1.74	3.84	4.88	2.86
Composition of pulp				
Specific gravity of must	1.064	1.088	1.092	1.077
Water	82.46	77.03	75.31	78.93
Fermentable sugar	14.09	13.83	19.55	16.95
Pot. bitartrate	0.62	0.56	0.67	0.35
Free acids (tartaric acid, malic acid, etc)	0.39	0.37	0.23	0.61
Nitrogenous substances	0.27	0.22	0.46	0.09
Mineral substances	0.13	0.14	0.06	0.05
Insoluble woody substances	0.43	0.36	3.36	2.74
Substances not determined	1.61	3.49	3.36	2.74
Composition of skins				
Water	76.8	73.47	67.30	74.63
Tannin	1.27	1.33	1.53	0.30
Woody substances	20.10	23.73	29.90	24.08
Mineral substances	1.83	1.47	1.27	0.99
Composition of seeds				
Water	34.82	32.80	29.54	36.61
Oil	6.92	7.25	7.98	4.96
Volatile acids	0.57	1.81	1.04	0.93
Tannin	2.56	0.72	4.17	4.67
Resinous substances	4.45	3.35	5.40	3.77
Woody substances	48.82	52.38	50.12	47.64
Mineral substances	1.86	1.69	1.75	1.42
Composition of stalks				
Water	79.66	53.83	45.46	76.64
Tannin	1.23	0.61	0.34	1.64
Resinous substances	1.07	1.05	0.91	0.80
Woody substances	15.71	40.62	50.95	19.58
Mineral substances	2.33	3.89	4.24	1.34

N.B.—The pot. Bitartrate and free acids in the skins and stalks are not given separately

Grapes with small berries contain a larger proportion of stalk (by weight) than those with big berries. Thus Lewis found the weight of the stalks with Hermitage to be 1.97 per cent and with Riesling 3.91 per cent of that of the bunch. Pacottet,[1] according to an analysis by Bouffard, quotes a stalk weight of 6.6 per cent for Jacquez. The wild American vines possess small berries and well-developed stalks, which necessarily result in a high stalk percentage by weight.

The skins generally form 6–10 per cent of the weight of the berries. According to Girard and Lindet's analyses just quoted, the fresh skins contain (in the four cases quoted) 67.3-76.8 per cent water, hence 23.2-32.7 per cent dry material. According to Mach,[2] quoted from Babo u. Mach, *l.c.* i. p. 368, fresh skins contain on an average 26.1-34.9 per cent dry material. In the skins we find nitrogenous

---

[1] P. Pacottet, *l.c.* p. 2.
[2] Mach, Die Gärung und Technologie des Weines, Wien, 1884.

substances (proteids, etc.), cellulose, fatty substances (e.g. in the bloom), colouring and aromatic substances, tannin, pot. bitartrate, calcium tartrate, malic acid, sugar. The sugar content of the skins is less than that of the pulp. Their tannin content varies greatly with different varieties and with the degree of ripeness. The skins of over-ripe grapes contain much less tannin than those of ripe grapes. The mean tannin content of ripe fresh berries is about 0.10-0.15 per cent, or about as much as that of the stalks; that of fresh skins about 1-2 per cent, and that of dry skins about 3.7 per cent.

We have already seen that the colouring substances of the ripe grapes occur almost exclusively in the skin. Armand Gautier has shown that the pigments of red and black grapes are formed in the colourless state in the leaves, whence they are conducted to the berries as soon as these begin to colour. Here they change into the coloured form (pigment) by oxidation. This process requires both air and a certain amount of heat. Thus grapes that hang more or less in the dark on a vigorously growing vine, or high above the ground, will sometimes not develop sufficient colour. I myself saw a case in point where Barlinka grapes were so perfectly covered in by the shoots and leaves that they hung in the dark, and had only a light pinkish colour, although they were sweet and ripe enough to have been perfectly black. Three days after I had removed sufficient leaves to expose them to direct sunlight, they were almost quite black. Before I removed the leaves, these grapes had been kept too cool. That it was not the light as such that brought about the change is proved by the fact that grapes hanging high above the ground will sometimes colour up badly although they are exposed to sunlight or diffused daylight. If we enclose a green bunch of a black variety in an atmosphere free from oxygen, it will not turn black, as the oxidation cannot proceed in the absence of oxygen.

The nature of the pigments varies greatly with the different varieties. A. Gautier calls these pigments enolic acids, and the different shades of colour of the grapes are related to the degree of neutralisation of these acids. In a strongly acid medium they are red. As the acidity decreases the colour passes into reddish violet, and into green in a faintly alkaline medium. The colour of the ripe grapes is intimately connected with this behaviour of the colouring substances. According to Ottavi-Marescalchi,[1] Armand Gautier found the formula $C_{42}H_{40}O_{20}$ for the red colouring substance, one of his enolic acids. According to these authors, Prof. Gombom has isolated the following pigments from black grapes and red wine: (1) a violaceous, very soluble, very unstable pigment; (2) a ruby red pigment; (3) two yellow pigments, of which the one dissolves with difficulty in water and crystallises, whilst the other is more soluble in water but cannot crystallise; (4) a green pigment; (5) an amorphous substance which seems to be a tannin. The red pigment which Comboni has isolated from the Rabosa grape had the composition $C_{21}H_{20}O_{10}$ (*i.e.* just half that of Gautier's). His violaceous pigment is differently constituted and does not contain any nitrogen or iron.

The quantity as well as nature of pigment in the skins and wine has been investigated by Girard and Lindet. The reader will find a good deal of information on this point in Pacottet, *l.c.* pp. 14-19. The pigments of the grape seem to be fairly closely related to its tannins. They are rendered insoluble and separated out by oxygen (hence air). This process is prevented by the presence of 10-15 grams tartaric acid per litre of solution. The pigment in solution is discoloured by sulphur dioxide which reduces it to its colourless form. The action of air will, however, bring it back to the coloured form. Hence coloured grape-juice or red wine that has been discoloured by the use of large quantities of sulphur dioxide or a metabisulphite will get back and keep its original intensity of colour, if it is well aerated.

The oxidation and consequent deposition of the pigment is caused by an enzyme, an oxydase, in the presence of air or oxygen. When ripe white grapes are crushed and exposed to the air, their light yellowish-green colour on the surface soon changes to a darker brownish colour. This too is brought about by such an oxydase.

The *aromatic* substances also are formed in the leaves and conducted thence to the berries, where we find them in the pulp and in the skin, whence they pass partly into the air. On walking through a block of ripe Ferdinand de Lesseps (on a hot day) one detects the peculiar aroma in the air. It is

---

[1] Ottavi-Marescalchi, *I principii della viticoltura*, 1909, p. 552.

in this grape, in Isabella, Muscat of Alexandria, and other Muscat varieties (e.g. Muscadel, Muscat Hamburg, etc.) that one finds a strong aroma. Next to these aromatic substances the grape still contains substances that are decomposed during the alcoholic fermentation, when they also yield aromatic substances. They are called *bouquet-giving* substances. Jacquemin[1] regards them as glycosides which, during fermentation, are split up by enzymes of the yeast cells into sugar and aromatic substances which we subsequently find in the wine and are characteristic of the variety (according to Babo u. Mach, *l.c.* i.p. 374). The bouquet of a wine therefore contains some aromatic substances that were present as such in the ripe grapes, and others that were formed during fermentation. According to Müller-Thurgau (quoted from Babo u. Mach, *l.c.* i.p. 374), the aromatic substances can be extracted by ether from the leaves and the grapes, and possess the characteristics of essential oils. The bouquet-giving substances are insoluble in ether and only slightly soluble in alcohol, and are odourless until they have been split up. By fermentation the same aromatic substances as are formed in the wine can be obtained from the leaves. Such characteristic bouquet substances are formed, for instance, in the Riesling wines of the Rhine and Moselle.

The leaves to a certain extent possess the characteristic bouquet of the wine, and industry has succeeded in making from the leaves preparations which, when added to a wine, give it a bouquet similar to that of the wine made from the grapes of the variety which had had its leaves used in making the extract.

According to Ordonneau (quoted from Guillon, *l.c.* p. 295), the bouquet substances of wines mainly consist of esters of fatty acids with higher alcohols; these esters are present in the must before the commencement of the alcoholic fermentation, and have probably been formed in the leaves. These esters, like the acids, tannins, and the sugar, can be oxidised under the influence of heat and intensive sunlight. This explains why the grapes, and hence their wines, do not develop or at least show their finest aromas or bouquets in the hot countries, but only in the

---

[1] G. Jacquemin, "Developpement des principea aromatiques par fermentation alcoolique en presence de certaines feuilles", *Comptes rendus*, cxxviii., 1899, p. 369.

cooler regions near the northern limit of the European vineyards, where their aromatic substances are not oxidised (or evaporated) so rapidly. This further explains why the most highly aromatic light white wines are produced along the Moselle and Rhine (especially in the Rheingau). The finely bouqueted red wines of the Medoc are produced in a district where, during and even before the vintage, it is fairly cool and frequently foggy till well into the forenoon. In South Africa the Constantia area produces such excellent dry red wines, owing to its cooler climate and consequent later vintage (*i.e.* not till March) than that in our other wine districts. The sweet wines form an exception. For these we desire plenty of sugar in the grapes and must, and generally use Muscat varieties (Muscat of Alexandria, Muscadel, Muscat de Frontignan, etc.), and sometimes also the very highly aromatic Ferdinand de Lesseps in the hot wine countries. These varieties produce highly aromatic wines in spite of the great heat.

The pulp contains the *must* or *grape-juice*, which is the sap of the large thin-walled cells of the pulp. According to Babo u. Mach, *l.c.* i. pp. 379-380, the fresh must contains the following constituents:

1. Water.
2. *Carbohydrates:* glucose, fructose, cane sugar (?),[1] pectic substances, pentosanes, inosite.
3. *Acids:* tartaric (racemic acid not present), malic (glycosuccmic, succinic, glycollic, glyoxylic, and formic acid: not definitely proved), acetic, oxalic, citric, salicylic (?) acid.
4. *Nitrogenous substances:* total nitrogen, proteid, albumoses and peptones (?), amido-acids, ammonium, organic nitrogen bases, nitrates, lecithine (not proved).
5. *Enzymes:* invertase, oxydase.
6. Tannin, phlobaphenes.
7. Red wine pigment (oenine).
8. *Green and yellow pigments:* chlorophyll *a* and *b*, carotine, xanthophyll, quercetine, quercitrine.
9. *Fats* and *waxes*.
10. Primary grape bouquet substances and bouquet-giving substances.
11. Unknown extractive substances.
12. *Mineral substances:* potassium, sodium, calcium, magnesium, manganese, iron, aluminium, (copper, arsenic, etc.), phosphates, sulphates, silicates, chlorides, borates.

We have seen already that the composition of the must

---

[1] 2012 Edition Editor's note: The questions marks on this page appear in the original text.

varies greatly (particularly quantitatively). In fully ripe grapes the sugar content is usually 10-30 grams per 100 c.c. must, though in most cases it will hardly exceed 28 grams per 100 c.c. In the Cape viticultural area it is usually 18-25 grams per 100 c.c., which is equivalent to a must with 19-25° Balling. Sweeter musts we get from over-ripe grapes. In ripe grapes dextrose and levulose are present in about equal quantities, although the levulose often slightly predominates. In 1903 Windisch[1] found in 60 musts from the Rhine, Nahe, Moselle, and Ahr, that the dextrose was more than the levulose in 13 cases, that they were equal in 13 cases, and that the levulose was more than the dextrose in 34 cases.

The total acidity of the must of fully ripe grapes varies from 4 to 20 grams per litre (calculated as tartaric acid). In the Cape viticultural area this acidity runs from 4g with White French to nearly 10‰ with Pontac (Teinturier).

The pure fresh juice of ripe grapes contains practically no tannin. But where it remains in contact for some time with the stalks, husks, and seeds, it dissolves some tannin which is increased if we press the must out of the crushed grapes.

The nitrogen content of fresh must runs from 0.18 to 1.37 grams per litre, and averages about 0.5 gram per litre. Its nitrogen content is increased by giving high nitrogen dressings to the vines. Grapes from moist, black, low-lying soils, rich in humus, contain considerably more nitrogen than those from slopes.

The ash constituents of the must consist mainly of potash (50-65 per cent), then phosphoric acid, which can amount to 26 per cent, then sulphuric acid, magnesia, and lime, each from about 3-5 per cent, then boric acid, manganese, etc., in smaller quantities. In order to be able to control an adulteration of must and wine, it is necessary to make numerous analyses of musts and pure natural wines during a number of years in the different wine districts and from the different varieties of grapes grown there.

The grape seeds can be completely wanting, as with Sultana, Currant, Black Monukka, and a few other seedless varieties, or in the seedless berries of Muscat of Alexandria, etc.; usually, however, they form from 1.70 per cent (Lewis

---

[1] K. Windisch, Die chemischen Vorgange beim Werden des Weines, 1905, p. 194.

for Hermitage) to about 4.88 per cent (Girard and Lindet for Pinot Noir) of the weight of the berries. This percentage will rise as the berries grow smaller in size and the seeds become bigger and more numerous. The water content of fresh grape seeds varies from 28 to 30 per cent (according to Babo u. Mach, *l.c.* i.p. 376). As the embryo, during germination, depends mainly upon the seed itself for its food supply (excepting air and water), it is not surprising that the seed contains the necessary food stored in the shape of proteids, carbohydrates, fat (oil), and all the mineral substances the germinating embryo requires for its development. The seed further contains the cellulose and lignin of its cell tissues, and tannins (tannin and phlobaphenes) as important constituents.

The fresh seeds have a nitrogen content of from 0.78 to 1.19 per cent, average 0.96 per cent, from which the nitrogenous substances are calculated (by multiplying by 6.25) at 4.88-7.44 per cent, average 6.00 per cent (according to Babo u. Mach, *l.c.* i.p. 376).

The *grape-seed oil* forms 10-18 per cent of the weight of the ripe dry seeds. Sometimes it may be less. The fresh seeds usually contain only 5-8 per cent oil. In Italy this oil is sometimes extracted by grinding the dry seeds to a powder and mixing it with water (25 litre per 100 litre seeds), heating to 60-80° C., pouring into small bags, and then pressing out the oil, which is burnt in lamps on the farms. According to Ottavi-Marescalchi,[1] 100 kg. air dry seeds give an average of 10 kg. oil, even after the seeds have been in the still with the husks. According to analyses made by Dr. Martinotti (see Ottavi-Marescalchi, *l.c.*—just quoted—p. 192), the dry seeds of the Sicilian grape, Inzolia, contain 12.93 per cent oil and those of Riparia 16.60 per cent oil. This oil solidifies at 16° C. Its specific gravity at 15° C. is 0.9202. It is odourless and yellowish in colour. It contains 95 per cent fatty acids, and on saponification with soda, potash, or ash, it gives a good soap.

The *tannins* of the grape seed consist of tannin and phlobaphene or tannic anhydride which, in Girard and Lindet's analyses, figure as "resinous substances". Together they form about 10 per cent of the weight of the fresh seeds.

---

[1] Ottavi-Marescalchi, *I residui delta vinificazione*, 1901, p. 194.

Girard and Lindet[1] have found that the weight of tannin and phlobaphene remains constant. When the tannin increases, the phlobaphene decreases, and inversely. If must comes into contact with the seeds, especially during the alcoholic fermentation, a good deal of the tannin is dissolved and gets into the wine.

The volatile acids in the seeds amount to 0.5-1.0 per cent, and they impart a sharp, unpleasant taste to the must if the seeds are broken whilst still in contact with the must. For this reason the seeds should remain whole during the crushing and pressing of the grapes. The seeds and grapes are physiologically ripe as soon as the seeds are capable of germinating and so propagating the vine. This occurs when the grapes are fully ripe. If we wish to harvest the grapes in the over-ripe stage, they will be ripe physiologically sooner than technically.

In its development the seed differs from the pulp in that its cells continue to multiply during the green period. At the veraison the seed is full-grown (according to Guillon, *l.c.* p. 295).

(c) **The Over-ripe Stage.**—During this stage important changes still take place in the grape. Through evaporation of water the concentration of sap in the berry gradually increases. The absolute quantity of sugar in the berries grows less, but the percentage of sugar in the must rises. Through respiration in the berry, a part of the sugar and acids is consumed (oxidised), so that the absolute quantity of both gets less. Where we desire to have a very sweet must (*e.g.* for making grape syrup or sweet wines), we allow the grapes to get over-ripe before they are harvested. The nitrogenous substances, pectic substances, and the pigments also undergo changes whereby they are rendered insoluble, at least to a certain extent. The colour of white grapes becomes golden yellow to brownish. The aromatic substances decrease somewhat.

Over-ripe grapes are easily attacked by fungi and insects as the skin becomes thinner and softer. Much rain will cause such grapes (like half-ripe and ripe grapes) to rot easily, mainly owing to the fungus of the grey rot, *Botrytis cinerea.* It is assisted in this work of destruction by the

---

[1] Quoted from Ottavi-Marescalchi, *I principii della viticoltura*, 1909, p. 531.

Penicilliums and other fungi. Where the climate is moderately moist and not too hot or too cold, the Botrytis causes the **Noble Rot** ("pourriture noble"), which helps in the production of the famous liqueur wines along the Rhine and Moselle ("Auslese-weine") and in Sauternes. The fungus grows in the skin and later also in the pulp. It causes fine cracks in the skin through which water evaporates readily, thus soon changing the berry into a noble-rot raisin. These berries are gathered individually, then crushed and pressed. The must is very thick and usually possesses about 35 per cent sugar. The berries with green mould (Penicillium) are kept separate and are worth very little.

The Botrytis causes important changes in the sap of the grapes. The nitrogenous substances and part of the sugar and acids are consumed by the fungus as food. Hence the yeast cells later on have difficulty in causing such a must to ferment properly. The lack of nitrogenous food and the high sugar concentration hinder their growth very much. In consequence, the fermentation is slow and usually drags on for a few years before coming to a standstill. In 1888 Müller-Thurgau[1] published his classical researches on the noble rot of grapes. By inoculating sterile must with pure cultures of *Botrytis cinerea*, he showed what influence the fungus had on the must. A must with 17.2 sugar, 13.4‰ total acidity, and 0.57‰ nitrogen, possessed only 11.6% sugar, 1.9‰ total acid, and 0.10‰ nitrogen, after the fungus had acted upon it for twenty-four days. Here no evaporation took place. The losses of sugar, acid, and nitrogen were respectively 32.5 per cent, 85.8 per cent, and 82.5 per cent of the quantities originally present. Hence we see that the fungus destroyed a great deal of sugar and relatively two and a half' times as much acid and nitrogenous substances.

Where the fungus develops spontaneously on the grapes in the vineyard, these substances are attacked in the same way, only, in this instance, a concentration of the sap takes place simultaneously through the evaporation of water. Müller-Thurgau also analysed the musts of fully ripe, over-ripe, and noble rot berries of the same variety of grape, and found:

---

[1] H. Müller-Thurgau, "Die Edelffäule der Trauben", *Landw. Jahrbücher*, 1888.

	Sugar gr. per 100 c.c.	Total Acidity (calculated as Tartaric Acid) gr. per litre
In must of fully ripe berries	18.2	6.9
In must of over ripe berries	20.6	7.1
In must of noble rot berries	33.5	10.5

The must of the noble rot berries had 84.1 per cent more sugar and only 52.2 per cent more acid than that of the fully ripe berries. This must has therefore been improved or rendered more "noble" by the fungus, and can give a very fine liqueur wine (*i.e.* a naturally sweet wine).

If the grapes are not yet very sweet when the fungus begins to grow on them, the must will be deteriorated rather than improved by it, whilst it will be greatly improved if the fungus begins its activity only when the grapes are already fully ripe and contain a fairly high percentage of sugar. This is proved by the following results of Müller-Thurgau's experiments:

	Sugar %	Acid %
The grapes with must of low sugar content originally had	13.4	11.6
The same after 8 days' action of Botrytis had	22.34	10.1
The same after 10 days' action of Botrytis had	20.8	7.8
The grapes with must of high sugar content originally had	23.67	11.6
The same after 10 days' action of Botrytis had	39.14	11.4

From this it is clear that the grapes with a low sugar content were improved only a little by the fungus, and in any case not sufficiently to make up for the loss in quantity whilst the grapes with a high sugar content were improved very much indeed, so much so, that their greater quality will more than balance the loss in quantity. This very sweet must will give a very fine and very valuable liqueur wine

Professor Laborde[1] has also shown that the Botrytis consumes relatively more acid than sugar, whilst the reverse is true of Penicillium. He further showed that Botrytis forms glycerine as an intermediate product out of the sugar of the must. It can destroy this glycerine again by oxidation. Therefore the must of noble rot berries can contain much glycerine if the action of the Botrytis has not been too prolonged. In such a case the berries will give a fine oily liqueur wine.

---

[1] Laborde, *Cours d'oenologie*, Bordeaux, 1908, pp. 229-231.

142

In a warm climate the noble rot cannot take place to any appreciable extent. One will find only a few noble rot berries here and there. The great heat will cause the berries to shrivel and become raisins too soon. It is usually too hot and dry for the fungus to develop properly. If abundant rains fall when the grapes are ripe, the Botrytis will cause the grapes to rot (grey rot) and may cause great damage to the crop, but noble rot will not be developed.

6. THE RIPENING OF THE WOOD AND THE DROPPING OF THE LEAVES

Some time after the shoots have acquired their full length and thickness, the previously mentioned layer of cork is formed in the outside layer of the bast. This suberification of the tissues cuts off all food supply from the external layers of the bast, which die and assume a reddish brown or light to dark brownish colour. This coloration of the bast begins from the internodes at the base of the shoot, and gradually proceeds towards the tip of the shoot. If a shoot keeps on growing very late, its front portion will remain green, even after the leaves have dropped. This green portion is not ripe, and during the cold winter it will die and become black. Only the properly ripened wood can pass through winter successfully. If a vine matures its shoots properly, they will assume a more or less brown colour almost to their tips. The brown colour shows that the shoot is ripening. The outer layers of cells are more and more lignified and reserve food begins to accumulate in the vine.

The reserve food consists mainly of starch grains which we find in the whole vine (roots, stem, and canes), and more particularly in the medullary rays, bast parenchyma, wood parenchyma, and wood fibres. Therefore, if we wish to know whether wood is properly ripe, we can see directly under the microscope, in thin sections of it, whether it contains a number of starch grains. Otherwise we can dip sections of the wood in an aqueous solution of iodine in potassium iodide, when the starch grains will be coloured blue-black and we can see with the naked eye whether the wood contains many starch grains. Well-matured wood is rich in starch and will therefore show plenty of blue-black coloration. As has already been said, we can recognise well-matured canes by their brown colour almost right up to their tips.

Kovessi[1] has made a deep study of the ripening of the canes of the vine, and he points out, amongst other things, that the canes of the vine are generally used for propagating it, and that only properly matured canes should be used for this purpose. This, is particularly important in selecting canes of American vines that are to be used as stocks for European vines. If, for this purpose, we make use of badly matured material, their grafted vines, according to Kovessi, will be very liable to diseases and perish after a few years of miserable existence.

He mentions the following external characteristics of canes which have ripened well and badly respectively:

A cane that is well matured sheds its leaves early; its wood is strong and well lignified and differs very much from the brown pith; the bast is coloured light or dark brown and can easily be torn off ; the cane is hard and breaks when it is bent.

A badly matured cane shows little difference between wood and bast; the wood is light green in colour; the cork is little developed in the bast, which is tinted brown with green patches; the cane is soft, and folds but does not break when it is bent.

As a result of his investigations Kovessi (*l.c.* p. 67) came to the following conclusions:

"The phenomenon of ripening consists, from the anatomical point of view, in a development and a differentiation of the tissues of the plant taking place after the appearance of the cork, brown coloration of the bast, development of the rings of wood and bast, thickening of the cell walls, formation of starch grains.

"The degree of ripening is measured by the intensity of the preceding phenomena; a well-matured cane always has the rings of wood and bast relatively more strongly developed, the pith very much reduced, more numerous and larger grains of starch, thicker cell walls. And in the same volume it contains a larger quantity of dry substance and a smaller proportion of water.

"The presence of a large quantity of water in the tissues of badly ripened canes explains why they stand the

---

[1] M. F. Kovessi, Recherches biologiques sur l'aoûtement des sarments de la vigne,1901.

cold of winter badly; the existence of a larger quantity of starch in the well-matured canes accounts for their greater suitability for grafting and for planting out as cuttings.

"All the external factors which influence the development and differentiation of the tissues, for this very reason also act on the ripening: light, heat, and dry weather favour a good ripening; shade, cold, and humidity have the opposite effect."

Ravaz[1] has also shown that canes ripen badly in the shade. Where he tied all the shoots to one stake, those in the middle grew mainly in the shade and matured only at their base, whilst their tops dried up without ripening.

Where vines are trained on pergolas or on a fairly high trellis the canes ripen badly in the shade.

It is of the utmost importance to remember that the leaves are needed to ripen not merely the grapes but also the canes. Hence we should keep the leaves on the vines as long as possible after the grapes have been harvested. Where vines are grown under irrigation, it will therefore generally be found advantageous to irrigate the vines after the vintage is over and if the soil is rather dry. This will lengthen the period of carbon assimilation by the leaves and result in better ripening of the canes (wood). It must be remembered that the success of the following crop will largely depend upon the manner in which the vine ripens its wood. If the wood ripens well, and the vine therefore contains plenty of reserve food, it will bud vigorously in the next spring and subsequently develop well, if the weather is not very unfavourable.

Vines that over-bear will usually ripen their wood badly, particularly if they also suffer from drought, and in the next spring they will either be dead or grow badly, and frequently even die later. Over-production is therefore very detrimental to the ripening of the wood and to the vine's whole future.

If, as sometimes happens, farmers are foolish enough to drive sheep and other leaf-eating animals into their vineyards once the grapes have been harvested, these animals will soon eat up all the leaves, after which the vines will not be able to ripen their wood any further. This results in a gradual weakening of the vine, which may end in the

---

[1] L. Ravaz, L'lnfluence, des opérations culturales sur la végetation et la production de la vigne, 1909, p. 48.

disappearance of the vineyard. I cannot too strongly condemn this malpractice. In the discussion on the nutrition of the vine, we shall see that more than half the nitrogen which the vine absorbs from the soil is present in the leaves. If the leaves are eaten up, there will be a great loss of nitrogen which has to be made good by the application of a nitrogenous fertiliser, which, however, cannot improve the badly ripened wood. The habit of some farmers of giving their vines a preliminary pruning (whereby only those canes that are to serve as bearers are left on the vine) shortly after the vintage and when the leaves are still green, will not do great harm to vigorous vines on fertile soil, and will help to make such vines bear more if their crop has been small through excessive growth. Where the vines are not over-vigorous, it is a bad practice to remove some of their shoots while their leaves are still green, seeing that these leaves could have formed more reserve food, which would have helped to keep the vine strong and healthy.

Where a vine's leaves are damaged or destroyed by diseases or pests, the vine naturally suffers. Hence the farmer ought to do his very best to keep the leaves healthy and on the vine as long as they are green. When the vine matures, the reserve food is stored in the whole vine. Thus the roots are supplied with the necessary organic food when they resume their growth at the end of their winter rest. For this reason a badly ripened vine will bud badly and runs the risk of being damaged and even killed by phylloxera.

Dropping of Leaves.—In autumn the leaves gradually lose their green colour; this means that the formation of reserve food is ceasing. The leaves now assume their autumnal colours, as has been referred to already under the external morphology of the leaf. The fact that the leaves of white varieties now become yellow and later on brown, but never red, whereas those of red and black varieties now usually turn yellow with red spots or red, proves that the colour of the grapes originates in the leaves.

As the canes are by this time as ripe as they will ever be, the leaves have now completed their task, and therefore drop after a while. Under normal conditions this happens in the Western Province of the Cape in April-May. In France it takes place in October-November (according to Guillon).

This closes the vine's *growing period* for the season, which is reckoned from the time the buds begin to open until the leaves drop. Guillon, *l.c.* p. 302, gives the following data for the growing period of Aramon at Montpellier:

Year	From Budding to Bloom. (Days)	From Bloom to Ripeness. (Days)	From Ripeness to Falling of Leaves. (Days)	Total Growing Period. (Days)
1889	61	79	65	205
1890	71	70	74	215
1892	64	78	66	208
1893	54	93	49	196
1895	64	57	78	199
Average	63	75-76	66-67	204-205

According to Mr. J. C. van Jaarsveld's previously quoted figures, the period from budding to bloom at the Paarl Viticultural Experiment Station was as follows:

Variety	1914	1915	1916	1917	Average
White Greengrape	71	60	60	67	64-65
White Hanepoot	56	55	60	65	59
Hermitage	73	52	48	60	58

As the grapes of these varieties at Paarl are well ripe about 30th January, 15th February, and 25th January, the period from bloom to maturity will last about 85, 100, and 84 days respectively, which is 1-2 weeks longer than Guillon quoted for Aramon at Montpellier. The leaves of Greengrape at Paarl drop about the beginning of May, so that its whole growing period there will be about 8 months or 240-250 days, which is 1-1½ months longer than that of Aramon at Montpellier.

The leaves fall when the cold weather sets in in autumn. At Montpellier this takes place about 16th November (Kovessi, *l.c.* p. 39), which is on an average fifteen days after the first frost in autumn. In the colder northern wine districts of Europe the leaves fall earlier, and about eight to ten days after the first autumn frost. In the neighbourhood of Paris and in the Champagne the first frost in autumn falls about 1st October, and the leaves of the local varieties drop about eight to ten days later, whilst those of southern varieties and of the American vines, *e.g. V. rupestris* and its hybrids, do not drop till four to eight days later. The first autumn frost falls in the South of France when the mean daily temperature has sunk to 10° C. (according to meteorological data of 1872-96). In parts where sudden changes in the weather take place, it can suddenly get very cold in autumn, so that the leaves drop before the mean daily temperature has sunk to 10° C.

For Rupestris du Lot, Kovessi (*l.c.* p. 43) gives the following periods of growth at the different places mentioned:

Locality.	Beginning.	End.	Length of Period of Growth.
Montpelier	March 16	November 16	244 days
Dijon	April 15	October 15	183 days
Nancy	April 15	October 13	180 days
Paris	April 18	October 14	179 days

The longer period of growth in Montpellier is due to the warmer climate experienced there during the vine's period of growth. For the four places mentioned, the sum of the mean daily temperatures during the period of growth was respectively 4362° C., 2979° C., 2796.4° C., 2763.4° C. If we divide these totals by the number of days that the period of growth lasted in the various places, we get respectively 17.9°, 16.3°, 15.5°, and 15.4°, which represent their mean daily temperatures during the whole period of growth. From this we see that this mean daily temperature was higher in the south than in the colder more northerly centres. In a hot climate the period of growth lasts longer than in a colder climate. In the equatorial parts it is so warm throughout the year that the vine there does not enjoy a winter rest. Its period of growth extends over the whole year and its product is worthless.

CHAPTER IV

GENERAL AMPELOGRAPHY

1. HISTORICAL

THE word ampelography is derived from the Greek words *αμπέλου* (vine) and *γραφή* (drawing or description), and therefore stands for the science dealing with the description of the vine. This term was used for the first time by D. Sachs of Breslau in his *Ampelografia*, Leipzig, 1661, which deals with this subject in Latin (see Viala-Vermorel, *l.c.* i. p. 4). After Sachs, Don Simon de Rojas Clemente y Rubio published an important work on ampelography in 1807 under the title *Ensayo sobre las variedades de la vid commun que vegetan en Andalucia*, in which he especially described the varieties grown in Andalusia. The Spanish Government had a second edition of this work published at Madrid in 1879.

In 1841 Count Odart's *Essai d'ampelographie, ou Description des cepages les plus estimés dans les vignobles de l'Europe de quelque renom* was published. Six editions of this work were published under the later name of *Ampélographie universelle, ou Traité sur les cepages les plus estimés dans les vignobles de quelque renom*. After Odart, J.-L. Stoltz published his *Ampélographie rhénane* in 1852, and Victor Rendu his *Ampélographie française* in 1854. In 1877 Count Giuseppe di Rovasenda in Italy published his *Saggio di una ampelografia universale*, of which a French edition appeared in the same year, 1877, and a second edition in 1887 under the title *Essai d'une ampélographie universelle*. In 1878 Hermann Goethe's *Handbuch der Ampelographie* was published, whilst his *Ampelographisches Wörterbuch* had already appeared in 1876.

From 1879 to 1890 the *Ampelografia italiana* of the Central Committee for Ampelography was published in Italy in 7 sections, in which 28 Italian varieties were described with as many large chromolithographic plates. It is an ornamental work of which unfortunately only 600 copies were printed, which were soon sold out.[1] From 1876 to 1887 the Italian Minister of Agriculture published the *Bolletino Ampelografico* (according to Molon, *l.c.* xxxviii.). From 1874 to 1879 Mas and Pulliat's beautiful work, *Le Vignoble*, appeared in 3 vols. containing the description of 288 varieties with as many plates of these ripe grapes in their natural sizes and colours. V. Pulliat further wrote an important little work, *Mille variétes de vignes*, of which the third edition appeared in 1884.

In 1883 G. Foëx and P. Viala's *Ampélographie américaine* appeared, and J. Roy-Chevrier's *Ampélographie rétrospective* in 1900. In 1891 Henri Marès published his *Description des cépages principaux de la région méditerranéenne de la France* with 30 beautiful chromolithographic plates of the 30 varieties in natural size and colour.

In 1906 Prof. G. Melon's very important work, *Ampelografia*, 2 vols., was published in Milan. In 1908 appeared the best and most comprehensive work on the grape varieties that are grown mainly in North America, and are indigenous or were bred in this country. I refer to the beautiful work of U. P. Hedrick, *The Grapes of New York*, in which 23 North American and 1 European (V. *vinifera*) species of the genus *Vitis* are described, also 202 American and 2 European varieties under the heading "The Leading Varieties of American Grapes", and short descriptive notes on 1207 less important varieties under the heading "The Minor Varieties of American Grapes". The work contains 102 large colour plates, of which 2 represent European and 94 American varieties in their natural colours and sizes, and is Part II. of vol. iii. of the 15th Annual Report of the Department of Agriculture of the State of New York. For a serious study of the North American varieties that are grown for their grapes, this undoubtedly is the best and most complete work to refer the reader to.

---

[1] According to Prof. G. Molon, *Ampelografia* (Milano, 1906, 2 vols.), pp. xxxviii-xxxix.

From 1901 to 1910 the largest and best work on ampelography was published by Masson et Cie of Paris. It was published under the able guidance of P. Viala with V. Vermorel as general secretary and 84 collaborators out of most of the wine countries, and is called *Traité général de viticulture ...AMPÉLOGRAPHIE.* It consists of 7 large quarto volumes with 3145 pages, 70 large white and black plates of 58 American species and their hybrids, 10 Asiatic species, and 2 wild types (European and Turkestan) of *Vitis vinifera*, and 500 coloured plates of as many varieties in their natural size and colour. The work contains about 24,000 names or synonyms for 5200 varieties, of which 627 are described in detail in vols. ii.-vi., each of which contains 100 coloured plates. Vol. i. deals with general ampelography, and amongst other things contains a description and discussion of 32 species of the genus *Vitis*, 13 varieties of some of these species, 32 hybrids, and fossil vines. Vol. vii. is called *Dictionnaire ampélographique*, and, in addition to the alphabetical list of species and varieties that are described in the other 6 vols. and short descriptions of varieties not yet described, contains a classification of the principal grape varieties according to the order in which they ripen, and an extensive ampelographic bibliography. It is thus truly a standard work on this subject. Had it not been for Monsieur V. Vermorel's great financial support, this monumental work could never have been published.

In the preceding pages mention has been made of the principal works that have been published purely on ampelography. The list naturally is not complete, and works like Portes and Ruyssens' *Traité de la vigne et de ses produits* (1886-89), which treats of Viticulture and Vinification, have been omitted, although in vol. i. of this work 719 grape varieties are mentioned or described.

A fairly complete *ampelographic bibliography* we find in the works of Molon, Viala-Vermorel, and Hedrick.

Ampelography is concerned with the study of the different *Ampelidaceae* or *Vitaceae*, their genera, species, and varieties with regard to their structure, characters, mutual relations, distribution, culture, and value as cultivated plants. From the practical point of view, the genus *Vitis* is by far the most valuable. For this reason ampelography is more particularly concerned with the study of the various species and varieties of the genus *Vitis*.

During the international exhibition in Vienna in 1873, an International Ampelographic Commission of 19 members was appointed for promoting the study of ampelography, when they drew up the following programme of action (according to Molon, *l.c.* p. xxxv):

(1) To start laying the foundations that can lead to a nomenclature for grapes that is understood and recognised by all, *i.e.* to indicate the variety by its main name, that is to say, that one which it has in its place of origin or in the district where it is most largely grown; all translations of this name and all other names are then added to its main name as synonyms.

(2) To study all the imported and new varieties and to conduct experiments with them in order to determine whether they are separate varieties or only sub-varieties of those already known, and to determine their value to the viticultural industry.

(3) To make known all the varieties of very poor quality, to advise against their being planted in future, and to recommend better varieties in their stead.

(4) As suitable for attaining the three objects just stated, and with the collaboration of all the members of the Commission, to publish a *general ampelographic catalogue* which can then be augmented by successive additions and corrections as the investigation progresses.

(5) To arrange the final catalogue of the varieties according to either a natural system based on family, or an artificial system based on characters, in order to be able easily to recognise and determine the varieties.

The Commission met for the eighth and last time in Geisenheim a.Rh. in 1880, and did a good deal of work. The fruits of its labours appeared in 15 reports in the *Bolletino ampelografico*; further in the *Rivista di Viticoltura ed Enologia* (1879, p. 503), and finally in the publications of H. Goethe, which I have already mentioned. Also in the *Ampelographische Berichte* reports are made about the work of the Commission.

2. CLASSIFICATION

(a) Classification of the Ampelidaceae (Vitaceae)

The following is Planchon's classification of the Ampelidaceae (Vitaceae) in 10 genera:

1. VITIS Tournefort.	6. LANDUKIA *Planchon.*
2. AMPELOCISSUS *Planchon.*	7. PARTHENOCISSUS *Planchon.*
3. PTERISANTHES *Blume.*	8. AMPELOPSIS *Michaux.*
4. CLEMATICISSUS *Planchon.*	9. RHOICISSUS *Planchon.*
5. TETRASTIGMA *Miquel.*	10. CISSUS Linné.

Of these ten genera thus far only the genus *Vitis* has given useful crop plants. Representatives of the genus *Ampelocissus* occur in Abyssinia, the Sudan, Cochin-China, and in South Africa (the Transvaal, etc.). They bear large bunches of grapes that weigh up to 10 lbs., and might be of some value in tropical countries. They give a very light wine containing 4-6 vol. per cent alcohol. These grapes have a very poor taste. Some of them possess thickened roots, which grow to 2 in. broad and 10 in. long, or even larger.

The genus *Parthenocissus* includes *Parthenocissus quinque-folia*, formerly known as *Ampelopsis quinquefolia*, which is the Virginia creeper.

The genus *Rhoicissus* includes *Rhoicissus capensis*, formerly known as *Vitis capensis*, Thunberg, and as *Cissus capensis*, Wildd. It is indigenous at the Cape and grows along the slopes of Table Mountain. A further South African representative of this genus is *Rhoicissus Thunbergii*, Planchon, which occurs in the Transkei and Natal.

The genus *Vitis* stands on a different plane from the other genera of the Ampelidaceae, when we consider its species and their varieties from the practical point of view, as they are the only ones that have given rise to the great viticultural industry.

(b) Classification of the Genus *Vitis*

Viala in Viala-Vermorel, *l.c.* i. pp. 111-112, divides this genus into the following 32 species:

I. MUSCADINIA *Planchon*

V. rotundifolia, Michaux (United States of America: south-east).
V. Munsoniana, Simpson (United States of America: Florida).

II. EUVITES *Planchon*
America

V. caribaea, De candolle (tropical America).
V. coriacea, Shuttleworth (Florida).

V. *Bourgaeana*, Planchon (Mexico).
V. *Blancoii*, Munson (Mexico).
V. *californica*, Bentham (California).
V. *arizonica*, Engelmann (U.S.A.: Arizona and New Mexico).
V. *Berlandieri*, Planchon (U.S.A.: Texas).
V. *monticola*, Buckley (U.S.A.: Texas).
V. *candicans*, Engelmann (U.S.A. : south and middle-south).
V. *Lincecumii*, Buckley (U.S.A.: middle and south).
V. *rupestris*, Scheele (U.S.A.: south and middle-south).
V. *cinerea*, Engelmann (U.S.A.: middle and middle-east).
V. *cordifolia*, Michaux (U.S.A.: middle and middle-east).
V. *aestivalis*, Michaux (U.S.A.: middle).
V. *bicolor*, Leconte (U.S.A.: middle).
V. *rubra*, Michaux (U.S.A.: Missouri).
V. *riparia*, Michaux (U.S.A.: north and north-east).
V. *Labrusca*, Linne (U.S.A.: north ; Canada : south).

Asia

V. *Coignetiae*, Pulliat (Japan).
V. *Thunbergii*, Sieb. et Zucc. (Japan, Corea).
V. *flexuosa*, Thunberg (Japan, Corea).
V. *amurensis*, Ruprecht (Japan, China).
V. *Romaneti*, Romanet du Caillaud (China).
V. *Davidii*, Carriere (China).
V. *Pagnuccii*, Romanet du Caillaud (China).
V. *Retordi*, Romanet du Caillaud (Tonkin).
V. *Balansaeana*, Planchon (Tonkin).
V. *lanata*, Roxburgh (China, India).
V. *pedicellata*, Lawson (India: Himalaya).

Europe, Africa, etc.

V. *vinifera*, Linne (Europe, North Africa, Central Asia).

The subdivision of the genus *Vitis* into Muscadinia and Euvites is also adopted by Foëx (*l.c.* p. 28), Ravaz,[1] Molon (*l.c.* p. 4), and others. The American species of the Euvites are grouped in the following scheme by Foëx (also adopted by Molon):

Series 1. Labruscae : *V. Labrusca*.
Series 2. Labruscoideae americanae: *V. californica, V. caribaea, V. coriacea, V. candicans*.
Series 3. Aestivales: *V. Lincecumii, V. bicolor, V. aestivalis*.
Series 4. Cinerascentes: *V. cinerea, V. cordifolia, V. Berlandieri*.
Series 5. Rupestres: *V. monticola, V. rupestris, V. arizonica*.
Series 6. Ripariae: *V. rubra, V. riparia*.

---

[1] L. Ravaz, Les Vignes américaines, porte-greffes et producteurs directs, 1902.

Ravaz classifies them as follows:

I. Vines with *continuous tendrils*: *V. Labrusca.*
II. Vines *with intermittent tendrils*:

A. *Canes with a round or oval cross-section*:
(*a*) Leaves wedge-shaped: *V. riparia.*
(*b*) Leaves kidney-shaped: *V. rupestris.*
 Doubtful species: *V. treleasei,* Bailey.

B. *Canes with an angular cross-section*:
(*a*) Leaves truncated and shining strongly on the lower surface:
V. aestivalis, V. bicolor, V. Lincecumii.
(*b*) Leaves wedge-shaped, broad: *V. coriacea, V. arizonica.*
(*c*) Leaves round, downy pubescence: *V. californica, V. monticola.*
(*d*) Leaves cordate: *V. cordifolia.*
(*e*) Leaves cordate, truncated: *V. rubra.*
C. Canes ribbed:
(*a*) Leaves cordate or round: *V. candicans.*
(*b*) Leaves wedge-shaped: *V. Berlandieri.*
(c) Leaves cordate, truncated: *V. cinerea.*

Hedrick[1] classifies the North American species of *Vitis* as follows:

A. Skin of mature berry separating freely from the pulp.
 B. Nodes without diaphragms; tendrils simple.
  1. *V. rotundifolia.*
  2. *V. munsoniana.*

B.B. Nodes with diaphragms; tendrils forked.
C. Leaves and shoots glabrous at maturity and without bloom; tendrils intermittent.
(*V. cinerea* and *V. arizonica* are partial exceptions and might be looked for under C.C.)
D. Leaves thin, light, bright green, generally glabrous below at maturity except perhaps in the axils of the veins (*V. champini* an exception), with a long or at least a prominent point, and usually long and sharp teeth or the edge even jagged. (*V. bicolor* might be looked for here.)
E. Leaves broader than long; petiolar sinus usually wide and shallow. (*V. treleasei* might be sought here.)
  3. *V. rupestris.*
E.E. Leaves ovate in outline; petiolar sinus usually medium to narrow.

---

[1] U. P. Hedrick, *The Grapes of New York,* 1908, pp. 107-108.

F. Diaphragms thin; young shoots not red.

> 4. *V. monticola.*
> 5. *V. riparia.*
> 6. *V. treleasei.*
> 7. *V. longii.*
> 8. *V. champini.*

F.F. Diaphragms thick; young shoots bright red.

> 9. *V. rubra.*

D.D. Leaves thickish, dull-coloured or greyish green, often holding some close dull pubescence below at maturity; shoots and leaves nearly always more or less pubescent when young; the teeth mostly short; the point mostly rectangular and conspicuous.

E. Plants strong, climbing, with stout, persistent tendrils.

F. Young shoots cylindrical, glabrous or very soon becoming so   10. *V. cordifolia.*

F.F. Young shoots angled, covered the first year with tomentum or wool

> 11. *V. baileyana.*
> 12. *V. Berlandieri.*
> 13. *V. cinerea.*

E.E. Plants scarcely climbing; tendrils perishing when without support

> 14. *V. arizonica.*

D.D.D. Leaves orbicular, scallop-shaped; species of the Pacific Coast   15. *V. californica.*

C.C. Leaves rusty or white tomentose or glaucous blue below, thick or at least firm. (*V. cinerea*, *V. arizonica*, and possibly *V. californica* might be sought here.)

D. Leaves flocculent or cobwebby or glaucous below when fully grown {i.e. not covered with a thick, dense, felt-like tomentum except sometimes in *V. doaniana*).

E. Shoots white-tipped; ends of the growing shoots and the under surface of the leaves whitish or grey.

> 16. *V. girdiana.*
> 17. *V. doaniana.*

E.E. Shoots rusty - tipped; the unfolding leaves and (except in *V. bicolor*) the young shoots distinctly ferruginous; mature leaves either rusty or bluish below, or sometimes becoming green in *V. bicolor.*

> 18. *V. aestivalis.*
> 19. *V. bicolor.*
> 20. *V. caribaea.*

D.D. Leaves densely tomentose or felt-like beneath throughout the season ; covering white or rusty white.

E. Tendrils intermittent   21. *V. candicans.*
                                               22. *V. simpsoni.*

E.E. Tendrils mostly continuous   23. *V. labrusca.*

A.A. Skin and pulp of mature berry cohering. (Old World.)
                                               24. *V. vinifera.*

(*c*) Classification of the European Varieties (*V. vinifera*)

These varieties all belong to the one species, *V. vinifera*, and include all the varieties of this species whether grown in Europe or elsewhere. The classification of these varieties is a matter of great difficulty. It has been attempted in the past by many ampelographers, but thus far there is no perfect classification in existence. The trouble is that the characters of the different varieties are not sufficiently constant to prevent doubt and errors from occurring. Take, for instance, the shape of the berry. Gros Colman passes for a grape with spherical berries, yet appreciable deviations from this shape frequently occur. The shape of berry gradually passes from spherical to elongated (acorn - shaped) with the different varieties, and varies considerably with one and the same variety. We find the same happening with the other characteristics. This explains why an exact classification is extremely difficult, if not impossible.

The first attempt to classify these varieties was made in 1777 by S. Helbling,[1] who divided the grapes into three groups according to their colour, namely blue, red, and white; every group he divided into two sub-groups according to the shape of the berries: round or oval (quoted from Molon, *l.c.* pp. 137-138). Although he has been long forgotten, the basis of his classification to-day still renders us useful service.

In 1876, during the fourth session of the International Ampelographic Commission in Marburg, H. Goethe proposed to take the angle made by the two main lateral nerves with the middle nerve (Ravaz's $a + \beta$) as a basis of classification, and to divide the varieties into four classes according as this angle measures up to 95°, 96-100°, 101-120°, and more than 120°. His brother, R. Goethe, after careful investigation,

---

[1] S. Helbling, Beschreibung der in der Wiener Gegend gemeiwn Weintrauben, Prague, 1777.

found that this value remained constant with one and the same variety. In 1902 Ravaz, *l.c.* p. 21, proposed the same thing in a modified form. He made ten classes for values up to 70°, 71-80°, 81-90°, 91-100°, 101-110°, 111-120°, 121-130°, 131-140°, 141-150°, 151-160°. This, however, does not yet give us the desired ampelographic classification.

Since 1876 V. Pulliat[1] declared himself in favour of a classification which rests upon the time of ripening, and against any classification which rests upon the organs of the vine. He recommended and applied his system in his works *Le Vignoble, Mille variétés de vignes*, and *Cours de viticulture et d'ampélographie*. As starting-point he chose Chasselas doré, and made the following grouping:

1. *Early varieties, i.e.* those ripening more than six days before Chasselas doré.
2. Varieties of the *first period* of maturity, *i.e.* those ripening not more than five or six days before or after the Chasselas doré.
3. Varieties of the *second period* of maturity, *i.e.* those ripening twelve to fifteen days after those of the first period.
4. Varieties of the *third period* of maturity, *i.e.* those ripening twelve to fifteen days after those of the second period.
5. Varieties of the fourth period of maturity, *i.e.* those ripening after those of the third period.

In 1877 the International Ampelographic Commission adopted at Florence the following classification:

Class I.	Berries Round	Order 1. Leaves glabrous on lower surface.
		Order 2. Leaves with felt-like pubescence on lower surface.
		Order 3. Leaves with woolly pubescence on lower surface.
Class II	Berries elongated	Order 1. As before
		Order 2. As before
		Order 3. As before
Class III	Berries with indefinite shape	Order 1. As before
		Order 2. As before
		Order 3. As before

H. Goethe applied this classification in the first edition of his *Handbuch der Ampelographie* (1878). In the second edition (Berlin, 1887) he applied the following classification

---

[1] V. Pulliat, Mille varietés de vignes, 3rd ed., 1888, pp. xix-xx

by himself and Oberlin, which was adopted by the International Ampelographic Commission at its congress in Budapesth in 1879:

Class I. With round berries.
Class II. With elongated berries.
Class III. With berries of indefinite shape.
Every class is divided into three orders, as follows :
Order 1. Leaf nearly glabrous on lower surface.
Order 2. Leaf felt-like on lower surface.
Order 3. Leaf woolly   "   "
Every order is divided into three sub-orders, as follows :
Sub-order 1. Leaves with open petiolar sinus.
Sub-order 2.   "   closed   "
Sub-order 3.   "   irregular "

Since 1893 G. Molon[1] as proposed the following classification:

			Period of Maturity	Taxonomic Sign
Class B Grapes white	Order *r*, Berries round	Sub-order 1, simple flavour	Early	B. r. 1 p.
			I. Period	B. r. 1 I.
			II. "	B. r. 1 II.
			III. "	B. r. 1 IIII.
			IV. "	B. r. 1 IV.
		Sub-order 2, Muscat flavour	Early	B. r. 2 p.
			I. Period	B. r. 2 I.
			II. "	B. r. 2 II.
			III. "	B. r. 2 IIII.
			IV. "	B. r. 2 IV.
	Order *o*, Berries oval	Sub-order 1, simple flavour	Early	B. o. 1 p.
			I. Period	B. o. 1 I.
			II. "	B. o. 1 II.
			III. "	B. o. 1 IIII.
			IV. "	B. o. 1 IV.
		Sub-order 2, Muscat flavour	Early	B. o. 2 p.
			I. Period	B. o. 2 I.
			II. "	B. o. 2 II.
			III. "	B. o. 2 IIII.
			IV. "	B. o. 2 IV.
	Order *l*, Berries long		Early	B. 1 p.
			I. Period	B. 1 I.
			II. "	B. 1 II.
			III. "	B. 1 IIII.
			IV. "	B. 1 IV.

Class R.—Red grapes ......   as before with R for B.
Class N.—Black grapes ...   as before with N for R

---

[1] G. Molon, *l.c.* pp. 200-201.

159

Molon himself is not quite satisfied with his classification, but wishes to retain it until he may some day be able to work out a better one.

For discussing the species and varieties of the genus *Vitis* that are of practical value, I group them as follows:

A. *American vines*	I. Used a stocks
	II. Grown for their grapes.
	I. Grown for wine.
B. *European vines*	II. " table grapes.
	III. " raisins.
I. is subdivided into	1. White wine varieties
	2. Red "
	1. White table grape varieties
II. is subdivided into	2. Red "
	3. Black "
III. is subdivided into	1. Varieties for raisins with seeds.
	2. " seedless raisins
C. Self-bearers or Direct Producers	I. Old
	II. New

For the numerous other systems of classification of the European vines I refer the reader to Molon, *l.c.* pp. 137-201.

3. DESCRIPTION OF THE VINE

(*a*) **General**

It is of the greatest importance to the practical man to find out which varieties it will pay him best to grow under his conditions. In order to find a proper solution for this problem, he should have at his disposal experience gained with varieties grown under conditions similar to his own. It is also of the greatest importance that there should be no confusion of names. Hence we must study the various varieties under their true names under varying circumstances. This study does not merely comprise a description of the various parts of the vine, but should extend also to the geographical distribution of the variety, its physiological behaviour (e.g. when it buds, blooms, ripens its grapes; its susceptibility to diseases and pests; whether it is a regular and good cropper with short or long pruning; how it answers on the various

American stocks; use and usefulness of its grapes, etc.), its origin and history, and in this connection a list of all the various names under which it is known, how and under what conditions it can be successfully grown.

The first point in connection with the description of a grape variety is its *name*. Most varieties are known under different names in the various countries and regions where they are grown. The writer of a treatise on this subject therefore usually chooses the name best known in his own country and adds to it, as synonyms, all the other names under which the variety is known elsewhere. Thus Muscat of Alexandria is known in France as Muscat d'Alexandrie, in Germany as Muskat Damascener, in Tuscany as Salamanna, in Sicily as Gerosolimitana bianca and Zibibbu, at the Cape as White Hanepoot, etc. One of the important points to settle is which names are synonyms of one and the same variety, and which refer to separate though closely related varieties.

In South Africa I have satisfied myself that *Hermitage* is the same variety as *Cinsaut* of the South of France; *White Greengrape* is almost certainly a slightly differing type of Semillon; *Old Stein* is to all intents and purposes the same as Riesling; *Stein* (woolly leaf) is closely related to Sauvignon blanc; *White French* is almost if not entirely the same as *Palomino*, the basis of the famous sherry wines; *Pontac* is the same as *Teinturier mâle*.

(b) Characters to be described

In order to prevent misunderstanding and confusion, we should, in describing a vine, try to give an accurate and complete description of all its important characteristics and properties, so that this description can enable us to recognise this variety with certainty wherever we meet it. Here photographic reproductions of the shoots, leaves, and grapes in their natural colours when the grapes are ripe, can render us very valuable service.

In describing a vine we should take into account all its parts at the different stages of development, although all are not of the same value for diagnostic purposes.

According to Pulliat, *l.c.* p. ix. the following characters are the most constant and reliable for descriptive purposes, in the order of their importance:

1. Period of *maturity*.
2. Shape and size of the *bunch*.
3. Shape, colour, flavour, and quality of the *berry*.
4. "*Bourgeonnement*" or rudimentary shoot.
5. Main characters of the leaf.

He then proceeds to describe the points to be noted in this connection.

H. Goethe[1] mentions the following points to be noted in describing a vine:

(a) General Properties of the Vine: Vigour, fertility, susceptibility to diseases and pests.

(b) One-year-old Wood (i.e. ripe canes): Thickness, hardness, colour, thickness of pith, smoothness. Internodes are short if not exceeding 4 cm. where bunches usually occur, medium long if 4-7 cm., long if over 7 cm. Nodes more or less thickened, round or flattened, differently coloured from internodes or not. Eyes small or large, pointed or rounded, prominent or not, glabrous or woolly, special colour if any.

(c) Green Shoots: Thin or thick, smooth or rough, glabrous or pubescent, straight or bent. Colour, bloom, tendrils. Growing tips (including the tips of the shoots and the laterals in a state of active growth) with green, red, brownish, or whitish colour, glabrous or pubescent, shining or dull, shape and dentition of uppermost leaves, etc.

(d) General Characters of the Leaves: Small, medium or large, thin or thick, like leather or like paper, stiff or flaccid, smooth or rough, folded or bullate, with much, little, or no gloss, dropping early or late.

(e) Shape of Leaves: Lobes, sinuses.

(f) Dentition of Leaves: Small or large, pointed and long or broad and rounded, even or uneven. Terminal tooth large or small, long and pointed or broad and dome-shaped.

(g) Colour of Leaves: Colouring of leaves, especially in autumn.

(h) Pubescence of Leaves: Felt-like, woolly, brush-like.

(i) Leaf-nerves: Standing out on upper or lower surface, thin or thick, glabrous, woolly or with brush-like hairs, colour, etc.

(k) Leaf-stalk or Petiole: Shorter or longer than middle nerve, thin or thick, pubescence, colour, etc.

(l) Flower: Opens early or late, influence of weather (non-setting). Length of blooming period. Form of flower's parts: length of stamens relative to that of pistil, shape of pistil.

(m) Bunch and Stalk: Shape of bunch: small, medium - sized, large; long or short, cylindrical or pyramidal (tapering), single (without large stalk divisions) or divided, with berries loose or compact, berries even or uneven.

---

[1] H. Goethe, *Ampelographisches Worterbuch*, Vienna, 1876, pp. 7-10.

(n) Stalk or Peduncle: Short or long, thick or thin, soft and brittle or tough, green or lignified, with or without a lateral bunch at the node.

(o) Pedicel: Short, long, thin, thick, smooth, warty, etc.

(p) Berry. Shape: Small, medium, large; decidedly long (ovoid, oval, acorn-shaped) when the long diameter of the loose berries is always bigger than the cross one; or longish (longish round or passing into spherical) if the loose berries are neither decidedly long nor round, or round (spherical or flattened spherical) if the long diameter is just as long as or shorter than the cross diameter. Skin: thin and soft or thick, cracking and leathery, with much or little pigment. Colour: green, white, yellow, red, blue, grey. Stigma or its mark at the apex (*i.e.* point opposite the base or point of attachment of berry to pedicel) small or large, prominent or more or less sunk, black, grey, or brown. Berry-content (pulp): its firmness or softness, taste, aroma. Seeds: number, colour, shape. Period of maturity, early, normal, late, or very late.

Molon, *l.c.* pp. 288-300, mentions more or less the same points for a description of the vine, perhaps in somewhat greater detail and with an interesting discussion of the points mentioned.

4. DISCUSSION OF CHARACTERS

Ravaz[1] discusses the characters of American vines in his work referred to at the bottom of this page. His remarks are of general application, and therefore I wish to give a short résumé of them in the following pages.

(*a*) **The Roots**.—"They can be thin or thick, hard or fleshy, more or less numerous, with few or numerous fine roots, spreading or plunging; have a yellow, red, grey, etc., colour. But most of their characters depend upon the state of the soil, its physical and chemical composition, its dryness, etc. . . and, with grafted vines, the variety used as scion. The colour varies noticeably with the soil; the same applies with regard to the *direction* of the roots. A certain plant will 'plunge' here and 'spread' elsewhere. And it is only in one and the same region, in one and the same soil, that the root system furnishes definite characters. ... In any case, the specific characters are more distinct than the characters of the varieties; they are therefore influenced less by the soil ... and they can everywhere be stated precisely. A root of

---

[1] L. Ravaz, Les Vignes américaines, ports - greffes et producteurs directs, 1902 pp. 5-30.

V. riparia in no way resembles a root of *V. aestivalis* or of *V. rupestris*. It will nevertheless be necessary to exercise much care in the diagnosis of the root-system."

The root-system possesses little value from the descriptive point of view, and in practice it will seldom if ever enable us to determine the identity of a grape variety. Its study is, however, valuable, as some of its characters give us some idea about the usefulness of the vine. Thus the fleshy roots of *V. coriacea*, etc., indicate a high resistance against drought; thin roots indicate that the vine requires a loose and fertile soil, etc.

(*b*) **The Stem**.—For ampelography the stem with its attachments constitutes the most important part of the vine. Ravaz attaches value only to the living organs of the stem. He discusses in succession the trunk, shoots (canes), leaves (young and full-grown), tendrils, and bunch.

The Trunk: Its thickness is of some value where vines occur in a collection. It gives an indication of the vigour and rigidity of the vine. Ravaz does not attach any value to a description of the dead bark on the trunk.

The Ripe Canes: Their characters possess an appreciable value for determining the identity of varieties. In different species we have to do with clearly differing characters of quality or kind, which persist after the bark has died. Here, therefore, we can proceed with much greater certainty than where we have to distinguish between different varieties of the same species where we have to deal with quantitative differences in the characters.

The Green Shoots: On these the outstanding characters of the ripe canes are more accentuated. "The shoots can be thin or thick, short or long, more or less erect or creeping, with short or long internodes, cylindrical or flattened; even, angular, ribbed; smooth or rough (rugose); glabrous or woolly or with brush-like hairs and even thorny; with a grey, brown, mahogany, red, violet, fawn-coloured, etc., colour on a green background."

Dimensions: The length and thickness of the shoots are influenced by the soil, etc. They therefore only possess relative value. Ravaz expresses the *slenderness* ("gracilite") of a cane by dividing its length (measured at the end of the growing season) by its thickness, which is always measured in the middle of the internodes from the 9th to the 12th nodes, and in the direction of the greatest diameter. The length of the internodes is variable and largely depends upon the conditions of growth. It may, however, be expressed by an average value.

Direction: This is subject to the vigour of the vine, and therefore to the soil and its cultivation. It is of little descriptive value.

Diaphragm, Wood, Pith: The diaphragm has been made use of by Millardet to characterise some species and varieties of vines. With *V. rotundifolia* it is wanting; with the other species its hardness and thickness depend very much on the progress of the vine's growth: they are therefore, on the whole inconstant. The same applies to the hardness of the wood and to the thickness and colour of the pith.

Form: The form or contour of the shoot (cane) is constant; it supplies excellent specific characters, especially on the green shoots. One should note it on the internodes when they have just acquired their full length. According to their cross-section through the middle of the internode, we can divide shoots into three groups: even, angular, ribbed.

Colour: This is a character of the bark, which is dead on the ripe canes. It varies with the species and with varieties in the same species. *V. aestivalis*, for example, can easily be distinguished from *V. Berlandieri* and *V. rupestris* by the colour of its canes. Varieties also can be thus distinguished, although the indications in this case are less precise and one must proceed with caution. The colour of the shoots, with a few exceptions, possesses only relative value and is difficult to describe precisely.

Pubescence: Green shoots as well as ripe canes can be glabrous, woolly, or covered with brush-like hairs. The brush-like hairs are straight and stand at right angles to the surface of the bark. They have been formed out of the epidermis, and, according to Ravaz, perish soon again. (This is not so with Rip. × Rup. 3306; A. I. P.) After the hairs have dropped, the marks where they have been attached to the bark can still be seen under the microscope. The same applies to the woolly hairs.

The glabrousness or pubescence can thus be always observed and provides valuable characters of quality. It is difficult precisely to indicate the intensity of the pubescence. Where a microscope is not used, Ravaz recommends that only the young, still growing parts of the shoots should be used for this purpose. He consequently classifies the pubescence of the green shoots as follows:

1. *Cobwebby* or slightly pubescent if the pubescence extends only over the first five internodes reckoned from the tip of the shoot.
2. *Downy* or fairly pubescent if it extends over the first ten internodes.
3. *Woolly* or strongly pubescent where it extends over more than the first ten internodes.

Usually pubescent shoots have pubescent leaves and glabrous shoots glabrous leaves. Some species have cobwebby shoots and glabrous leaves (*e.g. V. monticola*).

The presence of bloom on the shoots possesses only specific value; club-shaped and stinging hairs characterise only subdivisions of species.

The Eyes: Except in case of a few species, the eyes offer us not a single constant character. They contain a variable number of buds.

(c) *The Leaf.*—"The leaf provides very good ampelographic characters; and it will perhaps be possible to recognise each variety by the characters of the leaf only."

The Tip of Growing Shoots: This is really a developing bud. It possesses characters that have been used by all ampelographers. They can best be seen in the early stages of growth, and are less definite later on, particularly when the growth comes to standstill. (Molon, *l.c.* p. 288, recommends that the growing tips be described just shortly before or during the bloom, when they possess very constant characters. —A. I. P.)

Its glabrousness or pubescence usually recurs on the leaves and therefore does not assist us further. The colour of the unfolding leaves is more valuable. The most constant colour tints are green with its various shades, red, rose- and bronze-coloured. Sometimes the colour is confined merely to the borders of the leaves, sometimes along the nerves, and some-times it extends over the whole leaf. Sometimes only the hairs (down) are coloured and sometimes only the parenchyma, sometimes both. Thus the growing tips of the shoots of the varieties of *V. aestivalis* are all crimson-red, but only on the pubescence; the parenchyma under it has a pretty green colour. Hence the colour of the parenchyma and the pubescence should be described separately. As these characters are merely transitory, they serve in actual practice merely to differentiate between varieties that resemble each other fairly closely.

The dimensions and colour of the stipules are useful specific characters.

FIG. 43.—Reniform leaf. From Ravaz, *Les Vignes américaines*, etc., 1902. Coulet et fils, Montpellier; Masson et Cie, Paris.

Young Leaves: The young growing leaves continually change. Their colour and the manner in which they unfold themselves at the terminal bud give useful indications which can serve to characterise either the species or the varieties.

Full-grown Leaves: Once the leaf is full-grown, its form remains constant. As the leaves have developed one after the other, and hence under varying conditions, they do not precisely resemble each other. The lowest, middle, and upper-most leaves on a shoot can differ appreciably. The first mentioned are sometimes entire when the last named may be lobed. For comparative purposes Ravaz makes use *only of* those leaves that correspond, with the period of greatest growth of the vine: these are the 9th-12th leaves from the base of the shoot.

Leaf-stalk (Petiole): According to Petit (Ravaz, *l.c.* p. 13), the leaf-stalk offers excellent specific characters, and every species can be distinguished by making a cross-section of the leaf-stalk near the leaf-blade. Millardet has made use of the shape of the petiole for determining different species of vines. Its colour and pubescence are here taken into account.

FIG. 44.—Round or orbicular leaf. From Ravaz, *Les Vignes américaines*, etc., 1902. Coulet et fils, Montpellier; Masson et Cie, Paris.

It forms an angle with the leaf-blade (lamina) which varies with the position of the leaf. Ravaz does not attach much value to the character of the petiole.

Shape or Form: The general form of the leaf depends upon the leaf skeleton; it is therefore a function of the relative lengths of the primary or main nerves expressed by the ratios $\frac{nI^1}{nI}$ and $\frac{nI^2}{nI^1}$, and also by the angles α and β, which the first lateral nerve (I^1) forms with the middle nerve and the first (I^1) with the second (I^2) lateral nerve respectively.

I, I¹, and I² are respectively the lengths of the middle, first lateral, and second lateral primary nerves.

In the *reniform* leaf (Fig. 43) the lateral nerves are very long compared with the middle nerve and the angles α and β very acute. Let *a* and *β* now grow bigger, *i.e.* let the lateral nerves open whilst retaining their lengths constant; we now have a *round* or *orbicular* leaf (Fig. 44). Now let the first lateral nerve alone get shorter, and we get a heart-shaped or *cordate* leaf (Fig. 45). Let both lateral nerves now get shorter in the same proportion to their length, then the wedge-shaped or *cuneiform* leaf appears (Fig 46). Let only the lateral nerve I² get shorter, and we have the truncated leaf (Fig. 47.) There are thus five types of leafs forms: reniform, orbicular, cordate, cuneiform, truncated.

All these types which characterise the species or specific groups are united by numerous transition forms which can serve to distinguish between the varieties, and especially the hybrids. The general leaf form is no way affected by the lobe formation.

Some leaves are entire or nearly so, others are more or less deeply cut up into 3, 5, or 7 lobes; some varieties are even laciniated. The

FIG. 45.—Cordate leaf. From Ravaz, *Les Vignes américaines*, etc., 1902. Coulet et fils, Montpellier; Masson et Cie, Paris.

teeth are rounded or acute, and more or less short. The contour of the leaf varies a good deal with the variety, if we compare the upper with the lower leaves. If we take them, however, at the same height, 9th-12th nodes, their contour is sufficiently constant to constitute one of the best ampelographic characters.

Asymmetry: All the leaves are irregularly asymmetric. The half that always resembles itself is that which corresponds with the normal or resting bud in the leaf-axil; it is also the larger half, and *the ampelographic characters should be observed on it only*.

Length of Leaf: The length, L, of the leaf is measured from the terminal tooth of the middle nerve along the same to the top of the highest tooth of the petiolar or bottom lobe.

Width of the Leaf: .The width, *l*, is double the width of the half corresponding to the normal eye.

FIG. 46.—Wedge-shaped leaf. From Ravaz, *Les Vignes américaines*, etc., 1902. Coulet et fils, Montpellier; Masson et Cie, Paris.

Relative Dimensions: This is expressed by the ratio $\frac{L}{l}$, which expresses an important specific character.

Size of Leaf: Ravaz draws attention to its variableness and attaches only relative value to it. He considers a leaf large if the average length of the four leaves at the 9th-12th nodes is equal to one and a half times the average length of the 9th-12th internodes; medium where they are of the same length; and small where it is shorter than the internode. But he made little use of this factor.

170

Lobing of the Leaf: The leaf is entire when the upper sinus is zero, *i.e.* when its value is equal to the length of the terminal tooth of the corresponding lobe. It is *trilobed* if the upper sinus (i.e. that between the middle nerve and the first lateral nerve) has the value of at least two teeth. It is five-lobed if the lower sinus (*i.e.* that between the two lateral nerves) is clearly marked. The sinus is deep if the incision is equal to half the length of the nerve; very deep if three-quarters. The sinuses are often more or less closed near their opening.

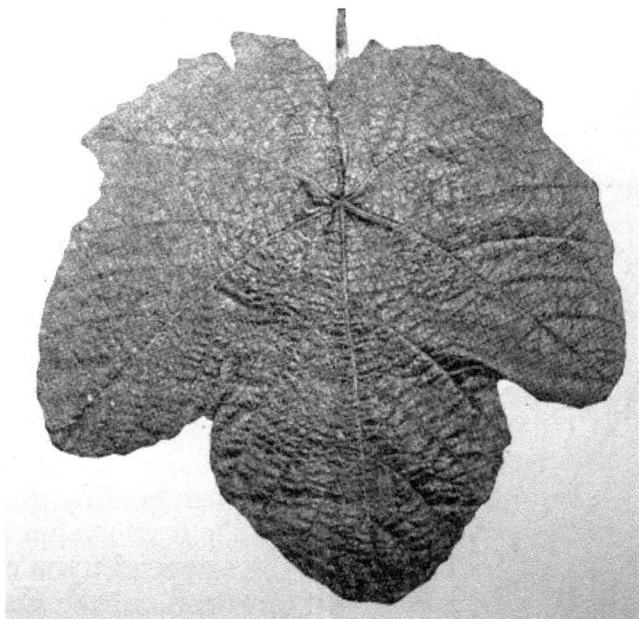

Fig. 47.—Truncated leaf. From Ravaz, *Les Vignes américaines*, etc., 1902. Coulet et fils, Montpellier ; Masson et Cie, Paris.

(With Cabernet Sauvignon their openings are frequently closed, as the edges of the lobes overlap, in which case we frequently see the five sinuses as five more or less round holes. See Fig. 10—A. I. P.)

Petiolar Sinus: The petiolar sinus is formed and bounded by the petiolar lobes. Great ampelographic value is attached to it. Its nature is determined by the direction of the main or primary nerves of the petiolar lobes, by a marginal zone of parenchyma, by the direction and degree of development of a few tertiary and quaternary nerves, by the length (l^2) of the first part of the petiolar nerves (see Fig. 48), and by the more or less vigorous growth of the leaf.

That the shape of the petiolar sinus is connected with the value of the angles a and β will no doubt be clear to every one. The angles γ, δ, ε (see Fig. 48) are formed by the secondary, tertiary, and quaternary nerves with the nerve out of which each one arises in its turn. The nearer to the petiole the secondary nerve (forming the angle γ) arises, the more the petiolar sinus will tend to be closed. Some times this secondary nerve arises out of the end of the petiole, in which case the petiolar sinus is closed.

Ravaz maintains that the opening and shape of the petiolar sinus are of no ampelographic value, as they depend upon climate, soil etc. He, however, makes use of the following values: a, β, γ, and the relative length[1] of the base of nerve l^2, which is constant with the same variety.

Under "Classification" we have already seen that Ravaz classifies the leaves in ten groups according to the value of $a + \beta$

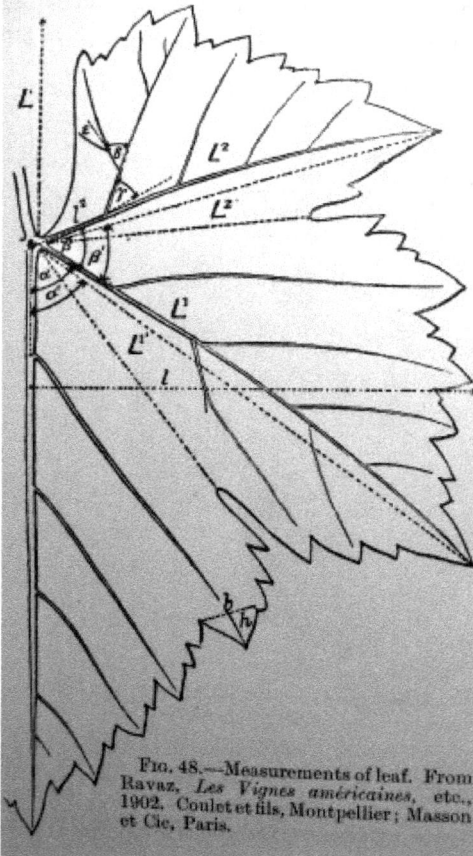

Fig. 48.—Measurements of leaf. From Ravaz, *Les Vignes américaines*, etc., 1902. Coulet et fils, Montpellier; Masson et Cie, Paris.

---

[1] NOTE.—Although Ravaz (*l.c.* p. 20) does not explain it any further, Molon (*l.c.* p. 294) says that it is $\frac{l2}{l1}$, where l^1 is the depth of the petiolar sinus, and hence = length of leaf minus length of middle nerve = L -I, and l^2 is the distance from the end of the petiole to where the secondary nerve arises out of l^2.

He points out that each of these groups can again be divided into three groups according to the value of γ, which is *independent* of the value $a + \beta$. The angles δ and ε may sometimes be made use of, but are only of secondary value.

Teeth: They are acute or rounded off. Their dimensions possess only relative value, which is expressed by $\frac{h}{b}$ the ratio of the height of the tooth to its breadth.

Teeth are very narrow when the ratio is   > 1
"   narrow   "   > 0.75
"   broad   "   > 0.50
"   very broad   "   > 0.25
"   almost nil   "   < 0-25

(Ravaz, *l.c.* p. 20, in the last instance, also uses the sign >, but this should obviously be <.—A. I. P.)

Nerves: The primary or main nerves arise directly out of the petiole just where it is attached to the leaf. The angles they make with each other have already been discussed. Their *relative* lengths also give useful indications. Their width or thickness is difficult to determine, but the relative thickness of each of the primary nerves can be made use of as specific characters. Their colour shares that of the petiole and that of the shoot, is of little value and changeable, and must therefore be described just when the leaf is full-grown.

Colour of Leaves: This varies with each variety and is only of secondary value. It runs from pale green to dark green; with some varieties the parenchyma of the leaf sooner or later assumes a red colour, sometimes prior to the véraison, *e.g.* varieties with red juice sometimes before, during, or after the vintage. The leaves of white varieties turn yellow in autumn,

The surface of the leaf may be glossy or dull. In a dry climate the leaves shine more than in a moist climate. The leaf surface can be even or uneven. In fertile soil the parenchyma of the leaf develops more strongly than in poor soil; therefore it will be more uneven in the former than in the latter case. The leaf is folded ("tourmentée") if the unevenness affects the whole of the parenchyma between the primary nerves; crinkled ("gaufrée") if the raised parts are numerous, nearly all of equal importance, and measure 1 cm. in diameter;

bullate ("bullée") if the unevenness affects the parenchyma between the last branchings of the nerves. The leaf is undulating ("ondulée") when it is raised between the nerves and parallel to their direction.

Glabrousness and Pubescence: There are no glabrous leaves. All have at least a few hairs, which may soon disappear but leave their marks behind. There are long, twisted, *woolly* hairs and straight or brush-like hairs of variable length and consisting of one or more cells. The brush-like hairs arise on all kinds of nerves, and the woolly hairs also arise at first on the nerves, but, where they are very numerous, they occur particularly on the parenchyma. The leaf is called glabrous (according to Ravaz) if it clearly has no woolly hairs, and *woolly* or *tomentose* when the woolly hairs are clearly visible. According to Ravaz it would seem that the straight hairs are not sufficiently constant for great importance to be attached to them. Both kinds of hairs appear on both surfaces of the leaf, but it is particularly the woolly hairs on the lower surface of the leaf that should be studied. In agreement with Simon Rojas de Clemente y Rubio and the International Ampelographic Commission, Ravaz divides the woolly leaves into two classes:

1. *Downy* ("duveteuses") leaves, if the hairs can easily be rubbed off with the finger and do not cover the whole parenchyma.

2. *Felt-like* ("cotonneuses") leaves, if the hairs resist rubbing with the finger and cover the whole parenchyma.

According to Ravaz a leaf is "pubescente" if it has straight hairs. He therefore makes the following classification of leaves according to their pubescence or hairiness:

A. Leaves glabrous or glabrous with brush-like hairs.

B. Leaves woolly (tomentose) on both surfaces,

C. Leaves woolly on lower surface.

c^1. Leaves downy or downy with brush-like hairs.

c^2. Leaves felt-like or felt-like with brush-like hairs.

c^1a.Leaves nearly glabrous or cobwebby.

c^1b. Leaves downy.

Ampelometry: Here Ravaz shows how one can express the most important ampelographic characters of the leaf by means of numbers. The pubescence he expresses by 0, 1, 2, 3, 4, 5, where 0 stands for glabrous leaves, 1-4 for woolly leaves with an increasing number of hairs per square cm., and 5 for a felt-like pubescence. The intensity of the brush-like pubescence is expressed by the number of nerves on which brush-like hairs occur. For his ampelometric description he makes use only of the *9th-12th leaves on the uppermost shoot of the bearer.*

Example.— Full-grown leaf: 136,48 = 184,45; - (54,112); - 0.89, 0.7;- 0.23; -0.70, 0.90; 0.77*a*; 0-; 0.90; 1,2. This means : $a+\beta=136°$, $\gamma=48°$, $a+\beta+\gamma=184°$, $\delta=45°$; a^1 and β^1 (see Fig. 48) = resp. 54° and 112° (the values of a^1 and β^1 are valuable for the construction of the leaf) ; 0.89 and 0.75 give the relative lengths of the nerves $\frac{l1}{l}$ and $\frac{l2}{l}$; 0.23 expresses the relative length of the base of I² hence $\frac{l2}{l1}$; 0.70 and 0.90 indicate that the upper sinus is clearly marked and the lower sinus nearly nil [here Ravaz evidently gives the ratio of the lengths of the dotted lines which run on Fig. 48 from the point or end of the petiole to the upper (on the illustration it stands below!) and lower sinuses, to the lengths of the corresponding lateral nerves. It is just the reverse of what he previously recommended for expressing the depth of the sinuses]; 0.77*a* indicates that the teeth are acute (*a* stands for acute) and narrow; 0 indicates that the leaf is glabrous; 0.90 is the value of $\frac{L}{l}$, in other words, the leaf is broader than it is long; 1, 2 states that the leaf is large. [This is not clear. Had the 2 stood by itself, there would have been no difficulty. I wish to remind the reader that Ravaz expresses the size of the leaf average length of leaf by the ratio $\frac{average\ length\ of\ leaf}{average\ length\ of\ internode}$.]

Such ampelometric formulae are handy and can render good service. In his own descriptions of the American vines, Ravaz uses numbers only for the values of the angles between the primary nerves (omitting the values for a^1 and β^1) and for the relative lengths of the nerves, as in his scheme that I have quoted. The rest is expressed in ordinary language.

(*d*) **The Tendrils.**—They are simple, bifurcated, or trifurcated. (Only on *V. rotundifolia* are they simple.) Their position on the cane is important. It can be continuous (characteristic of V. *Labrusca*), or *discontinuous* (indicating a Labrusca hybrid, according to Ravaz), or *intermittent* (with all other species).

(*e*) **The Bunch.**—Most American stocks are sterile and must therefore be recognised by their vegetative characters, which, however suffice. Where bunches are formed their characters should be made use of in distinguishing between the varieties. The bunch is a repeatedly branched shoot (see Ravaz, *l.c.* p. 26). He calls berries

Small: if their average diameter is		≤ 10 mm
Medium:	" "	≤ 15 mm
Large:	" "	≤ 20mm
Very Large:	" "	> 20 mm

Finally Ravaz points out that the seed possesses valuable specific characters.

5. FACTORS WHICH INFLUENCE THE USEFULNESS OF A STOCK

(*a*) **Soil and Climate.**—These two factors should be discussed together on account of their mutual influence on the vine. This is best illustrated by the sensitiveness of a vine *V. riparia* towards chlorosis, where it is caused by to much lime in the soil. In fairly dry soils the danger is much less than in moist soils. Thus vines grafted on Riparia thrive at San Michele (Tyrol) in a soil with 75 per cent calcium magnesium carbonate, because this dolomitic limestone is hard and the soil is fairly dry, whilst it often suffers from lime chlorosis in moist soils with 10-12 per cent calcium carbonate

In *poor, sandy* soils Rupestris varieties will answer much better than Riparia varieties, whilst the latter do well in *deep, moist, fertile soils* with little lime (less than 10 per cent calcium carbonate). *Heavy clay* soils are undesirable for a vineyard and it is difficult to find a suitable stock for these soils. It is very difficult to provide them with the necessary amount of air and water, and for the rest their cultivation is costly. *Broken* soils, especially decomposed granite, medium to light loams, and sandy soils, are much more suitable for this purpose, and there is no great difficulty in finding suitable stocks for them. In compact soils one must use stocks with stick roots as they can penetrate such a soil much better than those with thin roots. Therefore we use in this case 1202, 420A, 333, etc rather than the Riparia-Rupestris hybrids.

Humidity is a very important soil factor, which, however depends also very largely upon the climate. If the soil is of such a nature as to be able to absorb a large quantity of water and to retain it for a long time, particularly because the top soil is loose and can easily be kept loose at the surface, we can retain enough water in the soil to enable the vine to grow well *during its whole period of growth*. The important point is that during this period there should never be either too little or too much water in the soil.

As we have previously seen, the most favourable climate for the vine, as regards rainfall, is that which brings an abundant winter rainfall with little rain in spring and early summer, followed by dry weather in late summer and autumn. Where most of the rain falls in winter but the annual rainfall is insufficient (say 11-12 inches per annum), the vineyards have to be irrigated, and here there is the danger that the soil might become too dry between two irrigations. With a stock like Jacquez, which has a relatively low resistance to phylloxera, it can be attacked and damaged by the phylloxera during the intervals of drought when the vine is weakened. For this reason we should be very careful in selecting a stock for soils that are to be irrigated, particularly if they are of a clayey nature, and for shallow soils, and generally for soils which might at times become too dry. For such soils we should in any case select a stock which is highly resistant to phylloxera. *It should be the farmer's main object always to have sufficient moisture in his soil*—never too little and never too much during the vine's period of growth.

Fertility is a very important and desirable character in a soil for a vineyard. Whether a fertile soil will also be highly productive, depends upon various factors, and not the least upon the soil's moisture content and the climate. Later on I shall have more to say on this point. Meanwhile it will no doubt be clear that a fertile soil, under otherwise favourable conditions, will ensure more vigorous vines, greater crops, and a longer life to the vines than a poor soil. According to the fertility of the soil we shall have to choose our stock.

Lime is a very good ingredient in a soil, but where we have to grow vines on American stocks, it can sometimes cause us great trouble. Before we decide on a certain stock, we ought to know how much lime (it is always present in the soil as carbonate, and bicarbonate) the soil contains, or certainly whether it contains more or less than 10 per cent calcium carbonate. In the Cape viticultural area the soils are mostly poor in lime. In the sweet soils of Worcester (parts), Robertson, and Montagu the vineyard soils contain a fair amount of lime, but seldom more than 10 per cent calcium carbonate. In Montagu (whole district) and Robertson (part of district) I took 161 samples of vineyard soils and determined their lime content[1] with Bernard's calcimetre. Subsequently Crawford made more complete analyses of these samples, and these, together with a discussion of the results, were published by Perold and Crawford.[2] It was found that the fertile river (alluvial) soils nearly always contained less than 10 per cent carbonate of lime, and only a few of the samples had more than 10 per cent lime. Hence lime offers no difficulty in reconstituting the Cape vineyards on American stocks.

Where soils contain more than 10 per cent lime we must be careful in the choice of a stock. At the same time we should take note of the hardness or softness, coarseness or fineness of the lime, whether it is mixed with clay or sand, whether the soil is dry or moist, as with each of the first alternatives the chlorosive effect of the lime is much less than in the alternative cases. Where the soil contains more than 10 per cent lime we should use stocks that are only slightly sensitive to lime.

Chancrin[3] classifies the principal American stocks with regard to their *resistance against lime* (*i.e.* the lime passing through a sieve with 1 mm. mesh) as follows:

Stocks for soils with up to 50-60 *per cent lime* $CaCO_3$): the Berlandieris (Berlandieri Resseguier Nos. 1 and 2, Berlandieri Lafont, Berlandieri Mazade), Chasselas × Berlandieri 41B of Millardet.

Stocks for soils with up to 40-50 *per cent lime:* Mourvèdre × Rupestris No. 1202, Aramon × Rupestris Ganzin Nos. 1 and 2, Bourrisquou × Rupestris Nos. 601 and 603 of Couderc, Colombaud × Rupestris No. 3103 of Gamay Couderc, Cabernet × Rupestris No. 33A, A[1], A[2].

---

[1] See Dr. A. I. Perold, *De Vernieuwing van Wingerden in Kalkgronden op geschikte Amerikaanse Stokken,* Union Department of Agriculture, Bulletin No 48 of 1913.

[2] See Dr. A. I. Perold and D. C. Crawford, "Some Preliminary Investigations into the Chemical Composition of certain Vineyard Soils in the Montagu and Robertson Districts", *South African Journal of Science,* June 1915.

[3] E. Chancrin, *Viticulture moderne,* 1908, p. 105.

Stocks for soils with up to 30-40 *per cent lime*: Berlandieri × Riparia 420A and 420B of Millardet and de Grasset, Berlandieri × Riparia No. 157[11] of Couderc, Rupestris × Berlandieri 301A and 219A of Millardet and de Grasset.

4. Stocks for soils with up to 25-30 *per cent lime*: Riparia × Rupestris Nos. 101-14, 3309, and 3306, Rupestris du Lot, Rupestris Ganzin, Rupestris Martin. Stocks for soils with up to 15-25 *per cent lime:* Solonis, Solonis × Riparia No. 1616.

6. Stocks for soils with up to 10-15 *per cent lime:* Riparia gloire de Montpellier, Riparia grand glabre, Riparia à bois violet (with violet canes).

7. Stocks for soils with up to 1-5 *per cent lime:* Vialla.To Group 1 we can undoubtedly add the Tisserand or Cabernet × Berlandieri No. 333 E.M. of Foëx. Further, in connection with this classification, my previous remarks about the chlorosive power of the lime under different conditions must always be borne in mind.

(b) **Resistance against Drought**.—Where a vine grows on its own roots there exists an intimate relation between the roots, stem, and leaves as regards the circulation of water in the vine. Leaves that are woolly and possess a fairly strongly developed cuticle will lose relatively little water by transpiration. The vessels of such vines will then also be correspondingly narrow, and allow relatively little water to flow to the leaves. As soon as we graft the vine we change its leaves. We now get a vine of which the root system was not intended for its foliage system. Our European vines nearly all possess leaves that evaporate much water. I shall revert to this question when discussing the influence of grafting.

The width of the vessels in the roots of a stock will inform us fairly well of its capacity for conducting water to the leaves. On the whole, a vine with thick fleshy roots will stand drought better than one with thin, hard roots. Further, a stock with plunging roots will stand drought well in deep soils if the roots can penetrate into the subsoil. Such a root system shows that the vine is afraid of drought and hence sends its roots deep into the soil. Where the subsoil is almost impenetrable and the top soil is rather shallow, we should use a stock with a strongly developed and branched root system, *i.e.* many roots and rootlets, although they are thin and hard, and although they may not penetrate deep into the soil. The best way of determining a stock's resistance to drought is to make careful experiments with it for at least ten years. In this connection we should also bear in mind the number of square feet of soil which every vine has at its disposal, as well as its development above ground. If we are afraid that the vines may suffer from drought, we should plant them a fair distance apart and keep them relatively small. By tying up the shoots over the vine we diminish the loss of water by transpiration, and hence the danger of drought.

(c) **Resistance against Diseases and Pests other than Phylloxera**.—The aerial part of the American stocks are of value to us only in so far as it concerns their propagation, for as soon as they have been grafted only their root system remains. Most of these stocks are highly resistant against fungoid diseases. Jacquez is somewhat susceptible to anthracnose, but in South Africa it is usually not necessary to protect the American vines in the mother plantations or in the nurseries against fungoid diseases. In some European countries and elsewhere it may be necessary to spray them with Bordeaux mixture against *Plasmopara viticola.* Leaf-eating insects may sometimes be troublesome, when they have to be combated.

The roots of the American stocks are attacked by phylloxera and eel worms to a greater or lesser extent. I shall treat phylloxera separately on account of its outstanding importance. Eelworms are to be feared on 1202, Aramon Nos. 1 and 2, Constantia Metallica, etc., in moist or wet soils, especially if of a sandy nature. The European vines are far less susceptible to this pest than the American vines mentioned above.

In order to determine the resistance of the various American stocks to the diseases and pests, we must grow them in places where these are bad.

(d) **Ease of Propagation.**—Here we are concerned only with the vegetative propagation, so that the two points of interest here are the *rooting powers* and the *knitting of grafts.* It is well known that canes of the pure Berlandieris strike root very badly or with great difficulty when planted out, and this explains why more extensive use is not made of them in viticultural practice. The cuttings of Jacquez also strike root badly, but much better than those of *V. Berlandieri.* Usually varieties with hard canes strike root badly. Their buds open long before the cutting has formed any roots, exhaust the stored food supply, and frequently die again before the cuttings have developed decent roots. The varieties of *V. rupestris, V. riparia*, and many others strike root much more easily. The following results I have obtained on the Stellenbosch University Farm in comparative tests for determining the rooting power of the stocks used.

	1202	101-14	3306	3309	125-1	1616	Rip. gloire.	Rup. du Lot.	Jacquez.	333 E.M.	420A.	Aramon No. 1.	Rupestris Seedling H.	Rupestris Seedling F.
Number of cuttings planted on 13.8.20 .	980	988	273	239	231	181	129	157	130	100	257	148	222	214
No. of rootlings dug up on 29.6.21 . .	826	817	210	200	200	135	100	138	106	75	170	115	150	172
Percentage of cuttings rooted	84·3	82·7	76·9	83·7	86·6	74·6	77·5	87·9	81·5	75·0	66·1	77·7	67·6	80·4

These cuttings on the whole rooted well. The percentage for Jacquez is unusually high. We notice the Riparia × Berlandieri 420A rooted worst of all and Rupestris du Lot best of all. Unfortunately I do not possess any pure Berlandieris, otherwise they would have given a much worse result than any of the others in the experiment. In 1910 I imported pure Berlandieris from Montpellier. Out of 200 cuttings not a single rooted plant was obtained.

The *knitting on grafting* is a very important point. If we wish to make comparative tests with regard to it, we must select properly ripe cuttings from the same mother plantation, which should stand on a uniform piece of land; we must have all cuttings grafted with the same variety, by the same man, and as nearly as possible in the same manner, and we must treat all the grafted vines alike. In this way I made grafting experiments on the Stellenbosch University Farm, and now give the results so obtained (see Table, p. 182).

Jacquez gave the highest and Aramon No. 1 the lowest percentage of first-class (*i.e.* well knitted also) grafted vines. Jacquez generally gave the highest percentage of first-class grafted vines. The average percentage of 45.2 must be considered as quite good. Only vines with good joints were counted. The cuttings were bench grafted by hand, then stratified in the open, and later on planted out in the nursery.

181

GRAFTING EXPERIMENT WITH STEIN AS SCION
(Grafted July 1920, planted out 23rd August, dug up 28th and 29th June 1921.)

Stock	Number of Cuttings grafted	Number of first-class grafted Vines	Percentage of these	Remarks
1202	954	486	50.9	Grew and knitted well
101-14	1054	487	46.2	„ „ „
3306	177	70	39.8	„ „ „
3309	120	36	30.0	Grew poorly, knitted well
125.1	44	28	63.6	Grew fairly well, knitted well
Rip gloire	2434	1024	42.1	Grew best, knitted well
1616	26	10	38.5	• •
Rup du Lot	434	232	53.4	Grew worst, knitted well
Jacquez	809	536	66.2	„ well, „ „
333 E.M.	370	126	34.1	„ poorly, „ „
420A	292	100	34.2	„ „ „ „
Aramon No 2	338	114	33.7	Grew fairly well, knitted rather well
Aramon No 1	352	98	27.8	Grew less well, knitted rather well
TOTAL	7404	3347	Average 45.2	

Grafting Experiment with Cabernet Sauvignon.—The cuttings were grafted in August 1921, stratified in vineyard soil in the open till 22nd September, when they were planted out. This was very late, and the buds of the scions were then already mostly open, whilst the American cuttings had developed many roots.

Stock.	Number of Cuttings grafted.	Number of well-knitted grafted Vines.	Percentage of these.
1202	500	279	55.8
101-14	500	238	47.6
Aramon No. 1	450	81	18.0
Aramon No. 2	200	29	14.5
420A	690	179	25.9
333	682	158	23.2
Jacquez	285	105	36.8
Riparia gloire	265	51	19.2
TOTAL	3572	1120	Average 31.4

The general result in this case was merely fair. The percentages with 1202 and 101-14 were excellent: those on

Jacquez and Riparia gloire were decidedly poor, the more so as Jacquez usually gives a much higher percentage with practically any European variety; those with the Aramons again were the lowest, as usually happens.

(e) **Fertility.**—In order to decide whether a certain stock causes vines grafted on it to bear more than they do on other stocks, we must graft these different stocks with the same variety of grape, prune some vines long and others short, and for the rest keep all the conditions as nearly identical as possible. We shall later on see that some stocks increase the fertility of the vines grafted on them, whilst others decrease it. Some stocks may be suitable for varieties that give medium crops, but unsuitable for heavy croppers. This point will be discussed later on at greater length.

(f) **Vigour.**—Very vigorously growing stocks have a greater chance of success than those of little or medium vigour. In making our choice of a stock, we look for a vigorous one. But too vigorous a stock may cause a poor crop both in quality and quantity. The berries will be far more inclined to set badly and to be soft and even watery than where a less vigorous stock has been selected. Where the soil's water supply is rather limited, a vine grafted on a very vigorous stock will sooner be in danger of suffering from drought than one grafted on a suitable stock of somewhat less vigour.

In order to determine the relative vigour of a stock we plant it side by side with other stocks and compare results after four to five years, and preferably even longer. Some varieties and species, during the first few years, develop above ground much more than under, and *vice versa.* After some four to five years a state of equilibrium is reached, and it is only then that we ought to compare the various stocks with regard to their vigour. Ravaz, *l.c.* p. 43, gives the following relative weights which he found with one-year-old seedlings that had grown side by side.

Weight of Dry Material of One Plant.

	Roots.	Stem.
V. riparia	15.84	18.43
V. Berlandieri	18.52	4.32
V. monticola	10.91	3.82
V. Pagnuccii	12.47	6.30
V. vinifera	7.13	1.56

Here we thus see that *V. riparia* was the only one that, after one year's growth, had developed a greater weight above ground than underground. With *V. Berlandieri* and *V. vinifera* the weight of the root system was about four and a half times that of the stem portion of the plant. If we graft the same variety on the various stocks, we can determine their relative vigour under the local conditions and when grafted with the particular variety. As a rule, vines grafted on *V. Berlandieri* and on its hybrids with a pronounced Berlandieri character, will develop less strongly than when grafted on most of the other stocks. After about five years they are equal to the others, and sometimes even excel many of them.

A vigorously growing stock can, *ceteris paribus,* resist the phylloxera better than one with little vigour.

(g) **Affinity.**—According to Teleki[1] it was Couderc who, at the Viticultural Congress in Macon in 1887, first pointed out that some varieties of the European vines grow well on all American stocks, while other varieties thrive with difficulty on any of these stocks. He pointed out the importance of this in reconstituting the European vines on American stocks. A few years later Couderc published his observation that, in soils rich in lime, many European varieties, grafted on one and the same stock, answer well, whilst others suffer from lime chlorosis; further, that a certain stock can be used on certain soils only when grafted to certain European varieties. In this way Couderc brought up the question of affinity. *Affinity is simply the behaviour of the European vine towards the American stock in the grafted state.* Therefore I shall revert to this matter when discussing the influence of grafting. Meanwhile, however, I wish to discuss certain aspects of the matter.

Teleki, *l.c.* p. 155, says: "Good affinity is shown by an initial moderate fertility which increases up to a certain age limit. Then the size of the crops should remain constant and only the quality of the wine should improve with increasing age."

Concerning the question of affinity, Ravaz[2] says that, where scion and stock are of the same variety, the grafted

---

[1] Andor. Teleki, *Die Rekonstruktion der Weingärten.* p. 153; Vienna and Leipzig, 1907.

[2] L. Ravaz, *Porte-greffes,* etc., pp. 36

vine behaves just as if it had never been grafted. We shall see later on what Prof. Lucien Daniel has to say on this point. Ravaz then quotes the case of *V. rotundifolia,* which gives a first-class joint when a European vine is grafted on it, but the scion cannot continue to thrive on it, and soon dies. He attributes this to the great physiological differences which exist here between scion and stock, but not in the first instance. "Between these two extremes there naturally exists a large number of intermediate stages with regard to the stock as well as the scion; and we can conclude that two plants that are grafted on each other will suffer less according as they correspond more closely or possess more 'affinity' for each other." He then proceeds to argue that the European vines should therefore possess more affinity for the Vinifera-American hybrids than for the pure American stocks, although he admits that in practice no proof of this has been forthcoming. I shall soon prove that, in two instances at least, viticultural practice has proved the opposite. Ravaz further tells us that the external characters of an American vine do not give us any indication of its affinity for European varieties. He finally states that the differences in the affinity of the various American stocks for the European varieties are very small, if they exist, seeing that their effect is so small as not to be noticeable in the vineyards. He further maintains that the influence of the stock on the scion is just like that of the soil on the ungrafted vine; that the question of affinity is one of adaptation to the soil; and, finally, that we ought to attach very little value to the differences of affinity of the various stocks for the European varieties.

From my own experience with Muscat of Alexandria and Inzolia[1] in South Africa I am obliged to disagree with Ravaz. By a *bad affinity* or *lack of affinity* of a European variety for an American stock I mean that the one does not, *in the long run,* grow well or at all on the other; we can be more certain about a bad affinity, if some variety does not thrive on one stock whilst it does well on another stock next to the former, and also if another variety does well on the same stock on which the first variety did not answer in the same soil. Then we can rightly say that the first variety has a bad affinity for the first stock.

---

[1] See under the description of Inzolia in the next chapter.

This is precisely what has happened to our Muscat of Alexandria (Hanepoot) grafted on Aramon × Rupestris Ganzin Nos. 1 and 2, which lasted only three to five years. At the Paarl Viticultural Experiment Station, in a good soil of decomposed granite, the Hanepoot on the two Aramon stocks began to decline in vigour during their third year, when they formed only short shoots which ripened badly. The following year the European vine was dead, whilst the Aramon stock threw out new shoots. The Hanepoot grafted on Jacquez, next to these dead vines, is still flourishing after fourteen years. On the other hand, other varieties grafted on the Aramon stocks near by are still doing well after fourteen years.

On the Stellenbosch University Farm my experience with the Hanepoot grafted on Aramon Nos. 1 and 2, in a sandy loam (alluvial soil), has been still worse. The stocks were planted out in 1919 and grafted on the spot in 1920. During the 1921-22 vegetative period, *i.e. within two years,* the Hanepoot on the Aramon began to decline in vigour and the Hanepoot part of some vines died. In 1922-23 this continued till nearly all were dead or languishing. During the winter of 1922 I grafted over some of the languishing vines with Greengrape *in the Aramon part* of the vine *(i.e.* below the original joint), and in 1923 I grafted Hanepoot on the Greengrape, so that I now have Hanepoot on the Aramons intergrafted with Greengrape, which does well on the Aramons. It remains to be seen in how far this device of intergrafting will overcome the difficulty experienced with Hanepoot on Aramon.

This experiment has been made in my ampelographic collection of 200 European varieties on the Stellenbosch University Farm. Each variety is grafted on Jacquez, Rip. gloire, Aramon No. 1, Aramon No. 2, 1202, 101-14, 106-8, Rup. du Lot, 420A, 333 E.M. There are 20 vines of every variety and 2 on every stock. I have side by side one row of each of the following varieties: Muscat of Alexandria (imported from Montpellier in 1910 and identical with our White Hanepoot), White Hanepoot, Le Roux Hanepoot (a bud variation of our White Hanepoot), Canon Hall Muscat (a seedling of Muscat of Alexandria), and Gordo bianco. In the summer of 1923 the 20 (5×2×2) vines in the five rows were nearly all dead or nearly so *on the two Aramons,* while the 80 vines in the five rows on the other eight stocks were doing well. All these vines have been treated alike, and those on Aramon have not borne more grapes than the rest. *The failure can decidedly not be attributed to over-production.*

During the summer of 1911-12 Perold and Tribolet[1] investigated an interesting case of White Hanepoot grafted on a Rupestris seedling (Apricot Leaf Rupestris) in a decomposed granite soil along Paarl Mountain at the north end of Paarl (town). The vineyard was then nineteen years old. On the fertile hillocks (rich in humus) the vines were then still big, strong, and healthy. On the surrounding, less fertile land the vines were not so vigorous and some were languishing. From time to time, from about eight years prior to the investigation, some of such languishing vines had been grafted over with Greengrape in the Rupestris part, and they had again given big and vigorous vines on the same roots on which the Hanepoot had previously been languishing. From this we conclude that the affinity of Hanepoot for Apricot Leaf Rupestris is good enough in fertile soil, while the affinity of Greengrape for the same stock was sufficient to produce a good result in the same soil where Hanepoot started to decline on it after about ten years. Hence Greengrape has a better affinity for this stock than Hanepoot has.

We further learn from this that affinity is not something absolute, but is influenced by soil conditions. Experience, however, proves that it is wrong to attach the greatest value to the latter, as Ravaz does. Affinity, therefore, rests on certain subtle properties of the two vines that have to live together in the grafted state; properties which cannot yet be detected beforehand, but which nevertheless certainly exist (as is apparent, *inter alia,* from the cases quoted above), and which are influenced by soil conditions to a minor extent.

Ravaz's assertion that differences of affinity are not noticeable in a vineyard is refuted at the same time, and we have had two examples of Vinifera-American hybrids Aramon

---

[1] Dr. A. I. Perold and I. Tribolet, "American Stocks for Cape Vineyards. Report of an Inquiry into the Suitability of the American Stocks on which the vineyards in Cape Province have thus far been reconstituted". *Agric. Journal for the Union of South Africa,* July and Aug. 1912.

× Rup. Ganzin Nos. 1 and 2) possessing a worse affinity for Hanepoot than the pure American stocks. *Affinity is most certainly a practical reality, and one which we must face in the vineyard itself.*

(h) **Resistance against Phylloxera.**—As the European vines are grafted on American stocks to prevent them from being destroyed by phylloxera, the resistance of such a stock against phylloxera is one of its essentials. *Why does one variety of vine resist phylloxera better than another?* No conclusive reply has yet been given to this question, although various theories have been advanced. In order to be in a position to understand them, we must consider briefly what the phylloxera does to the roots.

In 1878 Max Cornu[1] published a detailed study of the phylloxera and its action on the vine. The insect causes *nodosities* and *tuberosities* on the roots (see Fig. 49).[2] The nodosities are formed on those parts of the young roots that are still growing in length. They have been carefully studied by Cornu, and he states that the nodosity at first is yellow, then golden yellow, and afterwards brown. Where the insect sits there is a depression and the rootlet swells a great deal, bulging out on the side opposite the insect and curving somewhat around it. The irritation arising from the puncturing of the tissues and the sucking by the phylloxera causes a hypertrophy or stronger growth of the cells. Towards the end of summer these parts of the rootlet die. Cornu, *l.c.* pp. 180-181, attributes the depression of the young rootlets beneath the insect to the sucking of root-sap by the insect, not to the puncture made by it or to an irritating liquid excreted by the insect into the wound. The vine dies because the young roots die and are then no longer able to absorb the vine's mineral food contained in the soil solution.

Prof. Millardet[3] states that Cissus, Ampelopsis, and *Vitis rotundifolia* are immune against phylloxera, whilst all the other species and varieties of the genus *Vitis* are susceptible to the formation of nodosities and tuberosities on

---

[1] Max Cornu, "Etudes sur le Phylloxera vastatrix", *Mémoires présentées par divers savants à l'Academie des Sciences de l'Institut National de France,* tome xxvi. No 1 pp. 1-357, mdccclxxviii. Imprimerie nationale, Paris.

[2] 2012 Editor's Note: Fig 49 is shown on page xiii. (PFM)

[3] Prof. A. Millardet in *Comptes rendus de l'Academie des Sciences,* 1878; and *Pourridié et Phylloxera* (1882); and "Altérations phylloxériques" in *Revue de Viticulture,* 1808, vol. ii. pp. 692-698, 717-722, 753-758.

their roots by phylloxera. The latter arise on thin and thick roots that have ceased to grow in length.

He believes that the formation of nodosities can weaken a vine without killing it, while the tuberosities are much more dangerous and can easily bring about the death of the vine. Millardet made a careful study of the tuberosities. According to him they are hemispherical, up to 3 mm. high on Vinifera roots, and not higher than 1 mm. on very resistant roots. With *V. vinifera* they occur on roots of all ages; with *V. riparia, V. rupestris, V. cordifolia,* and other resistant vines they occur only on the younger roots of one or at most two years old, and then rarely.

If the phylloxera punctures a periderm (layer of cork), an increased growth of the cambium cells gives rise to the formation of a tuberosity. On young roots passing into their secondary structure, the tuberosities sometimes do not penetrate the suberised endodermis, which then protects the parts of the root inside it. They are not very dangerous. As long as the outermost periderm is still the primary periderm, the tuberosities that penetrate the periderm are far more dangerous than those previously mentioned, *i.e.* the more superficial tuberosities. With *V. vinifera* it is only during the third and even fourth year that inside the primary periderm a secondary periderm is formed which can cut off the tuberosities. Millardet maintains that the fissures formed in the outer cell layers of the nodosities and tuberosities, in consequence of the high pressure caused by the hypertrophy, are, particularly, the places where the organisms causing putrefaction enter the root. These latter pass from the tuberosity through the medullary rays into the deeper parts of the root, and can thus cause the whole root to rot at the seat of the tuberosity, which naturally means the death of its whole further extension. Young rootlets die in one year, and roots $1/5$ in. thick after more than three years. This explains why the tuberosities are so much more dangerous than the nodosities.

As soon as rotting starts in a hitherto sound tuberosity, we notice a *layer of cork* being formed underneath the rotting part. This protective layer of cork is sometimes large enough and sometimes too small to cut off the rotting part and thus to protect the root. It is only the irritation set up by the rotting of the sound tissues of the tuberosity and of the bast which causes the formation of this layer of cork—

a sound tuberosity never possesses any such layer. If the rotting gets past the first cork-plate, more than two cork-plates (one inside the other) are very seldom formed with *V. vinifera.* With Jacquez, Herbemont, etc., three and very occasionally four such plates are formed. With *V. riparia, V. rupestris,* and *V. cinerea* the secondary periderm is formed in the roots of the year, so that the sub-periderms can cut off the tuberosities sometimes a few weeks after they have appeared. Millardet does not consider this, namely the formation of the secondary periderm, as of primary importance to the stock's resistance against phylloxera, since *V. aestivalis, V. cordifolia,* and *V. Berlandieri* are highly resistant against phylloxera, although they, like *V. vinifera,* form a secondary periderm only in the third to fourth year. Some punctures of the phylloxera are so irritating as to prevent the formation of secondary periderm under the tuberosity. On the older European roots (about four years old) the cork-plates formed are thicker and more complete and stop the rotting more often than on the younger roots. To sum up, the tuberosities are most dangerous to European vines if they are formed beneath or inside the primary periderm on one- to two- (perhaps even three-) year-old roots. With American vines *(V. riparia, V. rupestris, V. cordifolia, V. aestivalis, V. cinerea, V. Berlandieri)* the normal tuberosities are formed almost exclusively on roots of only one year old, so that those occurring on older roots date back from the first year. Here they are also lower and smaller than on European roots. On the roots of Jacquez, Herbemont, etc., we do find numerous tuberosities, but the cork-plates here are more numerous and denser than on *V. Vinifera.* Hence they are still fairly resistant to phylloxera, particularly where they grow well.

From the preceding abstract from Prof. Millardet's work, we see that he attaches great value to the formation of the cork-plates for isolating the rotting in and under the tuberosity from the sound parts of the root, and thus he regards this as one of the principal indications of a high resistance against phylloxera.

Prof. Ravaz[1] has conducted extensive experiments on the

---

[1] Prof. L. Ravaz, "Étude sur la résistance phylloxérique", *Revue de Viticulture* vii. (1897), pp. 100-114, 137-142, 193-190, and viii. (1807), pp. 688-604.

resistance of vines to phylloxera on his own method, which I shall describe later on, and has come to the conclusion that Riparia gloire, the Rupestris varieties (Rupestris Martin, etc.), and most Americo-American hybrids are more resistant against phylloxera than the Vinifera-American hybrids [*e.g.* Axamon × Rup. Ganzin Nos. 1 and 2, 1202, 33A (Cabernet × Rup.), etc.]. The former also develop tuberosities, but it is particularly on the youngest roots and not on the roots of two years old and older that they are formed. Where they here occur on two-year-old roots, they are superficial. With the Vinifera-American hybrids they occur on the thick roots just as frequently as, if not more often than, they do on the thin ones. This they have inherited from their Vinifera parent, which is attacked simultaneously on all its roots. Again, the tuberosities were much deeper on the roots of the Vinifera-American hybrids examined than on those of the pure Americans or of their intermediary hybrids that were examined.

Ravaz says that the resistance of a vine to phylloxera is unknown to us until we see that it is not resistant. He further maintains that the attractiveness of a vine's young roots to phylloxera has nothing to do with its resistance to phylloxera, and that the wounds on them (*i.e.* the nodosities) are of no or almost no importance—a view which by the way Millardet also holds. What we should know, according to Ravaz, is only the attractiveness to phylloxera of those roots on which the tuberosities are formed, as these alone are really dangerous.

Dr. L. Petri[1] reminds us that, according to Max Cornu's view, the rotting of the nodosities and tuberosities depends upon physiological processes which are independent of microbes in the soil, whereas, according to Millardet, these hypertrophies will not rot in a sterile soil but will continue to exist just as long as the normal tissues. Petri has grown seedlings from sterilised seeds in sterilised soil in sterile glass tubes and then infected these with sterile phylloxera. The result was that the nodosities and tuberosities were formed without any rotting taking place, but did not remain alive as long as the sound tissues.

---

[1] Dr. L. Petri, "Nodositätenbildung auf den Rebenwurzeln durch die Reblaus in sterilisiertem Mittel", *Centralblatt für Baku* Abt. II. vol. 24 (1909), pp. 146-154.

In 1911 Petri[1] discussed the influence of the acidity of the root-sap on the vine's resistance against phylloxera. He agreed with Comes[2] that the cultivated plants contain less acidity in their cell-sap than the corresponding wild plants. Comes maintains that American vines under cultivation will gradually lose their resistance against phylloxera. He further believes that a certain amount of variation takes place in the phylloxera resistance, particularly of Americo-Vinifera hybrids, but this is very limited. Comes further maintains that the more acid the root-sap contains, the more rapidly the formation of the protective cork-plate, after the root has been wounded by phylloxera, takes place. Dr. Averna Saccà[3] has come to the conclusion "that the degree of resistance of the American vines against phylloxera is closely related to the acidity of their cell-sap".

Petri attributes the degree of resistance against phylloxera to the following factors:

Degree of attractiveness of the roots to the insect (taste of root-saps).

Degree of irritability of the root-tissues on being punctured by the phylloxera (specific properties of irritability and reaction of the tissues).

3. Degree of resistance of these tissues against rotting (more or less early loss of primary periderm, formation of cork-plates, formation of positively or negatively chemotropic substances in the tissues directly or indirectly altered by the phylloxera).

Ravaz's Method for determining a Vine's Resistance against Phylloxera.[4]—On the hypothesis that only the tuberosities are dangerous, Ravaz arranged his experiment in such a way that the phylloxera could live only on the one- and two-year-old roots, and could therefore form only tuberosities. He takes a pot into which he puts a fairly thick layer of fine sea-sand. On this and in the centre only he throws a handful of ordinary soil in which the phylloxera can thrive.

---

[1] Dr. L. Petri, "L'acidité des sucs et la résistance phylloxérique *Revue de Viticulture,* vol. xxxv. pp. 487, 505, 544.

[2] Prof. Comes, Del fagiuolo comune (Phaseolus vulgaris), Napoli, 1909; and Giornale di Viticoltura ed Enologia, Avellino, xvii.

[3] Dr. Averna Saccà, "L'acidità dei sucohi nelle viti americane in rapporto alla loro resistenza alia filossera", *Giorn. di Vit. ed Enol., Avellino,* xvii. (1909), p. 350.

[4] See Ravaz, *Porle-greffea,* etc p. 42.

On this soil he now places the vine that is to be tested together with one of known resistance against phylloxera, putting their roots side by side on the soil and sand. He evidently takes two-year-old vines. For two-thirds of their length, reckoned from the stem, the roots are now covered with a large handful of the same soil. The pot is now filled with pure sea-sand. The phylloxera is placed in the soil when the vines are planted or when they are already growing. The phylloxera is now obliged to live on the older roots, and is kept away by the sea-sand from the young rootlets developing at the ends of the roots. Also, the phylloxera will now live on the vine which is not so resistant against it. The number of tuberosities and their nature (deep or superficial, etc.) on the two root systems give us an idea of their relative resistance against phylloxera.

However interesting a discussion of the natural inherent resistance of a vine against phylloxera may be, it is in itself not of much practical value to us unless we bear in mind the varying conditions under which the vine will have to live in actual practice. They undoubtedly have a strong influence on the vine's resistance against phylloxera, as they do with other parasites. Professor Lucien Daniel[1] expresses this more generally when he says: "Physiology and pathology teach us that every cause weakening a living being *ipso facto* diminishes its resistance against and increases its attractiveness to parasites. This is a fundamental truth which no one now will doubt." Further discussion of this point may be found in Chapter VII.

In 1892 Viala and Ravaz expressed the resistance of a large number of vines against phylloxera in numbers from 0 to 20. This list is quoted by Foëx, *l.c.* pp. 758-759, and appears also in the English translation[2] of Viala and Ravaz's work, *Les Vignes americaines,* Montpellier, 1892, as in the original French edition. I do not consider this list of sufficient value to reproduce it here. Ravaz himself does not quote it in his *Porte-greffes et producteurs directs.*

---

[1] Prof. Lucien Daniel, *La question phylloxérique, le greffage et la crise viticole,* p. 189; Bordeaux, 1908-11 (3rd fascicule appeared only in 1919).

[2] P. Viala and L. Ravaz, *American Vines: Adaptation, etc.,* translated by Dubois and Wilkinson (Melbourne, 1901) from the French edition of 1892.

CHAPTER V

SPECIAL AMPELOGRAPHY

A. AMERICAN VINES

UNDER American vines I here include species and varieties of the genus *Vitis* which are indigenous in North America, and their natural arid artificial hybrids amongst each other and with *V. vinifera,* in so far as the last-named hybrids can serve as suitable stocks. Those that are used as direct producers or self-bearers are discussed in Section C of this chapter. For practical purposes I shall divide the American vines into two classes, namely those that serve as stocks and those that are grown for their grapes.

I. AMERICAN STOCKS

Here I shall discuss only the more important stocks. Where a species has not itself given varieties that are used as stocks, but occurs in useful hybrids, it will also be briefly discussed. First the pure species and their varieties will be treated, then the hybrids, which are again subdivided into hybrids exclusively with American blood (Americo-American hybrids) and those with American and European blood (Americo-Vinifera hybrids).

(a) Pure American Species and Varieties

1. **Vitis rotundifolia,** *Michaux.*
   Synonyms.—V. Taurina, V. Vulpina, V. Muscadina; Muscadine, Bullet, Fox grape, etc.
   Geographical Distribution.—S. Delaware, west through Tennessee, S. Illinois, S.E. Missouri, Arkansas (except the N.W. parts),

to Grayson County, Texas, as a northern and western boundary, to the Atlantic Ocean and the Gulf of Mexico on the east and south, *i.e.* largely the S.E. States.

Culture.[1]—Varieties of this species grow best on light sandy or alluvial soils. As their canes possess no diaphragms, they can only be pruned towards the end of summer or in early winter, as they will otherwise bleed too much. They are mostly trained on trees or on pergolas (arbours) and not pruned, except for removing dead wood. Their roots are not attacked by phylloxera; their grapes and leaves are immune against *Plasmopara viticola* or downy mildew, black rot, oidium or powdery mildew; and generally they are surprisingly resistant to the attacks of all insect pests and fungoid diseases. Their cuttings grow only with difficulty; layers succeed better. They possess practically no affinity for European varieties. Where the scion grows and a good joint results, the European portion dies after a while. Here the physiological difference between stock and scion is too great. As a stock, *V. rotundifolia* is consequently worthless. Munson, according to Hedrick,[2] has bred a few Rotundifolia hybrids.

The Rotundifolia varieties require a warm climate for successful growth, and are very sensitive to lime in the soil. Their grapes ripen very late and unevenly. The ripe berries drop from the bunch, hence "the common method of gathering the fruit of this species is to shake the vines at intervals so that the ripe berries drop on sheets spread below the vines".[3] They give a poor wine with a musky aroma (foxy taste).

Description (according to Hedrick, *The Grapes of New York,* p. 109).—

Vine: usually very vigorous, climbing high, sends down aerial roots in the shade.

Shoots and Canes: wood hard, bark smooth, not scaling off except in old age, with prominent warty lenticels; shoots short-jointed, angled; diaphragms absent; tendrils intermittent, simple.

Leaves:[4] below medium in size, broadly cordate or roundish; petiolar sinus rather wide, usually shallow; margin with obtuse, wide

---

[1] Under "Culture" I shall discuss the cultivation of the vine, its most important properties, its use, etc.

[2] U. P. Hedrick, *The Grapes of New York,* 1908, p. 112.

[3] U. P. Hedrick, *Manual of American Grape-growing,* 1919; Macmillan Co., New York

[4] The leaves described in this chapter are adult leaves.

teeth; not lobed; dense in texture, rather light green colour, glabrous above, glabrous or sometimes pubescent along veins below.

Fruit: cluster small (6-24 berries), loose; peduncle short; pedicels short, rather thick. Berries large, globular, or somewhat oblate, black or greenish-yellow; skin usually thick, tough, and with a musky odour; pulp rather tough; ripening unevenly and dropping as soon as ripe. Seeds two to four, very large to medium, shaped something like a coffee-berry, somewhat flattened, shallowly and broadly notched; beak very short; chalaza rather narrow; slightly depressed with radiating ridges and furrows; raphe a narrow groove. Leafing, flowering, and ripening fruit very late.

Varieties

Flowers (syn. Black Muscadine) is a very late, black variety.

Scuppernong is a widely spread, well-known white variety. It grows and bears well. According to Hedrick it is pre-eminently the grape of the South, and is the chief representative of the species *V. rotundifolia*. In the past it was mainly valued for wine. Hedrick states that several very good wines are made from it. Its fruit and wine have a musky or foxy flavour. It is generally grown on arbours, is not pruned, frequently not cultivated, and occurs nearly everywhere in the South Atlantic States. The growth of the vine is prodigious. Thus there are accounts of vines of this variety over a hundred years old, which cover over an acre of land and which bear 500 bushels of fruit and make 2000 gallons of wine (see Hedrick, p. 401).

2. **Vitis Munsoniana**, *Simpson.*

Synonyms.—V. Floridana, V. peltata, Mustang grape, Bird or Everbearing grape, Florida grape.

Geographical Distribution.—Central and Southern Florida and the Florida Keys. It is said to be the only grape growing on these Keys.

Culture.—It appears to be a variation of Rotundifolia, fitted to subtropical conditions. It can easily be propagated by means of cuttings, is immune against phylloxera, and of no practical value.

Description (according to Hedrick).—

Vine: slender grower, usually running on the ground or over low bushes.

Canes: angular, internodes short; tendrils intermittent, simple.

Leaves: smaller and thinner than Rotundifolia and rather more circular in outline; not lobed; teeth open and spreading; petiolar sinus V-shaped; both surfaces smooth, rather light green.

Fruit: cluster of same size as, but with more berries than, rotundifolia. Berry one-third to one-half the diameter, with thinner and more tender skin; black, shining; pulp less solid, more acid, and without muskiness. Seeds about one-half the size of those of Rotundifolia, similar in other respects. Leafing, flowering, and ripening fruit very late.

N.B.—The reader is reminded that *V. rotundifolia* and *V. Munsoniana* belong to the Muscadinia, while all the other species now following belong to the Euvites.

3. **Vitis riparia**, *Michaux.*

Synonyms.—V. Vulpina L., V. odoratissima, V. cordifolia, Riverside grape, etc.

Geographical Distribution.—"It is the most widely distributed of any American species of grape. It has been discovered in parts of Canada north of Quebec and from thence southward to the Gulf of Mexico. It is found from the Atlantic coast westwards and as far as, say, the Rocky Mountains, according to most botanists. Usually it grows on river banks, on islands, or in upland ravines" (Hedrick, *l.c.* ii. p. 315).

Culture.—Riparia answers best in a temperate climate. Hence the climate of most European wine countries as well as that of the Cape suits it quite well. According to Viala-Vermorel, *l.c.* ii. p. 419, most varieties of Riparia which were imported in large numbers into France came from the fine alluvial, grey to dark-coloured, very fertile soils along the banks of the Mississippi. They have a depth of up to 7-8 metres or 23-26 feet. Riparia thus prefers the *deep, moist, fertile* river soils or alluvial soils, otherwise it must be irrigated and heavily fertilised. In poor, sandy soils it is worthless. Heavy clay soils and shallow soils do not suit it either. In broken, deep, cool, fairly *fertile* soils it is a very good stock, provided that the soil does not contain over 10 per cent lime; more lime, especially soft, finely divided lime, and in moist soils may make it suffer from lime chlorosis.

I wish to quote two instances of vines grafted on Riparia gloire in soils that one would not have considered suitable for this stock, but are answering very well.   The first is a block of Cabernet Sauvignon at Elsenburg in the Stellenbosch district, planted 5' × 10' rectangular, over twenty years ago, in a broken (sandy-clayey) soil with a lumpy clay as subsoil. This subsoil was formed out of the Malmesbury shale into which it gradually passes.   The total depth of soil is only 24-30 in. at most, and the vineyard is not irrigated, but the average annual rainfall (mainly in winter) is about 26 inches.   The vines are big and vigorous, and produce heavy crops.

The second is a block of Muscat of Alexandria (White Hanepoot) grafted on Rip. gloire in a fairly stiff, red Karroo loam, 5-10 in. deep, resting on a hard calcareous layer (hardened clay and lime) near the town of Robertson.   In 1921, when I saw these vines, they were twelve years old, very vigorous, and laden with fruit.   The vines are all irrigated in this part, and this particular block had (about a month before the vintage) a darker green colour than the adjoining vines grafted on Aramon and Jacquez, although those on Rip. gloire had then been a longer time without irrigation. In suitable soil, European vines grafted on Riparia usually withstand drought well, and better than the ungrafted Riparia, as its leaves are not adapted to dry surroundings.

Sometimes the complaint is raised that Riparia gives grafted vines with thin stems that cannot stand upright without support.   This happens only where Riparia is planted in soil not quite suitable.   Where it can grow vigorously, this difficulty is not experienced, and in any case it usually ceases after the first three years of growth.   Where necessary, the young vines should be supported by stakes. The reason for the above is the fact that the Riparia stem remains thinner than that of the European vine grafted on it, particularly if the stock does not grow vigorously.   In deep, cool, fertile decomposed granite in Bovlei (above Wellington, Cape), big, vigorous vines of Greengrape and Hanepoot grafted on Rip.gloire and over twenty-five years old have been and still are regularly giving heavy crops and ripening their grapes They certainly have strong stems and have never been supported in any way.

Riparia cuttings grow easily and develop numerous, roots at all the nodes. With bench grafting they usually give a high percentage of successful grafts. Riparia has a good affinity for most European varieties, including Muscat of Alexandria. It is highly resistant to phylloxera, and one usually finds only very few phylloxera on its roots. According to Ravaz, *l.c.* p. 82, nodosities are sometimes fairly numerous on its rootlets, while tuberosities on the thicker roots generally occur very rarely and then are only superficial. At the Cape its roots are generally free from phylloxera and perfectly sound. In any case it is always sufficiently resistant to phylloxera.

Ungrafted Riparia is highly resistant to most cryptogamic diseases. It is hardly ever attacked by *Plasmopara viticola,* black rot, and anthracnose. Oidium affects it slightly more, especially towards the end of the growing period.

Amongst the Riparias we find varieties with male and others with hermaphrodite flowers. The clusters (bunches) are small, and there are usually three to four on one shoot. Even water-sprouts from the old wood bear fruit. The fruit *ripens very early.* In the collection of the School of Agriculture and Viticulture at Montpellier there is a variety of Riparia the grapes of which ripen (according to Ravaz) one month before the earliest European variety. This property of great fertility and early ripening the Riparia stock largely communicates to the European vine grafted on it. Thus we know that vines grafted on Riparia regularly give good crops, ripen their grapes early, and develop a high sugar content in them. Muscat of Alexandria grafted on Riparia ripens its grapes fully a week earlier than when grafted on Jacquez or on 1202.

For reconstituting the French vineyards on American stocks, Riparia was at first used very extensively. In 1902 Ravaz, *l.c.* p. 83, still stated that *three-fourths* of the reconstituted vineyards in France were grafted on Riparia, Teleki *l.c.* p. 18, states (1907) that about 80 per cent of the Austro-Hungarian vineyards have been reconstituted on Riparia, and that about one-half of these should never have been grafted on Riparia. This latter statement applies particularly to cases of lime chlorosis in the highly calcareous soils, and of poor growth in otherwise unsuitable soils (poor soils and heavy clay soils that might become too dry). I would remind the reader that Riparia possesses a freely-branched root system with numerous fine roots which spread in the soil and do not penetrate it deeply, especially in very heavy soils.

In unsuitable soils Riparia was a failure. This prejudiced the farmer with a superficial judgement against Riparia, even where it would have answered well. In South Africa too little use has thus far been made of Riparia. Unfortunately we find in the Cape vineyards, next to the good Riparia varieties, very poor ones (the so-called Old Riparias) which frequently brought Riparia into disrepute. I wish to warn my readers against using Riparia in poor soils and in very compact clay soils. In deep, cool, rather fertile soils it answers well.

Riparia ripens its wood early and sheds its leaves early in autumn and before most other American vines. It also starts budding early.

Description.—
Bourgeonnement uniform pale green colour.

Shoots cylindrical or oval, usually smooth, thin, long, reddish violet in early summer, bark chestnut colour in autumn; diaphragms thin; pith fairly thick ; tendrils intermittent, usually bifid.

Leaves with large colourless stipules (about 1 cm. long); leaf-blade medium to large, thin, entire or 3-5 lobed, cuneiform, usually longer than broad; sinuses shallow, angular; petiolar sinus broad, usually shallow; angles made by primary nerves fairly small; teeth narrow and sharply serrate, those at the ends of the middle and lateral primary nerves being particularly sharp and long; glabrous above, usually glabrous but sometimes pubescent on primary and other nerves below.

Fruit, cluster medium to small, usually compact and shouldered, short peduncle; berries small to medium, black with dense blue bloom, round, ripening early (after Kavaz); according to Hedrick their taste and time of ripening vary a good deal; seeds 2-4, notched, short, plump, with very short beak.

Vine vigorous to medium vigorous grower, with a thin stem; roots very thin and tough, yellow, numerous; shoots rest on the ground unless they have a chance of climbing.

Riparia Varieties

Professor Millardet was the first to draw the attention of the French wine-growers to the probable value of Riparia as a stock (1874), although it had been grown in France in botanical gardens since the eighteenth century. In 1875 and 1876

Bush and Meissner sent to France a large number of Riparia cuttings which they had collected along the banks of the Missouri and Mississippi. These gave rise to the numerous plantations of Riparia in Europe (see Viala-Vermorel, *l.c.* i. p. 414).

Ravaz, *l.c* p. 79, divides the Riparia varieties into four groups, according as their young shoots are glabrous, woolly, entirely covered with brush-like hairs (on the whole shoot), or partially covered with brush-like hairs (only on the nodes). Group 1 includes Rip. gloire de Montpellier, Rip. Grand glabre, Rip. Baron Perier, Rip. Meissner Nos. 2, 6, 7, 8, 9, 10, 13, and generally the best varieties. Group 2 contains Rip. des bords sableux. Group 3 contains Rip. Meissner Nos. 5 and 12. Group 4 contains Rip. Meissner No. 4, Rip. belle Souche (Despetis), Rip. pubescent bleu (Despetis), Rip. pubescent blanc, Rip. a bourgeons bronzes (Fitz-James), etc.

Riparia gloire de Montpellier (see Fig. 46), also known as Rip. gloire, Rip. Portalis, Rip. Michel, etc., is the best among the Riparias. Its leaves are large, longer than they are broad, 3-lobed with shallow sinuses, three long sharp teeth at the ends of the middle and first lateral ribs (primary nerves), teeth uneven and sharp-pointed, $a + \beta = 98°$, $\gamma = 46°$, $a + \beta + \gamma = 144°$. Leaf bullate, light green, glabrous except the straight brush-like hairs on the nerves below. Petiolar sinus deep like an open U; leaf-stalk long and strong. Shoots glabrous and reddish violet in early summer; long, medium, thick to thin, long internodes, creeping. Plant bears only male flowers. Louis Viala first noticed it on the estate of Portalis belonging to Mr. Michel, and he selected, described, and drew public attention to it.

It is a vigorous grower in suitable soil and a splendid stock for producing wines of high quality. It can be strongly recommended for most European varieties, including Muscat of Alexandria, provided the soil is suitable. It is specially valuable for varieties grown for quality, *e.g.* Cabernet Sauvignon, which are inclined to give only light or moderate crops.

The *Old Riparias* of the Cape are not worth much. They produce smallish grafted vines. Their leaves are small and serrate, their shoots very thin with short internodes.
The *Blue Riparia* of the Cape is a strong grower and about as good as Rip. gloire. Its leaves are larger, longer than

broad, and its young shoots are almost covered with straight brush-like hairs (very striking feature).

4. **Vitis rupestris,** *Scheele.*

Synonyms.—Sugar grape, Sand grape, Rock grape, Bush grape, Mountain grape.

Geographical Distribution.—"This species is an inhabitant of south-western Texas, extending eastward and northward into New Mexico, southern Missouri, Indiana and Tennessee to southern Pennsylvania and the district of Columbia. Its favourite places are gravelly banks and bars of mountain streams or the rocky beds of dry watercourses" (Hedrick, *l.c.* ii. p. 313).

Culture.—*V. rupestris* drew the attention of the grape-growers long after *V. riparia.* Planchon[1] in 1875 merely wrote about *V. rupestris*: "Unknown in viticultural practice, this species will probably merit being planted in the gardens as a curiosity". This passage occurred in a description of his tour through North America. He therefore did not give much encouragement. Numerous successful experiments with Rupestris varieties in France, however, made them gain increasing popularity, so that more and more they began to supplant Riparia as a stock.

V. rupestris likes a warm climate. Its name, from Latin, *rupes,* a stone, reminds us of its common occurrence in stony soils in its natural habitat. From this it must not be inferred that this species is particularly resistant to drought, for actually the Rupestris varieties are usually fairly sensitive to drought. Under similar conditions *V. rupestris* will suffer from drought sooner than *V. riparia.* Hence it should not be planted in dry soils unless the subsoil is easily penetrable. Where the subsoil is firm and impenetrable to its roots, it should not be planted. It gives the best results in deep, *moist* soils, even *though they be poor.* Hence it is preferable to Riparia in moist, poor sandy soils, and also in other poor soils where it will not suffer from drought. In stony, fairly dry soils which are deep, it still does fairly well.

It is very sensitive to lime. Its purest varieties are often more sensitive than Riparia. Those Rupestris varieties that

---

[1] Planchon, J.-E., Led Vignea américaines, Montpellier, 1875.

202

are less sensitive than Riparia, *e.g. Rup. du Lot, Rup. Metallica*, possess blood of other species (see Ravaz, *l.c.* p. 108). The vines show a shrubby growth which is due to the strongly developed branching of the shoots. Its leaves sometimes resemble the leaves of the apricot tree more than those of the vine (hence the name Apricot Leaf Rupestris used at the Cape for a certain Rupestris seedling). Its shoots usually are short, thick, and numerous. Every eye generally develops several shoots. It has a more upright habit of growth than the creeping Riparia.

Its cuttings readily strike root. The roots arise on the stem at various depths and form numerous rootlets. Where European varieties are grafted on Rupestris, the stem of the stock grows nearly as fast in thickness as that of the scion. This species is *very highly resistant to phylloxera.* The rootlets show numerous nodosities, but the roots rarely show any tuberosities. According to Ravaz it is more resistant than Riparia. Its varieties show marked differences amongst themselves, and some resemble other species somewhat. This is the result of its blooming after Riparia, and simultaneously with several other species. The different Rupestris varieties also differ appreciably in their resistance to phylloxera and drought, and as regards the soils to which they are well adapted. The varieties that are related to other species are less highly resistant to phylloxera than the pure Rupestris varieties, but they too are usually good stocks. *V. rupestris* is practically immune to downy mildew, black rot, and powdery mildew or oidium; it is, however, attacked to some extent by anthracnose. It soon perishes in soils that are too wet or too dry. When budding takes place, one sees practically only a mass of young inflorescences or clusters. Where it thrives, it forms very vigorous grafted vines which sometimes set their fruit very badly ("coulure"), though this can usually be overcome by long pruning.

Description.—
Bourgeonnement usually yellowish green, slightly cobwebby.

Shoots glabrous, red when unripe, brown and striped when ripe, thick, and sometimes thinner, much branched, few tendrils; internodes short, pith and diaphragms thin.

Leaves small, reniform, much broader than long, entire or 3-lobed (rarely), glabrous on both surfaces or with just a few brush-like hairs on the lower surface only, thick, soft, bluish green: the first lateral nerve 1′ is just as long as and often longer than the middle nerve 1, angles between nerves small (70-105°) ; young leaves copper green, glabrous or cobwebby, glossy; leaf-stalk shorter than leaf; petiolar sinus usually wide and shallow.

Fruit small, loose clusters with small, round, black berries which are characterised by much pigment under the skin and hence coloured sap when fully ripe, sprightly taste free from foxiness. Seeds small, plump, short beak.

Roots thin or somewhat fleshy, reddish, numerous, plunging.

The varieties have either male or hermaphrodite flowers. "Leafing, blossoming and ripening early" (Hedrick, *l.c.* ii. p.313)

Rupestris Varieties

Rupestris du Lot. Synonyms.—Rup. Monticola (in Germany, Austria, and Hungary it is most commonly known by this name), Rup. Richter, Rup. Lacastelle, Rup. Sijas, Rup. St. George (in California), etc.

Culture.—In the sweet Karroo soils of Worcester, Robertson, and Montagu it is a most vigorous grower. It wants a deep soil and subsoil that its plunging roots can easily penetrate. Under such conditions it can resist drought fairly well. If the soil is somewhat moist, it is very subject to root-rot, and dies where Jacquez still does very well. Where the soil is *deep,* of medium fertility and sufficiently moist without being too moist, it forms big, strong vines of which the grapes are liable to set badly, especially with varieties like Muscat of Alexandria, Malbec and others that are very susceptible to coulure.

In poor sandy soils it forms smallish vines and is much inferior to Jacquez. In a deep, heavy clay soil at Elsenburg it gave vigorous vines that did well until they were cut out after ten years, when the experiment was abandoned. In shallow soils it is worthless.

At the Cape this stock has not yet been used to any large extent. In California and Australia,[1] as in several European wine countries, it has been largely used with great success in suitable soils and with varieties not very subject to coulure or non-setting of the berries. It is very highly resistant to phylloxera.

---

[1] F. de Castella, *Resistant Stocks,* pp. 6-7, reprinted from the *Journal of the Dept. of Agriculture of Victoria,* May and August 1921.

Its cuttings strike root well and graft well. It seems to have a good affinity for most European varieties, including Muscat of Alexandria, and can withstand 25 per cent lime in the soil.

Description.—

Bourgeonnement strongly red. The vine now seems to be covered with red inflorescences (undeveloped).

Shoots glabrous, cobwebby near growing point, violet red, branched, usually rather short, with very short internodes. Most shoots grow upright. Eyes very small.

Leaves entire with *typical open and shallow petiolar sinus* (see Fig. 51), $\alpha+\beta= 71°$, $\gamma =25°$; small, about as long as broad, thick, glabrous on both surfaces, dull green with (somewhat) silver-leafcolour, teeth sharp and narrow, nerves violet-red near base and on upper surface. Male plant.

FIG. 50.—Leaf of **Rupestris du Lot.** From Ravaz, *Les Vignes americaines*, etc., 1902. Coulet et fils, Montpellier; Masson et Cie, Paris.

Rupestris Mission.—It is a vigorous grower in sandy-clayey soils, very sensitive to lime in the soil. Little known at the Cape and not much used anywhere nowadays. Its petiolar sinus is also shallow and open, but not as much as that of Rup. du Lot. $a + \beta = 82°$, $\gamma = 29°$. Leaf entire, much broader than long, teeth short; those of 1 and 1' are fairly long. Shoots glabrous, creeping. Male plant.

Rupestris Metallica.—This vine is sometimes called French Metallica at the Cape, where it is little known and has never been used to any extent. Vigorous grower, nearly everywhere sufficiently resistant to phylloxera, $a + \beta = 91°$, $\gamma = 33°$. Leaf as long as broad, entire, glabrous, grey fatty green colour with strong gloss. Shoots pale green, cobwebby, growing nearly upright.
Male plant.

Rup. Constantia Metallica.—This variety was raised from seed and selected by the late Mr. J. P. de Waal at Groot Constantia (Cape) on account of its great vigour. He also propagated it. It is a vigorous grower, strikes root, and grafts very well, and vines grafted on it start bearing early and give uniform heavy crops. In cool, deep, loose, well-drained soils which become neither too wet nor too dry, it is an excellent stock. In moist sandy soils it is destroyed by eelworms, and in somewhat shallow, dry soils its grafted vines, after about eight years, become smaller and smaller until they do not pay and must be chopped out. Between 1900 and 1912 it was about the most commonly used stock at the Cape, except for Muscat of Alexandria, but is now no longer used. It is unsuitable for heavy clay soils. It has been imported into Australia and was called Metallica Cape about the beginning of the present century. F. de Castella, *l.c.* p. 7, who gives this information in 1921, states that "Shiraz grafted on this stock at the Viticultural Station nearly twenty years ago still yields excellent crops". In view of our experience with this stock at the Cape, he sees no reason for propagating it further in Australia.
Its leaves are glossy, light green, and later in summer it has a whitish-yellowish green colour; teeth sharp and well developed; leaf entire, medium size to small. Young shoots reddish, with numerous loose hairs; ripe canes reddish brown.
Male plant.

Rupestris Martin,—According to Ravaz this is a very vigorous grower, very highly resistant to phylloxera, and one of the most robust stocks. In sandy-clayey, stony soils, and in somewhat moist soils, it is a splendid stock. It is sensitive to lime. Its shoots are long and thick, strike root readily, and grow fast. Its roots spread and plunge. Its stem thickens almost as rapidly as that of the European vine grafted on it. It is a good stock, but has the great drawback of grafting very badly. I do not know whether we have it at the Cape. The information here given was taken from Ravaz.

α + β = *105°*, γ=40°. Leaf entire, broad, teeth very broad and rounded off; glabrous on both surfaces, dark green, very glossy. Petiolar sinus an open V. Shoots glabrous, violaceous.

Male plant.

Rupestris Ganzin. — This plant was selected and propagated by V. Ganzin. Moderate vigour, very highly resistant to phylloxera. Gives good, vigorous grafted vines, but has been replaced by better varieties ; important as one of the parents of Aramon x Rup. Ganzin Nos. 1 and 2. Male plant with open petiolar sinus, α + β = 98°, 7 = 30°. Leaf entire, broad, glabrous, pale green, teeth broad and pointed. Shoots glabrous, yellowish green, afterwards light reddish.

Rupestris No. 9 (H. Goethe).—[According to Teleki, *l.c.* p. 28] This variety was grown and selected by the famous ampelographer H. Goethe, in 1889, out of seed which he obtained from America (Missouri) through Prof. Millardet of Bordeaux. It is a vigorous grower, highly resistant to lime chlorosis and phylloxera. Leaves fairly large (the largest among the Rupestris varieties), thick, dark green, glabrous, nerves strongly developed below, teeth broad, entire; growing tips glabrous, glossy green with faint reddish tint. Shoots thick with relatively few tendrils and laterals, internodes medium long, eyes strikingly small. Buds and blossoms eight days after Riparia. Canes strike root and graft easily. It is very popular in Styria (Austria). Evidently a very good stock, but not known to me personally.

5. **Vitis aestivalis,** *Michaux.*

Synonyms.—V. sylvestris, V. Labrusca var. Aestivalis, Summer grape, Bunch grape, Pigeon grape.

Geographical Distribution.—From Long Island to Central Florida, Pennsylvania, Missouri, Mississippi (after Ravaz).

(*Note.*—Hedrick gives the following division of the original species: *Vitis aestivalis,* Miehaux; *Vitis aestivalis Lincecumii,* Munson; and *Vitis aestivalis Bourquiniana,* Bailey.)

Culture.—This species is important, since Herbemont and Jacquez contain some of its blood and characteristics. According to Ravaz they are natural hybrids of Aestivalis-Cinerea-Vinifera. It is a vine for warm regions. In France its wood and fruit ripen well only in the south. According to Bush it endures the severest drought. The fine green bloom on the lower surface of the leaves would make one suspect this. Its thick, hard cuttings with dense wood *generally root badly,* although better than those of *V. Berlandieri* and *V. cinerea.* Its roots are not numerous, and at first grow slowly. Afterwards they soon begin to accelerate their growth. The vines therefore grow slowly during the first year. The roots of well-developed Aestivalis vines are thick and fleshy, spread and also plunge, have few but thick rootlets; they *grow very well in a compact, hard soil.*

It is more sensitive to lime than Riparia. In compact soils it does better than Riparia and Rupestris. In sandy soils it also does well; the same applies to Jacquez and Herbemont. Ravaz states that its resistance to Phylloxera is on the whole insufficient. Hedrick *(l.c.* ii. p. 320) says: "The hard roots of Aestivalis enable it to resist phylloxera, and varieties with any great amount of blood of this species are seldom seriously injured by this insect". In good, deep, cool soils it will do, and in moist and sandy soils it is a good stock. Its grapes produce a good, very dark red wine. In suitable soils it can serve as a self-bearer.

Description (according to Ravaz and Hedrick).—

Bourgeonnement woolly, crimson or rusty.

Shoots ribbed when very young, smooth or pubescent (woolly), pale green, with bloom, short, thick, ripe canes violaceous red, distinctly striped, almost cylindrical.

Leaves truncated (see Fig. 47), angles between primary nerves open, teeth hardly noticeable; leaf-blade large, thick; petiolar sinus deep, usually narrow, frequently overlapping; usually 3- to 5-lobed; upper surface dark green and finely crinkled; lower surface with more or less reddish or rusty pubescence which, in mature leaves, usually shows in patches on the ribs and veins; petioles frequently pubescent.

Young leaves with woolly pubescence, white or reddish or rusty.

Bunch more or less loose with round, black berries (10-15 mm. diameter), juicy, sweet, with high acidity. Seeds two to three of medium size, plump, smooth, not notched; chalaza oval (according to Hedrick) and round (according to Ravaz), distinct; raphe a distinct cord-like ridge. Leafing (budding) and ripening fruit late to very late.

Roots thick and fleshy with well-developed vessels.

6. **Vitis aestivalis Lincecumii,** *Munson.* (After Hedrick.)

Synonyms.—Post-oak grape, Pine-wood grape, Turkey grape.

Habitat—Eastern Texas, Western Louisiana, Oklahoma, Arkansas and Southern Missouri on high sandy land, frequently climbing post-oak trees, hence the name Post-oak Grape, by which it is locally known.

Culture.—H. Jaeger and T. V. Munson domesticated Lincecumii. "The qualities which recommend it are: First, vigour; second, capacity to withstand rot and mildew; third, hardiness and capacity to endure hot and dry summers without injury; fourth, the large cluster and berry which were found on certain of the wild vines. The fruit is characteristic because of its dense bloom, firm yet tender texture, and peculiar flavour. The cultivated varieties have given satisfaction in many sections of the Central Western and Southern States. Like Aestivalis, it is difficult to propagate from cuttings." This species requires a hot climate and long pruning. Ravaz makes particular mention of Early Purple and Jaeger's No. 43 as varieties with pretty, large bunches.

Description (after Hedrick).—

Vine vigorous; canes cylindrical, much rusty wool on shoots; tendrils intermittent. Leaves very large, almost as wide as long; entire 3-, or 5-, or (rarely) 7-lobed; lobes frequently divided; sinuses, including petiolar sinus, deep; smooth above, and with more or less rusty pubescence below. Fruit small to large, usually larger than typical Aestivalis, usually black, with heavy bloom. Seeds larger than Aestivalis, pear-shaped; chalaza roundish.

7. **Vitis Bicolor,** *Leconte.*

Synonyms. — Blue grape, Northern Summer grape, Northern Aestivalis.

Hedrick says that "Bicolor is readily distinguished from Aestivalis by the absence of the reddish pubescence and by blooming slightly later. The habitat of Bicolor is to the north of that of Aestivalis, occupying the north-eastern, whereas Aestivalis occupies the south-eastern quarter of the UnitedStates ... The horticultural characters of Bicolor are much the same as those of Aestivalis... Like Aestivalis, Bicolor does not thrive on limy soils, and it is difficult to propagate from cuttings." Its clusters are of medium size and the berries are small, black with much bloom, acid but pleasant tasting when ripe. For a description see Hedrick, *l.c.* ii.p. 322.

8. **Vitis Coriacea**, *Shuttleworth.*

Synonyms.—V. candicans var. Coriacea (Bailey), V. cari-baea (Chapman), Leather Leaf.

Habitat: South and Central Florida.

There exists a difference of opinion among botanists as to whether it should be regarded as a different species or merely as a variety of *V. candicans.* Leaves thick, broad, bullate, strongly woolly, teeth hardly developed, entire or 3- to 5-lobed. Growing tips strongly woolly and red. Bunches loose, medium size; round, medium-sized berries; ripens very late. Roots very thick and very fleshy. Its cuttings strike root with as much difficulty as those of *V. Berlandieri* and *V. candicans.* It is fairly resistant to phylloxera, and requires a very warm climate.

9. **Vitis Arizonica**, *Engelmann.*

This species is spread over the arid regions of the Western United States (W. Texas, New Mexico, Arizona, S.E. California), and is very highly drought resistant. It is a weak grower, and is of no practical value.

10. **Vitis Californica**, *Beniham.*

This species occurs in northern California and southern Oregon. It is a vigorous grower with a thick stem; thick, fleshy roots; small bunches with small, round, black berries with a good taste; leaves fairly orbicular, soft, woolly below, and remind one much of *V. vinifera.* Its cuttings root without difficulty (according to Hedrick). It has no practical value, and is scarcely more resistant to phylloxera than *V. vinifera,* but is of interest as being an indigenous vine in a part of America whither the phylloxera penetrated only in recent times.

11. **Vitis monticola**, *Buckley.*

According to Hedrick *(l.c.* i. p. 116) monticola inhabits the limestone hills of central and south-western Texas.

It is highly resistant to heat and prolonged droughts.

Hedrick says that "it is found to be very resistant to phylloxera" while Ravaz *(l.c.* p. 139) states that "its resistance to phylloxera does not seem to be very high". It is most highly resistant to lime chlorosis, even more than *V. Berlandieri.* It might have been a splendid stock but for the following bad properties which it also possesses: poor aerial growth, cuttings strike root with great difficulty (like *V. cordifolia* and something better than *V. Berlandieri)* and graft very badly.

Description (after Ravaz).—

Bourgeonnement woolly, rose-coloured.

Shoots angular, cobwebby, short, branched, violaceous green; ripe canes striped and reddish-brown.

Leaves orbicular, about as broad as long; angles of nerves not very wide, α + β = 110° and less, but γ goes up to 70°, average 60°, with the result that the petiolar sinus tends to be closed; characteristic strong gloss, glabrous (sometimes with a few hairs below), pretty green colour, thick, brittle. The parenchyma of the leaf consists of layers of palissade cells above and below with a dense spongy tissue in between: the stomata are sunk deeply into the lower epidermis, and the leaf is covered with a thick cuticle. The leaf of Monticola shows all the characteristics of leaves adapted to hot and dry climates. The leaf is entire, and has sharp teeth with equally long sides.

Bunch small with small, round, red to black berries. Seed with circular chalaza, raphe prominent. Its shining green leaves and shrubby growth are very characteristic.

12. **Vitis cordifolia,** *Michaux.*

Synonyms.—V. Pullaria (Leconte), V. Vulpina var. Cordifolia (Regel), True Frost grape, Chicken grape, Winter grape, etc.

Geographical Distribution—According to Ravaz it occurs in the woods and along the river banks in Pennsylvania, eastern Kansas, south-western Florida, and in Texas. It frequently occurs together with Riparia, with which it has frequently been confused. In warm, dry regions it grows well in limestone soils.

Culture—It is a species of hot rather than temperate climates. It grows well in all kinds of soil (sand, clay, alluvial and hill-soils) but is very susceptible to lime chlorosis in limestone soils unless they are very dry. It is splendid for compact, dry soils.   In the South of France it remained green during very dry summers when Riparia, Rupestris, etc., lost some of their leaves and visibly suffered from drought. Cordifolia × Rupestris, and particularly Cordifolia × Riparia hybrids, remain green and healthy in the driest soils.   Its roots plunge more than those of Riparia, *i.e.* they have a smaller angle of geotropism.   In sandy-clayey *(i.e.* broken) soils it gives vigorous grafted vines that bear well, but sometimes set their fruit badly on account of excessive vigour.   The Cordifolia portion of the stem thickens just as rapidly as the Vinifera portion.   It is not used as a stock, because its cuttings strike root only with great difficulty. This has been overcome in its hybrids, which are sometimes used to great advantage, *e.g.* Rip. × Cordif.—Rup. No. 106-8.   Cordifolia is very highly resistant to phylloxera. It has a powerful root-system. The roots are thick, fleshy, and yellowish.   It is a most vigorous grower.   Its stem reaches a diameter of 45 cm. or 18 in. (Ravaz, *l.c* p. 143).

Description (after Hedrick).—

Vine very vigorous, climbing.

Bourgeonnement almost glabrous (Ravaz).

Shoots rather slender; internodes long, slightly angular, usually glabrous, sometimes slightly pubescent; diaphragms thick; tendrils intermittent, long, usually bifid.

Leaves with short, broad stipules; leaf-blade medium to large, cordate, entire or sometimes indistinctly three-lobed; petiolar sinus deep, usually narrow, acute; margin with rather coarse angular teeth; point of leaf acuminate; upper surface rather light green, glossy, glabrous; glabrous or sparingly pubescent below.

Clusters medium to large, loose, with long peduncle. *Berries* numerous and small, black, shining, little or no bloom. Seeds medium in size, rather broad, beak rather short; chalaza oval or roundish, elevated, very distinct, raphe a distinct, cord-like ridge. Fruit usually sour and astringent, and frequently consisting of little besides skins and seeds. Leafing, flowering, and ripening fruit very late.

13. **Vitis candicans**, *Engelmann.*

Synonym.—V. mustangensis, V. caribaea var. Coriacea, Mustang grape, Mustang.

Geographical Distribution.—Hedrick (*l.c.*) writes: "The habitat of this grape extends from southern Oklahoma, as a northern limit, south-westerly into Mexico. The western boundary is the Pecos River. It is found on dry, alluvial, sandy or limestone bottoms or on limestone bluff lands, and is said to be especially abundant along upland ravines."

Culture.—The white woolliness of its bourgeonnement, shoots, adult and young leaves, inflorescences, give it a whitish aspect, hence the name Candicans. It is a vigorous grower and climber, and is adapted to hot, dry regions. It thrives best in deep, moist, fertile soil. In dry, shallow soils it does not develop much vigour. It is highly drought-resistant, but very susceptible to lime chlorosis. Hedrick says: "Candicans grows well on limestone lands, enduring as much as 60 per cent of carbonate of lime in the soil". But this only happens in the very dry soils, where the lime is not dangerous. It is fairly resistant to phylloxera. Its great drawback is its *very poor rooting power.*

Description (after Ravaz and Hedrick).—

Bourgeonnement very woolly, red around the borders.

Shoots densely woolly, ribbed, thick, long, little branched; diaphragm thick.

Leaves dense felt-like pubescence; with adult plants they are entire, cordate, bullate, with the borders curved in towards the lower surface; in young shoots and on young vines and sprouts usually deeply 3- to 5- or even 7-lobed; teeth shallow, sinuate; petiolar sinus shallow, wide, sometimes lacking; thick.

Fruit clusters small. *Berries* medium to large, black, purple, een, or even whitish, thin blue bloom or bloomless. Seeds usually three or four, large, short, plump, blunt, notched; chalaza oval, depressed, indistinct; raphe a broad groove.

Vine very vigorous.

Roots thick, fleshy, soft, dark grey.

14. **Vitis Berlandieri**, *Planchon.*

Synonyms.—V. monticola (Millardet), V. Montana (Buckley), Mountain grape, etc.

Geographical Distribution.—Berlandieri is a native of the lone hills of south-west Texas and adjacent Mexico. It grows in the same region as *V. monticola,* but is less restricted locally, growing from the tops of the hills down and along the creek bottoms of these regions.

Culture.—*V. Berlandieri* is a vine for a hot climate. It is late, and its wood ripens fully only in hot regions. Its *cuttings root only with great difficulty,* especially when not fully ripe. This is its greatest drawback.   It can serve as a stock in parts where it would not ripen properly in the ungrafted state.   As it is most highly resistant to lime in the soil, more so than any other American species (Hedrick, *l.c.* ii. p. 318), *it is an excellent stock for soils very rich in lime.*   In fairly dry soils it will also answer well.   Its canes are hard, and strike root with very great difficulty (5-10 per cent against 90 per cent with *V. riparia).*   This is due to the fact that the eyes develop and sometimes form shoots 6 in. long when the roots only just begin to form.   Thus it is worse than Jacquez in this respect.   It is for this reason that the pure Berlandieri varieties have been used so little in viticultural practice. Their hybrids with good rooting qualities have, on the other hand, been very considerably used, especially on soils very rich in lime.

During the first couple of years Berlandieri vines grow much more underground than above ground (18-52 : 4.32 in the first year, according to Ravaz, *l.c.* p. 156).   Its stem fed shoots therefore remain somewhat weak during the first couple of years, although its root system develops well. At first its grafted vines remain a little backward compared with vines grafted on most other stocks.   After some five years they are about equal, when they are fine, vigorous grafted vines that bear well and ripen their grapes well.   The Berlandieri stem remains rather thin, and much thinner than the stem of the Vinifera variety grafted on it—as happens with Riparia.   In a lecture given by P. Viala in Algiers in 1911, he expressed the opinion that Berlandieri and its hybrids should be the stocks of the future.   Berlandieri has thick, fleshy roots which are highly resistant to phylloxera. "Leafing, flowering, and ripening fruit very late" (Hedrick).

Planchon described this species in 1880 and called it *Vitis Berlandieri,* according to herbarium material collected by the Swiss botanist Berlandier in Texas in 1834, and according to living plants which had been imported into France.

Description (Ravaz and Hedrick).—

Bourgeonnement woolly white or coloured red.

Shoots long, thin, green to reddish-violet, more or less angled (and ribbed) and pubescent. The ripe canes are coloured light grey or more or less dark brown, mostly with short internodes and thick diaphragms.

Leaves fairly large, broadly cordate or cuneiform (much like the Eiparia leaf), broad but rather shallow teeth, cobwebby above and below, sometimes with brush-like hairs (especially below), bullate ; thick, dark green and glossy above, green below; entire or shortly 3-lobed; petiolar sinus almost closed or closed at the top (Kavaz), and, according to Hedrick, rather open, V- or U-shaped; short, straight hairs make the leaves rough to the touch.

Clusters large, compact, compound, with long peduncle. Berries small, black, with thin bloom, juicy, rather tart but pleasant tasting when thoroughly ripe. Seeds few, small, short, plump, oval or roundish with short beak; chalaza oval or roundish, distinct; raphe narrow, slightly distinct to indistinct. Leafing, flowering, and ripening fruit very late.

Varieties of Berlandieri.—

Of the numerous Berlandieri seedlings that have been selected and propagated, Berlandieri No. 2 (Resseguier), B. Mazade, B. Lafont, B. Richter are amongst the best. In order to make them root better, P. Viala recommends cutting the canes only in winter, stratification in moist sand, planting out late in spring (after other varieties are in leaf) in well-manured and irrigated nursery soil, after having removed a piece of the bark on the last internode near the bottom end of every cutting. In the vineyard only grafted, rooted vines should be planted.

(b) **Hybrids.**

In the hybrids we find some if not all the characteristics of the parents. Sometimes those of the one parent predominate and *vice versa*. Thus Rip. × Rup. 101-14 is predominatingly Riparia in character, while Rip. × Rup. 3309 is mainly Rupestris in character. The object in hybridising is to combine as many of the useful characters of the parents and to eliminate as many of their bad characters as possible in the hybrids. Thus *V. Berlandieri* is very difficult to propagate from cuttings, but is very highly resistant to lime, while *V. riparia* roots very easily but is sensitive to lime.

When these two species were crossed the hybrids were selected with a view to combining good rooting power with high resistance to lime. This we find, for example, in Rip. × Berl. 420A.

Crosses between American species are called Americo-American hybrids, while those between American species and the European species (*V. vinifera)* are called Americo-Vinifera hybrids. Those hybrids that are only of value as direct producers, I shall discuss under these. Here I shall discuss only those hybrids that are used as stocks.

1. Americo-American Hybrids.

Riparia x *Rupestris* 101-14.

It was produced by Millardet and de Grasset by crossing *V. riparia* with *F. rupestris* (father).

Description.—
Bourgeonnement pale green, glossy.
Leaf distinctly cuneiform, 3-lobed, glabrous above and below α + β = 94°, γ = 37° ; the middle nerve ends in a fairly long, pointed tooth; light yellowish-green.
Shoots glabrous, smooth, reddish-violet, ripe canes light to fairly dark brown and striped, fairly thick, with fairly long internodes, ripen early and shed leaves early, which then have a yellow or brown colour.
Bunches small, with small, round, black berries.

Culture.—It is a vigorous grower in most soils, even when somewhat poor. Its resistance to lime is about the same as that of Riparia. Its root system is strongly branched. The roots are thin, resembling more closely those of Riparia than those of Rupestris, and are highly resistant to phylloxera. Its cuttings root easily and graft well. It has a good affinity for most Vinifera varieties, including Muscat of Alexandria. It forms big, robust grafted vines which give good crops and ripen their grapes well and early. Where Jacquez is unreliable (in heavy loams and clay soils), it and 420A are good stocks for Muscat of Alexandria. It is *one of the best stocks* for the whole of the Cape viticultural area. At Alphen (Constantia) Hanepoot (Muscat of Alexandria) sets better and ripens a fortnight earlier on it than on 1202. In the poor sandy-gravelly and sandy-clayey soils of Helderberg (Stellenbosch district, Cape), where Stein and Hermitage (=

Cinsaut) have so miserably perished on Aramon Nos. 1 and 2, it and 1202, after twelve years of experiment, proved to be the two best stocks. The vines grafted on them were then still big and strong, and gave good crops. In Australia 101-14 does not seem to be looked upon with great favour, whilst at the Cape it is now one of the most favoured stocks. Prof. Bioletti[1] thinks that it is the most generally useful stock for the hotter and interior irrigated regions of California.

Rip. ×Rup. 3306 and 3309:

They have been bred by Couderc, who thinks that they must also be related to Monticola on account of their shining leaves.

Description.—
Their shoots tend to grow flat on the ground, are much branched, long, and rather thin. Their leaves are shining, dark green—those of 3309 more so than those of 3306 — medium size, glabrous on both surfaces excepting those of 3306, which have straight brush-like hairs on the nerves below, as well as on the leaf-stalk and on the shoots. Their growing tips are respectively yellowish-green and pale green, shining and glabrous. Both are male plants. The shoots of 3309 are glabrous, while those of 3306 are covered with a brush-like pubescence (short, straight hairs), which is very characteristic.

*Culture.—*They are fairly resistant to lime chlorosis and can stand up to 20 per cent carbonate of lime in the soil. Their cuttings root and graft well. They have a good affinity for most European varieties, evidently including Muscat of Alexandria. At the Cape their leaves sometimes become burnt during spells of great heat in January and February. In moist, fertile soils 3306 is excellent, and gives vigorous grafted vines that bear well. In stony and fairly stiff, deep clay soils that become somewhat dry, 3309 is a good stock, and better than 3306. They are both highly resistant to phylloxera. At the Cape they have not been made much use of in the past, but more attention should be paid to them in future. In Australia and California they have been largely used, with, on the whole, good results. Prof. Bioletti[2] considers 3309 the most generally useful stock for the cooler parts and the coastal regions of California.

---

[1] According to a letter to the author dated Berkeley, June 25, 1925.
[2] In his letter just quoted.

Riparia x *Monticola* 1 R.

It was bred by Ravaz in 1892. Its shining, thick, dark green leaves with sharp teeth are very characteristic. The terminal teeth of the middle and first lateral primary nerves are long and sharp pointed. At the Cape it has not yet given any particularly good results, and is of moderate vigour.

Riparia × *Cordifolia* 125-1 (Millardet and de Grasset).

Description.—

Leaves 3-lobed, medium, broad, crinkled, thick, light green; teeth narrow and sharp; nerves about red near base; $\alpha + \beta = 94°$, $\gamma = 72°$, $\delta = 41°$, glabrous. The leaves show no Cordifolia characters. Male plant. Root-system strongly developed and almost exclusively Cordifolia in character; roots fleshy and yellowish. Ravaz is of opinion that its Riparia parent must have contained Rupestris blood also.

Culture.—It is one of the first vines to begin budding. It is a vigorous grower, highly resistant to phylloxera and drought, but still more susceptible to lime chlorosis than Riparia. In sandy, clayey, or broken soils, poor in lime and subject to drought, whether stony or not, whether poor or fertile, it still answers well. It roots and grafts well, and forms vigorous and fertile grafted vines. The parts of the stem above and below the joint thicken about equally fast, and no appreciable knob is formed at the joint.

Riparia x *Berlandieri* 34 E.M.

In 1899 G. Foëx, Principal of the School of Agriculture in Montpellier, crossed *V. riparia* (father) with *V. Berlandieri Ecole,* which gave rise to 34 E.M. amongst others.

Description.—

Shoots angular, brush-like pubescence, green with red stripes, thick, long.

Leaves 3-lobed, $\alpha + \beta = 91°$, $\gamma = 41°$; teeth narrow and sharp, much longer than broad; the middle nerve in particular ends in a long pointed terminal tooth; light green, glossy, glabrous except for brush-like hairs on main nerves below; also shows strongly Berlandieri character, *e.g.* thick leaf, etc. Male plant.

Culture.—Vigorous grower, especially in moist, light, deep or shallow soil. Of all the Rip. × Berlandieri hybrids it is the most highly resistant to lime. It is therefore particularly valuable in moist limestone soils. Its cuttings root well (60-80 per cent in the nursery) and graft well. Its root system is strong, and much branched and somewhat fleshy. It is practically everywhere sufficiently resistant to phylloxera. Its grafted vines set their fruit well, are fertile, and ripen their grapes well. So far little used at the Cape, but promising.

218

Rip. × *Berl.* 157-11.

It was produced by Couderc by crossing Berlandieri Las Sorres with Riparia Gloire (father).

Description.—

Shoots cobwebby, angular, green with light red tint, thick, long.

Leaves 3-lobed, $\alpha + \beta = 99°$, $\gamma = 40°$; teeth narrow, sharp; glabrous excepting a few brush-like hairs on main nerves below; dark green with high gloss, somewhat bullate, large; much Riparia character in shape of leaf and Berlandieri character in its thickness, colour and gloss. Young leaves yellowish green and woolly.

Bunches small, with round, black, fleshy berries. Thick stem and strongly developed root system, about intermediate in character between its two parents. Roots somewhat fleshy and strongly branched.

Culture.—According to Viala, in his discourse in Algiers in *1911* (previously quoted), it is more Riparia than Berlandieri in character, and cannot stand more than 30 per cent carbonate of lime in the soil. With the possible exception of stony and very dry soils, Rayaz thinks it can be cultivated successfully in all soils. It is a very vigorous grower and highly resistant to phylloxera. It grafts well, but roots rather poorly, especially when the canes have not ripened well. Its mother vines should thus be grown on a warm, somewhat dry soil. Its grafted vines are vigorous, fertile, little subject to coulure, and ripen their grapes well. It is therefore a very valuable stock. Unfortunately its cuttings strike root only with great difficulty. This is its greatest drawback. It has so far hardly been used at the Cape, or, I think, in Australia and California.

Rip. x *Bed.* 420A.

It was produced by Millardet and de Grasset by crossing Berlandieri with Riparia (father), (probably Rip. Gloire).

Description.—

Shoots angular, green with red stripes and violet on the nodes; ripe canes fairly dark brown.

Leaves 3-lobed, $\alpha + \beta = 105°$, $\gamma = 36°$; teeth somewhat rounded off, narrow; brushlike hairs on primary and secondary nerves below; dark green, glossy, thick, broad. The middle and first lateral nerves are terminated by three well-developed, sharp teeth. The leaf shows more Berlandieri than Riparia character.

Male plant. Root system strong and intermediate between those of the parents.

Culture.—Vigorous grower in nearly all soils. Its shoots grow long and ripen early. They root well (70-85 per cent in the nursery) and graft quite well, particularly when rooted vines are grafted on the spot, a practice which has become popular in Victoria (Australia) with this stock (according to F. de Castella, *l.c.*). It possesses great toleration for lime (quite 35 per cent), and is fairly drought resistant in deep, free soils. It has a good affinity for most Vinifera varieties, including Muscat of Alexandria. In heavy loams and clay soils it is one of the best stocks for Muscat of Alexandria. It is everywhere sufficiently resistant to phylloxera. As it does not like wet feet, the soil must be well drained where it is grown under irrigation. At first its grafted vines are backward when compared with those on most other (non-Berlandieri) stocks, but after about five years they have equalised the difference and are then strong, fruitful vines, that ripen their grapes well and tend to raise their quality. I consider it to be one of the best all-round stocks, and farmers should not be prejudiced against it on account of its initial slower growth above ground.

Rip.× Berl. 420B is also an excellent stock of this group. It is highly resistant to lime chlorosis and still more resistant to phylloxera than 420A (according to Ravaz, *l.c.* p. 237).

Rip. x Berl. Teleki.

Historical.—Siegmund Teleki of Hungary has grown this stock from seed which he had obtained from M. Resseguier in France in 1896. At the same time he also made other selections out of the large number (about 40,000) of seedlings which he obtained from the imported seed. For details I refer the reader to Teleki, *l.c* pp. 65-76.

Description (after Teleki).—

Growing tips light to dark bronze coloured, strongly glossy, frequently somewhat woolly.

Shoots angular, strongly pubescent; at first green with a reddish tint, red in summer, the ripe canes are brown to ash-grey internodes very long; canes thick with thin pith.

Leaves remind one much of those of Rip. Gloire, 3-lobed, long, pointed tooth at end of middle nerve, teeth sharp and narrow; primary and secondary nerves slightly pubescent below, with light rose colour near the petiolar sinus; glossy, dark green, glabrous, thick, leathery,

220

slightly longer than broad, parenchyma raised somewhat between the nerves, *very large*. Teleki says that he has measured leaves up to 48 cm. or 19 in. long.

Root system strongly developed, much branched, yellow, slightly fleshy, grows fairly deep.

Culture.—The most striking thing about this vine is that it ripens its wood very early in Hungary, three to four weeks before that of Riparia Gloire, and six to eight weeks earlier than that of Rupestris varieties and Rupestris hybrids. In spring it is the first to commence budding. It forms relatively few shoots, which, however, grow fast and become thick. It is a very vigorous grower. Its cuttings strike root very well. It answers about equally well in dry and in wet marly and highly calcareous soils; only soils that are too dry and shallow do not suit it. It is almost as good in poor as in fertile soils. It is highly resistant to cryptogamic diseases; only anthracnose affects it slightly.

In an experiment where Welschriesling had been grafted on Aramon × Rup. Ganzin No. 1 and on Rip. × Berl. Teleki standing side by side, the vines on the latter bore heavier crops, ripened their grapes a fortnight earlier, and showed 2 per cent more sugar in the must than those on Aramon No. 1. In both cases the grafted vines were vigorous.

Grape varieties that are susceptible to coulure, set their berries well when grafted on Rip. × Berl. Teleki. In soils with 40-50 per cent readily soluble carbonate of lime, its grafted vines do not suffer from lime chlorosis. It is an excellent stock for quality wines and table grapes. Wherever it grows well it is sufficiently resistant to phylloxera. It is particularly valuable for limestone soils and for the northern viticultural areas on account of its early maturity. It seems to have a good affinity for most Vinifera varieties. Up to 86 per cent of good grafted vines have been obtained on bench grafting it; Thus it grafts splendidly. Male plant.

This stock seems to merit being tested out in various parts.

Rupestris × *Berlandieri* 219A (Millardet and de Grasset).

This is a vigorous grower that is evidently sufficiently resistant to phylloxera. It will answer well in stony soils that are dry on top but can be penetrated by its roots. It is very highly resistant to lime chlorosis. Its cuttings root fairly well and graft well. Its grafted vines are vigorous and fertile. It possesses a strong root system. The roots plunge, are fleshy and well branched. Its bunches are very small, with small, black, round berries.

Cordifolia x *Rupestris de Grasset No. 1* (Millardet).

This is a natural hybrid from the Indian Territory in Arkansas. It is a vigorous grower with thick canes that root fairly easily; root system strong, yellow, somewhat fleshy; everywhere sufficiently resistant to phylloxera; gives a rather low percentage of successful grafts, which is one of its gravest defects. Its grafted vines grow very vigorously and become big. It is a good stock for dry or sandy-clayey (broken) soils with little lime.

Rupestris × *Cordifolia* 107-11 (Millardet and de Grasset).

It is a very vigorous grower, with thick, long shoots. Its cuttings strike root satisfactorily, and graft well, lit is very little attacked by phylloxera. In soils poor in lime, in stony or in stiff sandy-clayey soils it forms big and fertile grafted vines. It is a valuable stock. Its bunches are large and loose, with small, black, round berries.

Riparia × *Cordifolia-Rupestris* ᵣ 106-8 (Millardet and de Grasset).

This vine was obtained by Millardet and de Grasset in 1882 by crossing Riparia with Cordifolia-Rupestris de Grasset. It thus contains 50 per cent Riparia blood and is predominatingly Riparia in character.

Description.—

Leaf medium to small, 3-lobed, with sharp narrow teeth, *crinkled,* dark green, nerves below with violet coloration and brush-like pubescence.

Shoots glabrous, violaceous, rather thin.

Bunches small, short, with small, black, round berries, $\alpha + \beta = 93°$, $\gamma = 47°$.

Culture.—It is a vigorous grower that roots well and highly resistant to phylloxera but very sensitive to lime. It is a good stock for sandy clay soils which soon become hard and dry after rain, and for stony soils poor in lime. In stiff, dry, shallow soils it is one of the best stocks. In hot dry regions it may still become a very good stock, so long as the soil is poor in lime, and should therefore be experimented with in California, Australia, Algeria, and at the Cape.

Solonis.

According to Ravaz it contains Riparia, Rupestris, and Candicans blood. It is characterised by its grey whitish leaves with copious pubescence on both surfaces, and narrow, sharp, long teeth. Shoots woolly. Clusters small with small, round, black berries. Until about 1908 it was used fairly extensively near Vienna and elsewhere in Austria as a stock. It gave grafted vines of fine appearance. Latterly it has been abandoned owing to an insufficient resistance to phylloxera. As a stock it is worthless.

Riparia × *Solonis* 1616 (Couderc).

This is a fairly strong grower; sufficiently resistant to phylloxera and fairly resistant to lime chlorosis. It is a good stock for soils that do not contain much lime, are somewhat wet and contain some salt brack (white). It is especially valuable for such brackish soils, and may therefore be found useful in California, Australia, and at the Cape, where irrigation is practised and the drainage is not too good. In compact, moist soils it is better than Riparia. Its grafted vines are quite vigorous, fertile, and ripen their grapes early and well. It reminds one of both parents. Its leaves are glossy, dark green, 3-lobed, with narrow and very sharp teeth, pubescent on the nerves. Growing tip white, wholly. Shoots cobwebby, green-rose-coloured, rather thin. Bunches small, with small, round, black berries.

2. Americo-Vinifera Hybrids.

Aramon × *Rupestris Ganzin* No. 1. This vine was obtained by M. Ganzin in 1879 by crossing Aramon (Vinifera) with Rup. Ganzin.
Description.—
Bourgeonnement: cobwebby, red.

Shoots green, with, violaceous stripes, and shining red growing tip with a few loose hairs; ripe canes ash grey.

Leaves medium, 3-lobed; α + β = 101°, γ = 30°; sharp, fairly-broad teeth; dark green; nerves above violaceous near the base; autumn colours yellow and red. Male plant with numerous inflorescences.

Root-system very strong, fleshy, somewhat reddish.

Culture.—Very vigorous grower in most soils; bad affinity for Muscat of Alexandria but good affinity for most other Vinifera varieties. It roots well but grafts only moderately well (about 30 per cent with bench grafting). Ravaz *(l.c.* p. 261) is wrong in saying that all Viniferas do well on it. On account of its great vigour and great adaptation to different kinds of soils, it has been largely used in many wine countries. Ravaz *(l.c.* p. 260) says: "The resistance of this vine to phylloxera is fairly high; it is not that it is little attacked by the insect; on the contrary, it frequently has numerous and voluminous tuberosities on its roots which sometimes penetrate them very deep; the rootlets are attacked slightly; but its great vigour enables it rapidly to repair the alteration of the tissues thus brought about. One sees at once that it will develop well, even in presence of phylloxera, in cool, sandy or fertile soils, and that it runs the risk of being insufficiently resistant in dry or shallow soils."

Some years ago failures of vines grafted on Aramon No. 1 were reported from Sicily. It was then pointed out that this happened after very hot, dry summers. For over twenty years Aramon Nos. 1 and 2 have been used everywhere as stocks at the Cape with conspicuous success. When Jacquez failed in certain cases the Aramons succeeded well. From about 1912 certain vineyards, Stein and Hermitage, grafted on these two Aramons began to decline in the Helderberg area (Stellenbosch district). Now hundreds of acres of these vineyards have been uprooted and replanted with 101-14 and 1202. The vines on Aramon were evidently killed by phylloxera. In some cases the vines died when only two to three years old and before they had given any crop worth mentioning. They died also in sandy soil where Jacquez still succeeded well! This dying of vines grafted on Aramon has gradually spread to other districts, so that to-day it would be rash to graft any more vines on Aramon Nos. 1 and 2 for Cape soils. Cape wine-growers are therefore *strongly advised to stop using Aramon as a stock.*

In the soil at Helderberg where the trouble first started, I planted an experimental plot in 1912 with Aramon No. 1 and No. 2 next to 1202, 101-14, etc., inside the vineyard on Aramon. The American vines were grafted on the spot with Stein, Hermitage, and other varieties in 1913. After some four years the vines on Aramon No. 1 became weak and miserable, those on Aramon No. 2 following suit the next year. After twelve years the vines grafted on 101-14 and 1202 were vigorous and the best of all, while those on the Aramons had died long ago. The soil is sandy clayey, and in parts sandy with loose clay ironstone gravel and clay in the subsoil. It is a poor soil and becomes rather dry in summer. The original vineyard on Aramon gave splendid crops and grew well during the first eight to ten years. Now vines grafted on Aramon and planted in the same and better soil near it, start declining after two to three years. And vines grafted on Aramon in good soils have begun to decline fast. The Aramon roots of the suffering vines show numerous nodosities and tuberosities to almost the same extent as Vinifera vines. Why the Aramons should have collapsed so badly is not yet known. I am inclined to believe that a *new biological race of phylloxera* has evolved on the Aramon roots in Helderberg, where Aramon has been the almost exclusive stock. If this assumption is correct it will explain the gradual spreading of the trouble to other districts and soils. In this connection I refer the reader to Chapter VIII., where, under the discussion of the Biology of Phylloxera, mention is made of Borner's work and the *pervastatrix* race of phylloxera. At the Cape a race might have been evolved which has a special fancy for the Aramon roots.

Aramon × Rup. Ganzin No. 2.

It issued from the same cross as No. 1, but is usually an even more vigorous grower than No. 1. In the Karroo soils of the Cape they are about equally vigorous, the preference being perhaps slightly in favour of No. 1. In other soils, where they are mixed, all the biggest and most vigorous vines are invariably found to have been grafted on Aramon No. 2. Hence, prior to the bad failure of the Aramons, No. 2 was recommended in preference to No. 1. Now neither of them is safe, as both suffer and die.

It differs from No. 1 in having a yellowish rather than reddish growing tip, which has numerous woolly hairs, like those which occur on the young shoots. By this character alone they can readily be distinguished. Further, the leaves of No. 2 turn yellow but *never* red as do those of No. 1 in autumn. Both are male plants.

Mourvèdre × Rup. 1202.

This vine was produced by Couderc by crossing Mourvèdre (Vinifera) with a Rupestris (father).

Description.—

Bourgeonnement woolly, light reddish tint.

Shoots thick, fairly upright, violaceous green; loose hairs on shoots and leaf-stalks; young shoots also fairly pubescent.

Leaves medium to rather small, 3-lobed, $\alpha + \beta = 110°$ $\gamma = 45°$; teeth acute and broad; cobwebby threads on primary and secondary nerves below; uneven bullate; light green with red spots in late summer (especially on moist spots) and strongly red with yellow in autumn.

Bunches very numerous, rather small, with small, black, round berries of poor quality, but with dark-red juice when fully to over ripe.

*Culture.—*It is everywhere our most vigorous stock at the Cape, giving big strong vines in nearly all soils. In moist sandy soils it is easily killed by eelworms. It is everywhere still sufficiently resistant to phylloxera, although attacked by it. If its grafted vines should suffer from drought or become weakened through other causes, it might be destroyed by phylloxera. No such case has, however, been observed hitherto at the Cape. Its tremendous vigour is its salvation.

Its stem thickens as fast as that of the Vinifera vine grafted on it. It has a good affinity for most Vinifera varieties excepting Muscat of Alexandria, odd vines of which grafted on 1202 die suddenly during the first four years. Where a block of Sultana had been grafted on it next to Muscat of Alexandria, the vines of the former all did very well. After four years, the vines of Muscat of Alexandria then still flourishing, do not die but continue to flourish. As the result is an uneven vineyard, I do not recommend it as a stock for Muscat of Alexandria. Its root system is very strongly developed, numerous fleshy roots which arise anywhere on the root-stump. It is highly resistant to lime chlorosis.

According to Ravaz it can stand 30 per cent carbonate of lime in the soil. In the Karroo soils of the Cape it is very vigorous. Its grafted vines are big, vigorous, fertile, though inclined to set their berries badly (coulure). Its cuttings root and graft very well. I therefore recommend it as a good stock. But care must be taken that its grafted vines remain vigorous, by proper manuring, cultivation, and irrigation where necessary, whilst it had better not be used on moist sandy soils on account of the eelworm danger.

4401 (Couderc).

This is a hybrid produced by Couderc in 1884 by crossing Chasselas rose with Rupestris. It is a vigorous grower with a strong, fleshy root system, insufficiently resistant to phylloxera (according to Ravaz), and more of a direct producer than a stock. Not to be recommended.

Chasselas × Berlandieri 41B. 11

This vine has been bred by Millardet by fertilising flowers of Chasselas with pollen of *V. Berlandieri.*

Description.—

Shoots angular, with loose hairs, violaceous green; growing tip woolly, rose-coloured.

Leaves fairly large, 5-lobed, $\alpha + \beta = 93°$, $\gamma = 45°$; teeth broad and angular; shining, dark green (later in summer light yellowish-green); young leaves shining, woolly, copper-coloured; glabrous except for cobwebby pubescence on nerves 1, 2, 3 below.

Bunch rather small, with small, round, black berries.

Root system more Berlandieri than Vinifera.

*Culture.—*Vigorous grower, but at first slow above ground like the Berlandieris; especially good in soils very rich in lime, to which its use is practically limited. It is fairly resistant to phylloxera. Its cuttings root fairly well, forming four to five roots, mostly at the lowest node. It gives vigorous grafted vines, though not as vigorous as those of 333.

Tisserand or 333 E.M.

Foëx bred this vine at the School of Agriculture of Montpellier by crossing Cabernet Sauvignon with pollen of *V. Berlandieri.*

Description. —

Bourgeonnement woolly, crimson-red.

Shoots light reddish-green, with cobwebby pubescence, often bent, *angular,* growing tip woolly, red. Ripe canes a lighter brown colour and usually thicker than those of 420A. The young shoots tend to grow upright and are easily blown off by strong winds.

Leaves 5-lobed, sinuses fairly deep as with Cabernet Sauvignon, where vine vigorous and leaves big, otherwise almost round and hardly lobed; dark green, shining, thick, cobwebby below. Male plant.

Culture.—Very strong and rapid grower, with a strongly developed root system ; cuttings root very easily and graft very well; forms big, vigorous grafted vines that bear well and ripen their grapes well; good affinity for almost all varieties, including Muscat of Alexandria. In my ten-year-old experimental plots it still does well everywhere and is very promising. It is extremely resistant to lime chlorosis, and therefore a capital stock for soils with a very high carbonate of lime content. It also does well in soils poor in lime, in loose and stiff soils, in poor and fertile soils. It will probably still turn out to be a good stock for all kinds of soils. Owing to the presence of tuberosities which frequently occur on its roots, Ravaz doubts whether it will be sufficiently resistant to phylloxera in too shallow and too dry soils. It seems to me to be sufficiently resistant to phylloxera wherever it grows well.

White Herbemont.

This vine has been raised by Malègue from Herbemont seed, and contains Aestivalis, Cinerea, and Vinifera blood just as Black Herbemont and Jacquez do. It is a vigorous grower in deep, moist soil, corresponding largely with Black Herbemont, except for its white grapes.

Herbemont or Black Herbemont.

Synonyms.—Herbemont's Madeira, Neil Grape, Warren, Warrenton, etc.

Origin.—Unknown, but has been cultivated since 1878 from an old vine in South Carolina; is named after Nicolas Herbemont, who grew it in America before 1834. According to Munson and Bailey it came from Madeira. According to Millardet it is a natural hybrid of Aestivalis, Cinerea, and Vinifera (see Ravaz, *l.c.* p. 346).

228

Description (after Ravaz).—

Bourgeonnement woolly, red.

Shoots glabrous, violaceous green, covered with bloom, long, thick, short internodes ; ripe canes reddish-brown.

Leaves large, 5-lobed, α + β = 120°, γ = 48°; pale green, glossy, with brushlike hairs on nerves, 1, 2, 3, 4 below; teeth broad and somewhat rounded off; growing tip woolly, red (my Herbemont is more yellow than red).

Bunch medium, with small, round, reddish-black berries, compact, sweet, juicy.

Culture.—Vigorous grower, formerly cultivated in France as direct producer and as stock, but now abandoned. In deep, cool soils it is usually sufficiently resistant to phylloxera, like Jacquez. It is sensitive towards lime. It possesses a good affinity for almost all Vinifera varieties. It seems to be a particularly good stock for Hermitage (Cinsaut). Its cuttings root very badly (owing to its Aestivalis blood) but graft well. It does well in soils where Jacquez answers well. Elsewhere it too is not sufficiently resistant to phylloxera.

Jacquez.

Synonyms.—Lenoir, El Paso, Jack, Segar Box.

Origin.—The Segar Box of Longworth (Ohio) is considered to be the same as Jacquez or Jack. Jacquez is named after a Spaniard Jacques. Munson is of opinion that Herbemont and Jacquez came from Madeira to America. It certainly contains Vinifera blood, as is evident from its susceptibility to fungoid diseases, *e.g.* downy mildew and anthracnose, and its high resistance to lime chlorosis. Its shoots, grapes, seeds, etc., clearly show characters of *V. aestivalis*. It is regarded as a natural hybrid of Aestivalis-Cinerea-Vinifera.

Description.—

Bourgeonnement red, woolly.

Shoots glabrous, green, fairly upright, covered with bloom, short, thick; short internodes; ripe canes reddish-brown, smooth, hard. Growing tip red, woolly.

Leaves large, 5-lobed, α + β = 120°, γ = 52°; teeth broad and somewhat rounded off; woolly below, glabrous, shining, dark green above.

Bunches fairly large, long, loose, with small, round, black berries, which are juicy and contain a great deal of pigment.

Culture.—It is a fairly vigorous grower in suitable soil. It forms a thick stem that grows in thickness as rapidly as the Vinifera stem of vines grafted on it. It has a splendid affinity with all Vinifera varieties, including Hanepoot. Its cuttings root with some difficulty but graft splendidly, up to 90 per cent being successful, well-knitted grafts. The uppermost eyes of the ungrafted cutting, when planted out, start budding before the cutting has formed any roots, and die again if roots have not been developed soon. The scion on the grafted Jacquez cutting buds later and gives the latter a better chance of forming roots. The bad rooting qualities of the Jacquez cuttings are thus no drawback for their use as a stock in bench grafting. The cuttings develop only two to four roots, and these nearly all near or on the bottom node. Its roots are fleshy and little branched.

Its rootlets are not much attacked by phylloxera, but its thicker roots are much attacked. In France it was formerly grown on a fairly large scale as direct producer and as a stock for Vinifera varieties, but has now been almost completely abandoned. *In moist or cool, deep, fertile or sandy soils it answered well in France as a direct producer.* It gave there, on fertile soils, 100-120 hl. wine per ha., or 890-1068 gallons wine per acre, or 14½-17¹/₃ leaguers wine per Cape morgen. Its wine is very dark and alcoholic (12-14 vol. per cent alcohol or 21-24½ deg. proof spirit), and is valuable on account of its intense colour. In order to obtain and keep the maximum amount of colour in its wine, one should add 100-150 and even up to 300 grammes tartaric acid per hl. (equivalent to 18-27 and even 54 oz. per leaguer) of its must in the fermenting tank. This addition of tartaric acid prevents the colouring substances from separating out afterwards.

With us at the Cape Jacquez is one of the most commonly used stocks. Here it also was frequently disappointing, as it had at first been planted indiscriminately on all sorts of soils. *In shallow soils that become too dry and in very heavy loams and clay soils Jacquez should not be planted.* It was precisely in such soils that vines grafted on Jacquez have deteriorated and sometimes perished at the Cape. *Good Jacquez soils are all deep, loose, cool soils that never become too dry.* Further, care should be taken that vines grafted on Jacquez do not over-bear during the first three years, as this has, to my knowledge, in the past weakened such vines and threatened their whole future.

I conclude this discussion of Jacquez with the following quotation from the "Report of the Commission of Investigation of Grafted Vineyards,"[1] of which Commission I was a member:

"*Jacquez*—Where there is sufficient rainfall and irrigation is not practised, and where the soil is deep and open, this stock gives satisfaction. On the deep sandy soils of Goudini, Jacquez is by far the best stock for all kinds of grapes. On stiff clay soils, on shallow soils, and on all soils which become too dry during the summer, it is unsatisfactory and very often unsuitable, and therefore must not be used.

"In areas where irrigation is practised, Jacquez must only be used on deep open soils. It is essential that the soil should take up water readily and that the irrigation should be sufficient for the vine not to suffer from drought during the summer. The stock should also not be planted on stiff Karroo soils, which usually do not take up water readily."

What American Stocks should be used in the Future?

According to past experience and in the light of what has been said in the preceding pages about the better-known American stocks, I wish to recommend the following stocks for the Cape viticultural area:

1. *For poor, moist (cool) sandy soils* : Jacquez.
2. *For deep, loose, cool (moist), fairly fertile soils*: Jacquez, Riparia gloire de Montpellier, 101-14, 1202, Rup. du Lot, 333, 420A, 3306.
3. *For broken sandy-clayey soils* which are fairly deep and do not become too dry: 101-14, 1202, 420A, 3309, Jacquez.
4. *For rather dry, stiff loams and clay soils*: 420A, 3309, 106-8,333.
5. *For red Karroo soils under irrigation:* 1202, Rup. Du Lot, 101-14, 420A, 333.

Further experiments should be conducted to test the value of 157-11, 34 E.M., Rip. × Berl. Teleki, and other Berlandieri hybrids, our best Rupestris seedlings occurring in old grafted vineyards, and possibly a few other stocks.

---

[1] Reprint No. 26,1920, from the *Journal of Union Dept. of Agrio.*, Sept. 1920

It is impossible for me to say how far this classification will apply to California and Australia, as soil *and climate* must both be considered in this connection.

II. AMERICAN GRAPE-VARIETIES

Here I wish to discuss some of the grape varieties that are grown *for their grapes* in the eastern states of the United States. Whereas the European or Vinifera varieties are grown in California, this is impossible or at least unremunerative in the eastern states on account of the warm moist climate that is experienced there during the vegetative period of the vine, and the consequent extraordinary development of dangerous fungoid diseases (black rot, downy mildew, etc.) and insect pests.

The principal varieties grown are Labrusca seedlings and artificial Labrusca-Vinifera hybrids, though other hybrids also occur. They are consumed as table grapes or used in the manufacture of grape-juice (free from alcohol) or of "juice", *i.e.* wine to be made and consumed in the home. Prior to prohibition a considerable portion of these was used for the manufacture of champagne.

Here I shall briefly discuss the following varieties, according to information obtained from Hedrick's standard work, *The Grapes of New York,* and his *Manual of American Grape-growing,* which I have previously quoted:

1.   *White* varieties: Diamond, Niagara, Triumph.
2.   *Red* varieties: Catawba, Delaware, Goethe.
3.   *Black* varieties: Concord, Barry, Herbert, Isabella.

Diamond (Labrusca-Vinifera).

It is a white or green grape, originated by the late Jacob Moore of Brighton, New York, about 1870, from Concord seed fertilised by Iona, a red Catawba seedling. In quality and beauty it is surpassed by few other American grapes. It ripens early, bears well, is a vigorous grower, and quite as resistant to fungoid diseases as Concord. In quality it surpasses Niagara, which is regarded as the first American green grape. The vines resemble Concord very much. Its grapes have no foxy taste, and pack, carry, and keep well.

Description.[1]—

Vine vigorous, hardy, productive. *Canes* short, brown with a slight red tinge; nodes enlarged; internodes short; tendrils intermittent, bifid.

Leaves thick; upper surface light green, dull, smooth; lower surface light bronze, downy; lobes three in number, indistinct; petiolar sinus very shallow; teeth shallow.

Flowers self-fertile, open early, stamens upright.

Fruit early, keeps well. Clusters medium to short, broad, blunt, cylindrical, often single-shouldered, compact; pedicel short, thick, with a few inconspicuous warts; brush slender, pale green. *Berries* large, ovate, green with a tinge of yellow, glossy, covered with thin bloom, persistent, firm; skin thin, tough, adherent, astringent; flesh pale green, transparent, juicy, tender, melting, fine-grained, aromatic, sprightly; very good. Seeds free, one to four, broad and long, sharp-pointed, yellowish-brown.

Niagara (Labrusca-Vinifera).

This is the leading American green grape. Among grapes of this colour it holds the rank that Concord has among the black varieties. It was raised in 1868 out of Concord seed that had been fertilised by Cassady (Labrusca-Vinifera). In vigour and fertility it equals Concord in places where they answer equally well. Like Concord it has much of the foxy taste and aroma of the wild Labrusca, which some regard as distasteful while others consider it an asset. "The foxiness of Niagara is most marked just after the fruit is picked, and it is usually better flavoured after having stood for a few days. The flavour is not at its best unless the grapes be fully ripe." Both bunches and berries are larger and better formed than those of Concord. The skin of its berries does not crack as easily as that of Concord. Both the vine and its fruit are more susceptible than those of Concord to fungoid diseases, especially to black rot, which proves a veritable scourge with this variety in unfavourable seasons and localities. It will no doubt continue for some considerable time to be regarded as the leading green table grape, especially if the buyers want a pretty grape without concerning themselves much about quality.

Description.—

Vine vigorous, lacking in hardiness, very productive.

---

[1] The descriptions of these grapes are those of Hedrick.

Canes long, thick, reddish-brown, deepening in colour at the nodes, which are enlarged and slightly flattened; internodes long, thick; tendrils continuous, long, bifid or trifid.

Leaves large, thick; upper surface glossy, dark green, smooth; lower surface pale green, pubescent; lobes three to five, with terminus acute; petiolar sinus of medium depth and width; basal sinus shallow, wide, often toothed; lateral sinus wide, frequently toothed ; teeth shallow, variable in width.

Flowers self-fertile, open in mid-season; stamens upright.

Fruit mid-season; keeps well. Clusters large, long, broad, tapering; frequently single-shouldered, compact; pedicel thick, with a few small, inconspicuous warts; brush pale green, long. Berries large, oval, pale yellowish-green with thin bloom, persistent, firm. Skin thin, tender, adherent, astringent; flesh light green, translucent, juicy, fine grained, tender, foxy, good. Seeds free, one to six, deeply notched, brown.

Triumph (Labrusca-Vinifera)

When quality, colour, shape and size of berry and bunch are all taken into account, Triumph is one of America's best table grapes. "At its best it is a magnificent bunch of golden grapes of highest quality, esteemed even in Southern Europe, where it must compete with the best of the Viniferas." In the south of France it also made a good impression, but its foxy taste and insufficient resistance to phylloxera were against it. As its fruit requires a long season for proper development, this variety justifies its name only in the south, and particularly in the south-west of Concord, its Labrusca parent.

Its habit of growth, vigour, productiveness, and foliage characters are, in general, those of Concord, its Labrusca parent. In hardiness, resistance to fungoid

Fig. 51.—Triumph. From Hedrick, *The Grapes of New York*, 1908. Albany: J. B. Lyon, State Printers.

234

diseases, and earliness of maturity, it falls short of Concord. Its grapes have a less pronounced foxiness than those of Labrusca, while its skin, seeds, and pulp show more the agreeable properties of a European grape. It gives a good white wine. Generally speaking, it is the best artificial hybrid of *V. labrusca* and *V. vinifera*. George Campbell of Delaware (Ohio) produced it in 1860 by fertilising Concord with pollen of Chasselas doré.

Description: $a + \beta = 129°$, $\gamma = 52°$ (after Ravaz).
Vine vigorous.
Canes long, dark brown with much bloom; nodes enlarged; tendrils intermittent, long, trifid, sometimes bifid.
Leaves large; upper surface light green, dull, rugose; lower surface greyish-white, pubescent; leaf usually not lobed, with terminus obtuse; petiolar sinus deep, narrow, often closed and overlapping; basal sinus absent; lateral sinus shallow and narrow when present; teeth deep, wide.
Flowers self-fertile, late; stamens upright.
Fruit very late. Clusters very large, long, broad, cylindrical, sometimes single-shouldered, compact; pedicel slender, smooth; brush short, yellowish-green. *Berries* medium in size, oval, golden yellow, glossy with heavy bloom, persistent, firm. *Skin* thin, inclined to crack, adherent without pigment, slightly astringent; flesh light green, translucent, juicy, fine-grained, tender, vinous; good to very good. Seeds free, one to five, small, brown.

Catawba (Labrusca-Vinifera).

In many respects Catawba is the most interesting of the American grapes, it possesses great elasticity of constitution, and therefore great powers of adaptation. It consequently grows over a vast area. Its origin is unknown. Catawba was the first great American grape, and after a century is still one of the four leading varieties of grapes cultivated in eastern America. Its high quality and attractive appearance give it intrinsic value as a table grape and for making wine. It is the chief of all northern varieties for wine-making. It is still the leading grape along the shores of Lake Erie in northern Ohio, and about the Central Lakes of New York. "Of all the commercial grapes grown in New York Catawba is the best keeper, lasting until March or later."

It possesses great vigour and productiveness. Its greatest drawback is the susceptibility of its foliage and fruit to fungoid diseases, especially downy mildew.

Its botanical characters, its adaptation, and its susceptibilities, all indicate a cross between *V. vinifera* and *F. labrusca*. Catawba is the standard red grape in the eastern American markets. It is practically free from foxiness. About 1823 John Adlum of the District of Columbia introduced this grape. Up to the introduction of Concord in 1854, Catawba was the most popular American grape. It is late in ripening, and was on this account superseded in many places by the earlier-ripening Concord. It is only slightly resistant to phylloxera, and was a failure in France.

Description.—
Vine vigorous, hardy, productive.
Canes numerous, thick, dark brown; nodes enlarged; tendrils continuous, bifid or trifid.
Leaves large; upper surface light green, dull, smooth; lower surface greyish-white, heavily pubescent; lobes sometimes three, terminal one acute; petiolar sinus deep, narrow; basal sinus often lacking; lateral sinus narrow, teeth shallow, narrow.
Flowers self-fertile, open late, stamens upright.
Fruit late, keeps well. Clusters large, long, broad, tapering, single-or sometimes double-shouldered, loose; pedicel with a few inconspicuous warts ; brush short, pale green. *Berries* of medium size, oval, dull purplish-red with thick bloom, firm. *Skin* thick, adherent, astringent; flesh green, translucent, juicy, fine-grained, vinous, sprightly, sweet and rich; very good. Seeds free, frequently abortive, two, broad-necked, distinctly notched, blunt, brown.

Delaware (Labrusca-Bourquiniana-Vinifera).
Hedrick opens his discussion of this grape as follows: "Delaware is the American grape *par excellence*. Its introduction raised the standard of quality in our viticulture to that of the old world, for there is no variety of *Vitis vinifera* more richly or more delicately flavoured or with a more agreeable aroma than the Delaware. This variety is rightly used wherever American grapes are grown as the standard whereby to gauge the quality of other grapes." The origin of this variety is uncertain. It has been traced back to the garden of Paul H. Provost in New Jersey, and was named Delaware by A. J. Downing in 1849 after the town from which the samples of its grapes had been sent to him by Abram Thompson, editor of the *Delaware Gazette* of Delaware, Ohio. "Delaware is the best American table-grape, and as such commands a premium in all of the markets,

selling oftentimes for double the price of Concord.   It is also much sought for by wine-makers both for Delaware wine and for blending in making champagne or other wines of high quality." It ripens a few days before Concord, looks pretty, keeps well on the vine and during transportation, and is highly resistant to black-rot.   Its faults or drawbacks are: small vine, slow growth, susceptibility of its leaves to downy mildew, its small berries.   On account of the first two points its vines should be planted closer than other varieties.   "It succeeds best in deep, rich, well-drained, warm soils, but even on these it must receive good cultivation, close pruning, and in some cases the fruit must be thinned."   On being pruned on the double Guyot system it gave good crops in the South of France.   Its wine was of excellent quality: very fine, high alcohol content, and a very pleasant bouquet.   It is only very slightly resistant to phylloxera.

Description.—
Vine weak, hardy, productive.

Canes short, numerous, slender, dark brown; nodes enlarged ; internodes short; tendrils intermittent, short, bifid.

Leaves small; upper surface dark green, dull, smooth; lower surface pale green, pubescent; lobes three to five in number, terminal one acute; petiolar sinus narrow; basal sinus narrow and shallow when present; lateral sinus deep, narrow; teeth shallow.

Flowers self-fertile, open late, stamens upright.

Fruit early, keeps well. Clusters small, slender, blunt, cylindrical, regular, shouldered, compact; pedicel short, slender, smooth; brush light brown. *Berries* uniform in size and shape, small, round, light red, covered with thin bloom, persistent, firm; flesh light green, translucent, juicy, tender, aromatic, vinous, refreshing, sweet; best in quality. Seeds free, one to four, broad, notched, short, blunt, light brown.

The following additional points are taken from Ravaz *(l.c.* p. 308): $a + \beta = 105°$, $\gamma = 44°$. Leaves thick, glossy on upper surface. Young leaves yellowish-green, shining. Bourgeonnement pale green and woolly.

N.B.—According to Ravaz it contains Labrusca-Aestivalis-Vinifera blood. Hedrick agrees with this view, except that he substitutes Bourquiniana for Aestivalis.

Goethe (Vinifera-Labrusca).

It was first mentioned in 1858 under the name of *Rogers' No.* 1 (Rogers bred his hybrids out of a Labrusca, Carter, or Mammoth Globe, with Black Hamburg and White Chasselas as fathers), but was named Goethe by Rogers in

1869. "Of all Rogers' hybrids, Goethe shows Vinifera characters most, resembling in appearance the White Malaga of Europe, and not falling far short of the best Old World grapes in quality." In the north it ripens with difficulty (considerably later than Concord). Where it does not, it is unexcelled in its class. It is a most vigorous grower. It does best in gravelly or sandy soils. It is fairly immune to downy mildew, rot, and other fungoid diseases. It should be pruned short. "Where it succeeds, the vines bear so freely that thinning becomes a necessity." For New York it ripens too late, hence it is mostly grown in the Middle and South Atlantic States and in the valleys of the Ohio and Missouri. It pays best as table grape, although it produces excellent wine. It is very little resistant to phylloxera.

Description.—
Vine vigorous, hardy.

Canes short, dark brown; nodes enlarged, flattened; internodes short; tendrils continuous or intermittent, long, bifid to trifid.

Leaves irregularly round, thin; upper surface light green, glossy; lower surface pale green, pubescent; leaf usually not lobed, terminus broadly acute; petiolar sinus narrow, closed and overlapping; basal sinus usually lacking; lateral sinus shallow, often a notch; teeth shallow, narrow.

Flowers partly self-fertile, open in mid-season; stamens upright.

Fruit late, keeps well. Clusters short, broad, tapering, frequently single-shouldered, usually two bunches to shoot; pedicel long, thick with numerous conspicuous warts; brush long, slender, yellowish-brown. *Berries* very large, oval, pale red, covered with thin bloom, persistent. *Skin* thin, tender, adherent, faintly astringent; flesh pale green, translucent, tender with Vinifera flavour; very good. Seeds adherent, one to three, large, long, notched, blunt, brown.

Concord (Labrusca).

"The Concord is known by all. The most widely grown of the grapes of this continent, it also represents the dominant type of our native species, and with its offspring, pure-bred and cross-bred, furnishes 75 per cent or more of the grapes of eastern America. In New York, approximately 75 per cent of all the grapes grown are Concords alone."

It was grown by Mr. E. W. Bull of Concord (Massachusetts) in 1843 out of seed from a wild grape. It was introduced in 1854, and its growth in popularity from this date was phenomenal. It is a very vigorous grower, which bears heavily and regularly. Concord succeeds in more kinds of soils than any other variety. This largely accounts for its great popularity. In calcareous soils it is subject to lime chlorosis. In the north it is highly resistant to fungoid diseases and insect pests. It is little resistant to phylloxera, and can only be grown on its own roots in fertile, deep, cool soils. It is a table grape for the masses. Concord is, *par excellence,* the grape used for the manufacture of grape juice. Under prohibition it is also used for making wine in and for the home, besides being used as a table grape. The grape looks attractive, but is not of high quality. The skin is inclined to crack and the berries to shell after picking. "The seeds and skin of Concord are objectionable, the seeds being large and abundant and difficult to separate from the flesh, and the skin is tough and unpleasantly astringent." It buds and flowers fairly late in spring, and the grapes are not easily injured by late frosts and hang well on the vine.

Description.—
Vine vigorous, hardy, healthy, productive.

Canes long, thick, dark reddish-brown; nodes enlarged, flattened; internodes long; shoots pubescent; tendrils continuous, long, bifid, sometimes trifid.

Leaves large, thick; upper surface dark green, glossy, smooth; lower surface light bronze, heavily pubescent; lobes three when present, terminal one acute; petiolar sinus variable; basal sinus usually lacking; lateral sinus obscure and frequently notched; teeth shallow, narrow.

Flowers self-fertile, open in mid-season; stamens upright.

Fruit mid-season; keeps from one to two months. Clusters uniform, large, wide, broadly tapering, usually single-shouldered, sometimes double-shouldered, compact; pedicel thick, smooth; brush pale green, *Berries* large, round, glossy, black with heavy bloom, firm. *Skin* tough, adherent with a small amount of wine-coloured pigment, astringent; flesh pale green, translucent, juicy, fine-grained, tough, solid, foxy; good. Seeds adherent, one to four, large, broad, distinctly notched, plump, blunt, brownish.

According to Ravaz: $\alpha + \beta = 110°$, $\gamma = 34°$.

Barry (Labrusca-Vinifera).

This is one of Rogers' hybrids between Labrusca and Black Hamburg. It was at first known as *Rogers' No.* 43. Hedrick says: "The Barry is one of our best black grapes, resembling in berry and somewhat in flavour and keeping quality its European parent, Black Hamburg". It is an attractive black grape with a delicate and sweet flavour, tender flesh, thin skin, and unobjectionable seeds. It keeps splendidly. Barry is a good table grape, good grower, and fairly resistant to fungoid diseases, but susceptible to mildew, fully ripe after Concord, commonly grown in all the gardens in eastern America, little resistant to phylloxera.

Description.—

Vine vigorous, hardy, productive, susceptible to mildew.

Canes long, numerous, thick, dark brown with heavy bloom; nodes flattened; shoots glabrous; tendrils intermittent, bifid or trifid.

Leaves large; upper surface light green, glossy, smooth; lower surface pale green, pubescent; lobes one to three, terminus acute; petiolar sinus deep, narrow, sometimes closed and overlapping; basal sinus usually lacking; lateral sinus shallow, narrow, teeth shallow.

Flowers open in mid-season, self-fertile, stamens reflexed.

Fruit mid-season; keeps well. Clusters short, very broad, tapering, often subdividing into several parts, compact; pedicel with small warts. *Berries* large, oval, dark purplish-black, glossy, covered with heavy bloom, adherent. *Skin* thin, tough, adherent; flesh pale green, translucent, tender, stringy, vinous, pleasant-flavoured; good. Seeds adherent, one to five, large, deeply notched, with enlarged neck, brown.

Herbert (Labrusca-Vinifera).

This grape was first known as *Rogers' No.* 44, but in 1869 Rogers called it Herbert. It is a beautiful grape. "In all that constitutes a fine table grape Herbert is about as near perfection as we have yet reached in the evolution of American grapes." It is a good grower and bearer, ripens at the same time as Concord, but keeps much longer and stands transportation well. As it is self-sterile, it should be interplanted with other varieties to be fertilised. It deserves to be grown.

Description.—

Vine very vigorous, productive.

Canes long, numerous, thick, dark brown; nodes enlarged, flattened; internodes long; tendrils intermittent, long, bifid or trifid. *Leaves* large, round; upper surface dark green, dull, smooth; lower surface pale green with some pubescence; leaf entire, terminus obtuse; petiolar sinus deep, narrow, closed, overlapping; basal and lateral sinuses lacking; teeth shallow.

Flowers self-sterile, open in mid-season; stamens reflexed.

Fruit mid-season; keeps well. Clusters large, broad, tapering, two to three clusters per shoot, heavily single-shouldered, loose; pedicel thick with small russet warts; brush yellowish-green. *Berries* large, round-oval, flattened, dull black, covered with thick bloom, persistent, firm. *Skin* thick, tough, adherent, astringent; flesh light green, translucent, juicy, tender, fine-grained, very good. Seeds adherent, three to six, large, broad, notched, long with swollen neck, blunt, brown with yellow tips.

Isabella (Labrusca-Vinifera).

For half a century after its introduction about 1816, Isabella and Catawba were the main varieties grown in eastern America. At the present time it has practically only historical value. In the north Concord has replaced it because it ripens earlier, is more productive, and hardier. Isabella now really occurs only in collections of amateurs. Its flavour is good, but the thick skin and muskiness in taste are objectionable. It has also a foxy taste. Its pretty green leaves and great vigour make it a very suitable and ornamental vine for pergolas. It was obtained from Mrs. Isabella Gibbs of Brooklyn, New York, by William Prince of Long Island in 1816, and called Isabella by him.

The foliage is fairly susceptible to powdery mildew or oidium and the grapes to black-rot. Hedrick is of opinion that it contains some Vinifera in addition to Labrusca blood, while Ravaz regards it as a pure Labrusca.

It was imported into France in 1820 as an ornamental grape, and has been grown to a considerable extent in France and Italy since 1853, when the oidium or powdery mildew wrought such havoc among the grape crops of these countries, while its grapes remained sound. It is only slightly more resistant to phylloxera than Vinifera varieties. It is erroneously called Catawba in the Transvaal, and is grown on pergolas and trellises in several parts of South Africa with summer rainfall.

Synonyms.—Raisin du Cap, Black Cape, Constantia, Strawberry grape, Framboisier, etc.

Description.—

Vine vigorous, hardy, productive.

Canes short, numerous, with heavy pubescence, thick, light brown; nodes enlarged, flattened; internodes short; tendrils continuous, long, bifid or trifid.

Leaves thick; upper surface dark green, smooth, glossy; lower surface whitish-green, heavily pubescent; lobes three when present, with terminal lobe obtuse; petiolar sinus shallow, narrow, often closed, overlapping; basal sinus usually wanting; lateral sinus shallow, narrow, frequently notched; teeth shallow, wide.

Flowers self-fertile, open in mid-season; stamens upright.

Fruit late, keeps and ships well. Clusters large, cylindrical, frequently single-shouldered; pedicel slender, smooth; brush long, yellowish-green. *Berries* medium to large, oval, black with heavy-bloom, persistent, soft. *Skin* thick, tough, adherent, astringent; flesh pale green, translucent, juicy, fine-grained, tender, meaty, some foxiness, sweet; good. Seeds one to three, large, broad, distinctly notched, short, brown with yellow tips. (In my experience the dead ripe berries drop easily from the bunch.—A. I. P.)

B. EUROPEAN GRAPES *(Vitis vinifera)*

I. WINE VARIETIES

(a) White Wine Varieties

1. Greengrape.[1]

Synonyms.—Groendruif, Wyndruiwe *(i.e.* wine grape).

General.—Greengrape is one of the oldest varieties grown at the Cape, and it was probably amongst the varieties imported by Van Riebeek in 1653. It is to-day still the predominant white wine variety. Greengrape and Hermitage (for red wine) are the two leading wine varieties grown at the Cape as far as quantity is concerned, and they bulk about equally in the total wine crop. There is a White and a Red Greengrape differing only in the colour of their ripe grapes and in the autumnal colour of their leaves, those of Red Greengrape being red in both cases, while those of White Greengrape are greenish-yellow and yellow respectively. Among the White Greengrape we find two types differing from the normal. The one has thick leaves and large, round, soft berries (worthless), the other rather short bunches, with fairly large, round, yellowish, firm berries. The latter gives

---

[1] *2012 Editor's Note*: Greengrape was later identified as Semillon. PFM

decidedly smaller crops than the normal Greengrape, but sweeter must and better wine, and agrees very closely with the Semillon of Sauternes. I have not yet been able to determine the identity of Greengrape compared with varieties grown in Europe.

Culture.—It is a good grower and free bearer when pruned short, ripens mid-season, and gives musts with 19-26° Balling and 6.5-5‰ total acidity.[1]   Its ripe grapes are soft and juicy, therefore very good for children and old people. For light wines it should be picked when just ripe and the juice separated immediately from the husks.   If the must ferments in contact with the husks, the wine will have an unpleasant flavour *sui generis*.   For heavy, full-bodied wines, for brandy and for grape juice, it should be picked when fullyripe or slightly over ripe.   Its products are of only fair average quality.   It might to advantage be replaced by better varieties, *e.g.* White French, Stein, Semillon, Sauvignon blanc, and some Clairette blanche.

It is practically immune to anthracnose, and not very susceptible to oidium and sunburn.[1] In moist spots some of its berries develop noble-rot. It possesses a good affinity for practically all American stocks, and grafts well.

Description.—(A. I. P.)[2]

Bourgeonnement red, especially the low edges and lower surfaces of the little leaves, which show a woolly pubescence on both surfaces.

Canes thick and brittle, internodes medium long, pith fairly thick, reddish-brown when fully ripe. Growing tip white with little red, woolly.

Leaves entire, 3- to 5-lobed, medium large; petiolar sinus fairly open; at first light, then darker green, light pubescence below. Young leaves garnet reddish but soon becoming green, above slight, below heavy woolly pubescence.

Bunches medium size, conical, seldom double, berries roundish to round, greenish-yellow. Mid-season.

N.B.—Sometimes red and white bunches occur on the same shoot; sometimes red and white berries in the same bunch; sometimes red and white occur on the same berry.

2. **Stein.**

Synonyms.—Steen, Vaalblaar, Vaalblaar Stein.

---

[1] I always express the total acidity as "*tartaric acid*".

[2] Where not otherwise stated, the descriptions were taken mainly from Mas et Palliat, *Le Vignoble*.

General.—It is one of the oldest varieties grown at the Cape. In Paarl, Stellenbosch, and Constantia it is largely-grown; elsewhere much less. It corresponds more closely with Sauvignon blanc than with any other European grape, as far as I know, but is not identical with it.[1] When ripe its must shows 22-27° Balling and 7.5-6‰ total acidity, which is 0.5-1‰ higher than that of Greengrape. Its wine has a typical Stein flavour which becomes stronger when the must ferments in contact with the husks, and also after the wine has been in the bottle for a couple of years. On the whole it gives wine of a higher quality than Greengrape does.

Culture.—Its shoots adhere firmly to the stem and are not easily blown off by strong winds. For this reason it is preferred in parts that are subject to strong winds, especially violent south-easters. The old stem develops plenty of water-shoots, which are troublesome to remove during winter pruning, and should therefore be removed when young. It bears well with short pruning, and answers best on sandy soil or a light sandy loam, where its grapes assume a fine golden - yellow colour when fully ripe. It is immune to anthracnose but very susceptible to oidium—much more so than Greengrape. It should therefore be sulphured with special care. Its affinity for Aramon is rather unsatisfactory, but good for most other stocks.

Description.—(A. I. P.)

Bourgeonnement reddish-brown; little leaves brown above along the border and red below, densely cobwebby above and below.

Canes medium thick, inclined to be somewhat thin; internodes fairly short, tough, greyish-brown when ripe.

Leaves entire to 3-lobed, medium size; above, almost glabrous, below, woolly; petiolar sinus V-shaped, medium open. Young leaves with numerous long loose threads above, and strongly woolly below. Growing tip reddish-green with dense felt-like pubescence.

Bunches somewhat small to medium size, usually double, broadly shouldered. *Berries* small, distinctly oval, with well-developed spots of cork. *Skin* greenish to golden yellow. Mid-season.

3. **White French.**

Synonyms.—Fransdruif, Vaalblaar, Palomino (?).

General.—It, too, is one of the oldest varieties grown at the

---

[1] *2012 Editor's note:* Perold's protégée and successor at Stellenbosch University, Professor C J Orffer, finally confirmed it as identical to the French Loire variety Chenin Blanc in 1963. PFM.

Cape, and is practically identical with Palomino[1], which forms the basis of the Spanish sherries. Unfortunately it is grown much less than the preceding varieties. It is most common in Paarl, then on the sweet soils of Worcester (Wilgerivier, etc.), Robertson, and Montagu; also in Tulbagh and elsewhere on a smaller scale.

Culture.—It is a vigorous grower and good bearer when short pruned, 1-2 bunches per shoot. Its big bunches ensure a good crop. It prefers sandy soils to medium loam, and river soils. It ripens fairly early, and is also sold locally as a table grape. Its total acidity is low, being only 4-4.6‰. In Paarl, Stellenbosch, Tulbagh, and on other soils poor in lime, its must usually shows 19-21 per cent sugar, which rises to 27 per cent and more in the sweet soils of Worcester, Robertson, and Montagu. It is the basis of our popular Witzenberg wine, and is eminently suitable for *light* white wine, sherry, white jeripico with a neutral flavour, cognac brandy, and grape syrup.

It is susceptible to anthracnose in Constantia, Stellenbosch, and Paarl, but not in the drier climate of Worcester, Robertson, and Montagu. Its susceptibility to oidium is about the same as that of Greengrape. It is particularly susceptible to "curly leaf" (the leaves are covered below with cobwebby threads and curl in to the lower surface; where it is bad, the leaves remain small and the internodes short, shoots thin and sickly; probably due to a mite). "Curly leaf" has thus far not caused any damage, although it has long been noticed. Together with Stein and Canaan grape, it resisted the phylloxera longest when at first attacked. Its shoots resist strong winds fairly well, and it has hardly ever been known to have suffered from sunburn. *It deserves to be grown largely on account of its many good qualities.*

Description.—(A. I. P.)
Bourgeonnement white and green, little leaves strongly pubescent on both surfaces; soon becomes shining with less and loose threads on the upper surface.

Shoots medium thick to thick, internodes fairly long, with red stripes; when ripe, light colour.

Leaves 3- to 5-lobed; upper sinus deep, lower sinus shallow, petiolar sinus V-shaped, usually closed; teeth sharp, large, leathery, thick; almost glabrous above, dense woolly pubescence below. Growing

---

[1] *2012 Editor's Note:* Perold was correct: Greeengrape *is* Palomino.

245

tip greenish-white; dense woolly pubescence also on the shoot; young leaves white-yellowish-green, strongly pubescent above; dense felt-like, white pubescence below.   Leaf-stalk red.

Bunches large, broadly shouldered, long, conical (sometimes roundish). *Berries* spherical, medium size, white to light yellowish to golden yellow at full maturity, brownish when over-ripe. *Skin* thin, soft. Early mid-season.

4. **Canaan Grape.**

Synonyms.—Kanaandruif, Beliesdruif, Vaalblaar.

This grape was thus named because its bunches are larger than those of any other variety grown at the Cape, where it has now been grown for a long time. The vine is a most vigorous grower. It gives heavy crops on account of the tremendous size of its bunches, not because they are very numerous. The wine produced from it is usually light, of very ordinary and even poor quality, though good for distilling into brandy and spirits. It prefers sandy soils and light loams.

Description.—(A. I. P.)

Canes thick, reddish-brown, fairly long.

Leaves large; soft, dense, fine woolly pubescence below; above, more or less glabrous.

Bunches very large, roundish; berries spherical, medium size, very juicy, yellowish-green.

5. **Pinot blanc Chardonay.**

Synonyms.—Weisser Rulander, Weisser Clavner, Chardonnay, Pinot blanc Chardonnay, etc.

Culture.—It is grown mainly in Burgundy and Champagne, and produces the famous Chablis wine. It is also used in making champagne. It produces good wine. At the Cape it suffers very much from coulure and millerandage and thus gives poor crops. But it grows fairly well in poor soils. It must be long pruned.

Description.—

Vine with medium vigour, but more than Pinot noir.

Shoots medium, thin; internodes short, more yellowish than those of Pinot noir.

Leaves medium size, above glabrous, below slightly pubescent; sinuses little developed, as long as broad.

Bunch small, short, cylindro-conical, compact. *Berries* small, spherical, light green with a golden colour on the sunny side; pulp firm, juicy. Mid-season (twelve days after Pinot noir).

246

6. Sauvignon blanc.

Synonyms.—Sauvignon,   Sauvignon   jaune, Feigentraube.

Culture.—It is probably the finest grape for quality in Sauternes, where it is grown together with Semillon and a little White Muscadel (sometimes also a little Metternich, *i.e.* Riesling) to produce the world-famous Sauternes wines. At the Cape it also produces an excellent wine, but is unfortunately grown very little.   It is a vigorous grower, and sufficiently productive if pruned to at least three good eyes. With shorter pruning it produces too little. It gives a finely bouqueted wine with a pretty, golden colour if picked when fully ripe. In sandy soils, broken soils, and decomposed granite soils it does well.   It ought to be grown for a good light wine and also for a fine heavy or even a sweetish wine.

Description.—

Bourgeonnement woolly, light rose-coloured around the border and on the lower surface of the little leaves.

Canes long and inclined to be thin, yellowish-grey with brown spots, short internodes.

Leaf medium to rather small, 3-lobed, somewhat bullate, glabrous above and light cobwebby pubescence below; teeth broad, obtuse.

Bunch small, cylindrical to cylindro-conical, compact. *Berries* medium to small, short oval, with corky spots, and a reddish-golden tinge when fully ripe; pulp fairly firm, juicy, very sweet, with a special aromatic flavour. *Skin* fairly thick. Mid-season.

7. Sémillon.

Synonyms.—Colombier, Chevrier, Sémillon blanc.

Culture.—It is grown mainly in Sauternes, where it is the principal variety, and where its berries develop the noble rot ("pourriture noble"), whereby they are turned into half raisins which give the very sweet musts (over 30 per cent sugar) from which the world-famous liqueur wines of Sauternes are made.[1] Foëx considers it to be one of the most famous white wine grapes grown in France, while Mas and Pulliat[2] call it "le grand cepage blanc francais". It combines maximum quality with maximum yield. For good crops it must be pruned to at least three eyes, as the crop will

---

[1] Foëx, G., *Coura complet de viticulture,* 4th ed. 1895
[2] Mas et Pulliat, *Le Vignoble,* 3 vols. 1874-1879. Paris.

otherwise be too light.   Its leaves and canes resemble those of Greengrape very closely.[1]   It gives a very sweet must and a good wine.

Description.—
Vine fairly vigorous.

Bourgeonnement very woolly, violaceous white passing into yellowish-white.

Shoots thick and vigorous, long, internodes long, somewhat flattened, light reddish-brown when ripe, like those of Greengrape.

Leaves large, thick, 3- to 5-lobed, glabrous above, light cobwebby below.

Bunch thick, fairly compact, cylindro-conical, medium to rather small. *Berries* fairly large, nearly spherical, pretty golden colour when properly ripe, fairly firm, thick skin, special flavour, sweet and pleasant. Mid-season.

8. **White Muscadel** (Muscat blanc).

This is one of the oldest varieties grown at the Cape. It starts budding early and ripens early.   It grows and bears well with short pruning.   The must of its fully ripe grapes contains up to 30 per cent sugar, and thus it gives heavy and sweet wines.   It is mainly used for making sweet wines and jeripico with a strong muscat flavour.   These products are very valuable.   It is a nice table grape, but is not suitable for export, as its berries are too small and do not stand long transport well.   It answers well on broken soils, alluvial soils, and especially on the sweet karroo soils and river soils in Worcester, Robertson, and Montagu, where it is most productive, and yields the finest and most highly aromatic products, and is therefore grown most extensively. The *Red Muscadel* differs from it only in the colour of its ripe grapes. They are both fairly susceptible to oidium and anthracnose, and graft very badly on Aramon; on other stocks they succeed better.

Description.—
Vine vigorous, upright grower.

Bourgeonnement violet-red, dense mass of white threads. Little leaves strongly cobwebby (below more than above), but this soon becomes less intense when they are glossy and green.

Canes fairly thick and long; internodes long; whitish light brown with brown dots when ripe.

Leaf medium size, 5-lobed; glabrous above, slightly woolly below, especially on the nerves.   Leaf-stalk fairly long.   Growing tip

---

[1] 2012 Editors Note: Greengrape was later identified as Semillon. PFM.

whitish-green with little red; young leaves garnet red with green nerves, above a few threads, below fairly woolly. Upper sinus deep, lower one shallow, petiolar sinus narrow V-shaped.

Bunch medium size, sometimes fairly broad shouldered and moderately long, sometimes almost cylindrical. *Berry* roughly spherical, small to medium, light yellowish to somewhat brown on the sunny side when fully ripe. Very sweet with pleasant muscat flavour when fully ripe. Early.

9. **Muscat de Frontignan.**

Of this grape a *white* and a *pale red* variety are known. The latter is one of the oldest varieties grown at the Cape under the name of *Frontignac.* In the olden days it was mainly used for making the then world-famous sweet Constantia wine, also known as Cape Wine. In France both varieties are grown mainly in the neighbourhood of Frontignan and Lunel for the production of sweet wines. They closely correspond with White and Red Muscadel in their characters, except that they are less productive and that their bunches are smaller and shorter than those of Muscadel. They also ripen somewhat earlier, become sweeter, and possess a stronger and finer muscat aroma than Muscadel. Unfortunately Frontignac is now not largely grown at the Cape. Although of only moderate vigour, it is one of the finest grapes for the production of a sweet wine of the highest quality, and deserves to be grown for this purpose in fairly warm wine countries.

10. **Clairette blanche.**

Synonyms.—Clairette, Petite Clairette, Clarette, Blanquette, etc.

In France it is grown in the coastal region along the Mediterranean from Nice to the Spanish border. It usually does well where the olive thrives. It has been grown in France from time immemorial. It gives a good white wine which improves the quality, and especially the bouquet of other light white wines when blended with these. Its wine is light, and has a characteristic and very pronounced aroma, which makes a great impression when the wine is a year old. After about three years it becomes largely lost, and the wine retains little character. It should therefore be blended with other wine and consumed when about one and a half to two years old. This, at least, is according to my experience with

249

it at the Cape. In France it is sometimes grown to be sold as a late table grape, as it ripens late and keeps very-late in winter. As a table grape it is, however, inferior and not at all attractive.

During the last twenty years it has been grown on a fair scale in Groot Drakenstein (Paarl) for blending with the wine made from Greengrape and other locally grown white varieties. The vine is healthy and vigorous. It is highly resistant to oidium and anthracnose. With short pruning it bears well. It does well on practically all deep soils. Where the vines are exposed to strong winds in spring and early summer, many young shoots are blown off. Hence it cannot be recommended for such localities. It ripens rather late and gives many second crop bunches.

Description.—
Bourgeonnement strongly white woolly, violaceous-red around the borders of the developing little leaves.
Shoots medium thick to thick, upright growth, short internodes; leaves drop very late.
Leaves medium, dark green, glabrous above, strongly woolly below; 5-lobed, petiolar sinus usually closed.
Bunch medium, cylindro-conical, well branched and fairly loose (inclined to coulure); peduncle fairly long, medium thick. *Berry* short oval, below medium size; pedicels fairly long and somewhat thin. *Skin* thin, fairly tough, passes from greenish-white to white, with yellow patches at maturity. Pulp fairly firm, juicy, very sweet, pleasant taste. (My own experience with this variety has thus far been that its juice is not very sweet and has a fairly high total acidity. —A. I. P.)   Late mid-season.

11. **Folle blanche.**

*Synonyms.—*Fol, Enrageat, Plant de Dame, Enrage, etc.

This is the grape out of which the world-famous French brandy, Cognac, has mainly been made since the seventeenth century. In France it is a very vigorous grower and heavy bearer. It answers well in all kinds of soils, but soon exhausts itself, and weakens in light soils unless heavily manured. At the Cape it bears heavily with short pruning, but develops only little vigour when grafted. On Jacquez it develops most vigour. In the limestone soils of the Charentes it does well on 41B. At the Cape it has not yet been tried on this stock. Here its must shows 20°-22°

Balling, with a total acidity of about 8‰.   In Cognac its must shows about 17° Balling with a total acidity of easily 15 ‰, with up to 10‰ free tartaric acid, and its wine has a strength of 8-10 vol. per cent alcohol.   At the Paarl Viticultural Experiment Station I have distilled and matured a most excellent brandy out of its wine, according to the methods adopted in Cognac.   It resembles a good French Cognac very closely, and is of equal quality.

Description.—

Bourgeonnement reddish-grey, heavy woolly pubescence, tinted violaceous red before passing into yellowish-green. *Vine* very vigorous, very fertile.

Shoots thick, erect or half erect, internodes rather short.

Leaves below medium, glabrous and somewhat bullate above, tinted reddish on the nerves, covered with a fairly dense cobwebby pubescence below; upper sinuses deep, usually open; lower sinuses well marked; petiolar sinus open or slightly so; petiole short, medium thick, tinted slightly rosy ; teeth short, obtuse or rounded off, obtusely acuminate.

Bunch above medium, compact, cylindro-conical, peduncle short and thick. *Berries* medium, spherical when loose, with short and thick pedicels. *Skin* thick, fairly tough, light green colour passing into golden yellow on a good site when fully ripe. Pulp soft, very juicy, sweet, simple flavour. Mid-season

12. **Ugni blanc**.

Synonyms. — Trebbiano, Trebbiano fiorentino, Saint - Emilion, Maccabeo (in the Roussillon).

This is a very old variety. It is grown mainly in the Provence and in Tuscany. In the Charentes it is grown for Cognac in the deep, moist soils, where it gives a wine with 10-11 vol. per cent alcohol and a lower total acidity than Folle blanche. In Tuscany the Trebbiano is dried slightly on trays or mats, and then used for making *Vino santo* or for the *governo* in making the famous *Ohianti* wine. In Provence it gives a good, strong wine on warm sites. It is sensitive to oidium and grey rot *(Botrytis cinerea)*. It is a vigorous grower and good bearer with short pruning on good soil and with heavy fertilising.   Its grapes keep well.

Description.—

Bougeonnement white woolly, tinted slightly violaceous rosy.

Shoots of medium thickness, long; internodes fairly long; vine igorous, fertile, grows very old.

Leaves medium, yellowish light green, glabrous above, covered below with a striking and fairly thick cobwebby pubescence; lower sinuses deep, usually closed, often unequal or marked only on one side; secondary (lower) sinuses well marked, narrow, not attaining the length of the middle nerve; dentition uneven, fairly broad, sharp.

Bunch large, loose, somewhat winged, long cylindro-conical, ordinarily with a little bunch at the fertile node of the peduncle; peduncle fairly long or long, rather thin or thin. *Berries* medium, almost globular, pedicel long and rather thin. *Skin* fairly thick, tough, at first dull white, passing into more or less dark yellow according to degree of exposure, and often into light rosy along very hot and stony hills when ripe. Pulp fairly firm, very juicy, very sweet, simple flavour. Late.

13. **Riesling.**

Synonyms.—Weisser Riesling, Gewürztraube, Rheingauer, Kleinriesler, Kleinriesling, Metternich (in Sauternes), Rheingauer Riesling, White Rissling (Dr. Hogg), etc.

General.—The famous Rhine and Moselle wines [Hocks and natural liqueur ("Auslese") wines] are made mainly out of Riesling grapes, and sometimes also from Roter Traminer (mainly in the Palatinate). Along the Moselle the wine-growers distinguish between a yellow-berried Riesling with small bunches and a green-berried Riesling. In Germany it takes the place of Pinot blanc, Sauvignon blanc, and Semillon in France. In Sauternes it is grown to a limited extent under the name Metternich. In Austria it is known as Kleinriessler, Kleinriesling, and Rheingauer Riesling. At the Cape odd vines of a variety known as Old Stein or Green-leaf Stein occur in the Stein vineyards. This variety is very much like Riesling, if not identical with it. There also exists a red variety of Riesling which differs from ordinary Riesling in the colour of its ripe grapes.

Riesling produces everywhere a good wine, but it is only in the northern viticultural area (Rhenish Palatinate, Rheingau, and Moselle, with its tributaries Saar, Ruwer, etc.) that it produces the world-famous wines (Schloss Johannisberger, Rudesheimer, Steinberger, Rauenthaler, Liebfrauenmilch, Deidesheimer, Bern Casteler Doctor, Scharzhofberger, Niersteiner, Hochheimer, etc.) with their characteristic, strong, lovely bouquet or aroma, taking the first place among the world's light, still, dry, white wines. On favourable sites its grapes are allowed to develop *noble rot,* when, as in Sauternes, these noble rot berries (half raisins) are gathered one by one, and give musts with over 30-40 per cent sugar. They give, after a long fermentation extending over several years, the unexcelled glorious natural sweet wines (liqueur wines), called "Auslese Weine" in Germany, to which the best Sauternes wines approximate closely though with a different character. Its dry wines have 10-13 vol. per cent alcohol and 7-6‰ total acidity. It is a vigorous grower and good bearer with short pruning, not very susceptible to oidium and anthracnose, more so to downy mildew *(Plasmopara viticola).* It starts budding fairly late. At the Cape it gives good, fruity light wines, and also splendid, full-bodied heavy wines if picked very ripe. It should be grown where quality is aimed at, especially as it yields good crops. In warm wine countries, as also at the Cape, its wine does not possess the fine, strong bouquet so typical of its wines in the Rheingau, Rhenish Palatinate, Rhenish Hessia, and the Moselle Valley, although they still have a fine bouquet and high quality.

Description.—(After Babo u. Mach.)

Bourgeonnement woolly, reddish.

Shoots medium (to fairly thick), half erect, internodes medium; ripe wood light brown, striped, with strongly developed tendrils. Growing tips yellowish-green, slightly woolly, with faint reddish hue.

Leaves medium, more or less round, deeply lobed (upper sinuses broad and deep; lower sinuses usually slightly marked; petiolar sinus closed or almost closed. After Mas et Pulliat), teeth obtuse and broad, strongly bullate, above dark green and glabrous, below pale grey-green and fairly pubescent (long, flat, slightly cobwebby hairs on nerves below. M. et P.).

Bunches small, short and broad (cylindrical, conical. M. et P.), usually simple or slightly branched. Peduncle short (and fairly thick. M. et P.); pedicel short, stiff (thick. M. et P.).

Berries small, round. *Skin* thin, greenish, yellow, greenish-yellow, with brown spots and white bloom, passing into pale red when overripe. Pulp fairly firm, with a pronounced and characteristic flavour, reminding one of the wine's bouquet. Late mid-season.

14. **Roter Traminer**.

Synonyms.—Traminer, Gewürztraminer, Fränkischer, Savagnin rose, Savagnin (Jura), etc.

253

General.—White Traminer or Savagnin blanc differs from the red variety only in the colour of its ripe grapes. The red variety is the one most grown. The main region where it is grown is the Rhenish Palatinate and Franconia (Wurzburg), then in the Nahe and Moselle Valleys, and much less in the French Jura, Lower Austria, etc.

Its grapes have a typical, strong, pleasant aroma which we find again in its wine. It produces most excellent wines, *e.g.* Forster, Deidesheimer, Bocksbeutel, Klosterneuburger, etc. In deep, fertile sandy soil it grows and bears well with long pruning. It is highly resistant to oidium and arithracnose. Its must is less acid than that of Riesling; 5‰ total acidity corresponds with 20 per cent sugar, and when well ripe the sugar rises to 24-26 per cent with a corresponding total acidity of about 4‰. It ripens fairly early, and can be recommended for the production of a full-bodied, fairly strong wine of high quality.

Description.—
Bourgeonnement very woolly.
Shoots thin, long, creeping, short internodes. Eyes very small. Leaves drop late. Ripe canes have a dark brown colour.
Leaves below medium or small, almost round, finely hullate, dark green and almost glabrous above, grey-green and very cobwebby below; 3-lobed, but upper and lower sinuses almost zero; petiolar sinus more or less closed; teeth shallow and obtuse; petiole and nerves reddish at their point of meeting. Growing tip pretty, white woolly.
Bunch small, broad, branched, cylindro-conical, compact, peduncle fairly thick. *Berries* small, ellipsoidal; pedicels short, fairly thick. *Skin* thick, tough, light red to brownish when fully ripe. Pulp firm, juicy, very sweet, somewhat slimy, fine, strong aroma. Mid-season.

15. **Furmint.**

Synonyms.—Formint, Zapfner, Gelber Mosler, Szigeti.

General.—This is the national and most famous grape of Hungary, where it occurs in the best vineyards usually on the hills, especially those of the Hegy-Allya, where the world-famous Tokay wine is made from its over-ripe berries, and it is known as Furmint or Zapfner. In Styria it is widely grown under the name Gelber Mosler. It grows vigorously and bears well with short pruning. It can give a sweet must with a fairly high total acidity. At the Cape it has given splendid crops, but thus far not very sweet musts.

Description. —

Bourgeonnement very woolly, white, with a little faint red. *Canes* thick, erect, yellowish rusty with brown dots; nodes swollen, internodes fairly long.

Leaves medium, somewhat thick, glabrous or slightly woolly above; fine, dense, thick cobwebby pubescence below, especially on the old leaves; upper sinuses rather shallow; lower sinuses hardly marked; petiolar sinus almost closed; petiole somewhat violaceous, slightly pubescent, long or fairly long, medium thick.

Bunch medium and sometimes larger, rather compact when the berries have set well, peduncle of medium thickness and rather short. *Berries* medium or slightly larger, oval; pedicels long, rather thin, swollen bourrelet. *Skin* thick, passing from greenish-white to golden-yellow on a good site, somewhat subject to rotting. Pulp rather firm but juicy, sweet, pleasant simple flavour. Late mid-season.

16. **Roter Zierfahndler.**

Synonyms.—Rotreifler, Spätrot, Gumpoldskirchner.

General.—It is spread over a large area in Lower Austria, but is grown more particularly in Gumpoldskirchen near Vienna, where the popular Gumpoldskirchener wine is made principally from this grape. It is a very pleasant, aromatic table wine. It requires a fertile soil and a fairly warm climate. In 1910 I imported it to the Cape, where it does well, it grows and bears well with short pruning. It is highly resistant to oidium and anthracnose,but is susceptible to downy mildew *(Plasmopara viticola)*. Its grapes ripen rather late and do not rot easily.

Description.—(Babo u. Mach.)

Vine very vigorous grower.
Canes fairly thick, reddish-brown, long internodes.
Leaves medium, deeply lobed (5-lobed, orbicular, thick, leathery. A. I. P.), light green, somewhat bullate, teeth obtuse, fairly woolly below. Growing tip fine woolly, light green, somewhat bronze coloured. In autumn the leaves turn yellow with red patches.
Bunch large, cylindrical, compact. *Berries* medium, spherical, light red or flesh-coloured, remaining green only where shaded. *Skin* thin (with numerous corky dots. A.I.P.). Pulp very juicy, pleasant taste, without aroma. At 20 per cent sugar the must contains 7 ‰ total acidity. On favourable sites its must contains as much as 28 per cent sugar when fully ripe. Late mid-season.

17. Catarratto.

Synonyms.—Cataratto, Catarattu, Catarrattu.

General.—Catarratto is the principal and classical grape of Sicily, and has been grown there from time immemorial. It forms the basis of the white, yellow, and heavy wines of Sicily. In the vineyards of Palermo, Trapani, Marsala, Girgenti, and Syracuse, it occupies a prominent place, and it is Catarratto which has made the Marsala wines so famous. Untrellised and with short pruning, it yields 5-6 leaguers wine per morgen or 34.5-41.4 hectolitres per hectare. Its must easily reaches 25-27 per cent sugar. The untrellised vine should be given many arms and short bearers (spurs). It is highly resistant to oidium and drought. The vine is a vigorous grower and good bearer with short pruning. Its ripe grapes have sufficient acidity. It (like Inzolia) has so little affinity for 1202, 101-14, Aramon ×Rup. Ganzin No. 2, Rup. Martin, and Gamay Couderc, that it cannot be grown on them. This was F. Paulsen's experience in Palermo.[1] In my ampelographic collection on the Stellenbosch University Farm, both Catarratto and Inzolia died on 1202, 101-14 and Aramon No. 2 after a few years, thus confirming Paulsen's experience with these varieties in Sicily. On Jacquez, Rip. gloire, Aramon No. 1, Rup. du Lot, 106-8, and 420A, they have so far both done very well in my collection in a sandy loam. Paulsen mentions Rup. du Lot, Rip. gloire, Chasselas x Berlandieri 41B, and Aramon No. 1 as good stocks for Catarratto and Rup. du Lot, Rip. gloire, and Aramon No. 1 for Inzolia. With regard to these latter, my local experience has also been in agreement with that of Paulsen.

Description.—(Paulsen in Viala-Vermorel, *l.c.*)

Bourgeonnement white, woolly.

Shoots of medium length and thickness, few hairs on green shoots; when ripe they are glabrous and light chestnut-coloured; wood rather soft with little pith.

Leaves medium, somewhat longer than broad, roundish, 3-lobed, sometimes 5-lobed, thick and soft; lower sinuses deep, upper sinus slightly marked; petiolar sinus open in young leaves and usually closed in adult leaves; crinkled and undulating; glabrous, green above, covered with a dense, white woolly pubescence below; nerves strongly

---

[1] See Viala-Vermorel, *l.c.* vi. pp. 224-225.

developed and outstanding; two series of broad teeth. Petiole of medium length, thick, covered with some cobwebby hairs.

Bunches medium or larger, conical, generally winged, sometimes developing into two nearly equal bunches ; peduncle short, thick, yellowish-green, pedicels short. *Berries* round, medium. *Skin* hard, fairly tough, little subject to rotting. Pulp juicy, greenish-white; golden yellow colour on the side of the light when fully ripe, and light green in the shade. Seeds 1-2 of medium size. Late mid-season.

18. **Inzolia bianca.**

Synonyms.—Insolia bianca, Inzolia, Nzolia

General.—It is grown together with Catarratto for the production of Marsala wine, and also for dry wines in Palermo, in Calabria, etc. It gives good wines. It is a vigorous grower and very productive. Its grapes look pretty, and serve both as table grape and as wine grape. It can become very sweet, when its must easily contains as much as 25 per cent sugar with 5-6 ‰ total acidity. Its affinity for American stocks has already been discussed under Catarratto. According to Paulsen it also fails on 3306 and 3309.

FIG. 52-Inzolia bianca (near Palermo). Original

Inzolia nera is merely a black type of Inzolia bianca, and, like it, ripens late mid-season.

Description.—(Paulsen in Viala-Vermorel, *l.c.)*

Vine vigorous, thick trunk, upright; strong thick bark in strips.

Bourgeonnement somewhat pubescent, light rosy, passing into green, glossy, glabrous (A. I. P.).

Shoots long, thick, pale green, striped; when ripe light brown, rusty brownish, with darker zones; internodes irregular, usually long, irregularly and strongly striped; nodes hardly marked; pith not very thick; wood hard; tendrils discontinuous, strong, bifid.

257

Leaves medium, as broad as or broader than long, 5-lobed, sinuses very deep and almost elliptical; upper sinuses very deep (up to two-thirds of the main nerves) and little open, sometimes lobes even superposed; lower sinuses less deep but well developed, sometimes with secondary sinuses towards the petiole; petiolar sinus deep, little open, with parallel borders; teeth sharp and well developed; completely glabrous, yellowish light green above, very slightly pubescent on the nerves and greyish-green below, nerves prominent below. Petiole thin, long, reddish-brown.

Bunches large, long, conical, broad-shouldered, often branched; peduncle short, thick, lignified near cane; pedicels long, thin. *Berries* medium, slightly oblong, olive-shaped. *Skin* thick with white bloom; when rubbed off a yellow skin is seen: golden yellow on sunny side. Pulp crackling, little juicy, sweet, very pleasant to eat. Seeds rather large, usually 2.   Late mid-season.

19. **Pedro Jimenez.**

Synonyms—Pedro Ximines, Pedro Ximen, Pedro Ximenès.

General.—This grape is largely grown in Andalusia. In Jérez de la Frontera it is used in the making of Sherry wine, in San Lucar de Barramedes for making Manzanilla, and in Malaga for making the Malaga sweet wine. It is a good wine grape for *hot* wine countries, and is a very vigorous grower and heavy bearer with very short pruning. According to Don Simon de Rojas Clemente y Rubio and others (Mr. Gonzalez also told me this during my visit to Jerez in 1909) it is not a Spanish grape, but came originally from the Rhine and Moselle district in Germany, whither it was imported from the Canary Islands.

At Jerez its ripe grapes are allowed to shrink somewhat by being left for some days in the sun on mats on the ground, when its extra sweet must is used for making jeripico and grape syrup, which are blended with the dry wines to make sweet and sweetish sherries. It is fairly susceptible to *Plasmopara vitieola. Unless it is pruned very short it bears too heavily and its grapes do not become sweet enough.*

Description. —
Bourgeonnement slightly pubescent, yellowish-green, passing into brilliant green. *Shoots* fairly thick, half erect, thickened at the nodes; internodes somewhat short.

Leaves medium or larger, light green (to yellowish-green: A. I. P.), glabrous and smooth above; few loose hairs below; upper sinuses

258

closed; lower sinuses fairly deep, slightly open; petiolar sinus almost closed; petiole of medium length and thickness; teeth unequal, the strongest ones broad, somewhat obtuse and shortly acuminate.

Bunch above medium or large, branched, conico-cylindrical, fairly compact on young and vigorous vines, somewhat loose on old vines; peduncle long, more or less thick. *Berries* medium, somewhat ellipsoidal *(i.e.* short oval); pedicels long or fairly long and thin. *Skin* thin and yet tough, passing from greenish-white to golden yellow with bloom on a good site, at complete maturity. Pulp very juicy, sweet, rather soft, simple flavour, pleasant.   Late mid-season.

20. **Codega.**

This is probably the same grape as Malvasia grossa (Douro and Madeira) and Vermentino (Corsica). It is one of the white varieties grown in the Port wine district Alto Douro. It is a most vigorous grower, gives sweet must and good wine, but at the Cape it is so susceptible to millerandage and coulure that it is not worth growing.

Description.—(A. I. P.)
Bourgeonnement white woolly.
Shoots medium thick, internodes fairly long.
Leaves large, dense white woolly pubescence below, 5-lobed with deep sinuses; long, sharp, broad teeth.
Bunch medium, conical, loose. *Berries* medium, oval. *Skin* thick, pretty yellow colour when fully ripe. Fairly early.

21. **Verdelho.**

Synonyms.—Verdelho da Madeira, Gouveio.

In the Upper Douro it is known as Gouveio and in Madeira as Verdelho, where it is supposed to be indigenous and produces fine bunches, and is largely used for making Madeira wine. It answers well in poor, dry soils. It is a vigorous grower and good bearer with short pruning, but very susceptible to oidium, and should therefore be sulphured for the first time when the young shoots are only a couple of inches long. Mas and Pulliat regard it as a splendid grape both for the table and the press.

Description.—

Bourgeonnement with a violet tint and somewhat woolly
Canes thin, light red, internodes short.
Leaves medium or fairly small, cordate rounded off, glabrous and smooth above, almost glabrous below, only slightly woolly on the

nerves; upper sinuses shallow, lower sinuses zero or almost zero; petiolar sinus slightly open; teeth almost regular, somewhat narrow, short, obtuse or very shortly pointed.

Bunch small, conical, well branched and very loose; peduncle long, thin, with a violet tint. *Berries* small, regularly oval; pedicels very long and extraordinarily thin. *Skin* somewhat tough, very-transparent, at first bright green, then passing into slightly golden light green, with good white bloom. Flesh somewhat firm, sweet and pleasant flavour. Early mid-season.

22. **Sercial.**

Synonyms.—Sarcial, Cerceal, Cerceal branco, etc.

It is grown in Madeira and in Portugal (Beira alta and Estremadura). It contributes much to the quality of Madeira wine. It can be pruned short or long and gives good yields.

Description.—(J. Marques de Carvalho in Viala-Vermorel.)

Bourgeonnement woolly, light rose-colour on the young leaves.

Canes medium thick, long, bark fine, chestnut or dark cinnamon colour, finely striped, wood rather soft, pith fairly thick, diaphragm thick, horizontal.

Leaves medium, cordate, smooth, 3- or 5-lobed, dark green and glabrous above, dirty green and woolly below; upper sinuses well marked, fairly deep; lower sinuses shallower, petiolar sinus U-shaped; teeth in two series, regular, obtuse. Petiole long and thin, dirty green. Leaves drop late.

Bunch 2-3 per shoot, small or sub-medium, simple or slightly branched, very compact, cylindro-conical or conical; peduncle short, lignified, very dirty; pedicels short, pronounced bourrelet. *Berries* small, sub-ovoid. *Skin* thick, greenish, slightly amber-coloured on the sunny side when ripe, translucent and slight bloom; style point at the apex central. Pulp tender, juicy, high acidity; juice greenish to yellowish. Seeds 2 to 3, small, elongated. Late mid-season.

(*b*) **Red Wine Varieties**

1. **Pontac.**

Synonyms.—Pontak (Cape), Pontacq (Bohemia, after Stoltz), Teinturier mâle, Dix fois coloree, Faerbertraube, etc.

General.—This is one of the oldest cultivated varieties. At the Cape it has been grown for probably over two centuries. It is characterised by the dark red, nearly black juice of its ripe grapes and by the red colour of its leaves, which is noticeable before the grapes are ripe. I consider Pontac to be the same as the Teinturier mâle of France.

Culture.—Pontac is of medium vigour and gives moderate crops with short pruning. It should preferably be grown on good decomposed granite (Paarl and Stellenbosch, Cape) or on sweet fertile alluvial soils, as at Wilgerivier, Worcester(Cape). If grown on a fertile soil it can be pruned half long without the vine becoming exhausted. It is immune to anthracnose but susceptible to oidium. At the Cape its production of grapes is about $1/3 - 1/2$ that of the ordinary wine varieties per unit area. In the last century a fairly large amount of Pontac wine was made, which was blended with Muscadel wine to produce a Cape Port. This was not a real Port, but it was an excellent wine of this type. Pure Pontac is an excellent medicinal wine. Blended with other red wines, it acts as a colouring agent. For this purpose it has largely been replaced by Alicante Bouschet and Petit Bouschet. At the Cape well-ripe Hermitage grapes are sometimes boiled for obtaining a very dark wine to take the place of Pontac. This resulted in too low prices being offered for wine made from the Pontac grape, and has caused its almost complete disappearance. This is most unfortunate, as Pontac possesses valuable qualities not present in other grapes. Pontac gives a very dark, almost black, wine, which may deposit most of its colouring substances in about ten years' time, when the wine will assume a yellow-brown colour.

Its must, in addition to plenty of colour, possesses also a high total acidity (8-9.6 ‰) and sugar content (25-28 per cent) when well ripe. Its wine is very rich in colour, tannin, alcohol, and dry extract, and possesses a typical Pontac taste and aroma. It has, on the whole, little affinity for American stocks. Jacquez is about the best stock for Pontac.

Description.—(Mainly that of M. et P. for Teinturier mâle.)

Bourgeonnement strongly white woolly, passing into deep violet-red. *Canes* thin, internally the colour of wine lees, short internodes; white woolly eyes (A. I. P.).

Leaves small; coloured red or showing red patches at an early stage, in autumn almost completely red (A. I. P.); almost glabrous above, slightly pubescent below; upper sinus deep ; lower sinus deep or well marked; petiolar sinus closed or nearly closed; teeth short and fairly pointed; petiole medium long, thin, well coloured.

Bunch small, cylindrical, sometimes more or less conical (A. I. P.), compact; peduncle short and thin to medium thick (A. I. P.). *Berries* small, spherical or almost so; pedicels short and thin. *Skin* thick, tough, turns violaceous red shortly after setting (A. I. P.) and pitch black when ripe, some white bloom. Pulp firm; rather little juice, deep blood-red, sweet. Early.

2. **Petit Bouschet.**

This is a hybrid between Aramon and Teinturier du Cher (or Teinturier femelle) obtained by Louis Bouschet in 1829. Like Pontac it has a dark red juice and gives a very dark red wine. It is a vigorous grower and bears well with short pruning. Its wine is almost black. In order to retain its colour it should be blended early with wine containing much acid and alcohol. It does not form much sugar. Even in Algeria its wine has only 8-11.5 vol. per cent alcohol. It is fairly resistant to fungoid diseases, and does well on Jacquez, Riparia Gloire, 101-14, 3306, 420A.

Description.—
Bourgeonnement grey rusty, strongly woolly with a violaceous red tint.

Shoots vigorous, rather thin, somewhat creeping ; internodes fairly long.

Leaves medium, longer than broad, glabrous above, light cobwebby below; upper sinuses deep, well closed; lower sinuses well marked; petiolar sinus open; nerves tinted violaceous red; teeth fairly short, uneven; petiole rather short, medium thick, intensely wine red. The leaf has a blood-red colour around its border when the grapes are ripe; before dropping the leaves turn very dark red.

Bunch above medium or large, branched, winged, conical, rather loose; peduncle thin, fairly long, violaceous. *Berries* medium, roughly globular or sub-globular; pedicels thin, fairly long, violaceous like the peduncle. *Skin* thick, turning violaceous black at maturity. Pulp rather soft, juicy, juice very dark red; brush brilliant red. Early mid-season.

3. **Alicante Bouschet.**

Synonyms.—Alicante-Henri   Bouschet,   Alicante-Bouschet No. 2. [1]

---

[1] According to P. Viala in Viala-Vermorel, J.c. vi. p. 426.

This grape was produced by Henri Bouschet, son of Louis Bouschet, in 1865, by crossing Black Alicante (Grenache noir) with Petit Bouschet. It is a very vigorous grower and *very heavy bearer* with short pruning, having frequently four inflorescences on one shoot, of which only the two lower ones usually develop into normal bunches. In Algeria its un-trellised vines produce with short pruning up to 100 hl. wine per ha. or 14.4 leaguers per morgen or 890 gallons per acre. At the Cape it also produces big crops, and can be strongly recommended for the production of a good, very dark-coloured red wine for blending with other red wines. According to Leroux[1] it produces a good red wine in Algeria, which has a very intensive dark garnet red colour, 12-14 and even 15 vol. per cent alcohol, and 25-31 grammes dry extract per litre. Its wine contains free acid. It is just now very popular in California, where some Thompson's seedless vineyards are being grafted over to Alicante Bouschet on account of the very high prices paid for its grapes, which, under Prohibition, are excellent for making red wine for the home out of raisins and other suitable material.

It is fairly resistant to fungoid diseases, and possesses a good affinity for most American stocks. The stocks that suit it best are: 101-14, 3306, 3309, 106-8, 420A (according to Viala-Vermorel, *l.c.).* Jacquez and Rupestris du Lot usually cause it to grow too vigorously, with the result that both the quantity and the quality of the crop suffer. In my own collection it answered well also on these two stocks.

Description.—(P. Viala in Viala-Vermorel.)

Vine vigorous, almost creeping.

Shoots long, fairly thick, little branched, green with broad brick-red stripes; ripe wood yellow wine tone with purple coloration at the nodes; internodes medium.

Leaves medium to larger, orbicular, thick, soft, *curving round the borders downward (typical),* entire to 3-lobed with shallow sinuses, petiolar sinus deep V-shaped; dark green, glabrous, glossy above; paler green and cobwebby below ; when grapes ripen leaves have red spots, and in autumn crimson red patches until the whole leaf is dark red.

Bunches large, usually well branched, thick, truncate conical to roundish, never very compact; peduncle medium long and thick, light green, tough and hard but not lignified; pedicels thick and short

[1] Leroux, S., *La Vigne et le vin en Algérie et en Tunisie,* 2 vols., 1894. Vol. i. pp. 92-93.

with thick bourrelet, light dirty red. *Berries* fairly large, round, black, with strong bloom, lenticels at the surface and persisting style point in the centre, not prominent; fairly firm with fairly thick, elastic skin. Pulp melting, juicy, bright dark-red colour, refreshing sweet taste. Seeds 2-3 per berry, medium size.   Fairly early mid-season.

4. **Pinot noir.**

Synonyms.—Pineau noir, Pineau de Bourgogne, Noirien, Blauer Burgunder, Black Cluster (Dr. Hogg).

Mas and Pulliat[1] consider the grey, red, and white Pinots as variations of Pinot noir. It is one of the oldest cultivated varieties, and is considered to be indigenous in the Cote d'Or (Burgundy-Beaujolais), where it is largely grown, and mainly determines the character of the world-famous burgundies. In the Champagne district (Rheims-Epernay) the famous champagnes are made mainly from this grape. It answers also in the other cold wine districts of Europe (Moselle, Assmannshausen, Wurtemberg, Wadensweil, along the lake of Zurich, etc.) on account of its early maturity. It seems to produce the best wine on limestone soils and dry hills. According to Mas and Pulliat, it still gives a fine wine on granitic soils, but lighter, less aromatic (bouqueted), and loses a little of its pretty colour.

On the Stellenbosch University Farm it produces a beautifully coloured, strong, full-bodied wine with an excellent bouquet on red decomposed Malmesbury shales on a hill with a northerly aspect.   It is a wine of high quality. Un-trellised its yield is too low.   Therefore it should be trellised on one wire 16-18 in. above the ground, with two wires 8 in. apart and 11 in. above the bottom wire, and pruned according to the double Guyot system, or else with a long stem and half long bearers each with four eyes, when it will give satisfactory crops.   It flowers very early, and is very subject to millerandage, which can reduce the crop very considerably. It should therefore be grown on a warm and sheltered site. It is highly resistant to anthracnose and oidium but not to *Plasmojpara viticola.*   It possesses a good affinity for American vines.

In order to obtain the finest wine, it should be picked when somewhat over-ripe.   In Burgundy its must then has 20-22 per cent sugar. At the Cape it then has easily 25 per

---

[1] Mas et Pulliat, *l.c.* vol. ii. p. 121.

cent sugar. The yield in Burgundy is about 15 hl. per ha., or 2¹/₆ leaguers per morgen, or 133½ gallons per acre, but this is very excellent wine. Even in hot countries it gives excellent wines, and should therefore be grown where quality is aimed at. If combined with Petit Gamay, the average yield will be quite satisfactory when the high quality is taken into account.

Description.—

Vine vigorous and healthy.
Bourgeonnement early, white woolly.
Canes medium thick (inclined to be somewhat thin: A. I. P.), creeping, internodes short and dark grey-brown (A. I. P.), with long and numerous tendrils.
Leaves medium, nearly round on fertile shoots, more deeply lobed on the unfertile shoots, petiolar sinus ordinarily open, slightly rough and glabrous above, distinctly cobwebby below, with prominent nerves; teeth somewhat obtuse, short; leaves drop early.
Bunch small or below medium, usually cylindrical, sometimes shouldered; peduncle short or medium, brown or lignified above the node. *Berries* fairly small, round or almost round when pressed in the bunch. *Skin* thick, tough, rich in pigment, deep black with light bloom. Pulp juicy, very sweet, simple flavour. Early.

5. **Gamay noir or Petit Gamay**.

*Synonyms.—*Gamay noir, Petit Gamay, Gamai noir, Plant d'Arcenant, Lyonnaise, Bourguignon noir, etc.

According to Mas and Pulliat this variety is a native of Burgundy, where Pinot noir is grown along the hills for quality and Gamay noir on the fertile lower land for quantity. Farther south, in the Beaujolais, it gives better wines, because it too is grown on the better hill sites. Thus the famous wines of Beaune are made mainly from Gamay noir. It bears soon, much, regularly, and its grapes ripen early, about the same time as those of Pinot noir. On limestone soils its wine is darker and more full-bodied than on granitic and shale soils, but its finesse (bouquet, etc.) is less.

It is a better grower and bearer with short pruning than Pinot. In my experiments it gave sweet musts (easily 25 per cent sugar) and a first-class wine with a beautiful colour. In hot countries it should be associated with Pinot for producing dry red wines with much quality. In colder countries they are useful on account of their early maturity. The vine is vigorous and has an upright (erect) habit of growth. It bears well with short pruning and possesses a good affinity for American stocks.

Description.—
Bourgeonnement yellowish-green and slightly woolly.
Canes erect, internodes of medium length, tendrils short.
Leaves medium, somewhat longer than broad, tender green, glabrous and smooth above, almost glabrous below; upper and lower sinuses usually little marked; petiolar sinus open; teeth small, not very broad and short, obtuse or seldom somewhat sharp; petiole of medium length and thickness, faintly rose-coloured and almost glabrous.
Bunch medium, most often cylindrical and sometimes shouldered, somewhat compact; peduncle short and lignified. *Berries* medium, shortly oval; pedicels fairly short and medium thick. *Skin* thin, fairly tough, pretty black colour with bloom at maturity. Pulp soft, juicy, sweet, and without a particular flavour. Early.

6. **Cabernet Sauvignon.**

Synonyms.—Petit Cabernet, Petite Vidure, Vidure Sau-vignonne, Navarre, Bouchet, etc.

General.—The Cabernets (Cab. Sauvignon and Cab. franc) are among the oldest and best wine varieties of the Gironde. They are vigorous growers and good bearers with long pruning, but remain varieties of high quality, and produce only about two-thirds of the crop obtained from Hermitage (Cinsaut). The value of their crop is fully equal to that of Hermitage. In view of the present state of the wine position in the civilised world and of the likely position during the present generation at least, wine farmers will be well advised to aim at the production of *less wine* but of *wine of higher quality.* In the group of dry red wines of the claret class, this can best be attained by *growing at least one-third as many vines of the two Cabernets as of Hermitage and other similar varieties giving less quality but more abundant yields.* At the Cape the dry red wines of Constantia are the best and possess real quality. The soil and climate are favourable, but it is mainly the very appreciable amount of Cabernet Sauvignon wine blended with that of Hermitage and other varieties *(e.g.* Shiraz, Verdot, etc.) to a lesser extent, that has brought about this favourable result. At the Cape, as in the Medoc, Cabernet spells quality in the dry red wine. Pure Cabernet wine is expensive, too astringent, and takes too long to mature in order to pay well at the Cape. In the

Medoc, during my visit in 1908, the Cabernets constituted practically the whole vineyard at Chateau Lafitte, the "premier cru" of the Medoc. Cabernet gives a beautiful dark red wine with a peculiar and typical flavour. It needs two to three years' maturing in wood before it is fit to be consumed. Before this time it is too astringent. As table-grape Cabernet is worthless, as it can hardly be eaten.

M. d'Armailhacq[1] characterises the two Cabernets as follows: "*Cabernet Sauvignon* shows the greatest analogy with the *gros cabernet* (same as *Cabernet franc*: A. I. P.), and one must recognise that they belong to the same family, but *Cabernet-Sauvignon* is somewhat less vigorous" (p. 38). . . . "*Cabernet sauvignon* begins to bud a little later than the *gros cabernet*; it flowers almost as soon, but on the whole a few days later than the latter; and what is remarkable, is that it turns colour later and ripens a few days earlier. This grape has the great advantage of keeping without alteration for several days when perfectly ripe; it would require lots of rain to cause it to rot" (p. 39). . . . "The bunch, more cylindrical, is longer; it does not *branch* as much; the berries are round, slightly smaller than those of *gros cabernet*; their colour is likewise deep black, somewhat blue, but they are more velvety, and have more bloom, which gives them a whitish appearance" (p. 37).

In order to obtain the best results the Cabernets should be picked when fully ripe and trellised on one wire about 15-18 in. above the ground, with two wires 11 in. higher and 8 in. apart, between which the young shoots should grow, and should be pruned according to the double Guyot system, *i.e.* two halves each with one spur of two eyes and one rod about 18 in. to 2 ft. long. The rows should be about 7-8 ft. wide and the vines should be 3-4 ft. apart in the rows. Both Cabernets are very susceptible to oidium, much less to anthracnose. From personal experience I know that they answer very well on Riparia gloire, 101-14,1202, and Jacquez.

Description.—
Bourgeonnement woolly, tinted dark wine red; developing little garnet coloration obscured by a thick woolly pubescence.

---

[1] A. D'Armailhacq, *De la culture des vignes dans le Midoc,* 3mo ed. Bordeaux, 1867, pp. 38, 39, 37.

Shoots thick, with fairly long internodes, distinctly striped.

Leaves medium; almost as broad as long, dark green, somewhat bullate and glabrous above, with a cobwebby and in certain parts slightly woolly pubescence below; upper sinuses deep, lower sinuses little marked; petiolar sinus closed (see 5-lobed leaf of Cabernet Sauvignon in Fig. 10. It should be pointed out that the leaves of Cabernet Sauvignon are frequently 5-lobed with five deep sinuses showing five holes owing to the lobes being superposed: A. I. P.); petiole fairly long and not very thick; teeth very broad, not very deeply incised, obtuse and rounded off.

Bunch medium, somewhat conical, shouldered, somewhat branched, usually somewhat loose; peduncle fairly long, somewhat thin. *Berries* below medium, spherical or almost spherical; pedicels somewhat long and thin, tinted with red at maturity. *Skin* thick and very tough, a pretty black colour, white bloom at maturity. Pulp somewhat firm, fairly juicy, with a very pronounced flavour characteristic of all the Cabernets. Mid-season.

7. **Cabernet franc.**

Synonyms.—Cabernet, Carmenet, Cabernet gris, Vidure, Petite Vidure, Fer, Petit Fer; Arronya, Gros Cabernet (d'Armailhacq), etc.

In addition to what has already been said under Cabernet Sauvignon, it should be pointed out that Cabernet franc is somewhat subject to coulure. It is grown less extensively than Cabernet Sauvignon in the Medoc. At the Cape they occur mixed in the Cabernet vineyards, although Cabernet Sauvignon predominates. Cabernet franc is somewhat more productive than Cabernet Sauvignon, but also requires long pruning.

Description.—

Bourgeonnement light rusty colour, woolly, with a slight violaceous red tint around the borders and on the lower surface of the little leaves.

Canes vigorous, internodes fairly long, half erect. Leaves drop somewhat late.

Leaves medium, about as broad as long, glabrous and almost smooth above, slightly cobwebby below; upper sinuses fairly deep, closed or almost closed without leaving a very apparent opening; lower sinuses marked, somewhat open; petiolar sinus almost closed or slightly open, leaving an opening below; petiole fairly long or broad, somewhat thin; dentition uneven, neither very broad nor very deep, slightly sharp or somewhat obtuse, with a short fine point.

Bunch below medium, somewhat loose, cylindrico-conical, sometimes a little shouldered, peduncle of medium thickness and length. *Berries* below medium (but slightly larger than those of Cabernet Sauvignon: A. I. P.), globular; pedicels fairly long and fairly thick. *Skin* somewhat thick, tough, pretty bluish-black with a strong bloom when ripe. Pulp somewhat firm, juicy, fairly sweet, somewhat astringent, with a special flavour characteristic of all the Cabernets. Mid-season.

8. **Merlot.**

This is a good variety, of uncertain origin, and cultivated in the Medoc only since the nineteenth century. There one finds it in the best vineyards, though much less than the Cabernets. Its grapes are sweet, its wine softer and sooner fit for consumption than that of Cabernet. It suffers from coulure (less than Malbec), is little subject to oidium and anthracnose but susceptible to downy mildew, and its ripe grapes rot easily. It should be long pruned and trellised like Cabernet, when it gives good crops. The vine is vigorous and healthy, a regular and good cropper. It ripens a little earlier than Cabernet, but the two can usually be picked together. Blended with Cabernet it will make the blend sooner fit for consumption. The Cabernets blend well with Merlot and Malbec, both of which increase the yield relatively to Cabernet.

Description.—

Bourgeonnement somewhat woolly, greenish-white, tinted a light rose-colour around the borders of the little leaves.

Shoots half erect, of medium vigour and thickness; internodes fairly short.

Vine vigorous, productive, less lasting than the Cabernets.

Leaves medium or larger, of variable shape and size, glabrous and almost smooth above, with a woolly pubescence below; upper and lower sinuses more or less deep, slightly or not closed; petiolar sinus open; dentition uneven, rather shallow, fairly sharp; petiole fairly thick, of medium length.

Bunch above medium size, long, cylindrico-conical, much or fairly branched, peduncle long and thick. *Berries* medium or smaller, globular; pedicels fairly long, somewhat thin, often tinted red. *Skin* somewhat thin, little resistant, dark black, with bloom when ripe. Pulp juicy, very sweet. and pleasant, simple flavour. Early mid-season.

9. **Malbec**

Synonyms.—Malbek, Malbeck, Cot, Pressac, etc.

According to Mas and Pulliat, *l.c.* vol. ii. p. 5, it is the most widely spread and cultivated grape in France, especially in the south and middle-west. It is a vigorous grower and good bearer with long pruning. It should be pruned and trellised like Cabernet. It is fairly resistant to oidium, but sensitive to *Plasmopara viticola,* and much subject to coulure. In France it has been grown for a very long time. It gives a good dark wine, but not of the same high quality as that of Cabernet, with which it blends well. With long pruning it grows very vigorously and bears well. Its affinity for most American stocks is good.

Description.—

Bourgeonnement turning from light rusty to violaceous white, which becomes lighter as the leaves open.

Shoots medium thick, internodes rather short, visibly striped.

Leaves above medium, nearly as broad as long, intensely green; glabrous and largely bullate above, with a fairly abundant flocculent pubescence below; teeth short, narrow, sharp pointed; upper and lower sinuses little marked; petiolar sinus open; petiole fairly short and thick.

Bunch above medium, of variable shape, but usually conical, branched, fairly loose; peduncle fairly long or long, thick, reddish. *Berries* above medium, almost spherical or spherical; pedicels long and thick (usually coloured wine red at maturity: A. I. P.). *Skin* thin, fine, fairly tough, black with bloom when ripe. Pulp rather soft, very juicy, melting; very sweet and pleasant.   Early mid-season.

10. **Verdot or Petit Verdot.**

Synonyms.—Petit Verdot, Fer, etc.

In Viala-Vermorel, *l.c.* vol. vi. p. 19, Cazeaux-Cazalet distinguishes between Gros Verdot and Petit Verdot, of which the former is the more productive but the latter gives a better wine.   Both are very old varieties of the Medoc, where they seem to have originated.   Mas and Pulliat refuse to call them two varieties, as they are too closely related, differing mainly in the size of their bunches.   Verdot is grown in the Palus (low-lying fertile soils rich in humus near the Gironde) of the Medoc, where it gives a better wine than any other variety. Elsewhere it is not favoured.   It is a moderately vigorous grower and bearer; the bunches are situated high up the shoots, the grapes ripen rather late. It gives a dark, coarse, astringent wine with little to recommend its cultivation.

Description.—

Bourgeonnement rusty, turning white, very woolly. *Canes* fairly thick, little branched (A. I. P.), half-erect; internodes medium.

Vine vigorous and fairly productive in suitable soils.

Leaves medium, glabrous and bullate above, with a whitish, fairly thick woolly pubescence below; upper sinuses fairly deep, lower sinuses more or less marked; petiolar sinus almost closed or closed by the superposed lobes, leaving a heart-shaped opening; petiole thick, fairly long, often slightly woolly; dentition rather shallow, uneven, fairly sharp.

Bunch hardly medium, cylindrico-conical, somewhat branched, little compact; peduncle long and thin or somewhat thin, often with a small bunch at the node. *Berries* below medium or small, globular; pedicels fairly long and thin, assuming a reddish colour near the berries at maturity. *Skin* fairly thick, tough, a pretty black with bloom at maturity. Pulp fairly firm, juicy, simple flavour. Late mid-season.

11. **Shiraz.**

Synonyms.—Schiras, Syrah, Sirah, Syrac, Sirac, Petite Sirah, Marsanne noir, etc.

Owing to the name Schiras some ampelographers are of opinion that it originally came from Schiraz in Persia. It has in any case been grown now for a very long time in the Rhone valley near l'Ermitage (written Hermitage in English; strangely enough, the grape grown at the Cape as Hermitage is the Cinsaut of the South of France and not Shiraz, the grape of Hermitage), where it produces a fine, famous red wine on the hot dry slopes. In order to make its wine finer and to increase the strength of its bouquet, it is blended with a local white wine of the Hermitage made from the Roussanne or Marsanne grape. At the Cape its wine is sometimes blended with Hermitage and Cabernet for producing a good dry red wine. In Australia it is grown to a considerable extent. When grown in suitable soil, well manured, and propagated from well-selected cuttings, it gives good crops with short or half long pruning. In the Cote Rotie it is sometimes pruned with long rods. When pruned to three good eyes it usually bears well.

Description. —

Bourgeonnement white woolly, with a light wine-red border a the developing little leaf.

Canes fairly thick, long, creeping, with long internodes.

Leaves medium or smaller, about as broad as long, dark green, glabrous above, covered below with a cobwebby pubescence; upper sinuses deep, lower sinuses well marked; petiolar sinus open; dentition broad, somewhat deep and obtuse; petiole fairly long and thin.

Bunch medium or larger, conical, elongated, more or less shouldered, usually little compact; peduncle long, somewhat thin (usually yellowish and not lignified at maturity: A. I. P.). *Berries* medium or smaller, oval; pedicels long and not very thick. *Skin* fine, thin, fairly tough, turning black with white bloom at maturity. Pulp fairly firm, juicy, sweet. Mid-season.

12. **Mondeuse.**

Synonyms.—Mandouze, Persagne, Savoyanne, etc. There is a white as well as a black Mondeuse. The name Mondeuse stands for the black type, which differs from the other one merely in the colour of its ripe grapes. It is grown mainly in Savoy (S.E. France), and generally on warm southerly slopes. There it produces a good wine. It demands a warm climate, is a very vigorous grower and a very good bearer even with short pruning, although it can stand long pruning.

Description.—

Bourgeonnement covered with a white woolly pubescence.

Canes fairly thick, grey-reddish, long, not erect, and with long internodes.

Leaves large, longer than broad, glabrous above, with a light, flocculent pubescence on the nerves below; upper sinuses not very deep and closed ; lower sinuses more or less marked; petiolar sinus closed or nearly closed ; teeth not very long, very uneven and obtuse, while those terminating the lobes are long and sharp; petiole short and thin.

Bunch large, conical, strongly shouldered and branched, generally little compact; peduncle fairly long and thick. *Berries* medium, short oval or spherico-ellipsoidal; pedicels long and not very thick. *Skin* not very thick and yet very tough, at first dark purple, then turning violet-black with bloom at maturity. Pulp tender, melting, juicy, sweet, but retaining a characteristic astringent taste even when fully ripe. Mid-season.

13. **Aramon.**

Synonyms—Ugni noir, Plant-riche, Pisse-vin, Burchard's Prince (Dr. Hogg), Arramont, etc.

This is the most generally grown grape on the plains of the South of France, where it produces tremendous crops (up to and exceeding 300 hi. per ha. or 43$^{1}/_{3}$ leaguers per morgen or 2670 gallons per acre) under irrigation. The wine is of *very mediocre quality* and very light (14-15° proof spirit). It grows well, bears many and large bunches with large, round, reddish-black berries. Ought not to be grown where any quality is aimed at.

14. **Carignan.**

Synonyms.—Carignane, Crignane, Bois dur, Plant d'Espagne, Tinto, etc.

It is grown in the eastern French Pyrenees, the Aude, and the Herault. In Algeria it is regarded as one of the best

FIG. 53-Carignan (near Tunis). Original

red wine varieties, and is largely grown. It is very highly resistant to sunburning. Unfortunately it is very susceptible to fungoid diseases, e.g. oidium, anthracnose (during wet summers), and *Plasmopara viticola* (downy mildew). It originated in Spain, where it is known as Tinto. In good, fertile soil it grows vigorously and bears well with short pruning. It produces a good, strong, dark-coloured wine. This grape has probably been grown in the South of France since the twelfth century. In poor, dry, warm hill soils it still answers well. It possesses a good affinity for American stocks, and is a good grape for hot climates.

Description.—(Viala-Vermorel, *l.c.* vi. 335.)

Bourgeonnement late, white, woolly. *Vine* vigorous grower, fairly erect, fertile.

Canes light reddish-brown, hard, brittle, thick, with broad stripes, internodes short at the base, longer higher up, on the whole fairly short; nodes coloured, fairly thick.

Leaves large, thick, 5-lobed, upper sinuses deep, lower sinuses shallow, petiolar sinus deep and nearly closed V-shaped; teeth sharp ; fairly dark green and glabrous above, lighter green and fairly woolly below; petiole large, thick, with a red colour; develops red borders in autumn.

Bunch large, compact, cylindro-conical; peduncle lignified, thick, short, tough. *Berries* fairly large, round, black, *Skin* thick, astringent; dense bloom.   Pulp firm, juicy, not nice to eat. Mid-season.

15. **Mataro.**

Synonyms.—Mourvèdre (Provence), Mataro (Catalonia, Roussillon, Cape), Catalan, Espar and Spar (Herault and Gard), etc.

This is one of the most important varieties for the south of France. It is of Spanish origin. In stony, fairly heavy and deep hill-soils it answers best. Also in deep limestone soils and clay soils which are not too wet, it grows well and is very productive. On light soils it produces less. It begins budding late, is little subject to coulure, bears regularly and well, is highly resistant to anthracnose; its grapes resist oidium well and do not easily rot. It is therefore a robust, healthy vine. At the Cape it produces a good, full-bodied, strong, dark red wine, and might be planted to advantage, though less than the Cabernets.

Description.—

Vine vigorous and robust.

Bourgeonnement very woolly, with rose or violaceous tint. *Canes* thick, erect, somewhat swollen at the nodes.

Leaves medium, dark green, somewhat rough above, with a fairly thick white woolly pubescence below; petiole fairly long, reddish; upper sinuses somewhat deep and closed; lower sinuses only slightly marked; petiolar sinus open; dentition fairly broad, somewhat sharp.

Bunch conico-cylindrical, broad shouldered, generally compact; peduncle thick, short or very short, dark brown where attached to the cane. *Berries* medium or larger, globular, pedicels thick and short or somewhat short. *Skin* thick, tough, black with heavy bloom at maturity. Pulp fairly firm, juicy, sweet when well ripe, more or less astringent before completely ripe ; simple flavour. Late mid-season.

16. **Hermitage.**

Synonyms.—Cinsaut,   Cinq-Saou,   Bourdales, Boudalès, Plant d'Arles.

This grape has been grown at the Cape since about 1880 under the name Hermitage. By means of comparative tests I was the first to prove the identity of the Cape Hermitage with the Cinsaut of the South of France. Marès[1] describes it as an excellent wine grape. It is on the dry hills in the South of France that it gives such excellent wines with a production of 30-60 hl. per ha. or 4½ -9 leaguers per morgen or 267-534 gallons per acre. At the Cape it generally produces 10-15 leaguers per morgen without being trellised. On the farm Evertsdal, near Durbanville (Cape), it produces 42 leaguers per morgen or 291⅔ hl. per ha., where it is trellised on two wires, grafted on a Rupestris seedling, and not irrigated. It gives a thin, light red wine with little character on account of the extraordinary high yield, which probably is a world's record for this grape.

In the South of France it is grown both as wine and table grape. Its grapes ripen early, and are sent as a table grape from the South of France to Paris and other large cities in the North of France. At the Cape it is also grown as a wine and table grape, and is exported to London, where it sells well as an early table grape. It is, however, not an export grape of high quality. Its main function is to produce wine. On account of its high production it is the most popular red wine grape at the Cape, where fully 75 per cent of the red

---

[1] Mares, Henri, *Description des cépages principaux de la région méditerranéenne de la France* 1891, p. 59.

wines are made from it. In Constantia and Stellenbosch it produces a *good* light dry red wine on the hills, where it does not produce such heavy crops. On the fertile low-lying soils it produces big crops of rather indifferent quality. In the fertile, sweet Karroo soils it produces very heavy crops of poor wine, and should not be planted except perhaps for the production of spirits of wine. Where it produces good dry wines, these should be blended with Cabernet, as was previously indicated.

It is practically immune to anthracnose and not very susceptible to oidium. It should be pruned short, as it will otherwise exhaust itself through over-production, being but of medium vigour. It does well on Jacquez, Herbemont, 101-14, 1202, 3309, 106-8, etc.

Description.—
Vine of medium vigour, with rather thin shoots, somewhat creeping or half-erect.

Bourgeonnement very late, light rusty, woolly, passing into rose-colour, then green, almost glabrous.

Leaves above medium or large, glabrous, almost smooth or slightly bullate, with brush-like hairs on the nerves and a light pubescence on the parenchyma below; upper sinuses deep, narrow or slightly closed; lower sinuses somewhat deep, slightly open; petiolar sinus slightly open; petiole of medium length, slightly thin; dentition fairly broad, deep, sharp; leaves develop some red colour along their borders (also on rest of leaf: A. I. P.) before dropping, which is somewhat late.

Bunch above medium or large, conico-cylindrical, somewhat branched, generally loose or somewhat loose on dry soil; peduncle somewhat long, thin, breaking easily. *Berries* large, oval; pedicels long and thin. *Skin* somewhat thick (at the Cape somewhat thin on productive soils.—A.I.P.), fairly tough, a pretty black with bloom at maturity. Pulp firm, crackling (only on fairly dry, warm sites: A. I. P.); sweet, juicy, simple flavour. Early mid-season to fairly early.

17. Grenache noir.

*Synonyms.—*Grenache, Garnacho (Rioja, Spain), Alicante, etc.

This grape is different from the table grape, Alicante or Black Alicante (Dr. Hogg). It was imported into the South of France from Spain. In 1910 I imported it into South Africa from Spain (Rioja) as Garnacho and from Montpellier as Grenache noir. Both turned out to be identical.

It is also grown in the Alto Douro (port wine district). Grown on slopes, it produces a good red wine. In the Roussillon (eastern French Pyrenees) a natural straw-coloured sweet wine is made from its over-ripe grapes. With moderate crops its grapes become sweet and give wines of good quality. Heavy crops considerably reduce the quality of its wine. At the Cape it produces fully as much as Hermitage. In deep, broken, stony, clayey, calcareous, fairly dry soils it gives wines of good quality. It therefore answers well in decomposed granite, Karroo soils, and decomposed shales if they are fairly warm and dry, therefore preferably on slopes.

It is resistant to oidium and anthracnose, but very susceptible to downy mildew. Owing to its large, compact bunches, mealy bug sometimes infests these very badly. With short pruning it bears very well. Its affinity for most American stocks is good. It is grown at the Cape together with other port varieties to produce a Cape wine of the port type. In cool, fertile soils it produces too heavily and then lacks quality.

Description.—

Vine very thick, very vigorous, very productive; reaches an old age in suitable soils.

Bourgeonnement slightly or not at all pubescent, light rusty or yellowish turning into light green, then brilliant green.

Canes half-erect, thick; internodes short and swollen, reddish-yellow, lightly dotted.

Leaves medium or smaller, smooth and glabrous on both surfaces, light (to yellowish: A. I. P.) green; upper sinuses well marked; the lower ones less apparent; petiolar sinus more or less open; petiole fairly thick, of medium length; dentition fairly sharp, uneven.

Bunch above medium (to large: A. I. P.), compact, sometimes much branched, usually broad-shouldered (sometimes roundish: A. I. P.); peduncle thick, fairly long. *Berries* medium, short oval (spherico-ellipsoidal); pedicels fairly thick, fairly short or short. *Skin* delicate, somewhat subject to rot, not very dark black at maturity. Pulp rather soft, very juicy, very sweet; pleasant, simple aroma. Late mid-season.

Note.—There are also a white and red Grenache which differ from the black Grenache merely in the colour of their ripe grapes. They, too, give a good wine.

18. **Blauer Portugiesier** or **Portugais bleu.**

This is an early variety which, according to Viala-Vermorel, *l.c.* vol. ii. p. 136, has been grown from time immemorial in Austria and Hungary. It is grown also in Würtemberg, the Palatinate, and the Rhine districts on account of its early maturity. It produces a very dark red wine which is somewhat thin and poor in tannin and acid. It grows and bears well with short pruning. It answers best on dry soils and along hillsides. Its canes are very thick, cinnamon-brown; leaves large, 3- to 5-lobed, glabrous on both surfaces; bunch fairly large, berries medium, spherical.

19. **Blue Kadarka.**

Synonyms.—Blaue Kadarka, Kadarka-Keck, Schwarzer Skutariner, Kadarkas noir.

This is the best and most widely-spread grape of Hungary, from where it was imported to Croatia and Styria. In Hungary it is generally grown as bush vines *(i.e.* untrellised) and pruned short. It is very productive, and in good years gives musts with 18-20% sugar and 7-6‰ total acidity. Its berries also produce raisins for fine liqueur (sweet) wines. It produces a very fine and aromatic wine, and most Hungarian red wines are made mainly from it. It is a strong grower, and answers on most soils. It ripens rather late and requires a warm climate and dry autumn. Rain easily causes its berries to crack and rot. When grafted on a vigorous stock it suffers easily from coulure. Use in preference Riparia gloire, 101-14, 420A, etc., which are not such over-vigorous stocks.

Description.—(H. Goethe.)

Vine vigorous and robust, very productive.

Bourgeonnement grey rusty, woolly; light violaceous rose tint on under surface of little leaves.

Canes thick, erect, not very numerous, reddish-brown.

Leaves large, thick, 3- to 5-lobed, dark green above, white grey dense felt-like pubescence below; nerves strongly developed, light green, with white brush-like hairs; petiolar sinus closed or nearly closed; petiole thick and shorter than the mid-rib.

Bunch large, seldom branched, compact; peduncle short, brownish-red. *Berry* round, medium; pedicel yellowish-green with small bourrelet. *Skin* very thin, black with blue bloom at maturity. Pulp juicy, sweet, aromatic. Somewhat late.

20. **Barbera**

Like Freisa, Nebbiolo, Grignolino, and Dolcetto Nero, Barbera is a grape of Northern Italy, and is largely grown in Piemonte, particularly in the neighbourhood of Asti and Monferrato.   It is an excellent variety, and grows on al lsoils, but prefers ferriferous, clayey, deep soils with a warm sunny aspect.   It is fairly resistant to oidium but not to downy mildew. It grows and bears well with short pruning, and produces a good, dry, very full-bodied and dark red wine which is sold under the name Barbera, and is highly esteemed in Italy.   Its dry wine contains 11.5-14 vol. % alcohol, 10-7 ‰ total acidity, 25-30 gr. dry extract per litre. Its ripe grapes do not rot easily.   In 1910 I imported all the above-named Italian varieties into the Cape, where they answer well.   Barbera seems to do best on Rup. du Lot and 420A.

Description.—
Vine vigorous and productive. *Bourgeonnement* white woolly.
Canes cinnamon colour, thick, with long internodes, tendrils numerous and thick.
Leaves fairly large, as broad as long, light green, glabrous and almost smooth above, woolly below; upper sinuses broad and deep; lower sinuses well marked; petiolar sinus closed; teeth fairly long, rather broad and slightly sharp; petiole long and reddish.
Bunch of variable size, cylindrico-conical and not very compact, with green stalk; peduncle long and almost entirely herbaceous (not lignified). *Berries* fairly large, longish oval; pedicels of medium length and thickness. *Skin* fairly tough, violaceous black, with bloom. Pulp fairly firm, juicy, somewhat astringent, not pleasant to eat, sweet. Pedicels red when somewhat over-ripe.   Late mid-season.

N.B.—My information about these Italian varieties has largely been obtained from the works of Molon, *l.c.,* and Strucchi, *I migliori vini d' Italia,* Ulrico Hoepli, Milan, 1908.

21. **Grignolino.**

It is indigenous in the district of Asti. It is fairly subject to fungoid diseases. It prefers a fertile, light soil and a warm site, as it ripens rather late; it is moderately productive with short pruning; produces a good table wine— the best of Piemonte—with 11-13.5 vol. % alcohol, 7-6 ‰ total acidity, 18-28 gr. dry extract per lit. (according to Strucchi, *l.c.* p. 51). In my ampelographic collection at

Stellenbosch it does not thrive too well. It does fairly well on Jacquez and 420A.

Description.—(Molon.)

Vine good grower and bearer. *Bourgeonnement* white, woolly.

Canes robust, light brown or walnut colour, short internodes.

Leaves above medium, generally 3-lobed, glabrous and copper-green above, white woolly below; petiolar sinus nearly always closed.

Bunch above medium, cylindrical (long: A. I. P.), compact: peduncle fairly long and thick, lignified up to the node. *Berries* below medium, oval, violaceous black at full maturity, otherwise reddish-violet. *Skin* thin but tough. *Pulp* firm, somewhat crackling. Seeds 3-4 per berry. Late.

22. **Nebbiolo.**

This is one of the finest grapes grown in Piemonte. In Novara it is called Spanna. The well-known and famous wines Barolo, Barbaresco, Gattinara, etc., are made from this grape. It starts budding early, and is sensitive to frost; it sets its berries badly if rain falls during the flowering; it is fairly sensitive to oidium and downy mildew. It ripens late (at the same time as Grignolino); rain easily causes its ripe berries to rot. It requires a warm site and prefers a tuffaceous, limestone soil. On good, warm soils it grows and bears well with long pruning. At the Cape it does well. It possesses a good affinity for most American stocks.

Description.—(Molon.)

Vine very vigorous.

Bourgeonnement white, woolly.

Canes creeping, medium thick to somewhat thin, reddish-cinnamon colour, long internodes, striped (with numerous fine black dots: A. I. P.), numerous tendrils, narrow pith.

Leaves medium, longer than broad, often 5-lobed but usually 3-lobed; fairly dark green, rough, glabrous above; light woolly pubescence below; sinuses large and deep; petiolar sinus open; teeth medium long, sharp; petiole longer than mid-rib, reddish, with a few hairs.

Bunch medium to large, conical, sometimes somewhat cylindrico-conical, always branched, nearly compact; peduncle long, herbaceous, robust; pedicels somewhat short, greenish. *Berries* medium, roundish to short oval. *Skin* somewhat thin, tough, violaceous red, passing into black at full maturity, when covered with very thick white bloom. Pulp soft, fairly sweet, somewhat astringent; vinous. Seeds 2-3 per berry. Late.

23. **Dolcetto Nero.**

This grape is indigenous in Piemonte. It requires a good soil or heavy manuring, as it possesses only moderate vigour; it bears well with half long pruning (3-4 eyes), prefers loose to compact soils, ripens early. Its cultivation in Northern Italy is on the increase. In compact soils rich in lime it gives a very dark-coloured heavy wine. Its dry wine has 10-12 vol. % alcohol, 7-5 ‰ total acidity, and 22-27 ‰ dry extract. Its wine is known as "Dolcetto" and is one of Piemonte's best table wines, where it is considered a very pleasant table wine; its must is sweet, with little acid and tannin. It is soon fit for consumption. Of all the varieties grown in Northern Italy, it is probably grown most generally and most largely. It starts budding late, and resists frosts well.

Description.—(Molon.)

Bourgeonnement woolly, reddish.

Canes somewhat thin and short; internodes short, brown with numerous black dots.

Leaves medium and smaller, broader than long, 5-lobed or also 3-lobed, with somewhat deep, not always open sinuses; petiolar sinus usually open; glabrous above, very slightly cobwebby below; teeth small, sharp, red when the grapes are ripe; petiole medium long, rather thin, tinted *red* like the nerves—this is a constant characteristic.

Bunch somewhat above medium size, conical, shouldered, fairly loose and long; peduncle long, somewhat thin, red to brown when ripe. *Berries* medium or small, round, uneven, drop easily; pedicels somewhat thin, with wine-red colour. *Skin* fairly thin, tough, rich in pigment, passes from red to bluish-black at maturity, strong bloom. Pulp soft, juicy, sweet, red, simple flavour. Seeds 2 per berry. Early mid-season to early.

24. **Freisa** or **Fresia** or **Fresa.**

It is also known as *Freisa di Chieri,* since Chieri is the centre where it is grown most extensively. It is also indigenous in Piemonte, and is grown mainly round about Torino, Alba, Canelli, Casale-Monferrato, and Asti. It is being grown more and more, since it is resistant to fungoid diseases (especially oidium), and being very vigorous can easily be grown on poor soils, and is a good bearer with fairly short pruning. Its grapes become sweet on a warm site.

Description.—

Vine robust, fertile, vigorous.

Bourgeonnement somewhat reddish and slightly woolly. *Canes* light chestnut colour, fairly thick, internodes long rather than short.

Leaves medium or smaller, light green and glabrous on both surfaces; entire to 3-lobed with narrow sinuses; petiolar sinus always very open; teeth shallow and slightly sharp; petiole of medium length and thickness.

Bunch cylindrical, branched, more often loose than compact; peduncle fairly long, stalk green. *Berries* medium or larger, almost spherical or slightly ellipsoidal; pedicels fairly long, somewhat thin and reddish at complete maturity. *Skin* somewhat tough, little subject to rotting; light green with bloom before turning red, then passing into bluish-black with abundant, almost ash-coloured bloom at maturity. Pulp soft, somewhat astringent, not very pleasant, yet vinous. Mid-season.

25. **Sangioveto grosso.**

*Synonyms.—*Montepulciano, Sangiovese, Sangioveto chiantigiano, Sangioveto dolce, Sangioveto, etc.

This grape is extensively grown in Central Italy. In the districts of Florence and Siena the famous Chianti wine is composed of Sangioveto grosso $7/10$, Canajolo nero $2/10$, Malvasia or Trebbiano fiorentino $1/10$. It is a very healthy grape, and it is grown over a large area in Italy, as far as Puglia, the district of Naples, etc. There are two types of this variety: *Sangioveto grosso* or *dolce* and *Sangioveto piccolo.* The former has larger berries and is more productive than the latter; it is also more widely grown. It is fairly resistant to fungoid diseases, a fairly vigorous grower, and a regular good bearer. In fertile soil it can be pruned long, in poor soil it must be pruned short. Its ripe grapes do not easily rot owing to their thick skins. It gives an excellent dry red wine. At 20% sugar its must contains 6‰ total acidity.

Description.—(Molon.)
Vine vigorous and healthy.
Bourgeonnement early, woolly, light green.
Canes fairly thick, internodes short, light mahogany colour.

Leaves medium or larger 3-lobed and sometimes 5-lobed; petiolar sinus open; glabrous above, nearly glabrous below; teeth sharp and rather small; petiole rather short, thick.

Bunch medium, conical, shouldered, fairly loose. *Berries* above medium, oval; pedicels long, somewhat thin, light green; peduncle short and thick. *Skin* thick, tough, with bloom. Pulp soft, juicy, sweet, simple flavour. 2-4 large seeds per berry. Mid-season (somewhat late).

26. **Alvarelhao.**

Synonyms.—Alvarelhão pied de perdrix, Alvarelho, etc.

This is one of the oldest varieties grown in the Douro Valley. According to Villa Maior[1] there are two varieties of this grape, one with a green peduncle and the other with a red peduncle called Pied de perdrix or Pied rouge. The latter is valued most highly (quoted from Mas and Pulliat, *l.c.* i. p. 179). The famous Port wines are, to a large extent, made from it and from Bastardo. Its grapes do not rot easily in wet seasons and resist drought well in very hot summers. On heavy soils it is very productive. It is grown untrellised and pruned long. Oidium frequently attacks it; otherwise it is fairly resistant to fungoid diseases. Its ripe grapes give 62 lbs. must from 100 lbs. grapes, and the must contains 2‰ total acidity with 26.6% sugar. It is a grape of high quality.

Description.—
Vine of medium vigour, with thick adhering bark. *Bourgeonnement* white woolly.

Canes medium thick, half erect, reddish-brown cinnamon colour, with thin pith and hard wood; nodes not very prominent; eyes large.

Leaves medium or fairly large, light green, glabrous and almost smooth above, with some loose, white, flocculent pubescence below; upper sinuses deep, rounded off at the base and closed by superposition of the lobes; lower sinuses well marked; open or almost open; petiolar sinus closed above, leaving an opening at the base; teeth somewhat long, sharp ; petiole medium, thin, wine-red.

Bunch medium, slightly compact or loose, branched, conical; peduncle long, somewhat herbaceous, reddish; pedicels long and thin. *Berries* medium, very even, oval, dropping easily. *Skin* thin and yet tough, a pretty black with bloom. *Pulp* fairly firm, juicy, sweet, very pleasant acidity.   Early mid-season.

---

[1] Villa Maior, *Douro illustrado,* 1876, p. 168

27. Bastardo

There are several Bastardos, but the one under discussion here is the Bastardo of the Douro, of which I have imported two varieties, *Bastardo do menudo* and *Bastardo do Castelo.* They agree very closely. With short pruning they bear well, but as the bunches are rather small, the yield per acre is only moderate. The quality is excellent. They give very sweet must and excellent, dark red wine. Bastardo, together with Alvarelhão, Touriga, Souzão, and Donzellinho do Castello, are the principal varieties responsible for the quality of Port wine. Its leaves begin to develop a red colour when the grapes commence ripening, and turn a beautiful, intense red in autumn. On good soil or with good manuring it is vigorous and can be pruned to 3 eyes to increase the yield. It answers very well at the Cape.

Description.—(According to Villa Maior from Leroux.)[1]

Vine strong, bark thick, firmly attached, cracked, starts budding early.

Canes fairly numerous, erect, short (fairly thick: A. I. P.); short internodes, hard, little pith, grey coloured. Tendrils few and undivided.

Leaves small, regular, 5-lobed, shallow sinuses, petiolar sinus open, heart-shaped, teeth two series, slightly acute; glabrous and dark green above, slightly woolly below, nerves thin but prominent. Petiole short, reddish.

Bunches fairly numerous, usually small, cylindrical or cylindro-conical, very compact; peduncle short, greyish-green. *Berries* medium and even, ovoid, pitch black, firm. 100 berries give 51.8 gr. must, fine, light rose-coloured; s.g. 1.140 with 29.3 gr. sugar per 100 c.c. and 3.6‰ total acidity (as tartaric acid), when the grapes are dead ripe. Early mid-season.

28. Donzellinho do Castello.

A number of varieties are grown in the Douro and Traz-os-Montes under the name Donzellinho, of which some are black, some red, and some white. Of the black variety *Donzellinho do Castello* is the best. *Donzellinho gallego* closely resembles it, and no doubt is derived from it. Both have thick leaves which are dark green, glabrous above and slightly pubescent below. A grape of good quality, which is early and becomes very sweet.

---

[1] Leroux, S., *La Vigne et le vin en Algérie et en Tunisie,* 2 vols., 1894

Description.—(According to Duarte d'Oliveira in Viala-Vermoral, *l.c.* vol. iv. p. 189.)

Vine vigorous, thick, erect.

Bourgeonnement white, woolly.

Canes long, cylindrical, brittle, medium thick, chestnut-yellowish-red at maturity; internodes long (8-12 cm. or about 3-5 in.); nodes swollen, tendrils very numerous and thick.

Leaves large, broader than long, bullate; vaguely 5-lobed, almost entire; parenchyma very thick, above dark green, sometimes with reddish patches; greenish-white, woolly below, teeth short and somewhat obtuse. Petiole long, thick.

Bunches numerous, medium or small, cylindro-conical, compact; peduncle medium, upper portion lignifies, lower portion remains green; pedicels long, thin, bourrelet coloured wine-red. *Berries* medium, almost round, bluish-black. Pulp soft, juicy, sweet and aromatic. *Skin* hard, with very little pigment and a hard style-point.

29. Touriga.

This is an excellent variety for quality, and is largely grown in the Douro district. Villa Maior writes concerning it: "This variety is highly esteemed in the Alto Douro, because it bears regularly and gives the wine a pleasant fruity taste; together with Tinta Francisca and Mourisco tinto it preponderates in these vineyards." According to Odart it contributes most to the good quality, and especially the colour of Port wines. It must be pruned fairly long (7 eyes) to ensure good crops. It is fairly resistant to oidium.

Description.—(According to Duarte d'Oliveira in Viala-Vermorel, *l.c.* vol. v. pp. 21-22.)

Vine vigorous grower.

Bourgeonnement fairly woolly, whitish.

Canes very thick, internodes 3-4 in. long, somewhat pointed eyes, light brown.

Leaves medium large, 5-lobed, upper and lower sinuses deep, nearly U-shaped, petiolar sinus open to (sometimes) nearly closed; dark green and slightly cobwebby above, very woolly and whitish below; leaves soft; teeth short, obtuse, light yellowish. Petiole long, glabrous, reddish. In autumn the leaves turn yellow with very bright crimson-red stripes.

Bunches medium large, cylindro-conical, not very compact, short; stalk much branched, frequently with a secondary bunch at the fertile node which is almost as large as the main bunch; pedicels long, thin; peduncle very long, fairly thick and usually flattened. *Berries* nearly spherical, uneven, small rather than large, bluish-black. Pulp soft, juicy, aromatic. *Skin* hard, rich in red pigment. Mid-season and somewhat later.

30. **Tinta Francisca.**

Synonyms.—Tinta Franceza, Tinta de Franca. It produces a very fine wine of a dark colour and of excellent quality. After two years it is well matured. It will then have lost much of its colour, but retained its flavour and other good properties. It prefers a warm, sunny aspect in order to develop its highest qualities. It bears well when pruned half long on the Guyot system, It is fairly resistant to oidium and other fungoid diseases. It answers well on Riparia, 3309, Rup. du Lot, etc.

Description.—(According to Duarte d'Oliveira in Viala-Vermorel, *l.c.* vol. v. p. 252.)

Vine of medium vigour, semi-erect.

Canes vigorous, somewhat flattened; with crimson-red stripes and very glossy when still herbaceous; light walnut colour on the internodes and dark chestnut colour on the nodes when ripe; internodes short or medium; tendrils very long and thick, bifurcated.

Leaves medium or small, almost round; 5-lobed, upper sinuses deep and slightly open, lower sinuses irregular, petiolar sinus deep and nearly or completely closed; glabrous on both surfaces, slightly crinkled, light green above and yellowish-green below; teeth alternately large and small; nerves light yellow. Petiole thick, medium long or short, reddish.

Bunches numerous, large, cylindrical, sometimes with one to two long branches, fairly compact; peduncle short or medium, hgnified up to the first branching; pedicels short, thick, crimson-red, adhering well to the berries. *Berries* black with strong bloom, nearly spherical, medium large. Pulp soft, juicy. *Skin* hard with little pigment.

Note.—Teinturier male (or Pontac), which also occurs in the Douro vineyards and is also called Tinta Francisca, is sometimes confused with the real Tinta Francisca here described.

31. **Mourisco tinto.**

It is a very healthy vine and a very vigorous grower. It resists fungoid diseases well, and bears well with long pruning. On fertile, moist soil it suffers easily from coulure. It can yield heavy crops on account of the *large* size of its bunches. In Portugal it is very widely grown as a table grape. In the Douro district it gives also a good but rather poorly coloured wine. When the young shoots are topped (pinched), the laterals that subsequently develop bear numerous second crop bunches. The leaves look ornamental.

Description.—(According to Duarte d'Oliveira in Viala-Vermorel, *l.c.* vol. ii. p. 305.)

Vine extraordinarily vigorous.

Bourgeonnement almost glabrous, yellowish-green.

Canes somewhat flattened, very thick and long, semi-erect, light brown with wine-red stripes; internodes very long, 10-14 cm. or 4-5⅔ in.; nodes swollen; tendrils numerous and unusually thick and large.

Leaves 5-lobed, very large and pretty, dark green, glossy and glabrous above; yellowish-green and slightly pubescent below; upper sinuses more or less closed, lower sinus hardly marked, petiolar sinus very deep and open; thick; teeth large, sharp to obtuse. Petiole long, glabrous.

Bunch very large, long, conical, branched; peduncle thick and very long; pedicels long and thin. *Berries* very large, nearly spherical, ripening unevenly, so that some berries are pitch-black when others are reddish-black. *Skin* not very thick. Pulp firm, very juicy, sweet. Late mid-season.

32. **Tinto Cao.**

On the poor, decomposed shale soils of the Douro valley, with a warm, sunny aspect and without manuring, it produces wine of good quality, but the yield is limited. It starts budding early and its berries set well, even in adverse weather. Its wine is very aromatic, light in colour, and of high quality. The vine is of medium vigour, with long, rather thin canes with long internodes; leaves large, thin, 5-lobed, glabrous and yellowish-green above, yellowish and woolly below; teeth small, bunch medium, loose, conical; berries small and almost spherical, black; pulp not very juicy, soft; skin thin.

33. **Cornifesto.**

This grape plays an important part in the production of Port wine in the Douro district. It gives good yields with short pruning, and its wine is much darker than that of most other varieties, though it does not contribute to the "finesse" of the wine, which is usually made from some twenty different varieties which are interplanted in the vineyards. On a dry warm site it gives a good product, and its grapes may even become very sweet. Its wine is rich in tannin, has a very dark colour, and gives plenty of lees.

The vine is a vigorous grower, with medium large, somewhat thin leaves, glabrous above and slightly pubescent below; bunches large, compact, long, somewhat cylindrical, frequently bent; berries spherical, black, medium size; skin thick, hard, very rich in pigment.

34. Souzão.

This is one of the good varieties of the Douro and Minho districts. It bears well with short pruning. In fertile moist soils it becomes a yielder of quantity. On a warm site and in somewhat dry soils it gives a wine of good quality. Its juice has more or less the same colour as Alicante-Bouschet, and therefore it gives a very dark wine. It is resistant to fungoid diseases. It is valuable when blended with other varieties that introduce more quality into the wine.

The vine is a vigorous grower; leaves 3-lobed, thick, soft, large, yellowish-green, woolly on both surfaces, with red spots at maturity; nerves at first light yellow, then red; petiole long, woolly, striped red; bunches fairly large, long; berries spherical, medium, bluish-black; pulp very juicy and coloured red; skin tough and very rich in pigment.

35. Molar.

Synonyms.—Molard, Mollar noir, Mollar, Mollard.

According to Mas et Pulliat, *l.c.* vol. ii. pp. 177-178, this grape, which is grown in the Hautes-Alpes, is the same as the Mollar which is grown in Andalusia and in Portugal, and is of Spanish origin. Others doubt this, but, it would seem, they are in error. It gives a good wine in a warm district. It grows and bears well with short pruning, and produces a full-bodied, very dark wine if its grapes are picked when fully ripe. I recommend it for a relatively neutral, full-bodied, *very dark* wine to be blended with Hermitage and perhaps even with wines of the Port type. Its wine possesses a flavour somewhat resembling that of Cabernet Sauvignon.

Description. —

Vine vigorous on most American stocks.

Bourgeonnement whitish-green, somewhat pubescent, turning into yellowish-green.

Canes thick, somewhat creeping; internodes above medium.

Leaves large, almost even, glabrous and smooth above, slightly woolly below, entire or hardly lobed; petiolar sinus almost closed by the approaching lobes; dentition rather shallow, somewhat obtuse.

Bunch medium or above medium, cylindro-conical, sometimes slightly shouldered; peduncle medium thick and medium long. *Berries* medium or larger, spherical or slightly spherico-ellipsoidal; pedicels short, fairly thick. *Skin* thin, fairly tough, a pretty black with bloom. Pulp fairly firm, juicy, sweet, simple flavour. Mid-season.

II. TABLE GRAPE VARIETIES

(a) White Table Grape Varieties.

1. Muscat of Alexandria.

Synonyms.—White Hanepoot (Cape), Muscat d'Alexandrie (France), Moscatel gordo bianco (Spain), Uva Salamanna (Tuscany), Moscatellone (Italy), Zibibbu and Gerosolimitana bianca (Sicily), Malaga, Bowood Muscat, Muscat of Jerusalem, Muscat of Alexandria (England), Gordo bianco (Australia), etc.

This is perhaps the finest and most valuable grape taken all round. It is grown in all viticultural regions where the climate is suitable. As it requires a considerable amount of heat for its grapes to reach perfect maturity, we find it mainly in the warm wine countries, *e.g.* Spain, South of France, Sicily, North Africa, South Africa, California, Australia, etc. It is not grown much in hot-houses on account of the large amount of heat it requires. In the Central European wine districts the climate is on the whole not warm enough for it. In no other country does it bulk relatively so largely among the varieties grown as at the Cape, where it is used as a table grape, raisin grape, and wine grape. Although mainly a table and raisin grape, it can produce a nice light, dry, white wine and a better sweet wine (*e.g.* Constantia Berg). Its wine has a typical muscat flavour differing, however, from that of Muscadel. One can also make a good "Moskonfyt" (grape syrup made at the Cape by boiling must in an open pot until the syrup is thick enough) out of its must. It can therefore serve various purposes, and this explains why it is so largely grown.

The Dutch name *Hanepoot,* under which this variety is known in South Africa, is an intentional corruption of Hanekloot, *i.e.* cock's testicle, which the berry of this variety resembles somewhat in shape and size. "Hanepoot" has nothing to do with "cock's foot".

Owing to its large bunches and great fertility it produces very heavy yields. At Franschhoek (Cape) 3300 Hanepoot vines planted 5×6 ft. square (thus occupying 1.1 Cape morgen or just over 2 acres), grafted on Jacquez and trellised on one wire, have for several years in succession produced 55-57 leaguers of wine, *i.e.* 50 leaguers per morgen or 3080 gallons per acre or 346 hl. per hectare. This probably creates a world's record. The soil is good, fairly deep, and well fertilised, and the vines are not irrigated. Its average yield is about one-fourth to one-third of this extraordinary yield.

In 1910 I imported Muscat d'Alexandrie from Montpellier and planted it next to White Hanepoot. I find absolutely no difference between them. I had the same experience with *Gordo bianco. Bowood Muscat* is a seedling of Muscat of Alexandria raised in England. It is less subject to coulure, ripens earlier, but is less aromatic than Muscat of Alexandria. *Canon Hall Muscat* is also such a seedling. According to Dr. Hogg[1] it "differs from its parent, the Muscat of Alexandria, in having better set and more tapering bunches, and rather larger and rounder berries. The vine is of more robust growth, and the flowers have six, and sometimes seven stamens; but the fruit is not so highly flavoured as that of Muscat of Alexandria". It was raised at Canon Hall, near Barnsley, England.[2] In my Stellenbosch collection it is not very successful; its canes ripen badly; it is very subject to coulure, and the leaves and grapes are easily sun-burnt.

Among the White Hanepoot vines at the Cape one finds vines that grow more vigorously and bear fewer and smaller bunches than the normal vines. The bunches are fairly short and naturally loose, so that they do not require any pre-thinning for export, which is a great advantage. The berries are rounder than and as large as those of ordinary Hanepoot. As Mr. D. M. le Roux of Vredenburg, Paarl, was

---

[1] Hogg, Dr. Robert, *The Fruit Manual,* 5th ed., London, 1884, p. 370.

[2] *2012 Editor's Note*: The original vine, over 200 years old, still grows and can be seen at Cannon Hall, now a museum and park.

the first to grow it by itself, I named it *Le Roux Hanepoot.*
This, like Red Hanepoot, evidently arose out of ordinary
white Hanepoot as a bud variation. This type is now constant
and retains its distinctive characters when grafted on
different stocks. Its inflorescences are much smaller and less
numerous than those of ordinary White Hanepoot. Where
ordinary Hanepoot must be pruned very short, Le Roux
Hanepoot should be pruned somewhat longer, say 3-4 eyes,
in order to obtain satisfactory crops, as its bunches are only
about half the size of ordinary Hanepoot bunches.

Muscat of Alexandria is a regular and good bearer
with short pruning. With cold and rainy weather during
flowering and setting, it is very subject to coulure. It is very
susceptible to anthracnose in wet springs and summers.
This happens some years in the Cape, Stellenbosch, and
Paarl districts, whereas in the drier districts of Ceres,
Worcester, Robertson, and Montagu, anthracnose is
practically unknown. It is fairly resistant to oidium. It is
subject to sunburning during the latter half of December and
the first half of January (at the Cape), *i.e.* about six to eight
weeks before it is ripe.

It can be grown as a bush vine or on a trellis. It is
frequently used for pergolas, but is not an ideal grape for this
purpose, as it gives its best products where the permanent
stem is kept small. This is due to the fact that its
productiveness exceeds its vigour. Grafted on Aramon ×
Rup. Ganzin Nos. 1 and 2, it usually dies within a few years,
as we have already seen. On Jacquez, Riparia Gloire, 420A,
101-14, etc., it does well. On 1202 some vines die without
any apparent cause (barring an insufficient affinity) during
the first three to four years, after which this trouble It seems
to do well on 333, 3306, 3309, and 106-8 also. On Rupestris
du Lot it grows vigorously but is inclined to excessive vigour,
when its fruit tends to set very badly, as sometimes happens
on 1202. For export it should be trellised.

It is our principal white table grape at the Cape. Its
bunches and berries are large and pretty, and possess a
delicious sweet taste and a pleasant strong muscat flavour.
Its skin is somewhat thin, but when grown on suitable soil
on a dry site and packed with the necessary care, it stands
the long transportation to London very well, provided the
grapes are stored in suitable cool chambers during
transportation. On account of its many good properties it will

no doubt always remain a first-class table grape both for local consumption and for shipment to distant markets.

Description.—
Vine very productive and fairly vigorous grower.
Bourgeonnement garnet coloured and slightly woolly.
Canes fairly thick; internodes short, light brownish (A.I.P.).
Leaves medium or larger, as broad as long, glabrous on both surfaces; 5-lobed, upper sinuses deep and closed, lower sinuses well marked, petiolar sinus open; teeth long, narrow, numerous and sharp; petiole very long, fairly thick, glabrous and often violet-red.
Bunch large, long, shouldered, loose and sometimes even branched; peduncle fairly thick, long or fairly long. *Berries* large, oval, unequal in size, and with long, slender warted pedicels (Dr. Hogg) (Mas and Pulliat describe the pedicels as "thick"). "*Skin* thick,[1] generally greenish-yellow, but when highly ripened a fine pale amber colour, and covered with thin white bloom. Flesh firm and breaking, not very juicy, but exceedingly sweet and rich, with a fine Muscat flavour" (Dr. Hogg). Late mid-season.

2. **Rosaki.**

*Synonyms.—*Rasaki, Razaki, Rhosaki, Dattier de Beyrouth, Waltham Cross (?), etc.

In 1910 I imported Rosaki and Dattier de Beyrouth from Montpellier, and found them to be identical. According to Viala-Vermorel, *l.c.* vol. ii. p. 99, Eckerlin is of opinion that Dattier de Beyrouth is the same grape as *Razaki sari* or Yellow Rosaki which is grown in the neighbourhood of Constantinople and in Asia Minor (Smyrna, etc.). It is the Yellow Rosaki that I imported into the Cape.

Rosaki is one of the prettiest grapes we know. It forms large bunches with very large, long oval to elongated berries. The following data I obtained at Stellenbosch in February 1923 from berries of Rosaki (the numbers give the breadth and length in one-hundredths of an inch): 90 × 115, 83 × 115, 95 × 120, 87 ×120, 89 × 121, 82 × 124, 88 × 123. These were large berries, but larger berries can be found. Rosaki is very susceptible to anthracnose, but fairly resistant to oidium and downy mildew. It practically never suffers from sunburning. It is a most excellent table grape, stands long

---

[1] Mas et Pulliat fittingly describe the skin as "fairly thin and yet tough."

transport (in cold store) well, and fetches high prices on the English markets. It is best to trellis it and prune fairly short (2-3 eyes). Its bunches are naturally loose and need practically no pre-thinning in normal years. Rain and cold weather during the blooming period causes many small, seedless berries to be formed, and thus the bunch is sometimes perfectly worthless as a table grape. This, however, happens only under unfavourable weather conditions during flowering and setting, and is very exceptional at the Cape. The bunches are usually well set, and the berries are then very even and naturally loose. The fact that Rosaki requires practically no pre-thinning and very little trimming makes it one of our most valuable export varieties.

FIG. 54.—Rosaki (Paarl Vit. Expt. Station). Original.

In Asia Minor, particularly in the neighbourhood of Smyrna, it is extensively grown for the production of lye raisins. It is grown for this purpose elsewhere also, *e.g.* at the Cape, although to a much lesser extent than Hanepoot. It gives a beautiful lye raisin *when well ripe*. Its affinity for American stocks is good.

Description.—(A. I. P.)
Vine vigorous, productive.

Bourgeonnement slightly pubescent, light greenish or light yellowish.

Canes long, medium thick ; medium to short internodes, light brown reddish; nodes somewhat flat.

Leaves (young) bronze-reddish with green nerves; (adult) fairly large or medium, as long as broad; petiolar sinus closed, entire or 3-lobed; even, long, sharp teeth; fairly glossy, dark green, and glabrous above; below dull green, parenchyma glabrous, with some cobwebby pubescence on the strongly developed nerves; petiole medium thick and medium long, light rose-coloured green.

Bunch large to very large (sometimes medium), conical, loose, well branched; peduncle long and thin ; pedicels long and rather thin, with a strongly developed bourrelet. *Berries* large to very large, long oval to elongated, slightly thicker where the pedicel enters the berry; little bloom, golden-yellow when well ripe. Pulp firm, pleasant sweet. *Skin* fairly thick, stands transportation well. The berries adhere very firmly to the pedicels, which, however, break off rather easily when dry.

Mid-season to late mid-season (often eight days earlier than Muscat of Alexandria).

3. Waltham Cross.

According to Dr. Hogg this grape "was introduced by Mr. William Paul, of Waltham Cross (England), and received a first-class certificate from the Royal Horticultural Society in 1871" Dr. Hogg calls it "one of the largest, if not the largest grape in cultivation. It is an exaggerated Muscat of Alexandria, but has not the Muscat flavour." As Dr. Hogg gives no detailed information about its origin and merely states that it was introduced, it is quite possible that Mr. Paul imported this grape into England. I can in any case not definitely distinguish it from Rosaki. The general impression at the Cape is that it ripens somewhat later than Rosaki, and that its ripe berries are less yellow and more greenish-yellow, and more elongated than those of Rosaki. This might be so, but no one can with certainty distinguish between their grapes. Their vines, leaves, shoots, and canes, etc., also correspond so closely that one cannot distinguish between them either. Since 1912 I have studied them, but I am not prepared to say that they are two different varieties. If, in future, no definite points of difference can be found between them, we ought to drop the name Waltham Cross and retain only the better known name Rosaki for Rosaki, Dattier de Beyrouth, and Waltham Cross.

In Waltham Cross and elsewhere in England it is unobtainable to-day (at least several leading nurserymen in England could not supply me with cuttings of it last year, as they either did not know it or did not grow it any more; one or two stated that it had been grown in England some thirty years ago). Dr. Hogg describes it as follows: "Bunches very large, long and tapering, with strong stout stalks well set and well shouldered. Berry-stalk thick, stout, and warted. Berries very large, from an inch and a quarter to an inch and three-quarters long; oblong oval. *Skin* membranous, pale amber. Flesh firm and solid, with a sweet, brisk, and very pleasant flavour".

It has been grown under the name Waltham Cross for many (possibly some thirty) years at the Cape.

4. **Golden Queen.**

According to Dr. Hogg "it was raised by Mr. John Pearson, of Chilwell, Nottingham, from Alicante, crossed by Ferdinand de Lesseps, and was awarded a first-class certificate by the Royal Horticultural Society in 1873". He says of it: "This is a fine grape, and the constitution of the vine is very good. It requires a high temperature to ripen it properly".

Description by Dr. Hogg: "Bunches nine inches long, with a stout stalk, long, tapering, and well shouldered, like Muscat of Alexandria. Berry-stalks, rather long, but stout and warted. Berries, upwards of an inch, and sometimes an inch and a quarter long; oblong or oval. Skin, membranous, of a clear amber colour. Flesh, firm and crackling, very juicy, and richly flavoured".

As far as I have observed this grape at the Cape, where it is grown in the open, it is not as good as one might have expected from Dr. Hogg's description of it. Its leaves are very woolly below. It grows and bears well with short pruning. Its bunches and berries are large enough for export, but the berries are a good deal smaller than those described by Dr. Hogg. Its grapes have a rather sour taste until they are ripe, and then are not very sweet. They ripen fairly late, and easily become over-ripe, when the berries show brown spots and readily develop wrinkles. For the Cape, I do not value it much as a table grape, and am of opinion that we have at least a dozen varieties of table grapes that are to be preferred to it.

5. **Chaouch**

Synonyms.—Tsaousi (Greece), Tchaouch (Russia), Tcha-vouch userum (Constable-grape), Shavoah (Cape), Chavoush (Dr. Hogg).

The description given by A. Tacussel and E. Zacharewicz of this grape in Viala-Vermorel, *l.c.* vol. ii. pp. 199-203, has brought me to the conclusion that the grape known as *Shavoah* at the Cape is identical with the Chaouch there described. The only striking difference is that Chaouch is a hard grape that stands long transport well, whereas the Shavoah of the Cape is generally regarded as too soft for long transport. This will probably improve as the vines grow older.

Chaouch is grown along the coasts of the Mediterranean Ocean, especially in European and Asiatic Turkey, Greece, the Crimea, etc. Under unfavourable flowering conditions it is very subject to coulure. Artificial pollination with pollen from Aramon No. 1 and other varieties is wonderfully effective. According to C. Eckerlin, former Government Viticulturist for the Ottoman Empire, there are four sub-varieties of Chaouch: the white one with round berries, the white one with oval berries, the red, and the Muscat flavoured.

Eckerlin (from Viala-Vermorel, *l.c.* vol. ii. p. 202) gives the following information about this grape: "The Tchavouch is the king amongst the table grapes. It is successfully grown on American stocks, but the vines must be fairly old before they will produce grapes of very good quality. Grapes from eighty to one-hundred-year-old vines sell at double the price of grapes from young vines. The grapes keep well, and can remain on the vine in paper bags till Christmas, when they are sold for 4 francs per kg. or 1s. 6d. per lb. It demands a warm site and dreads dampness in particular. Rain during flowering can remove half the crop by coulure. It is advisable to interplant it with some other variety like Merlot in order to promote pollination, as it is very subject to coulure. It does well when trained as a cordon with short bearers of two eyes. It is a wild grower. Top its shoots at two leaves above the highest cluster."

At the Cape it also is very vigorous and productive, and very subject to coulure under adverse flowering conditions.

Its berries are short oval. In 1923 I measured a number of Shavoah berries in my Stellenbosch collection with the following results (the numbers are $^1/_{100}$ in. and the cross diameter or equatorial diameter is given first): 81 × 97, 81 × 100, 82 ×102, 87 × 97, 87 × 101, 91 ×106, 92 ×100, 92×104, 97 × 106.

Description.—(According to Viala-Vermorel.)

Vine very vigorous grower, thick stem. *Bourgeonnement* strongly white woolly.

Canes inclined to grow straight up, long and very long, thick, little branched ; internodes long with thick diaphragms ; wood hard, pith rather thin ; tendrils few, thin, bifid.

Leaves very large, 5-lobed, somewhat longer than broad, very much crinkled, thick, dense, woolly below; petiolar sinus almost closed, the other sinuses deep and closed; its leaves are about the last to drop; teeth sharp, long; petiole medium thick, long, green below and light reddish above.

Bunch large, conical, branched, long peduncle; pedicels short, medium thick. *Berries* large, round and ovoid, the poles being sometimes a little flattened; very compact (when not badly set. A.I.P.). *Skin* thick, white-greenish turning to amber-coloured white at maturity. Pulp plentiful and firm, juice sweet and colourless, with a very pleasant flavour. Late mid-season.

Note.—Dr. Hogg, *l.c.* p. 379, writes as follows about this grape: "Bunches, about nine inches long, very loose, tapering and shouldered. Berries, large and oval. Skin of a pale amber colour when quite ripe, thin, and adhering closely to the flesh. Flesh, firm, juicy, and agreeably flavoured. A second-rate grape, introduced from the Levant. It sets its fruit very badly, both when forced in this country, and also on the shores of the Mediterranean, where I have seen it in a very miserable condition, even when under the most advantageous conditions." For the Cape its value as an export variety has not yet been proved.

6. **Olivette blanche** (White Olivette).

The *Olivettes* were called thus on account of the olive-like shape of their berries. They belong to the oldest varieties known. It has been definitely established that they were grown before the birth of Christ. It is thus very difficult to fix their origin. In the South of France they have been grown for a very long time. There are white, red, and black Olivettes.

The black Olivette is the same variety as Malakoff Isjum (see later).

The *White Olivette* can be grown successfully only in a fairly warm district. It is grown as a table grape near Montpellier and along the lower Rhone. There it is grown in deep, moist, fertile soils and pruned long (according to Guyot or Cazenave). In dry soil it may set its fruit too badly. In Paarl and Stellenbosch, where I have observed it in my ampelographic collections, coulure was its great fault. With long pruning and increasing age of the vines, this will probably improve. Ring-barking here is worthless, but cross-pollination will help. Irrigation just prior to flowering might overcome this difficulty. In good soil, when trellised and long pruned, it produces many bunches and good crops if the fruit sets well. It ripens late and keeps well. This makes it very valuable, the more so as it stands long transport well. It is therefore worth trying, and will pay very well where it sets its fruit well. It is very highly resistant to rotting *(Botrytis cinerea)*. It possesses a good affinity for most American stocks.

Description.—(Viala-Vermorel.)
Vine very vigorous.

Canes long, medium thick, light yellowish-brown; internodes short or medium, wood hard, pith thin, tendrils fairly numerous, bifid.

Leaves fairly large, 5-lobed, glabrous on both surfaces, petiolar sinus U-shaped; petiole fairly long and at right angles to the leaf-blade (lamina).

Bunch large (smallish when badly set), long, branched, loose, weighs on an average 300 g. or nearly ¾lb.; peduncle fairly long and thick but brittle; green, turns brown only where it enters the cane; pedicels long and fairly thick but brittle. *Berries* large, olive-shaped (19 × 25 mm. and 20 × 30 mm. or 76 × 100 and 80 × 120 in $^{1}/_{100}$ inch), very hard, stick very well to the pedicels. *Skin* thick, tough, dull white, passing into amber-like yellow at full maturity, little bloom. Pulp firm and crackling with a very pleasant sweet flavour. Seeds pear-shaped and on an average two per berry. Late.

7. **Olivette Barthelet.**

This grape has been raised from seed by Antoine Besson in 1864 near Marseilles. When trellised it gives good crops with long pruning (Guyot, etc.). It is a variety which I can recommend as good for export. Its bunches and berries attain a good size, and it stands long transport very well.

It need not be thinned very much. Its affinity for American vines is good. It ripens about the same time as Muscat of Alexandria.

Description.—(Viala-Vermorel.)

Vine grows vigorously and erect.

Canes long, light brown with dark brown patches; wood hard, internodes medium long but very short at basis of canes; pith dense and thick, diaphragm thick.

Leaves medium, broader than long, orbicular, thick, soft; petiolar sinus sharp V-shaped and little open above ; glabrous above, light cobwebby below.

Bunches usually two per shoot, fairly large, branched; pedicels thick. *Berries* large, olive-shaped, very firmly attached to the pedicels. *Skin* thick with amber-yellow colour. Pulp firm, light yellow, pleasant flavour.

8. **Bicane.**

Synonyms.—Olivette jaune, Panse jaune, Grosse Perle blanche, etc.

This is an old, much-scattered variety, but nowhere is it grown on a large scale. *It is a very pretty grape, but unfortunately very subject to coulure and millerandage.* This accounts for its restricted cultivation. When trellised and long pruned (Guyot) it is very productive and grows vigorously. Coulure and millerandage will, according to some authorities, be considerably checked by keeping the shoots topped short, from when the flowers begin to open and for some time afterwards. Ringbarking, according to my 1924 experiments with this and other varieties, effectively prevents coulure, but merely results in a large number of small, round, seedless berries being formed, which are worthless. Cross-pollination will no doubt be a good remedy, but will have to be done by hand.[1] It is excellent for export, but *must be picked when just ripe.* When fully ripe it soon becomes rather soft for long transport. Its affinity for American stocks is good.

Description.—

Vine vigorous grower.

Bourgeonnement almost glabrous, light garnet tint.

Canes not very thick, internodes fairly short (long, frequently branched. A. I. P.).

---

[1] Mas et Pulliat recommend, as a sure remedy, the pinching of the extremity of the bunch and of its main branches during flowering. Under adverse weather conditions this is ineffective.

Leaves small (or medium, A.I.P.), longer than broad, glabrous on both surfaces, those near the base of the canes often slightly pubescent below; upper sinuses very deep and closed, lower sinuses little marked and narrow; petiolar sinus closed; petiole fairly short, thin; teeth long, somewhat broad, sharp.

Bunch large, much branched, conical, loose; peduncle of medium length and very thick. *Berries* very large, oval (the following measurements I made on its large berries in 1923 on the Stellenbosch University Farm: 91 × 109, 96 × 112, 98 × 116, 103 × 130, 107 × 134,108 × 132 in $1/100$ inch); pedicels short and thick. *Skin* thick, subject to rotting, turning to a yellowish-green to amber-coloured yellow at maturity. Pulp fairly firm, not very sweet, with a pleasant acidity when ripe. Early mid-season.

9. **Formosa** or **Diagalves.**

This is a very old Portuguese grape. It is grown everywhere in Portugal as a table grape, and most extensively in the neighbourhood of Lisbon, where it is known as Diagalves, probably a corruption of Diogo Alves. In the Douro valley it is mainly grown in gardens and along vineyard-paths under the name Formosa. From Lisbon its grapes are exported on a large scale to England and other European countries. In 1910 I imported it to the Cape, and some years later exported its grapes for the first time from South Africa to London. They arrived in good condition and fetched a good price. The grapes for export should be picked when just ripe. After that they develop brown spots and do not keep so well. The vine is a vigorous grower, bears well with short pruning, ripens about the same time as Muscat of Alexandria, is fairly resistant to oidium and other diseases, and has a good affinity for American stocks.

Description.—(Duarte d'Oliveira in Viala-Vermorel.)

Vine vigorous grower, thick and strong trunk, bark thick, fairly loose.

Canes fairly erect, numerous, nodes thin, eyes large and prominent, internodes medium, reddish or light reddish-grey (when herbaceous yellowish-green); tendrils long, thick, bifid.

Leaves medium large, as long as broad, thick, 5-lobed, deep sinuses, petiolar sinus deep and open V-shaped; glabrous and dark green above, lighter green and almost pubescent, with outstanding nerves below; two series of sharp teeth; petiole of medium length, violaceous.

Bunch large to very large, long; peduncle short, thick, whitish-green; pedicels short and thin. *Berries* large, short oval to oval, a somewhat golden colour when ripe, with light brown spots when dead ripe, sticking well to the pedicels. Skin thick. Pulp fairly firm, pleasant taste.

10. **White Prince.**

This is a very interesting grape, owing to its very large berries. As far as I know, it forms the *largest berries of all grape varieties.* In 1923 I measured some of its largest berries in my collection, but larger can be obtained. The following results were then obtained: 103 × 125, 103 ×130, 107 × 134, 108 × 132 in $^1/_{100}$ in. In Stellenbosch it is not very successful when grown on a short-pruned cordon. It bears well, but is not sufficiently vigorous. Its shoots remain somewhat short and do not ripen too well. To the best of my knowledge it was imported to the Cape from Australia, where it is also known as *Centennial.*

Description.—(A.I.P.)

Bourgeonnement fairly pubescent, red-brownish; little leaves with reddish borders, shining with little cobwebby pubescence.

Leaves fairly large, 5-lobed, glabrous on both surfaces, dark green, petiolar sinus open U-shaped, upper and lower sinuses deep and usually closed; growing tip whitish-green with a few threads.

Canes short and thick, at first green, then fairly dark brown.

Bunch large, pyramidal, long and striking stipules (bracteae) at each branching of the stalk, especially noticeable on the young inflorescences before flowering. *Berries extremely large,* oval, even. *Skin* medium thick, light yellowish at maturity with hardly any bloom. Pulp somewhat soft, pleasant taste. Mid-season.

Note.—Where this grape thrives well, it will pay to grow it for export, provided the vineyard is near the railway station and not far from the harbour through which it is exported, as it is rather soft for long transport. It has already been successfully exported from South Africa (Cape Town) to London, where it fetched high prices. This is always bound to be the case, so long as it arrives in good condition.

11. **White Cornichon.**

Synonyms.—Lady's Finger, Finger grape, Pizzutello (in Italy), Pizzutello bianco, Tetta di vacca, Doigts de Donzelle, Dedos de Doncella, Uva de vaca, Uva de Africa, Fourchou (Constantinople), Cornichidu, Cornichon blanc.

History and Distribution.—This grape must have been known from the oldest times. Columella, a writer on agriculture early in the first century of the Christian era, recommends, as an early grape, the *dactyli* grape (evidently this variety). Plinius also writes about the *dactyli* grape. Before the thirteenth century an Arabian writer gave an accurate description of this grape. In Tivoli, near Rome, it has been grown on pergolas from time immemorial. The grapes are sold mainly in Rome.

Culture.—Its cultivation is still limited to the gardens. It is a pretty and peculiar grape with its long berries with their *bent ends* (see Fig. 14, the lowest berry at the right), which is its most characteristic and most striking character, and which gave rise to the names Finger Grape and Cornichon, the French for the spur of a cock. There is also a *Cornichon violet,* which grows more vigorously and forms larger berries, which are coloured reddish-violet at maturity. The White Cornichon is of moderate vigour only, and should therefore be grafted on a vigorous stock, *e.g.* 1202 or Rup. du Lot. Its fruit sometimes sets badly. It should be pruned fairly long (5-6 eyes). In Tivoli it covers about 40 hectares. The pergolas have a height of 1.80-2.50 m. or 6-8⅓ ft. and are horizontal, as in Almeria. White Cornichon is fairly susceptible to diseases. It is a late grape.

Description.—

Vine of moderate vigour and fertility, subject to coulure.

Bourgeonnement glabrous or almost glabrous, yellowish - green; little leaves, smooth and shining above.

Canes somewhat thin, internodes fairly short, semi-erect, ripening badly in moist soil. Leaves drop late.

Leaves medium, almost as broad as long, glabrous and smooth, almost glossy above, completely free from pubescence below; upper sinuses ordinarily rather shallow, hardly marked on the leaves near the base of the shoot and fairly deep on the upper leaves; lower sinuses marked; petiolar sinus open; petiole long, smooth, fairly thick, equalling or usually exceeding the midrib in length; dentition long or fairly deep, slightly broad, sharp but slightly obtuse at the end.

Bunch medium, branched, loose, shortly conical, often truncated (Dr. Hogg describes them as "rather small, round, and loose"), subject to coulure; peduncle fairly long and fairly thick. *Berries* large or very large, very long and irregularly olive-shaped, always curved in on one side, rounded at the ends when reaching their full size, remaining sharp pointed, to the contrary, and more curved when not full sized (Dr. Hogg describes them as "very long, sometimes an inch and a half, and narrow, tapering to both ends, and just like very large barberries"), pedicels long, fairly thick, attached somewhat to the side of the centre of the lower end of the berry. *Skin* fairly thick, not very tough, passing from greenish-white to straw yellow at maturity. Pulp firm, very sweet, pleasant. Seeds few; usually only one large seed on the convex side of the large berries. Late.

12. **Servan blanc** or **Raisin blanc.**

Synonyms.—Servan, Servan blanc, Servant, Raisin vert, Raisin blanc (Cape), Late Syrian, etc.

About 1912 I obtained cuttings of *Late Syrian* from the late Professor Muller of Stellenbosch, which I grafted on American vines, and found to be identical with the *Servan blanc* which I had imported from Montpellier in 1910. The *Syrian* described by Dr. Hogg is probably a different grape from the one here discussed. He says that its berries are large and oval and its bunches very large, one bunch of it having weighed 25 lbs. 15 oz.—exhibited at Edinburgh, September 15,1875. It is also a good late grape.

The *Raisin blanc* grown for a considerable number of years at the Cape resembles the imported Servan blanc so closely that I cannot distinguish between them, and therefore regard them as the same variety.

Servan blanc has been grown for a very long time in the South of France. It is particularly cultivated on a fairly large scale near Thor (Department of Vaucluse and 12 miles east of Avignon on the Rhône), from which town over 300 tons of this grape are annually sent as a late table grape to Paris, Lyons, London, Switzerland, and Germany. Partly because it ripens very late, and especially because it keeps so long on the vines, its cultivation near Thor pays very well. In Constantia and the Hex River Valley it is grown as a late table grape. As such it is only valuable in places where the autumn is dry and fairly long. It is one of the varieties successfully exported from the Cape to London, but is not very suitable for this purpose, as its berries are not very large and are rather soft for long transport. When grown on a fairly dry, warm site, it gives fairly firm berries which travel well. On low-lying and on moist soils its grapes are too soft for this purpose.

It is a very good variety for growing on a pergola, where one wishes to let the grapes hang very late and to consume them on the spot or sell them locally. The vine is a very vigorous grower and a good bearer with short pruning.

As the bunches are situated rather high up the canes, it should be trellised, and the canes should somehow be attached to a wire to support them when the bunches grow out and become heavy.

It is somewhat susceptible to anthracnose, less to oidium, but easily becomes sunburnt, especially where it is not trellised. For export it must be severely thinned in order to get fairly loose bunches. This is a decided drawback to its cultivation. For local markets it is valuable where it can remain hanging late on the vines (as at De Dooms). Where it grows very vigorously (in moist, fertile soil) the bunches can become very large and assume a more or less round shape. The stalk then is very much branched and tender. It produces a light wine of poor quality.

Description.—(Viala-Vermorel.)

Vine very vigorous grower, thick trunk, very productive (A.I.P.).

Canes very long, little branched; young shoots first whitish-green (A.I.P.), then pale green; ripe canes thick, very long, creeping; internodes long, wood hard, bark finely striped, very light dull yellowish colour; diaphragms thick; pith thick; tendrils bifid; very thin and numerous near ends of canes.

Leaves medium or small, 5-lobed, thin, light green, glabrous and smooth on both surfaces; nerves thin but outstanding and green; petiolar sinus closed, upper and lower sinuses deep, closed and U-shaped; teeth short, sharp, generally in series of three; petiole long, thin, completely green, forms a very obtuse angle with the lamina.

Bunch large, branched, irregular (pyramidal to roundish), 1-2 lbs. each; peduncle thick, green, and brown only near the shoot at maturity; pedicels medium thick, short, green. *Berries* firm, fairly large, short oval to almost round, on an average 20 × 22 mm. or 0.80 × 0.88 in. *Skin* somewhat thick and tough, dull green colour and somewhat yellowish when very ripe; plentiful white bloom. Pulp somewhat soft, juicy, pleasant taste. Seeds few and small, fairly thick and round. Late.

13. **Saint-Jeannet tardif.**

As the name indicates, this is also a late grape. It has evidently been in France for a very long time, but was brought to the notice of the grape-growers by Barthélemy Michel only about 1865.

It is grown principally in Saint-Jeannet (Alpes-Maritimes). It is its lateness in ripening and its splendid keeping qualities which make it valuable. In order to enable it to bring out these qualities, it should be grown in a fairly warm and dry climate, especially during autumn. In Saint-Jeannet it is grown on trellises against walls and houses, where its bunches attain a weight of 1-2 lbs., ripen in October, and are picked from November to February. When it gets very late, the bad berries are removed and paper bags put over the bunches. When the vine has a permanent stem of 5-6 metres or 16 ft. 5 in. to 19 ft. 8 in., it bears well with short pruning; otherwise it must be pruned long to give good crops. It is sensitive to oidium, but fairly resistant to downy-mildew.

Description.—(Viala-Vermorel.)
Vine very vigorous grower, thick trunk.
Bourgeonnement slightly woolly, yellowish-green.
Canes erect, very straight, very long internodes, often flattened; ripe canes uniformly orange-tinted yellow; small prominent eyes.
Leaves medium or small, glabrous on both surfaces, 5-lobed, thin; nerves green and very slightly pubescent; petiole long, forms a right angle with the cane and an obtuse angle with the leaf-blade.
Bunch large and branched; very long, green peduncle; pedicels fairly long and thin. *Berries* large, short oval to nearly round, greenish, seldom yellow, covered with bloom. *Skin* fairly tough without being hard. Pulp pleasant, melting, not very sweet. Seeds extremely small, 2-3 per berry, heart-shaped, somewhat elongated. Late.

14. **Tribodo bianco** (Gros Verjus or Tribotu bianco).

This grape I imported to the Cape from Palermo in 1910. It is a most productive variety with short pruning. In fact, it easily overbears. Its grapes are very hard and suffer less from sunburning than any other variety in my collection. It ripens late and keeps very late on the vines. The berries adhere very firmly to the pedicels. It is a most excellent late grape for long transport. Outwardly the grapes resemble Muscat of Alexandria a good deal without having any Muscat flavour.

Description.—(A.I.P.)
Vine of medium vigour, extremely productive.
Canes medium thick, rather short, somewhat flat, light brown when ripe and with red stripes when green, pith thin, diaphragm thin.
Leaves much longer than broad, $\alpha + \beta \leq 90°$; petiolar sinus V-shaped, open to fairly closed, glabrous on both surfaces; petiole

strongly wine-red; growing tip whitish-green, woolly, young leaves reddish-brown parenchyma with green nerves, a few loose threads on the nerves; adult leaves glabrous above and below, shining, with a light garnet red hue in green.

Bunch large, long, pyramidal, more or less the shape of a Muscat of Alexandria bunch but usually slightly smaller, fairly compact. *Berries* oval, fairly large, *hard. Skin* greenish-white to somewhat yellowish when fully ripe, fairly thick, tough, with little bloom. Pulp firm and hard, crackling, not very sweet, pleasant, without Muscat flavour. Fairly late.

15. **Ohanez**.

Synonyms.—Almeria grape, White Almeria, Casta de Ohanez, Uva blanca, Uva de embargo (*i.e.* export grape).

This grape occupies a special place among all the grape varieties, as it ripens the *latest* of all and *keeps the longest on the vine, and more particularly after having been cut.* It is the *hardiest grape for long transport,* and the only one that can be shipped over long distances *without being in a cold store.* It therefore possesses extraordinary value. It is a very ancient Spanish grape. It is extensively grown in the coastal belt between Malaga and Valencia, and especially in the Almeria district, from where most of the Almeria grapes are exported. This fact gave rise to the name *Almeria grape,* generally applied to this variety in the fruit trade. Next to oranges, pineapples, and bananas, Almeria grapes constitute the fourth kind of fresh fruit in the fruit trade of the world.

From the harbour of Almeria about 30,000 tons of grapes are exported annually, the vast bulk of which consists of Almeria (Ohanez) grapes. It is shipped all over the world, but London takes the bulk. In South Africa, California, and Australia this grape is also grown, but on a much smaller scale than in Spain. The most important centres for the growing of this grape are: Almeria, Carthagena, Murcia, Gijona, and Alicante; but in the neighbourhood of Malaga and in the province of Granada it is also grown on a fairly large scale. The prettiest Almeria grapes are grown in Gijona. This locality lies 2200 ft. above and 11¼ miles from the Mediterranean Ocean. The vineyards stand on hills in a more or less semicircular basin, the slopes of which have been terraced with stone walls. Here the grapes keep well on the vines till February, whereas the ordinary export of Almeria grapes begins early in October. In Gijona the grapes are cut from the vines as they are sold. They then have a beautiful golden colour and are sold at high prices. In South Africa this grape is grown with great success in the Hex River valley from Orchard Siding to De Dooms.

Culture.—In the province of Málaga the Almeria grape is grown in a special way. The vines are planted 20 × 20 ft. square, and trained on an 8 ft. high, flat overhead trellis. Strong poles are planted 10 × 10 ft. square to support the weight of the trellis; every 20 in. a wire is fixed, and similarly across these, thus producing a network of wire with 20 in. mesh. The wires are plain galvanised, and Nos. 12 and 16 are generally used. According to Viala-Vermorel, *l.c.* vol. iv. p. 357, the cost of erecting such a trellis in Almeria is about £133-166 per morgen or about £66-83 per acre.

At the Paarl Viticultural Experiment Station I had such an Almeria trellis put up on a small scale (see Fig. 103), and although the crops have not been very abundant, they have been fair and better than with the ordinary low trellising. In the Hex River Valley (South Africa) this grape is trellised on three wires in a vertical plane, and gives very abundant crops. As the vine possesses very great vigour, it should be given a long permanent stem; in other words, a great development and long bearers in order to be able to yield heavy crops. Soil and climate exercise a tremendous influence upon its productiveness. In South Africa it yields its largest crops on *poor,* sandy or broken soils. In fertile soils it grows too vigorously and bears too little. Near Orchard Siding (Cape), it bears fully as much as any other variety ever bears anywhere, and there it grows on poor, dark, sour sandy soil. Near Prince Alfred's Hamlet (Cape) I am growing this variety, grafted on Aramon (Nos. 1 and 2 mixed), and planted 8 × 10 ft. rectangular, in a broken soil with a subsoil of lumpy clay passing into decomposed Bokkeveld shales. This soil is barely of medium fertility. These vines are now about seven years old, and for the last three years have shown about twenty-four inflorescences and more per vine, even on shoots arising on short bearers of two eyes. They are long pruned, the rods being bent over and tied above the vines, as they are not yet trellised.

I have not yet had a good crop from them *owing to coulure.*
Ringbarking them in 1924 gave the same result as with
Bicane (see Fig. 106). The climate here is cold in winter, with
winter rains (about 25 in. per annum) and dry in summer
and autumn. Here the dry summer and autumn have
favoured the formation of fruit buds, but self-fertilisation has
been unsatisfactory. Elsewhere at the Cape the greatest
difficulty usually is not to get the fruit to set, but to get
enough inflorescences on the vines. Cross-pollination will
most probably be the cure for my Almeria vines.

Coulure is one of the drawbacks of this variety in
Spain. This is the result of the abnormal structure of the
flower, the stamens of which are very short and bend
downwards as soon as the corolla is shed and the stamens
open out. The result is that the anthers with the pollen-sacs
are much lower than the stigma, and pollination is more
difficult. I noticed the same thing with the flowers of Bicane
and Gros noir des Beni-Abès, which usually set their fruit
badly. In Spain the shoots are topped just prior to or during
flowering to three leaves above the highest inflorescence, and
self-pollination is assisted by gently passing a soft bunch of
feathers or a woollen brush over the inflorescences on three
to four days in succession during flowering. Only in this way
do they obtain well-set bunches. At the Paarl Experiment
Station I tried this, but both the treated and the untreated
bunches were well set.

In order to obtain good yields from this valuable
variety, I recommend the following measures as most likely
to ensure success:

1.   Give the vine a long permanent stem, long
bearers (preferably laterals), and prune late. Plant the vines
at least 10 ft. apart in the rows and 6-8 ft. between the rows.

2.   Use scions only from vines that are productive.

3.   Plant it on relatively poor soil.

4.   Fertilise heavily with phosphates and some
potash, say 300-500 lbs. basic slag (16 per cent citrate
soluble P_2O_5) and 60 lbs. sulphate of potash (50 per cent K_2O)
per acre. Give nitrogen only when the vines are bearing well
and show signs of insufficient vigour. Then give about 100
lbs. ammonium sulphate per acre in addition to the
phosphates and potash.

5.   Top the shoots on the third leaf above the highest inflorescence just prior to or during flowering, and keep them well topped for the next three weeks.

6.   Use, if possible, Riparia gloire, 101-14 and 420A as stocks.

7.   Irrigate where possible during the summer months.

In Spain the Almeria grape vines are irrigated nearly everywhere. In the Hex River Valley, where it gives such heavy yields, it is also irrigated.

8.   If artificial pollination (with soft brushes, as is done in Almeria or with other pollen) is not practised, sufficient cross-pollination might be obtained by planting alternate rows of Ohanez and Molinera gorda, which flower at the same time.

Where it gives good crops, 7½ to 11 tons of grapes are obtained per acre. In Almeria it produces up to 11 tons per acre under the most favourable conditions and when trained on the Almeria trellis.

Picking and Packing.—In Almeria the picking of the grapes usually begins early in October, and is over when the rains commence to fall in November. In some parts the grapes are picked when they are yellow and fully ripe. Most of the grapes, however, are picked before they are fully ripe, as then they keep better. The picked bunches are immediately carefully examined, bunch by bunch, and all small and bad berries are removed by means of grape scissors. Then they are packed in barrels with cork dust. These hold about 50 and 25 lbs. of grapes respectively. They are made of oak. The large ones are 22 in. high, with a diameter of 12 in., and weigh gross about 77 lbs. when full and packed. The smaller ones are 14 in. high with a diameter of 10 in., and weigh gross about 44 lbs. when full.

The packing is done in *dry weather,* and immediately after the grapes have been picked and trimmed. On the bottom of the barrel a layer of cork dust is put and on it the first layer of grapes is packed; then another layer of cork dust on the first layer of grapes, and on it a second layer of grapes, and so on until the barrel is full. During packing the barrel is well shaken from time to time so that the cork dust can enter among the grapes, and that the whole can be tightly packed. When the barrel is full, its top end is again inserted on the top layer of cork dust. When thus packed, the grapes keep for six months, and can be sent all over the world without it being necessary to transport them in cold store. The foregoing information about picking and packing I have taken from an article by F. Richter on Ohanez in Viala-Vermorel, *l.c.* vol. iv. pp. 359-360.

At the Cape the Almeria grapes have thus far been packed like other varieties for export to London, where the boxes with 10 lbs. grapes net have repeatedly fetched £1 a piece on the market. It is, however, not necessary to wrap every bunch separately in paper. Owing to its lateness it is troubled by fruit fly, against which it must sometimes be sprayed two to three times with Mally's fruit fly bait. At the Cape it ripens late in March to April.

Description.—(F. Richter in. Viala-Vermorel.)

Vine most vigorous; trunk becomes very thick.

Bourgeonnement glabrous, starts budding fairly early; young leaves glabrous and bronze-coloured, glossy, thin; young inflorescences appear only at the sixth or seventh leaf and remain small for a long time; flowers small, pale green with very short stamens.

Canes long, thick; young shoots light green, very thick at the base and much thinner towards the top; much branched ; ripe canes light brown with medium long internodes; tendrils thick and strong.

Leaves large, 5-lobed, petiolar sinus open V-shaped; glabrous on both surfaces, light green; yellow in autumn; drop very late.

Bunches fairly large, branched; peduncle long, red near the cane, thick, green, brittle; pedicels fairly short, thick, with very large bourrelet. *Berries* oval to elongated, frequently somewhat flattened at the end; large, fleshy, crackling. *Skin* thick and tough with white bloom, yellowish-green to golden-coloured, with a pinkish tinge on the sunny side. Pulp firm, fairly or moderately sweet, little acidity. Extremely late.

16. **White Crystal.**

This is one of our oldest grape varieties at the Cape, where it is almost exclusively grown on pergolas. It is a very vigorous grower, demanding a long permanent stem in order to be productive, with late and short pruning; otherwise it requires long pruning. Untrellised it is very susceptible to anthracnose; otherwise a healthy vine. It is a pretty, fairly transparent grape that keeps fairly late on a pergola. It is too soft for long transport.

Apart from this its berries and bunches are too small for export purposes. It is therefore a good grape only for local use and for keeping fairly late. It is a splendid variety for a pergola.

Description.—(A.I.P.) *Vine* vigorous grower.

Bourgeonnement green with, reddish-brown tint and fair number of hairs; little leaves red-garnet coloured, somewhat pubescent.

Canes long, little branched, medium to fairly thick, distinctly flattened; internodes short to medium; diaphragm thick, light; pale reddish-brown.

Leaves smallish to medium, orbicular, shallow; 5-lobed petiolar sinus closed, glabrous on both surfaces with a few loose threads on the lower side of the main nerves, fairly glossy and green ; young leaves shiny, soon become glabrous, excepting the few threads on the main nerves below; growing tip practically glabrous, white with garnet-red ; little leaves strongly garnet-red with green nerves, glabrous on both surfaces.

Bunch small to medium, cylindro-conical, well filled. *Berries* spherical, medium, fairly transparent (hence the name Crystal), with white bloom (moderate), somewhat soft, juicy, sweet, white-yellowish with a fairly thick skin.   Late mid-season.

17. **Ferdinand de Lesseps.**

Mr. J. Pearson obtained this grape from Royal Muscadine (Chasselas dore) crossed by the Strawberry Grape (Isabella). It was awarded a first-class certificate by the Royal Horticultural Society in 1870. It has something of the Isabella flavour, on account of which it is sometimes called *Pineapple Grape* and *Honey Grape* at the Cape.

In Hungary it was grown on a fair scale prior to 1914, for export to Russia and Germany. It is an early table grape which stands long transport well, if it is picked before it is dead ripe. Its berries are too small for export. At the Cape it is grown as an early table grape and for the production of sweet wine, but so far only on a small scale. To those who like its flavour, it is a very nice, sweet table grape, and gives a delicious, highly aromatic sweet wine, which after a couple of years, however, loses a great deal of its typical aroma. It bears heavily with long pruning, especially if pruned according to the Royat system. With short pruning it bears fairly well, but not enough. It resists fungoid diseases well. Its grapes can become very sweet.

Description.—

Vine vigorous, erect grower.

Bourgeonnement whitish-green with reddish-violet along borders and on lower surface of young leaves, cobwebby pubescence; young leaves woolly above and below with light green-yellow nerves, not very thick.

Canes long, above medium thick; shoots light green-yellow, with cobwebby pubescence; canes dark brown, early ripe; internodes of medium length, short at the base; wood hard, pith well developed; diaphragm medium thick.

Leaves medium and larger, 5-lobed, petiolar sinus open and sometimes closed above; glabrous, dark green and somewhat glossy above; lighter green and strongly woolly below; petiole long and fairly red.

"Bunches [according to Dr. Hogg] about the size of those of Royal Muscadine, shouldered and tapering. *Berries* about the size of those of that variety, oval. *Skin* of a fine deep amber colour, membraneous. Flesh tender, juicy and melting, with a very rich and peculiar flavour, composed of a mixture of Muscat and strawberry." He adds: "This is a fine grape, and ripens well in a house without fire heat". It has a fairly strong white bloom. The skin slips off the pulp fairly easily. Seeds 1-2 per berry.   Early.

18. **Chasselas doré**.

*Synonyms.—*Royal (white, common, amber) Muscadine, Chasselas de Fontainebleau, Chasselas de Thomery, Gutedel, Weisser Gutedel, Fendant vert; those types with *parsley leaves* are called Chasselas Cioutat or Ciotat, Chasselas Persille, Petersilientraube, Parsley grape, etc.

It is one of the oldest and most widely grown grapes. In Algeria it is grown near the capital as an early grape for Paris. In Thomery it is grown on the Thomery Spalier against walls to be put on the market very late (up to May), *i.e.* six to seven months after ripening (it ripens in October). Along the northern shore of the Lake of Geneva it is the grape most extensively grown. There it is grown for wine under the name *Fendant roux* (also vert and blanc). It is fairly resistant to fungoid diseases, but least to oidium. It likes a fairly warm site and prefers sandy, light soils. The best stock for it is Aramon No. 1, then Riparia gloire and 1202.

It has given rise to a large number of sub-varieties which differ from the basic type in their sinuses, dentition, and colour of leaf. The Parsley grape or Chasselas Cioutat is a case in point.   The best training for it is the cordon with short pruning. In order to obtain the best quality it should not be allowed to keep too many bunches, and some of the leaves should be removed as soon as the grapes start to ripen. It keeps very well on the vine, and better than most other varieties. Where it can be kept very late on the vine, it is a valuable table grape. It is unsuitable for export to London, as its berries and bunches are too small.

Description.—

Vine vigorous, productive, somewhat creeping.

Bourgeonnement garnet-coloured, glabrous or almost glabrous.

Canes fairly thick; internodes medium long or fairly long.

Leaves medium or larger, slightly longer than broad; glabrous above, somewhat pubescent on nerves below; upper and lower sinuses more or less deep; petiolar sinus narrow or nearly closed; teeth broad, slightly deep, sometimes obtuse and sometimes shortly acute; petiole long, medium thick, and often rose-coloured.

Bunch medium or larger, conical, winged, sometimes compact and sometimes somewhat loose, according to the soil and the age of the vine; peduncle of medium length and thickness. *Berries* medium or larger, spherical; pedicels fairly short and rather thin. *Skin* thin and yet fairly tough; at first very light green, then passing into greenish-white with a yellow tinge and often with a golden rusty colour on the sunny side. Pulp sometimes crackling, sometimes soft; very juicy, sweet and pleasant flavour.   Fairly early.

19. **Madeleine Angevine.**

It is probably the earliest of all varieties. Raised out of mixed seeds, it bore fruit for the first time in 1861. It bears well with long pruning, and requires a large development. Its cuttings for propagation should be carefully selected, as it is very subject to coulure and millerandage. Ring-barking effectively prevents coulure, but merely causes a large number of *small,* seedless berries to be formed, and does not increase the number of full-sized berries. Coulure is the greatest drawback of this grape. Where it ripens very early, it is of some value. It is too soft for long transport, and its bunches and berries are too small for export. It can be valuable for local markets.   It keeps well on the vine.

The earliest grape known to me.

Description.—

Vine vigorous grower.

Bourgeonnement woolly, violet-red; this colour becomes very intense on the lower surface of the opening leaves also.

Canes of medium thickness, long, with long internodes and with a red colour whilst growing.

Leaves fairly large, broader than long, glabrous above, with light pubescence below, especially on the nerves; upper sinuses very deep and yet closed; lower sinuses well marked; petiolar sinus closed; teeth broad, uneven, long and sharp; petiole long, of medium thickness, somewhat pubescent and rough to the touch, soon coloured wine-red like the nerves below.

Bunch medium, short, loose, shouldered and branched; peduncle fairly long, thin, and generally coloured red. *Berries* medium or larger, slightly ovoid; pedicels long and thin. *Skin* thin, fairly tough, at first pale green, then turning to whitish-green, golden coloured on the sunny side when ripe. Pulp melting, sweet and pleasant. Extremely early.

20. **Madeleine royale.**

It was introduced into the trade by M. Robert of Angers in 1851. It is a vigorous grower and good bearer with short pruning on a cordon. Its grapes ripen very early, about eight days after Madeleine Angevine. When not trellised it should be long pruned. It is not subject to coulure, and gives much heavier and more regular yields than Madeleine Angevine. It is rather susceptible to oidium, and should therefore be sulphured early. It is a soft grape which is valuable for local markets, but worthless for export.

Description.—

Vine vigorous and productive.

Bourgeonnement of which the developing leaves are whitish-green above, sometimes slightly rose-coloured and woolly below.

Canes thick, vigorous, semi-erect; internodes medium long, greyish, and coloured brownish-red on the sunny side.

Leaves above medium or large, as broad as long, usually uneven and crinkled, glabrous above, covered below with a thick, dense woolly pubescence; upper sinuses deep; lower sinuses marked; petiolar sinus always closed; teeth broad, uneven, short and obtuse; petiole long or very long, thick and swollen where attached to the shoot.

Bunch medium or larger, conical, branched, rather short, sometimes compact, sometimes somewhat loose; peduncle hardly of medium length, and fairly thick. *Berries* medium, and a few larger, spherical or almost spherical; pedicels, fairly long and somewhat thin. *Skin* somewhat thin, tender, and liable to crack (not at the Cape. A.I.P.), slightly transparent; at first whitish-green, passing afterwards into yellow at maturity. Pulp soft, juicy, sweet, frequently with a slight musky flavour. Very early.

(b) **Red Varieties.**

1. **Red Hanepoot.**

This grape corresponds in practically every respect with Muscat of Alexandria (white), excepting that its ripe grapes have a red colour and that its leaves in autumn sometimes develop red patches here and there. As a table grape for export it is somewhat better than the white variety, as it usually fetches higher prices on the Covent Garden market than the latter, but then it must have a full red colour.

FIG. 55.—Red Hanepoot (Paarl Vit. Expt. Station). Original.

It does not everywhere develop a good red colour, and is therefore not everywhere superior to the white variety. Further, where Muscat of Alexandria makes a splendid raisin (Muscatels), the Red Hanepoot is unsuitable for this purpose, as its raisins are too dark in colour.

From a reliable source I was informed that a Mr. Cloete, fully one hundred years ago, one day in Constantia noticed a shoot with *red* grapes on a Muscat of Alexandria vine, marked this shoot, and afterwards propagated it. When it reached the fruiting stage, it bore only red bunches, and thus the Red Hanepoot was born, evidently as a vegetative sport, bud-variation. I myself have seen white and red

315

bunches on a Red Hanepoot vine, and even red and white berries in the same bunch (on vines of Red Hanepoot). The modification, therefore, does not seem to be *very* fixed, although such deviations occur only very rarely.

According to Mas and Pulliat, *l.c.* vol. ii. p. 151, a red and a black Muscat of Alexandria are known in Italy besides the normal white variety. I do not know how far the red variety in Italy is identical with our Red Hanepoot of the Cape, but it does not seem to have attracted much attention or to be grown to any extent, if it even exists there to-day. Strangely enough, Professor G. Molon makes mention of a black but not of a red Muscat of Alexandria. I therefore still believe that Red Hanepoot originated at the Cape.

White, Red, and Black Types of the same Grape Variety.

Mas and Pulliat, *l.c.* vol. ii. p. 119, are of opinion that a black grape cannot pass into a white type without a red or grey type being also formed. They point out that there is not a single black variety of which a white type is known which does not always have a grey or red type, whilst there are black varieties with no white types but with red or grey types. Further, there are red varieties with white types and no black ones. From this they conclude that the change from black to red or grey takes place more often and more easily than the change from black to red or grey and from red or grey to white; further, that the change from red or grey to white takes place more often and more easily than the complete change from black to red or grey and also to white. They quote the following examples:

(a) *Black, red or grey, and white*: Mondeuse, Calitor, Gamay, Pinot, Picpoul, Poulsard, Terret.

(b) *Black and red or grey, or red or grey and white:* Aspiran noir and A. gris, Chasselas rose and C. blanc, Clairette rouge and Cl. blanche, Marocain noir and M. gris, Peloursin noir and P. gris, Savagnin rose and S. blanc. To this one could add red and white Riesling, Greengrape, Hanepoot, and Sultana.

Here we deal exclusively with varieties which agree precisely in their ampelographic characteristics except in the colour of their ripe grapes, and which must therefore be regarded as types of the same variety.

2. **Flaming Tokay.**

Synonyms.—Flame-coloured Tokay, Red Rhenish, Red Taurida, Wantage, Lombardy, Red Lombardy.

This is a beautiful grape when well developed and fully coloured. The vine grows and bears well with short pruning, and the berries and bunches are very large. It is

Fig. 56.—Flame-coloured Tokay (Paarl Vit. Expt. Station). Original.

an excellent grape for export, being not only pretty but also very firm and hardy. Unfortunately it is very subject to sunburning, and does not always develop sufficient colour. At the Cape it burns least and colours up best in Constantia, where the climate is influenced by the sea and thus is not so very hot. It would appear that red varieties like Flaming Tokay and Red Hanepoot do not develop so much colour or such a fine colour on a very warm, dry site as they do on a less warm site, which must, however, be warm enough. In the fertile,

317

irrigated sweet karroo soils they usually colour up badly, and in these parts the summer is dry and hot. The fish-spine method of trellising, the pergola, and my own system of training table grapes, can be recommended as suitable to this variety and as minimising the danger of sunburning.

Description.—(A.I.P.)

Vine very vigorous grower.

Canes thick, long, internodes fairly long to long, reddish-brown, pith fairly thick, diaphragm medium. Young shoots fairly red.

Leaves fairly large, entire to 3-lobed; sinuses generally shallow, lower sinus sometimes distinctly visible, petiolar sinus usually open, sometimes narrow V-shaped and even closed; teeth large, broad, long, sharp, sometimes slightly rounded off, uneven; glabrous on both surfaces; leaves in autumn generally yellow with some red patches; petiole medium, red, glabrous.

Bunch very large, pyramidal, branched, compact, beautiful. *Berries* very large, oval, hard; the following measurements I made of its diameters on the Stellenbosch University Farm in 1923: 90 × 99, 90 × 113, 97 × 101 in $1/100$ in. These berries were not very large for this variety; the berry is often thinner near its end and the cross-section angular, while the end is often flattened; this is more general and more striking on the young berries. *Skin* pink to violaceous red. Pulp firm, crackling, sweet, pleasant flavour. Travels well. Late mid-season.

3. **Rosada.**

This is a pretty grape. I imported it to the Cape in 1910 together with Ohanez and Molinera gorda from Almeria. Rosada and Molinera gorda are two red varieties that are exported from Almeria on a fairly large scale before Ohanez, as they stand long transport very well. In order to give good crops, Rosada must be long pruned. It possesses much less vigour than Molinera gorda. It is probably a native of Almeria. I have not yet seen it described in any work on ampelography. It is a very pretty grape, and has fetched very good prices on the Covent Garden market in London. I can strongly recommend it for export, and consider it as still hardier than Molinera gorda. It should be grown on good soil. With short pruning it generally bears too little. It is decidedly late and has splendid keeping qualities.

Description.—(A. I. P.)

Vine of medium vigour to vigorous (on good soil), and medium fertility.

Bourgeonnement light green, with a fair number of loose threads.

Shoots green with light red stripes here and there, medium to somewhat thin, not long. *Canes* dark brown, internodes medium to fairly short, pith and diaphragm fairly thick; ends of canes ripen badly at Stellenbosch.

Leaves medium, 5-lobed, petiolar sinus open U-shaped; teeth sharp and uneven, glabrous on both surfaces; petiole whitish-green, glabrous; growing tip whitish-green, very slightly pubescent; young leaves yellowish light green, glabrous on both surfaces ; leaves show a fair amount of red in autumn.

Bunch medium large, pyramidal, moderately compact. *Berries* medium large, oval. *Skin* thick, tough, with a very pretty blush of light rose colour to violaceous red, a typical and peculiar colour of moderate intensity. *Berries* adhere very firmly to the pedicels. Pulp firm, hard and crackling, sweet, pleasant taste with little acidity. Late.

4. **Molinera gorda.**

Synonym.—Castiza.

In 1910 I imported Castiza and Meraviglia de Malaga from Montpellier; they turned out to be identical with Molinera gorda. As Meraviglia de Malaga is, however, a black variety, I evidently did not get it, and hence give Castiza as synonymous with Molinera gorda.

It is an excellent grape for long transport, as it is very hardy and keeps very well. It forms large to very large bunches, which must be very heavily thinned in order to obtain a loose bunch with large, even berries. It ripens fairly early, about the same time as Hermitage, but has a fairly long season. It is a most vigorous grower, and should be pruned long in order to produce good crops. It usually bears too little with short pruning. It is fairly resistant to fungoid diseases. Persistent rainy weather in spring and early summer, as happened at the Cape in 1923 and 1924, expose it to violent attacks by anthracnose, which can damage the leaves and shoots, and destroy almost the whole crop. Under fairly dry conditions it is not troubled by anthracnose. It can be strongly recommended for local markets and for export, particularly in hot districts and in localities exposed to strong winds, and where the grapes must travel a fairly long distance by road before reaching the nearest railway station. Its grapes are not easily bruised. At the Cape it readily colours up, and also in the hotter parts.

I strongly recommend it as the best all-round red grape for export. It possesses a good affinity for American vines.

Description (compiled).

Vine extremely vigorous grower.

Canes long, thick, glabrous, yellow, light brown, finely striped; diaphragm thick; internodes medium to fairly long and even long; wood hard.

Leaves medium to fairly large, 5-lobed, slightly longer than broad, glabrous on both surfaces, light green; yellow with red patches in autumn; petiole long, thin.

FIG. 57.—Molinera gorda (Stellenbosch Univ. Farm). Original.

Bunch large, pyramidal, sometimes long, frequently roundish, very compact, branched; peduncle thick and tough. *Berries* almost round to short oval. In 1923 I made the following measurements of its berries: 85 × 92, 87 × 95, 86 × 90, 81 × 85, 80 × 91, 86 × 96, 89 × 96, 87 × 95, 89 × 95 in $^{1}/_{100}$ in.; fairly large; adheres very firmly to the pedicels, which are long and somewhat thin. *Skin* thick and tough, red to fairly dark violet-red at full maturity, with fairly strong bloom. Pulp firm, hard, crackling, pleasant sweet taste with little acidity. Early mid-season.

5. **Purple Cornichon.**

Synonyms.—Cornichon violet, Cornichon noir, Black Cornichon, Pizzutello nero, Corniola nera, Galeta nera.

In the *California Grape Grower,* vol. vi. No. 1 (Jan. 1925) p. 6, L. O. Bonnet of the University of California has published an article on *Black Cornichon,* in which he gave Olivette noire and Malakoff Isjum (which are synonymous) as synonyms for Black Cornichon. This is contrary to the information given in the large Ampelography of Viala and Vermorel. In my own collection I have Cornichon violet and Malakof Isjum as two totally different grapes (these were imported from Richter, Montpellier, in 1910). Bonnet's description of the Black Cornichon grown in California agrees fairly well with that of Cornichon violet in Viala-Vermorel.

The Purple Cornichon is grown on a fairly large scale only in Tivoli, where the White Cornichon is also extensively grown. According to Bonnet, its culture increases gradually every year in California. He says further: "It is particularly a grape of the Sacramento Valley and of the west slopes of the Sierras. ... It thrives in deep red sandy loams. A certain amount of gravel in the soil seems to improve the quality of its grapes. In light sandy soils and under a hot climate the Cornichon does not colour well, hence the name of Purple Cornichon." It is a vigorous grower and good bearer with long pruning. The horizontal cordon with short and long bearers suits it very well. Overbearing can be guarded against by limiting the crop. It ripens fairly late and can be kept very long on the vines if the bunches are covered up in paper bags as soon as they are ripe. It has splendid keeping qualities and ships well. It has been successfully exported from the Cape to London, where it fetched good prices. Its large bunches and large berries with their strong bloom are very pretty. Its ripe grapes stand low temperatures well. As the bunches are naturally loose, it requires little pre-thinning for export. Where it answers well, it can be strongly recommended as a late table grape and for export.

Description.—(Viala-Vermorel.)
Vine very vigorous grower, thick trunk.
Canes fairly thick, of medium length, much branched; young shoots very green: ripe canes greyish-brown, smooth; internodes more short than long; wood hard, pith thin, diaphragm thick.
Leaves medium or fairly large, 5-lobed, dark green, glabrous on both sides, remaining green in autumn; petiolar sinus very open U-shaped, upper sinus very deep U-shaped, lower sinus very deep, both closed by superposition of the lobes; teeth long, sharp; petiole of medium length and thickness, green with a light crimson-red tint.
Bunch large, branched, cylindrical or cylindro-conical, fairly loose pedicels long, thin, green. *Berries* large to very large, 13 × 30, 17 × 31, and even 20 × 33 mm. or 52 × 120, 68 × 128, 80 × 132 in $^1/_{100}$ in.; shape very irregular, thick in the middle with a bent point (my

321

Cornichon violet has more ellipsoidal berries than are shown in Bonnet's illustration. A.I.P.). *Skin* very thick, tough, red-violet (or purple to black. A.I.P.), with abundant bloom. Pulp very firm, not juicy but fleshy, sweet, pleasant. Seeds 2-3, very large, pyriform, very long seed-beak. Late mid-season to late.

Note.—This grape I could have equally well, and perhaps better, described under the black varieties. I have in my collection at the Stellenbosch University Farm a *light red* variety with very large berries, medium to large bunches, which ripens very late and does not always develop enough colour; it evidently is a Red Cornichon. It grows vigorously, bears well, has berries that are often bent near the end with no seed on the inside of the bend and only one on the convex side. It is a very pretty grape. I got it under the name of Cornichon blanc.

6. **Laubscher's Gem.**

This is a natural hybrid, probably of Red Hanepoot and Waltham Cross, that has been raised by Mr. Laubscher of Graaff-Reinet (Cape Province) and named by Dr. C. Mally. Where it reaches perfection, is severely pre-thinned and grows out well, it is a pretty grape which keeps and ships well. It has been successfully exported to London, but is not as hardy as several other varieties. It bears heavily with short pruning. Grown on good soil, the vines are sufficiently vigorous. Unless the crop is limited and the bunches are severely pre-thinned, neither the bunches nor the berries will reach export size. In some localities it may be a paying variety for export. There are, however, at least a dozen varieties which have been tested out locally, that should be preferred to it for this purpose.

Description.—(A.I.P.)

Vine of moderate vigour, heavy bearer.

Bourgeonnement intense red-brown (especially the nerves) with some loose threads; little leaves intense red-brown, hardly any green.

Canes medium to thinnish; internodes short, brown; diaphragm thick, pith medium; growing tip yellowish-green, glabrous.

Leaves medium, glabrous on both sides, 5-lobed, open and deep sinuses, leaf serrate with red nerves; petiolar sinus open U-shaped; teeth sharp and narrow, alternately large and small: young leaves intense garnet-red, glabrous on both sides, glossy.

Bunch medium to fairly large, compact, pyramidal. *Berries* medium to fairly large, oval. *Skin* reddish-violet, thin, fairly tough. Pulp fairly firm, pleasant, sweet. Mid-season and later.

7. **Karroo Belle.**

This grape, it would seem, has originated in South Africa. Years ago it was introduced into the trade by Messrs. Smith Bros, of Uitenhage (Cape Province). It is a wild grower, and it is said that it can bear well, though I have not yet got it to bear a crop, even with long pruning. When pruned exceedingly late (very near the time of budding) and very long (rods of 3 ft. long), it gave me very moderate crops. The grapes are very pretty, and stand long transport well.

Description.—(A.I.P.)

Vine very vigorous grower but very poor bearer.

Bourgeonnement garnet-red around borders of opening leaves, which show a fair number of loose threads on both surfaces.

Canes fairly thick, long; internodes fairly long to long, yellowish-brown with numerous black spots; pith fairly thick, diaphragm thinnish and fairly soft.

Leaves entire to 3-lobed, lower sinuses sometimes slightly developed; petiolar sinus closed; dark green, glabrous on both sides ; growing tip white-yellowish-green, hardly any threads; young leaves glossy, light reddish hue in the green, glabrous on both sides.

Bunch longish, cylindro-conical, medium, not too compact. *Berries* medium, oval, firm, keep well. *Skin* fairly thick, tough, red to dark violaceous red, with fair amount of bloom. Pulp firm, sweet, pleasant flavour. Late mid-season.

8. **Chasselas rose Salomon.**

This is a hybrid of *Chasselas rose royale* with Fintendo. It is a very pretty grape. When fully ripe its berries are red and somewhat transparent. It is fairly resistant to oidium. It keeps well. For local markets it is a good grape.

Description.—(E. and R. Salomon in Viala-Vermorel.)

Vine fairly vigorous, good bearer with short pruning.

Bourgeonnement woolly, reddish, with red ends to the young inflorescences.

Canes medium thick, long, light reddish-brown and a strong purple colour at the nodes; internodes of medium length, smooth, striped,wood hard.

Leaves medium, fairly thick, soft, 5-lobed ; all the sinuses are deep, V-shaped and closed above, only the petiolar sinus is completely closed the leaf has a somewhat serrate appearance; light green, glabrous on both sides, even.

Bunches 2 and occasionally 3 per shoot, medium large, little branched, cylindro-conical, fairly loose; peduncle thick and fairly long and soft, yellowish-green; pedicels somewhat short, thick, adhere very well to the berries. *Berries* medium, spherical, very even, *Skin* thin, red, transparent, light bloom. Pulp somewhat greenish, juicy, pleasant flavour. Mid-season.

(c) Black Varieties

1. Gros Colman.

Synonyms.—Gros Colmar, Dodrelabi, Grosse Kolner, Pomeranzentraube (Orange grape), etc.

This grape originally came from the Caucasus. It was imported into France in 1858, and is there generally known as Dodrelabi. Mr. Rivers imported it into England in 1860 under the name Gros Colmar. According to Dr. Hogg both Gros Colman and Gros Colmar are corruptions of Grosse Kolner.

It is perhaps the king amongst the table grapes. In any case it fetches the highest prices on the Covent Garden market, 15s.-25s. per 10-lbs. box. It is indeed *a beautiful grape.* In the English, French, Belgian, and Dutch hothouses it is grown on a fairly large scale. In order to reach perfect maturity it, however, requires a great deal of heat, as much as Muscat of Alexandria. In fairly warm climates, as at the Cape and in Australia and California, it ripens perfectly in the open air. When its grapes are fully ripe they have an agreeable flavour and are sweet. It is only moderately resistant to oidium, but practically immune to anthracnose. Its great drawback is the *cracking of its berries* if rain falls after the grapes have begun to turn colour *(veraison).* It is grown on a fair scale for export at the Cape, and but for the cracking of its berries it would have been grown on a much larger scale. It bears well and regularly with short pruning, and stands long transportation well if carefully handled and shipped in cold store. If a berry rots, the thick black skin becomes a red bloody mass which easily soils the bunch and brings about further rotting. I strongly recommend it as a table grape for export. It must usually be pre-thinned pretty severely in order to obtain a loose bunch for export. Its thick blue bloom and very large spherical berries make it one of the very prettiest grapes known. Its affinity for American stocks is good. It prefers good, deep soil with sufficient moisture, on a warm site. Low-lying moist soils produce a rather soft and poor quality Gros Colman grape.

Description.—(Compiled.)

Vine of tremendous vigour, good bearer.

Bourgeonnement strongly woolly, tinted white, light rose-coloured.

Canes very thick, internodes long, brown; on young shoots and petioles there are numerous black dots.

Leaves extraordinarily large, broader than long, thick, almost entire, upper sinus shallow and narrow; petiolar sinus closed; teeth short, very broad, rounded off, dark green, glabrous above, strongly woolly below; when the grapes are ripe, and later on in autumn, the leaves get red borders and red patches; petiole thick and very long.

Bunch fairly large to large, generally with a smaller bunch at the fertile node (typical), usually two bunches per shoot; short, thick, shouldered, compact; peduncle stout and fairly long. *Berries* large to very large (in 1923 I measured the following diameters: 109 × 110, 110 × 114; 112 × 113, 114 × 123, 117 × 124 in $^1/_{100}$ in., but they grow to 4 cm. or 1.60 in.), spherical (the measurements just quoted prove that this term must not be interpreted too closely); style-point clearly noticeable and depressed. *Skin* thick and tough, adhering closely to the flesh; dark purple or black, covered with a pretty thick bluish bloom. Pulp juicy, fairly soft, sweet and agreeable flavour only when dead ripe. Mid-season.

2. **Gros Maroc** (Cape).

Synonyms.—Gros Ribier, Ribier, Gros Ribier du Maroc, Ribier du Maroc, Morocco, Black Morocco, Le Cœur, etc.

The Gros Maroc which has been grown at the Cape for a number of years and the Gros Ribier which I imported from Montpellier in 1910 proved to be identical. Hence our Gros Maroc is the same as Gros Ribier and Morocco (Dr. Hogg), but not the same as Dr. Hogg's Gross Maroc (Cooper's Late Black), in spite of the name, there is no proof that this grape came from Morocco. It is a beautiful table grape which, from the point of view of quality, is in every respect the equal of Gros Colman. Its bunches and berries are usually slightly smaller than those of Gros Colman, but only very slightly so, and this applies more to the bunches than to the berries. It looks just as pretty as Gros Colman, and has the advantage over it of standing long transport even better. On Covent Garden it usually fetches as high prices as Gros Colman. It has a further advantage over Gros Colman, inasmuch as its berries hardly ever crack.

Unfortunately, however, it suffers very much from sunburn at the Cape. This has thus far prevented its cultivation here on any important scale. With my system of training this trouble might be overcome, and it is overcome to a large extent by growing the Gros Maroc on pergolas 7-8 ft. high. It is rather susceptible to oidium, and should therefore be sulphured early and thoroughly. Anthracnose does not trouble it. In cool viticultural districts it is much subject to coulure. Dr. Hogg considers it as perhaps one of the very worst varieties to set its fruit. At the Cape it must usually be pre-thinned a great deal in order to give a loose bunch. Here it sets its fruit almost as well as Gros Colman. In contrast with Gros Colman it rarely forms a side bunch at the node of the peduncle. Where it can be grown successfully, it is one of the very best grapes for export. It keeps well (also on the vine), and ripens fairly late, usually after Gros Colman.

Description.—(Compiled.)

Vine very vigorous grower; good bearer with short pruning on a horizontal cordon.

Bourgeonnement white, woolly; the borders of the opening leaves tinted light reddish-violet.

Canes thick or fairly thick, internodes medium to short, chamois-coloured or white-brownish, bark smooth and finely striped, with numerous fine black dots; diaphragm fairly thick; pith thick.

Leaves medium to large, 5-lobed, *strongly crinkled,* roundish, light yellowish-green to light green; glabrous above, strongly woolly (somewhat cobwebby) below; petiolar sinus always closed, generally owing everywhere to the superposition of the borders of the lobes at the petiole; teeth broad, long and sharp, uneven, sometimes more or less rounded off; die a golden-yellow, often with light red patches.

Bunch medium to fairly large, loose to compact, short (truncate conical) and irregular; peduncle short and thick, pedicels thick, medium to fairly long, green, but at maturity coloured light-wine red where they enter the berries. *Berries* very large, short, oval to roundish (M. et P. call the berries of the Eibier "ellipsoids"). *Skin* thick and tough, pitch black at full maturity, with a dense white-bluish bloom. Pulp firm, crackling, juicy, sweet, pleasant flavour when fully ripe. Late mid-season to fairly late.

3. **Gros noir des Beni-Abès.**

Synonyms.—Gros noir du Beni-Abbès, Beni-Abès-Lekhal, Beni-Abbès, Gros noir.

This is the largest black grape of Barbary. It forms very large, beautiful, elongated (long oval) berries. I imported it to the Cape in 1910. In 1923 I made the following measurements of its berries in $1/100$ in.: 96 × 130, 103 × 126,106 × 117. It is grown most by the tribe of the Beni-Abès on the southern flanks of the Haute Kabylie (Kabylie is the mountainous coastal region to the east of the city of Algiers), where it is trained on high trellises and trees. The vine is a vigorous grower, and produces the best results when it is given a large development, long permanent stem. It is very subject to coulure, which is not improved by ringbarking, inasmuch as the result is a large number of small seedless berries but not a single additional large berry (according to my experiments of 1924). As it is *one of the prettiest table grapes,* and stands long transport well, it deserves to be grown wherever this can be done successfully. It keeps very late. It should be pruned late and with long bearers. With cordon training in Algeria it produces up to 13¾ tons per acre. It requires favourable weather during flowering in order to give well-filled bunches. These can sometimes grow fairly large but are usually only medium large, and sometimes rather small, particularly when the fruit has set badly. Graft it on Riparia gloire, 101-14, 420A, 106-8, etc., which will oppose its coulure, rather than such vigorous stocks as 1202, which will make it worse. It fetches very high prices in London.

Description.—(Compiled.)

Vine very vigorous grower.

Bourgeonnement rose-coloured, with white woolly pubescence.

Canes thick (growing shoots much thicker near base than near top); internodes fairly long, sinuous; herbaceous shoots fairly strong red, little branched, glabrous, not very long.

Leaves medium to fairly large, 5-lobed, yellowish-green, glabrous on both surfaces, petiolar sinus open V-shaped; petiole long ; leaves in autumn have a reddish hue and red patches.

Bunch large to medium, especially when poorly set, loose to compact, longish to roundish; peduncle thick and tough. *Berries* very large, elongated or long oval. *Skin* thick, tough, elastic, dark red-violaceous black to black with a strong blue bloom. Pedicels fairly short, medium thick, very firmly attached to the berry. Pulp firm, crackling, not very sweet (unless very ripe), somewhat aromatic and pleasant when fully ripe. Late mid-season to late.

See Leroux, *l.c.* vol. i. p. 329, and Viala-Vermorel, *l.c.* vol. iii. p. 322.

4. **Barlinka.**

This is a North African variety which I saw on my tour of viticultural investigation in 1909 in a small village called Novi, about 50 miles west of Algiers and practically on the Mediterranean coast. Its owner, Monsieur Leon Roseau, sent me cuttings of it with the aid of Professor Marès of Algiers, who carefully packed them and sent them to the Cape, where I planted them out in 1910. It struck me in Novi as a most promising table grape for export. At the Cape I propagated and studied it, and it has fully come up to my expectations. M. Roseau told me that he got this grape in Algeria from the natives. I have not been able to find its name in any work on ampelography or viticulture, nor in the list of the collection of Professor Marès of Algiers, which comprises 2000 varieties. To the best of my knowledge, mine is the first description ever given of this variety, which is fast becoming one of our most important export grapes.

For local markets, and particularly for long transport, hence also for export, it is one of the very best grapes known to me. It is very pretty, keeps much better than Gros Colman, and fetches about the same price on the Covent Garden market. On the 1st April 1925 I picked some of its grapes, packed them on the morning of the 2nd April in a 10-lbs. box as for export (using woodwool and wrapping each bunch separately in tissue paper). Mrs. Perold took the box with her on the same day, on the boat leaving for Hamburg; during the first five days it was kept in the cabin (the air being fairly warm) and then put in cold store; on the 6th May they were taken out of the cold store in Hamburg, and on the 8th May, *nearly six weeks after they were cut,* the box was opened and the grapes were still perfect. This was certainly a severe test, and proved its fine keeping qualities. It is one of the very best late varieties for a pergola. In my garden at Stellenbosch it keeps in muslin bags on the vines till late in June, after being exposed to the winter rains for six to eight weeks.

FIG. 58.—Barlinka (Stellenbosch Univ. Farm). Original.

In order to develop its finest eating qualities (*i.e.* sweetness and pleasant flavour), it should be picked when highly ripened or dead ripe. When just ripe its acidity is still rather pronounced. With short pruning it bears too little; *it should be pruned late and long,* when it gives very good crops. I train it on a horizontal cordon on one wire, the long bearers being bent down and tied to this wire and short bearers being pruned to provide future bearers.[1]

---

[1] Where the vines are not very vigororous, prune according to the double Guyot system.

The growing shoots are then tied to the two upper wires, as will be described under my own method of trellising and training table grapes. It should be pre-thinned, but much less than Gros Colman. It is fairly resistant to oidium, and practically immune to anthracnose. It is one of the best table grapes for localities exposed to strong winds. It wants a fairly warm site, particularly if the soil is deep and fertile, in order to become perfectly black. Where the vines grow very vigorously and the grapes hang permanently in the shade, they sometimes have a reddish-black colour when ripe. On exposure to the sun (by picking leaves) they soon turn black. This difficulty is not experienced in the ordinary culture of this grape, which easily turns a deep black under its blue bloom at maturity.

Description.—(A.I.P.)

Vine vigorous grower.

Bourgeonnement light reddish-green to brown, very slightly pubescent.

Canes a strong reddish-brown, bark smooth with numerous fine dots, long, inclined to spread, of medium thickness; internodes fairly long, wood hard, pith medium thick, diaphragm concave, thick, and firm; young shoots glabrous, glossy, with red dotted stripes along the bast fibres on both the sunny and the shaded side; growing tips with a few loose threads, green and light reddish mostly on the sunny side ; bunches fairly high up the canes, first bunch usually opposite the fifth leaf only, and seldom more than one bunch per shoot.

Leaves medium, 3- to 5-lobed, upper sinus fairly deep U-shaped, lower sinuses little developed, narrow U-shaped, petiolar sinus open, between U- and V-shaped; angles α, β, γ, δ well developed, but no ε; smooth and glabrous on both sides, fairly even, somewhat light green colour; during flowering a few odd leaves show red patches, sometimes a whole leaf is fairly red; when the grapes ripen the leaves develop small red spots; in autumn the leaves show numerous pretty red patches, and sometimes nearly the whole leaf dies red; broader than long, the lobes have no long terminal teeth but are rounded off; teeth uneven, obtuse, rounded off with a hard point. Young leaves very glossy, glabrous on both sides, 3- to 5-lobed; shallow sinuses, petiolar sinus very narrow, V-shaped to closed; many young leaves reddish, uneven, with a few loose threads, soon become completely glabrous, glossy, and bronze-coloured.

Bunch medium to large, fairly long, somewhat truncated pyramidal, shouldered, branched, fairly compact; peduncle thick, strong, lignified up to first branching. The young inflorescences up to flowering are very small. *Berries* fairly large to very large, short oval to almost round (the following diameters I measured on its berries in 1923: 90×103, 93×106, 95×110, 96×112 in $^1/_{100}$ in.). *Skin* very thick, tough, black, with strong blue bloom. Pulp firm, greenish tinge, sweet and pleasant flavour when fully ripe. Pedicels short to fairly long, thinnish to medium thick but much thicker near its insertion into the berry than at the stalk end; adhere very firmly to the berries. Seeds 2-3 per berry, elongated, flattened on the side of the raphe, beak end fairly broad, light brown, of medium size. Late.

5. **Henab Turki.**

Synonyms.—Henab, Henab Turqui, Aneb Turki, Haneb Turki, Olivette géante rouge, Trifere du Japon.

This grape has been grown for some time at the Cape. In 1910 I imported from Montpellier the Trifere du Japon, which turned out to be identical with our Henab Turki. The other synonyms are taken from Viala-Vermorel, *l.c.* vol. iii. p. 297. *Aneb Turki* means Turkish grape. It is grown mostly along the Mediterranean and in Algeria. As it is a very late grape, it requires a warm climate and a warm site to ripen up properly. At the Cape it answers well. It must be long pruned to give good crops. Every vine should be allowed to retain only a limited number of bunches (about one bunch for every 5 sq. ft. of soil the vine occupies), which must be pre-thinned in time in order to obtain fairly loose bunches for export that will ripen well.

When its shoots are topped, it develops laterals with numerous second-crop bunches. When the main bunches are ripe, one finds on the vine bunches (on laterals) which turn colour, are half-grown, are small, are flowering, and do not yet flower. This is a striking characteristic of this variety. If the main crop is sufficient, all the inflorescences on the laterals must be removed when small; where this is not the case, sufficient bunches of the second crop should be kept. They usually grow to a fair size and ripen up well under suitable conditions. It is rather sensitive to sun-burning, though it does not burn very badly. Where it can be grown successfully, it is one of the best and most valuable varieties for export and for local markets. It keeps very well and very late, and stands long transport very well. I can therefore strongly recommend it for export. It fetches some of the highest prices on Covent Garden market.

Description.—(Viala-Vermorel.)

Vine vigorous grower, thick trunk.

Canes fairly thick; fruit-bearing canes remain short, the others grow long, little-branched, bark smooth and striped, light brown, diaphragm thick and concave, pith fairly thick, wood not very hard, internodes fairly short.

Leaves medium, 5-lobed, glabrous on both sides, green ; nerves thin, green to reddish, especially near the petiole, which is long and thin; petiolar sinus V-shaped, closed; upper sinus shallow, lower sinus deeper; teeth obtuse in groups of two.

Bunch large to very large, somewhat short pyramidal, branched, compact; peduncle long, thick, tough, coloured red-violet, not lignified. *Berries* large to very large, can grow as large as those of Gros Colman (according to Viala-Vermorel, *l.c.* iii. p. 299, up to 96 × 120 in $^1/_{100}$ in.), oval to short oval. *Skin* thick, tough, reddish-black to pitch black when fully ripe, with a fair amount of bloom. Pulp firm, crackling, juicy, pleasant sweet taste when fully ripe. Late.

6. **Barbarossa** (Cape).

Synonyms.—Danugue, Gros Guillaume.

The Cape Barbarossa is a black grape, and is identical with Danugue and Gros Guillaume. It has been grown for a fairly long time at the Cape. In 1910 I imported the Danugue from Montpellier and found it to be identical with our Barbarossa. Dr. Hogg says: "The grape which has been grown in this country for some years [Dr. Hogg wrote this in 1884] under the name of Barbarossa is a totally different variety. Its correct name is Gros Guillaume, and it is black, while the Barbarossa is, as its name implies, a rose-coloured or grizzly grape". The Cape Barbarossa was no doubt imported from England to the Cape under this wrong name, which I retain simply because it has been grown at the Cape for a long time under this name, and has been exported to London and elsewhere under this name. Viala-Vermorel gives Gros Guillaume as a synonym of Danugue. Professor Molon in his *Ampelografia* makes mention of three main types of the true Barbarossa, of which the ripe grapes are all red. I shall not discuss them here.

The Cape Barbarossa is a very vigorous grower and an excellent grape for a high trellis or a pergola. It bears fairly well with short pruning; it is, however, better to prune it with some long bearers or at least to give the short bearers three good eyes. Its bunches can reach a tremendous size—in France bunches of 12 lbs. weight have been grown. It is a pretty and very hardy grape that keeps and ships very well. When dead ripe it is very sweet and pleasant to eat. It keeps very late on the vine. Its berries can attain a good size if the bunches are very heavily pre-thinned. In wet summers it is very sensitive to anthracnose.

Notwithstanding the fact that it requires very heavy pre-thinning in order to get a fairly loose bunch for export, and considerable trimming, it is grown to a considerable extent at the Cape, both for export and for local markets, as it travels well and fetches good prices. It is decidedly a late grape.

FIG. 59.- Cape Barbarossa (Stellenbosch Univ. Farm). Original

Description.—(Compiled.)

Vine extremely vigorous.

Bourgeonnement glabrous, light rose-coloured.

Canes long, thick, light red-brown, smooth, striped, diaphragm thick and concave, internodes long, wood hard, pith large.

Leaves medium large, 5-lobed, as broad as long, glabrous on both surfaces, light green; petiolar sinus very open U-shaped, upper sinus often hardly indicated, lower sinus fairly deep; teeth in several series, more sharp than obtuse; petiole long and thick, green, sometimes tinted red-violet.

Bunch large to very large, long, pyramidal, branched, very compact peduncle of medium length, very thick, lignified; pedicels of medium length, thick. *Berries* fairly large to large, spherical, firm. *Skin*

333

thick, tough, purplish-black to pitch black with a strong bluish bloom. Pulp firm, sweet, pleasant when highly ripe. Late.

7. **Prune de Cazouls.**

This beautiful grape I saw for the first time when 1 visited the Vivaio americano at Palermo in 1909, and I imported it to the Cape from F. Richter, Montpellier, in 1910.

Fig. 60.—Prune de Cazouls (Paarl Vit. Expt. Station). Original.

I propagated it and tested it out. To-day it is one of the leading varieties exported from the Cape to London. Although not very hardy, it stands long transport well if carefully handled and properly cold-stored. Its large bunches with their large berries are very decorative. Hence it is not surprising that it fetches very high prices in London. It is a good grower and a good bearer with short pruning. It needs little thinning, and this is rapidly done, as the young pedicels are brittle, not tough. It is a healthy vine. I can strongly recommend it for export. I cannot find it described in a single work on ampelography.

Description.—(A .I .P.)
Vine good grower, good bearer.

Bourgeonnement glabrous, reddish - brown; little leaves shiny, glabrous, red-brown nerves, reddish-brown borders, a few threads on nerves below, fairly red-garnet-coloured on the parenchyma which is raised between the nerves.

Canes long, fairly thick; internodes fairly long, pith well developed, diaphragm thick; reddish dark brown with numerous black dots; growing tip green, glabrous.

Leaves green with fairly red nerves, glabrous on both sides, medium large, generally longer than broad; thin, soft, even; 5-lobed, upper sinuses deep to fairly deep and open, lower sinuses generally shallow and sometimes hardly indicated, petiolar sinus open U-shaped to fairly closed; teeth sharp, long, irregularly large and small; yellow in autumn with very little red; petiole short, glabrous, red.

Bunch large, long, cylindrico-conical, fairly loose, branched. *Berries* large to very large, oval to long-oval; pedicels rather thin, fairly long, brittle, adhere well to the berries; peduncle long or fairly long, remains soft, *i.e.* does not lignify. *Skin* fairly thick, purplish- to pitch-black with a moderate amount of white bloom. Pulp firm, juicy, flavour poor. Seeds large. Late mid-season to late.

8. **Lady Downe's Seedling.**

Synonym.—Lady Downe's.

This grape was raised by Mr. Foster, gardener to Lord Downe, at Beningborough Hall, York, from Black Morocco *(i.e.* the Cape Gros Maroc, A.I.P.), crossed with Sweetwater, about the year 1835. It was from the same pot of seedlings as Foster's White Seedling was obtained (see Dr. Hogg, *l.c.* p. 394). Since 1858 it has been grown in hot-houses in England, on the Channel Islands, and in France. About 1912 Mr. Spencer of The Chilterns, Wynberg (now the property of Mr. J. B. Taylor), imported it to the Cape from England. At the Cape some growers have confused it with the Cape Barbarossa. A number of varieties were sold in the country as Lady Downe's Seedling without being this grape at all. It was not in South Africa until Mr. Spencer imported it.

In contrast with practically every other grape, its berries, especially when only half grown, are distinctly *ribbed* by three to four depressions running lengthwise across the berry. This is a very striking characteristic. It is a very vigorous grower and heavy bearer with short pruning. It is very subject to sunburning, and should therefore be kept in the shade as much as possible. It is a very valuable grape, and possesses splendid keeping qualities.

Dr. Hogg says: "I have seen bunches of this grape ripened in August, hang till March, and preserve all their freshness, even at that late season, when the berries were plump and delicious". At the Cape it has thus far kept less well on the grafted vine than Raisin blanc and Barlinka.

It is fairly resistant to fungoid diseases, but is rather susceptible to anthracnose and sunburn. Its affinity for American vines is none too good according to E. and R. Salomon in Viala-Vermorel, *l.c.* vol. ii. p. 377, who recommend Aramon No. 2, Riparia gloire, 101-14, and Rup. du Lot, to which I can add Jacquez and 420A. In my collection it forms very compact bunches if flowering takes place in favourable weather. Unfavourable weather (rain, cold) at this time causes the fruit to set very badly. Its bunches will usually have to be heavily pre-thinned. This has to be done early, so as not to disturb the bloom, and thus increase the danger of sunburning. It keeps very well, and can be recommended as a good variety for export and local markets. Unless it is grown in good, sufficiently moist soil, its bunches and berries will be rather small for export. For personal use it is an excellent variety to grow on a pergola, as its grapes keep very late and are delicious when highly ripe.

Description.— (E. and R. Salomon in Viala-Vermorel.)

Vine vigorous, very productive.

Bourgeonnement strongly white woolly, tinted red.

Canes long, little branched, thick; internodes short to medium, light pale reddish-brown; eyes large, compound, prominent; wood hard, bright green just underneath the bark; pith thick, diaphragm thick. Young shoots fairly woolly, covered with bloom, green, passing into light yellow before the grapes are ripe.

Leaves large, usually longer than broad, 5-lobed; upper sinus well developed, more or less U-shaped open, sometimes closed above ; lower sinus often little developed but always clearly visible, petiolar sinus usually wide open; dark green and glabrous above, below somewhat lighter green and strongly woolly with outstanding nerves; teeth long, sharp, large, alternating irregularly with smaller ones; petiole short to fairly long, woolly. Young leaves and stalks very woolly, above less than below; intense red-violet to garnet-red on the half-grown leaves, which later on turn yellowish-green and finally, when full-grown, dark green. In autumn they develop numerous wine-red patches. Dr. Hogg says: "The leaves die bright yellow".

Bunch medium to fairly large, cylindrical to conical, sometimes loose but usually very compact, 8-14 in. long; very frequently double,

2-3 per shoot; the bunches are inserted fairly low on the shoots; peduncle green, thick, tough, of medium length ; pedicels short, thick, conical; berries very firmly attached to them. The inflorescences appear completely covered with silvery white cobwebby threads. *Berries* above medium to large, short oval ("roundish oval", Dr. Hogg), ribbed. *Skin* thick, tough, "reddish-purple at first, but becoming quite black when fully coloured, and covered with a delicate bloom. Flesh dull, opaline white, very firm, sweet, and richly flavoured, with a faint trace of Muscat flavour, but not so much as to include it among Muscats. Seeds generally in pairs" (Dr. Hogg). Late mid-season.

9. **Bonnet de retord.**

This grape has been grown for some time at the Cape under the name of *Striped Berry Grape,* because, at maturity, the bloom runs in streaks over the berry from pole to pole. In 1910 I imported the *Bonnet de retord* from Palermo to the Cape. It proved to be identical with our Striped Berry grape, and the imported name is now used for it.

I know of no other black grape which is thus striped. When about half-grown, the green berries show faint red stripes where the stripes of white bloom will appear later on on the ripe berries.

Viala-Vermorel, *l.c.* vol. vii. p. 57, merely state that it is a peculiarly striped grape, and that it is grown by Besson in his collection near Marseilles. I cannot find it described anywhere. It is a vigorous grower but a poor bearer with short pruning. It should therefore be pruned long (Guyot or Cazenave system) to produce good crops. Its bunches reach a good size, and must be pre-thinned to get a loose bunch for export. The berries are very firmly attached to the pedicels, and this grape *stands long transport very well indeed.* In this respect it is somewhat better than most other varieties grown for export, Ohanez being naturally excluded. Its berries might have been larger, but they are large enough when the bunches are sufficiently pre-thinned and the berries can grow out well. If the grower's nearest railway station is some distance off, and if grapes have to be carried for a long distance by rail, it can be grown for export to greater advantage than softer varieties like Muscat of Alexandria, Rosaki, Gros Colman, etc. It obtains good prices in London and ripens somewhat late.

Description.—(A.I.P.)

Vine vigorous grower.

Canes fairly thick, long; internodes short to medium, whitish I somewhat dark brown, finely striped, very numerous fine black dots-bark smooth and glabrous; pith thick, diaphragm medium thick.

Growing shoots have a light green colour.

Leaves medium large, entire to 3-lobed, a few 5-lobed; petiolar sinus open V-shaped, upper sinus fairly deep and open, lower sinus usually wanting or slightly marked; teeth broad, shallow, more or less rounded off; glabrous on both surfaces, usually broader than long; in autumn yellow with wine-red patches; frequently the whole leaf is coloured light wine-red to blood-red.

Bunch fairly large to large, fairly compact, long, broadly shouldered, more or less pyramidal; peduncle soft and yellow at maturity. *Berry* fair-sized, oval, hard. *Skin* black with stripes of strong white bloom, thick, tough. Pulp firm, white, pleasant sweet. Late mid-season.

10. Grossa vivarais.

This grape I saw for the first time together with Prune de Cazouls in the Vivaio americano near Palermo in 1909, where it struck me as a pretty grape and one suitable for export. I imported it in 1910. In my diary I made the following note: "Grossa vivarais, black grape, compact bunch, fairly large; oval berries like Hanepoot (also as large) fairly thick skin, delicious taste; ought to sell well. *Import to the Cape.*" Both on the Paarl Experiment Station and the Stellenbosch University Farm it unfortunately grows poorly as a grafted vine. This is a great pity, as it is no doubt a grape of great value. In my collection it is grafted on Jacquez, Rip. gloire, Aramon No. 1, Aramon No. 2, 1202, 101-14, 106-8, Rup. du Lot, 420A and 333, as are all of the two hundred varieties grown. I am still experimenting with it. Where it can be made to grow well, it will be a first-class table grape for export and for local markets. It stands long transport well and sells well. I cannot find it described anywhere.

Description.—(A.I.P.)

Vine at the Cape hitherto a poor to moderate grower.

Canes reddish-brown, medium thick to thinnish; internodes very short; pith well developed; diaphragm often thick on the one side and thin on the other; on the whole medium thick.

Leaves medium, 5-lobed; upper sinus deep and often closed in front, lower sinus also deep and usually open, petiolar sinus open U-shaped, glabrous on both sides; petiole short and red; teeth uneven, small, sharp to somewhat rounded off.

Bunch fairly large, of medium length, compact, *Berries* large, oval. *Skin* thick, tough, black with white bloom. Pulp firm, delicious sweet taste. Late mid-season.

11. **Malakoff Isjum.**

Synonyms.—Olivette noir (France), Teta de Negra (Andalusia), Uva di Pergola (Italy), Pergoleze, etc.

In the Crimea it is known as Malakoff Isjum. In the South of France, Spain, and Italy it is also grown. As it is a very late grape, it requires a warm climate in order to come to perfection. In Italy it was grown during the time of Pliny. It ripens well at the Cape. It is a pretty and a hardy grape that stands long transport well. The best training for it is on a pergola, as the Italian name implies, otherwise on a horizontal cordon with *very long bearers.* With short pruning it bears very badly indeed. Its bunches and berries are large. So far it has not been grown to any extent at the Cape. In early localities it should answer very well as a late grape for export. It keeps and ships splendidly. It fetches good prices and need not be pre-thinned very much, but requires a good, deep, cool soil to produce very large berries. I imported it from Montpellier to the Cape in 1910.

Description.—(Viala-Vermorel.)
Vine vigorous grower.
Bourgeonnement glabrous, green.
Canes long, thick, striped, smooth, light reddish-chestnut coloured; internodes of medium length, wood hard; pith thinnish, diaphragm fairly thick.
Leaves medium, round, 5-lobed, flat, thin, glabrous on both sides, light green; petiolar sinus open V-shaped, upper sinus marked, lower sinus shallow; petiole short and thin.
Bunch fairly large, cylindro-conical, loose, irregular; peduncle green, rather short; pedicels fairly thin. *Berries* large, hard, elongated and of irregular shape and size; ends frequently somewhat pointed. *Skin* thick, black, with dense blue bloom. Pulp firm and crackling, colourless, sweet, pleasant flavour. Late.

12. **Black Alicante.**

Synonyms.—Alicante, Black Spanish, Black Lisbon, Black Portugal, Black Saint Peter's, Black Tokay, Meredith's Alicante, St. Peter's, etc.

After Frankenthaler comes Black Alicante as the second most important black grape grown in English, Belgian, Dutch, and North French hot-houses. The reasons for this are its great fertility, its pretty, large bunches and berries, and its further characteristic of keeping long on the vine and shipping fairly well. It has been exported successfully from the Cape to London, where it was sold at a satisfactory price. This variety is fairly susceptible to oidium and downy mildew. In my collection its bunches are loose and contain a fairly large number of small green berries (really only pistils), which should be removed a couple of weeks after the fruit has set. In the hot-houses it sets its fruit very well, and has to be pre-thinned very severely in order to produce a loose bunch. It bears well with cordon training and short pruning. To Cape growers of export grapes it is not of special value, as they have many better varieties to choose from.

Description.—(Viala-Vermorel.)

Vine vigorous grower.

Canes long, above medium, thick, pubescent when young and also later, dark yellowish; internodes of medium length but very variable wood fairly soft, pith very thick,

Leaves large, usually longer than broad, thick; petiolar sinus shallow and usually closed in front, other sinuses little developed; teeth very sharp, very woolly below, giving it a silvery appearance; nerves very prominent, and very woolly; petiole long, medium thick, reddish-green, woolly; in autumn yellow, sometimes with red patches.

Bunch large, often 3-4 per shoot, strongly branched of variable shape, always very compact; peduncle thick and short, green, tough pedicels short and thin, but adhering very well to the berries. *Berries* large, oval, or olive-shaped. *Skin* black with thin, bluish bloom, medium thick, tough. Pulp very tender and juicy, adhering a little to the skin; pleasant taste, slight Muscat flavour when highly ripe. Late mid-season.

13. **Black Hamburgh.**

Synonyms.—Frankenthaler, Hampton Court, Blauer Trollinger, Frankenthal.

According to Dr. Hogg, this grape was imported to England from Hamburg by John Warner in the early part of the eighteenth century, and hence the name Black Hamburgh. According to Barron it was in 1720. It is frequently met with in England, Holland, Germany, Austria, and Hungary, and is the favourite variety in English, Dutch, and Belgian hot-houses, as well as in those on the Channel Islands. In 1901 the 120-year-old vine of Black Hamburgh in Hampton Court still produced about 1500 bunches per annum (see Viala-Vermorel, *l.c.* ii. p. 128).[1]

It is a vigorous grower and should be given a large development. It bears well with short pruning, but Claassen, Hazeloop en Sprenger,[2] recommend long pruning for it, as they maintain that only from the 4th or 5th eye upwards (counting from the base of the cane) shoots are developed which bear good bunches. It gives very pretty grapes, but they are rather soft and not very suitable for long transport. To fungoid diseases it is very susceptible, and should therefore be grown on a warm dry site and on trellis. It has been successfully exported from Paarl to London.

Description.—(Compiled.)

Vine very vigorous grower, very productive.

Bourgeonnement light green, slightly pubescent; the youngest leaves frequently have slightly red borders.

Canes light yellowish, with reddish stripes, somewhat thick, long; internodes of medium length to long, very thick pith; wood fairly soft.

Leaves large, thick, 3-lobed; petiolar sinus deep, upper sinus irregularly developed but usually closed at end ; teeth shallow, uneven, rounded off; glabrous on both sides, light green, dying yellow; petiole long, above medium thick, light green with purple stripes.

Bunch fairly large to large, pyramidal, branched, compact; peduncle long, medium thick; pedicels long and thin, easily separating from the berries. *Berries* above medium large, almost round to egg-shaped ("roundish oval", Dr. Hogg). *Skin* thin, tough, blue-black, covered with blue bloom. Pulp rather firm but tender, very juicy, rich, sugary, and highly flavoured. Mid-season.

14. Madresfield Court.

On account of its Muscat flavour, it is sometimes also called

---

[1] *2012 Editors Note*: It grows still: now more than 240 years old *The Great Vine* produced in 2001 a bumper crop weighing 383 kilograms (845 lbs).

[2] Claassen, Hazeloop en Sprenger, *Leerboek voor de Fruit-teelt,* 2° druk, 1017, p. 126

Muscat Madresfield Court. "It was raised by Mr. William Cox, gardener to Earl Beauchamp, at Madresfield Court, Worcestershire, by hybridising Muscat of Alexandria with Morocco [the Cape Gros Maroc, A.I.P.]. It was awarded a first-class certificate by the Royal Horticultural Society in 1868" (Dr. Hogg). It is an excellent grape. It grows and bears well on a cordon with short pruning. Its bunches and berries are large enough for export. It has been successfully exported from the Cape to London, where it obtained good prices. It is fairly resistant to diseases. Once ripe it does not keep long on the vine at the Cape. It ships well. It requires fairly severe pre-thinning, and surplus bunches must be removed early.

FIG. 61.—Madresfield Court (Paarl Vit. Expt. Station). Original.

Description.—(Compiled.)

Vine fairly good grower.

Canes straight, of. medium length and thickness, unbranched; young shoots strongly woolly and remain thus for a considerable time, light reddish-brown; canes light brown; internodes short at the base of the cane, then of medium length, covered with bloom, with numerous fine stripes; pith and diaphragm thin; growing tip woolly and reddish-brown.

Leaves medium, round, thick, 5-lobed; petiolar sinus half-closed, dark green and almost glabrous above, yellowish-green with cobwebby pubescence below; petiole of medium thickness, long, woolly; turn a pretty dark red before dropping.

342

Bunch medium to large, long and tapering, well shouldered; peduncle short and thick; pedicels short and thick. "*Berries* large, oval or oblong, even in size. *Skin* tough and membranous, but not thick and coarse, quite black [at the Cape sometimes purplish-black, A.I.P.] with a fine bloom. Flesh greenish or opaline, tender, juicy, rich, and with an appreciable Frontignan flavour, though not so marked as in the Frontignans and Muscats" (Dr. Hogg). Early mid-season.

15. **Black Prince.**

This grape has been grown for some considerable time at the Cape, and is quite different from Frankenthal, for which Viala-Vermorel give Black Prince as a synonym. The Black Prince of the Cape is evidently the same as the grape described by Dr. Hogg under this name, and it was most probably imported to the Cape from England. For this grape Dr. Hogg gives the following

Synonyms.—Boston, Pocock's Damascus, Langford's Incomparable, Sir A. Pytches' Black, Stewards' Black Prince.

It is one of the earliest varieties grown at the Cape. It is indeed grown for the early market. Latterly it has been exported to London. It stands the voyage well, as it is a fairly hard grape. Its bunches, and especially its berries, are rather small for export and not very showy. It obtains moderate prices in London. It is a fairly good grower which bears well with short pruning. Its bunches are inserted high up the shoots and hang fairly bare. It is a fairly healthy vine. In early localities it can be of considerable value, especially for local markets.

Description.—(A.I.P.)
Vine fairly vigorous, erect grower.
Bourgeonnement intense red, little leaves garnet-red above, below violet-red in the border zone with a strong felt-like pubescence.
Canes long, thinnish, little branched; herbaceous shoots fairly intense red colour; ripe canes reddish to light brown; internodes long, pith fairly thick, diaphragm medium thick.
Leaves 5-lobed, green, glabrous above, woolly below, medium ; petiole fairly red, short, numerous brush-like hairs; young leaves fairly intense garnet-red, felt-like below, above at first strongly then slightly pubescent and soon shiny and glabrous, while the leaf still retains a reddish hue; growing tip whitish-green with little red, woolly; leaves with much red in autumn.
Bunch medium size; long, generally without shoulders, predominatingly cylindrical, moderately compact. *Berries* medium to above medium size, oval to short oval, fairly hard. *Skin* thick, reddish

to purplish-black with fairly thick blue bloom. Pulp white, soft, juicy, sweet, own mild flavour. Early.

16. **Muscat Hamburgh.**

Synonyms.—Snow's Muscat Hamburgh, Red Muscat of Alexandria, Black Muscat of Alexandria, Venn's Seedling black Muscat, Muscat Hambourg, Muscat Hambro.

This is a very ancient variety and had been grown in England for a very long time, when it nearly disappeared, as its pedicels dry up very easily when the berries turn colour, unless it is grafted on another variety. About 1860 Mr. S. Snow, gardener to Lady Cowper, Wrest Park, Bedfordshire, re-introduced it, but grafted on another vine. Dr. Hogg recommends Muscat of Alexandria as the best, and Black Hamburg as a good stock for it. Grafted on American stocks (for which it has a good affinity) it does very well and does not lose any berries.

It is one of the most delicious Muscats. It has a strong, exquisite Muscat flavour and a very pleasant sweet taste. Its flavour and taste are absolutely typical, and enable one to identify it without fail. It grows and bears very well with short pruning. Its berries are usually somewhat soft for long transport, and the berries are usually not very large, for which reasons it is not exported from the Cape. It is, however, quite possible that it may produce a suitable grape for export on dry, warm, deep soil, if the crop is suitably reduced and the bunches are slightly pre-thinned.

It might be connected with Muscat of Alexandria (possibly it is a natural hybrid between Muscat of Alexandria and Red Muscadel or some other dark Muscat variety), but shows very little signs of such a relationship, and has a Muscat flavour which is quite different from that of Muscat of Alexandria. It is therefore wrong to call it Red or Black Muscat of Alexandria, as it is not a differently coloured Muscat of Alexandria as is Red Hanepoot, which is truly a Red Muscat of Alexandria. It is one of the very finest grapes to grow for one's own use, and for selling locally. It possesses a very good affinity for American stocks.

Description.—(Compiled.)
Vine vigorous grower and very productive
Bourgeonnement woolly, reddish, passing into garnet-green with a light white woolly pubescence.

344

Canes fairly thick, long; internodes fairly long, pith thick, diaphragm medium thick, fairly dark brown.

Leaves fairly large, longer than broad, glabrous and dark green above, lighter green with short light woolly pubescence below; upper sinus deep and always closed, lower sinus well marked and slightly open, petiolar sinus open; teeth sharp, fairly long, uneven; die yellow, then brown.

Bunch fairly large to large, branched, pyramidal to long and variable in shape, fairly loose, never compact; peduncle fairly long and thin. *Berries* above medium size, oval to roundish oval, usually about 15 × 20 mm. or 0.6 × 0.8 in.; pedicels long, thinnish. *Skin* tough, but not thick, dark reddish-purple to black with a thin blue bloom. Pulp melting, very juicy, rich, and sugary, and with an exquisite Muscat flavour. Early mid-season to mid-season.

17. **Black Monukka.**

"This is supposed to be an Indian grape, and was first brought into notice by Mr. Johnson, gardener at Hampton Court, who sent it to the garden of the Royal Horticultural Society at Chiswick, where it is now to be seen growing in the large vinery in great perfection" (Dr. Hogg, *l.c.* p. 375). It is an interesting grape. It grows with tremendous vigour and bears well only when long pruned. A large development suits it best. Its bunches grow to a tremendous size. As it is seedless, and has a sweet and very rich flavour, it is a grape that can be eaten with pleasure. It produces excellent pitch-black seedless raisins, which taste somewhat like dried figs. For eating purposes its raisins are far superior to Sultanas and currants, and they would have a tremendous sale once they become popular. At present people are prejudiced against them owing to their black colour. By heavy sulphuring, this is reduced to a light pink and the raisins become semi-transparent, but unfortunately they lose all their flavour and acquire a somewhat sour taste. (This is the result of an experiment I have carried out. A.I.P.) They should be dipped in lye and dried without sulphuring. It is undoubtedly a grape with a good future, both as a table grape and for drying. The few growers who turn it into raisins at the Cape find a very good sale for its raisins locally. In California its vines have lately been in considerable demand by growers.

In the non-Karroo portion of the Cape viticultural area (Cape, Stellenbosch, Paarl, etc.) it is more productive and less subject to anthracnose than Sultana. Dr. Hogg, *l.c.* p. 375, briefly describes its grapes, otherwise I could not find a description of it. He also says that its leaves die "dull reddish-brown". In my collection they turn yellow, then brown.

Description.—(A.I.P.)

Vine extremely vigorous.

Bourgeonnement green (starts budding very early!).

Canes thick, long; internodes medium to short, light grey-brown, pith fairly thick, diaphragm medium thick; growing tip pale green, glabrous; young leaves glossy, glabrous, reddish yellowish-green.

Leaves 5-lobed; petiolar sinus fairly closed V-shaped, glabrous on both sides; teeth sharp, fairly broad, irregular.

Bunch large to very large, pyramidal, long, branched, fairly loose but full. *Berries* above medium, elongated ("obovate-oblong", Dr. Hogg). *Skin* very thin, fairly soft, purplish black ("a deep dull chestnut colour", Dr. Hogg) with blue bloom, "adhering closely to the flesh, which is firm, crisp, and very juicy, with a sweet and very rich flavour, more so than Black Hamburgh" (Dr. Hogg). Early mid-season.

III. RAISIN VARIETIES

1. **Muscat of Alexandria.**

This variety has already been discussed and described among the white table varieties. It is, however, one of the most important raisin varieties. In Spain it is used for making the famous Malaga stalk raisins and also the well-known Denia lye raisins. In California, South Africa, North Africa and Australia large quantities of raisins are made of this grape.

2. **Rosaki.**

Though not used to anything like the same extent as Muscat of Alexandria for turning into raisins, the larger berries of Rosaki can give the prettier raisin of the two, provided the Rosaki grapes are picked when fully ripe (sweet with a fine golden-yellow colour). Rosaki berries that have a greenish tinge and are not properly ripe, give a reddish and poor raisin. In Asia Minor, especially in the Smyrna district, large quantities of Rosaki raisins are made. During the last couple of years, quantities of Rosaki raisins of good quality have been produced at the Cape, mainly in the Worcester district. It produces very pretty lye stalk raisins.

3. **Sultana.**

Synonyms.—Sultanina, Thompson's Seedless (California), Sultanieh, Sultan, Sultani, Kechmish yellow with oblong berries (India), Couforogo (Greece).

Dr. Hogg describes this grape under the name Sultana, which is also the name under which it is grown at the Cape. It is a native of Anatolia, and was brought from Smyrna to Greece, where it throve remarkably well. It subsequently spread to every viticultural country on earth. It is mainly in Greece, Asia Minor, California, South Africa, and Australia that Sultana is grown for the production of Sultana raisins. It is also a nice eating grape, having no seeds. Its seedless-ness is one of its most valuable characteristics. This depends upon a defective development of its sexual organs, and more particularly of its pistil. I have never yet found a Sultana berry with seeds, not even where it is interplanted with other varieties, and cross-pollination is therefore rendered easy. It is not yet certain whether some sort of fertilisation is required by the Sultana vine to produce its normal berries. In 1924 I tied up a few of its bunches in paper bags just before the flowers began to open. These bunches were perfectly normal and well-set. Hence cross-pollination is at all events unnecessary. This year I shall emasculate some bunches to see whether self-fertilisation of a kind is needed or whether the normal Sultana berries are all formed purely partheno-carpically.

The Sultana vine is a very vigorous grower and is most highly productive in a dry climate where irrigation is practised, *e.g.* the Karroo districts of the Cape, and in California and Australia, where similar climatic conditions exist. In the Western Province of the Cape it bears too little and suffers so severely from anthracnose that its culture does not pay. In the drier parts it is not touched by anthracnose. It is, however, susceptible to oidium, and should therefore be sulphured. If grown where anthracnose is troublesome, it should be treated against this disease in winter. It requires long pruning. This is illustrated in Figs. 91 and 101. At the Cape, Sultana is usually not trellised but simply pruned, as shown in Fig. 91b, with this difference, that the vines are self-supporting and not staked.

It is best to trellis the Sultana vine and prune it as is shown on Fig. 101. H Maroger's system would suit it admirably in good soil, and would require two more wires above those shown on Fig. 101 to which the growing shoots can attach themselves or be tied. In good soil and under suitable climatic conditions the Sultana vine gives very heavy crops with long pruning. On the Stellenbosch University Farm it produced a heavy crop in 1924 (see Fig. 97), but these vines were then still young, having been grafted only in 1922 on nine-year-old Hanepoot vines grafted on Jacquez. Its fertiliser dressings should be rich in phosphates and potash where it is sufficiently vigorous. Once the grapes are ripe they rot very easily if rain falls for some time.

Description.—(Compiled.)

Vine very vigorous grower.

Bourgeonnement reddish-green, glossy, glabrous.

Canes thick, long, strongly branched; internodes of medium length; pith thick; diaphragm thick; light whitish to yellowish-brown, sometimes light reddish-brown, darker at the nodes, finely striped and with numerous black dots.

Leaves large, entire to shallow 3-lobed, petiolar sinus closed (sometimes slightly open V-shaped), upper sinus shallow but present; teeth broad, obtuse, rounded off, thin, glabrous on both sides, light yellowish-green to green; petiole long, glabrous, light rose-coloured.

Bunch fairly large to large, long, more or less conical, frequently well branched, compact; peduncle fairly long, medium thick, fairly soft and brittle; pedicels long and thin, fairly brittle. *Berries* smallish, sometimes fully medium size, oval to elongated. *Skin* thin, yellowish-green to fairly golden - yellow at full maturity. Pulp fairly firm, crackling, pleasant sweet taste with a fair amount of acidity when ripe. Seedless. Mid-season.

Note.—There is also a light *red* coloured type of Sultana. On a farm near Robertson Station (Cape) I saw red and white Sultana bunches on the same vine and on the same shoot. For turning into Sultana raisins the red type is not so valuable as the ordinary white or yellow Sultana.

4. **Black Corinth.**

Synonyms.—Black Corinth Currant, Zante, Patras Currant (England); Vitis Corynthiaca, Vitis Apyrina, Passa minor; Staphis, Kourenti, Passa (Greece); Corinto nero, Passerina nera, Uva Passolina nera (Italy); Corinthe noir, Cormthien, Raisin sans pepins, Raisin de Passe, Marine noir (France), etc.

There are three kinds of Corinth or Currant, namely, *white, red,* and *black.* For consumption as currants in cake, etc., practically only the black kind is used. It is this Black Corinth which produces the ordinary black currants. Currants are, however, also used for making into wine. In Greece, where it is indigenous, both the fresh and the dried grapes are used for this purpose. In England and other countries where only the dried currants are imported, they are the only form used for wine-making. Light wines, sweet wines, jeripico, and grape syrup are all made from the dried currants. They give an aromatic wine. At the Cape currants are too expensive for wine-making.

In Italy the white Corinth was once upon a time grown extensively in the districts of Asti and Canelli for the manufacture of sparkling wine, where it produces a very good wine. Since the devastation which oidium caused also in these parts from about the middle of the previous century, its culture has almost completely disappeared in these parts.

In its native land, Greece, it is grown most extensively on both sides of the Gulf of Corinth, the Gulf of Patras, southwards along the west coast of the Peloponnesus, and on the Islands of Zante, Cefalonia, Ithaka (Ithaki), and Leukas (Santa Maura). On Korfu it is not grown. In Asia Minor it is again grown on a fair scale at the entrance of the Gulf of Smyrna at Phokia and near Thyra and Vurla. The dried currants are exported from the harbour of Smyrna. In Australia it is grown to a considerable extent under the name of Zante Currant; it was imported from Australia to the Cape some years ago. At the Cape it is just being tried out in different parts. It yet remains to be seen whether it will supplant the Cape Currant.

The vine is a vigorous grower and a heavy bearer, and it desires a large development, although, like Sultana, it can be grown successfully untrellised. On trellises and with ordinary cordon training it does very well. It likes a stony, dry soil, which should, however, not be too dry. When not grown on level land, it prefers a westerly slope. In the sweet, fertile soils of the Cape where Sultana does so well, it, too, will no doubt produce good results, as the Cape Currant In Paarl and Stellenbosch it does very well on Jacquez in soils of decomposed granite and Malmesbury shale. It possesses a good affinity for American stocks. It is highly resistant to downy mildew, but fairly susceptible to powdery mildew or oidium, and should therefore be sulphured early.

As it must be *girdled* or *ringbarked* in order to produce really good crops, it is desirable to prune it with long and short bearers, and to remove a strip of bark near the base of the long bearers just when the flowers begin to open. This should be done with special ringbarking scissors. In this way the work is soon done, and the vine will not suffer appreciably. If, however, the girdling is done on the trunk of the vine, as is frequently done in Australia, the vine is bound to suffer in the long run. The intensity and extent of the damage done will depend on the width of the strip of bark that has been removed, the rapidity and perfection with which healing takes place, and the fertility of the soil (also intensity of fertilising and general soil treatment). It produces 2750-3666 lbs. or an average of about 11 tons of dried currants per acre (3000-4000 kg. per ha.), where it answers well and is ringbarked. In my collection the bunches of this grape shed their berries as they ripen, and they remain insignificant if ringbarking is not practised. Ringbarking overcomes this defect and causes much better sized berries to be formed.

Its ripe grapes are dried directly in the sun on trays, and are dry after about four to eight days. It is seedless. This also probably depends upon defective sexual organs. Sometimes one finds on a bunch berries that contain seeds, and are much larger than the seedless berries. According to Marès we have to do here with a case of millerandage that has become hereditary. See Viala-Vermorel, *l.c.* vol. iv. p. 289, where an experiment of M. Jurie with this grape is related, who obtained a lot of berries with seeds in a bunch of this grape by covering it up in a paper bag before flowering had begun, so that cross-pollination was impossible. M. Jurie did not succeed in raising seedlings from these seeds.

In Greece the currants are harvested in August, after having been ripe for about a month. Formerly the Greeks dried all their currants on ground floors, but now they are using trays more and more, which is of course the better and cleaner method of drying fruit.

Description.—

Vine vigorous and productive.

Bourgeonnement woolly, red-violaceous green, early.

Canes long, medium thick, fairly frequently branched, fairly dark brown, internodes short to medium, pith large, diaphragm thinnish; tendrils numerous, long, tough; growing tip and young leaves very strongly woolly and reddish.

Leaves medium large, longer than broad, 5-lobed, sometimes 3-lobed and very seldom entire, upper sinus well developed and open V-shaped, lower sinus well marked, petiolar sinus closed; leaf fairly thick and tough, dark green and glabrous above, light green and very woolly below; teeth uneven, short, sharp; petiole long, slightly thin to fairly thick.

Bunch small to medium, long, cylindrical, fairly compact, sometimes strongly branched. *Berries* small to very small, round, with some large ones interspersed. *Skin* thin, purplish-black to black, and covered with blue bloom. Pulp juicy, sweet, richly flavoured, and without seeds, with a fair amount of acid.   Early mid-season.

5. **White Corinth.**

*Synonyms.—*Corinthe blanc, Corinto bianco, Passerina bianca, Korinte-druif, Tramarina, Korinthe kleine weisse, etc.

It is grown much less than the black currant. Dr. Hogg describes it as "of no value". It is sometimes grown for the table, and in Greece it is also grown for turning into wine, etc. It differs from the black variety mainly in the colour of its ripe grapes, and so need not be described separately.

6. **Cape Currant.**

This is a quite different grape from the Grecian Corinth. It is evidently purely and simply a Red Muscadel, in which total millerandage has become constantly hereditary. Sometimes one finds on a bunch one or two large berries of the same size and flavour as Red Muscadel, and with seeds. This proves my contention about its origin. It has been grown for a long time at the Cape, where it may have originated. The vine, leaves, shoots, etc., look just like those of the Cape Red Muscadel. With long pruning (like Sultana) it bears well without ringbarking. It grows more vigorously than Muscadel, I suppose on account of its lighter crop of grapes. Its berries are dark red-violaceous black, like those of Red Muscadel (Cape), about as large as those of the Grecian currant, but has a sweeter and less acid taste, possesses a pronounced Muscat flavour, and is seedless.

Its dried currants sell as well as the true Grecian currants. It now merely remains to be seen which of the two will give the highest yields. In this respect I expect the Grecian currant to win. I have proved by experiment that the Cape Currant will give much heavier crops if ringbarked as is done with the Grecian Currant. Its ripe berries tend to drop just like those of non-ringbarked Grecian currants. Its affinity for American stocks is good and agrees with that of Muscadel.

Description.—(A.I.P.)
Vine vigorous grower.
Bourgeonnement, Canes, Leaves just like Muscadel.
Bunch small, long, cylindrical, moderately compact to somewhat loose; pedicels thin, tender, medium short. *Berries* small to very small, round. *Skin* thin, dark red-violaceous black, with light blue bloom. Pulp soft, juicy, sweet, with a pronounced Muscat flavour, seedless. Early to early mid-season.

C. DIRECT PRODUCERS

I. THE OLD DIRECT PRODUCERS

When the phylloxera began to destroy the vineyards in France, and grafting on resistant stocks was very uncertain as no lengthy experience with these was available, the French wine farmers began to grow certain American varieties which they knew were more resistant to phylloxera than their European vines. These American varieties were planted out directly without grafting (as was previously done with their European vines), and were therefore called *direct producers* or *self-bearers*. The varieties that were at first imported from America for this purpose are known as the *old* direct producers, as distinguished from the *new* ones that were only subsequently bred in France by manifold crosses between American and European varieties. At first the direct producers were mainly used because the phylloxera attacked the European vines. They included Jacquez, Herbemont, Concord, Isabella, Clinton, and a little later Othello, etc. Of these the first two and the last are still grown in some parts of France as direct producers for obtaining a very dark red wine, suitable for blending. They are grown only on cool, deep soils where they can successfully resist the phylloxera. Their wines have a wild, more or less foxy taste. At the Cape Othello was planted on a fair scale on some farms in the early days of our reconstitution. Jacquez has been used for this purpose only where large mother plantations of it existed for producing cuttings, which were used as stocks for our European varieties. Its grapes are not always used for turning into wine. To-day direct producers are practically not grown any more at the Cape, as there is no good reason why they should. As the old direct producers are not used anywhere (practically) now for this purpose, I shall refrain from discussing them any further.

II. THE NEW DIRECT PRODUCERS

As the fungoid diseases and insect pests increased in number and in severity in the European vineyards, the idea arose of breeding new hybrids between American and European varieties which would be sufficiently resistant to the enemies of the European vine, and yet produce grapes of good quality for eating or drinking purposes. Notwithstanding the progress that has been made by grape-breeders like Seibel, Couderc, Malègue, Bertille-Seyve, Baco, Jurie, Oberlin (in Alsace), etc. in France, the ideal direct producer has not yet been obtained. Where the vine is highly resistant to its enemies, the grapes are not of quite good quality. If the grapes are quite good in comparison with those of good European varieties, the vine is not very resistant to its enemies.

In course of time the objects aimed at has, in many instances, been limited to good quality grapes and resistance to enemies above ground, without paying much attention to resistance to phylloxera. This more limited object will no doubt be sooner attained than the ideal direct producer already mentioned. Monsieur Pée–Laby[1] indeed distinguishes between *"hybrides producteurs directs"* (i.e. the ideal direct producers) and *"hybrides greffons"* which should

---

[1] Pée-Laby, E., *La Vigne nouvelle,* 2nd ed., Paris, 1923, pp. 24-25.

be grafted on resistant stocks but are fairly resistant to enemies above ground, and produce grapes of good quality. Resistance to diseases and pests will no doubt always remain a matter of degree, and will not be equally great with regard to all diseases and pests. Resistance to diseases will probably be obtained sooner than resistance to pests.

As we shall see later on, Professor Lucien Daniel maintains that the resistance of the European vines to their natural enemies has been weakened by grafting them on American stocks, which have usually been chosen on account of their great vigour. He is interested most in the creation (by hybridisation) of vines that will be sufficiently resistant to phylloxera without requiring to be grafted on resistant stocks (i.e. real direct producers). The trouble so far has been that the grapes of such hybrids have a more or less wild taste, and do not equal good European grapes in quality at all.

For viticultural areas like that of the Cape, where diseases and pests are not bad, the whole question of direct producers is of little interest. Where the contrary holds good, as in Europe and North America east of the Rocky Mountains, it becomes a very important question. In many European wine-districts, owing to the numerous and dangerous diseases and pests, viticulture has become so costly and uncertain an undertaking, that it is threatened with extinction in some of these districts unless direct producers or something else bring relief. Under such conditions the desire for a good direct producer can be understood.

In order to show the reader a little of what has thus far been done in this direction, I shall briefly discuss some of the best direct producers as I find them described in Pée-Laby's book, La Vigne nouvelle, 1923.

Seibel's Hybrids.

(a) **Black:** Nos. 2, 128, 138, 1000, 1077, 2007, 4121, 6036,6042

1. *Seibel No.* 2 (Rupestris-Lincecumii × Vinifera).

It is a vigorous grower and good, regular bearer; with long pruning and in good soil it produces up to 22 lbs, of grapes per vine.   It is sufficiently resistant to fungoid diseases not to require any treatment for these. In deep, cool, sandy-clayey soils it does well on its own roots. It answers well on Riparia, 101-14, 1202, Aramon × Rup. Ganzin No. 1, and the Berlandieri hybrids. Bunch above medium size, fairly compact; berries medium, round; ripens mid-season; has a fairly pronounced wild (herbaceous) taste. Its wine is very dark violet-red and has a wild taste (not a foxy taste!) like the grape. It contains on an average 7.5 vol. % alcohol, 30.6 g. dry extract per lit., and 11.30‰ total acidity. A good wine for blending purposes.

2. *Seibel No.* 128 (Rup.-Line, × Vinifera).

It is one of the most widely grown black hybrids owing to its heavy crops, tremendous vigour on good soil and the good blending qualities of its wine. In deep, cool soils it does well on its own roots. For downy mildew it need not be sprayed except in very bad years. It is highly resistant to oidium. As it is sensitive to anthracnose, it should not be grown on moist low-lying land. Its bunches are of medium size or larger, fairly compact, longish, with round berries. Its grapes are sweet with a somewhat wild taste, and ripens mid-season. Its wine is fairly strong (for French notions of a dry natural wine), has a very dark red colour and a wild taste. Alcohol 10.5 vol. %, total acidity 7.9‰, dry extract 25.8‰.

3. *Seibel No.* 138 (Rup.-Line, × Semis.—a natural hybrid). It can be recommended for its great vigour and high resistance, and does equally well grafted or ungrafted. It can stand a good deal of lime, even when ungrafted. Although a slightly lighter cropper than No. 128, it bears more regularly and is highly resistant to anthracnose and to downy and powdery mildew. It produces satisfactory crops, and its grapes have only a slightly wild taste. Its wine has a pretty dark red colour, is excellent for blending purposes, and has very little of the wild taste. Alcohol 9.5 vol. %, total acidity 8.5‰, dry extract 25.2‰. It is a pretty vine with strong roots, highly resistant to diseases, and a good grower in all sorts of soils.

4. *Seibel No.* 1000 (Rup.-Line, × Aramon-Rup. No. 1).

Grown on suitable land, *i.e.* on a flat plateau or along a sandy slope, it is a splendid hybrid. It cannot stand wet, and must not be grown on compact soils, clayey lime-soils, or on the fertile deep soils of the plains. As it is a tremendously vigorous grower, it answers best on somewhat dry soils. On damp soils it regularly sets its fruit badly, and its inflorescences suffer from anthracnose. It does well on its own roots and when grafted. In South-east France it is grown extensively on account of its heavy yields and the good quality of its wine. It is fairly resistant to diseases, and need be sprayed for downy mildew only in bad years. It bears heavily with short pruning. Its bunches and berries are of medium size, berries round. It ripens early mid-season. Its grapes become sweet and are free from any wild taste. Its wine is usually regarded as good ; pretty colour. Alcohol 10-12 vol. %, total acidity 6.8%o, dry extract 17%o.

5. *Seibel No.* 1077 (Rup.-Line, × Aramon).

This is a very vigorous grower and a fairly good bearer with long pruning. It is somewhat sensitive to black rot; downy mildew attacks only its grapes, which, in bad years, will be well protected by spraying once or twice with Bordeaux mixture. Oidium does not attack it. Its bunches are above medium size, fairly long, sometimes winged, with medium size berries, sweet, without a wild after-taste. It stands rotting well and ripens mid-season. Its wine is unanimously recognised as an excellent wine for blending, both on account of its pretty colour and its good taste. It is one of the best, if not the very best, red wine made from a direct producer, so far known. Alcohol 10.5 vol. %, total acidity 6.9%o, dry extract 26%o. Excellent cepage for resistance to phylloxera and diseases, and for good wine.

6. *Seibel No.* 2007 (Rup.-Linc, × Aramon).

It starts budding very early and answers well on its own roots except on too strongly chlorosive limestone soils. On American stocks it does well also. Ungrafted it gives heavier yields than when grafted. It is satisfactorily resistant to fungoid diseases. In normal years it need not be sprayed. Its bunches are above medium, berries large, round, sweet, with little or no wild taste, ripen mid-season. Its wine is not too good, but is readily bought for blending purposes on account of its dark red colour and good alcohol content.

Alcohol 9 vol. %, total acidity 7‰, dry extract 21‰. Do not plant where frost is feared in spring.

7. *Seibel No.* 4121 (Seibel 52 × Terras No. 20).

This hybrid is being grown more and more, both on its own roots and grafted. In chlorosive clayey limestone soils it grows moderately, otherwise it is a most vigorous grower. It bears heavily with long pruning. It starts budding fairly early. Its bunches are of medium size or larger, and its berries can grow to above medium size in good soils. It ripens rather late, is sweet with a neutral taste and coloured sap. Its wine (9-10.5 vol. % alcohol) has a very dark colour and a good taste. Blended with white wines it gives a good ordinary wine. It gives 7-12 lbs. of grapes per vine, answers well on all soils that do not contain too much lime, and has already been well tested as direct producer (ungrafted). It is highly resistant to fungoid diseases.

8. *Seibel No.* 6036 (Seibel 29 × Seibel 2510).

This is a splendid bearer. It buds late, grows very vigorously, much more so when ungrafted than when grafted. Ungrafted it is also more resistant to oidium than when grafted. It does not fear downy mildew. It has pretty bunches with medium-sized or larger, somewhat short oval berries, ripens mid-season, becomes sweet, and has a fairly pronounced Cabernet taste. Its wine has a pretty, very dark colour, 9.5 vol. % alcohol with 5.6‰ total acidity. Cepage that can be strongly recommended: good crops and high resistance.

9. *Seibel No.* 6042 (Seibel 29 × Seibel 2021).

This is a still more interesting hybrid than the preceding one. It is as vigorous and bears as well as or more than Seibel No. 6036, but buds later, and has prettier bunches with larger berries, compact and long bunches. It is highly resistant to fungoid diseases and needs no treatment against these. It ripens mid-season. Its grapes are sweet and also have a pronounced Cabernet character. Its wine has a very dark colour and a Cabernet taste. It is a cepage with a future, interesting on account of its yield, resistance, and its wine.

(*b*) **White:** Nos. 4986, 4995, 5279, 5409.

1. *Seibel No.* 4986 (Seibel 405×Seibel 2007).

In good soils it is a good grower on its own roots. It is a good bearer and very resistant to downy and powdery mildew. Its ripe grapes rot easily, especially on damp spots, they become very sweet, are nice to eat, and have a pleasant after-taste. Its wine is valued very highly, has a good taste and finesse. Alcohol 12.5 vol. per cent.

2. *Seibel No.* 4995 (Seibel 2510 × 272-60 Couderc).

This is a very vigorous grower, whether grafted or ungrafted. It stands a good deal of lime in the soils without getting chlorotic. As it buds rather late, it escapes the spring frosts. It is highly resistant to downy mildew and black rot. In moist areas oidium attacks it. It can give fairly heavy crops with long pruning. Its bunches and berries are above medium size and have a neutral taste with a good acidity. Its wine is considered good.

3. *Seibel No.* 5279 (called the Aurora). (Sicilien × Clair, dorée Ganzin × Seibel 29).

It is a very early grape (hence called Aurora) which sells well as a very early table grape. Its grapes are too soft for long transport. It grows well, both grafted and ungrafted. Its thick, glossy leaves are highly resistant to fungoid diseases. It is very productive. When pruned too long, it easily suffers from coulure. It buds fairly early. Its bunches are of medium size or larger, berries round and of medium size, very sweet. It ripens long before Chasselas. Its skin is thin, pulp soft, and very juicy. Its must is very sweet with very little acid. Its wine is good. Alcohol 13.1 vol. % with 4.6‰total acidity (expressed as sulphuric acid). Its robust constitution and large yields make it increasingly popular.

4. *Seibel No.* 5409 (Seibel 867 × Alic. Ganzin × Seibel No. 1).

It possesses a practical resistance to fungoid diseases. Its yield is more than average. Its bunches are not very long, berries fairly large, crackling, very sweet and pleasant to eat. It buds late and ripens early mid-season. Its wine is considered the best of the white direct producers. It is oily, fine, with a pleasant taste. Alcohol 12-14 vol. per cent. It is not a very heavy bearer, but is considered very valuable.

Couderc No. 4101 (called the Madonna) and **Couderc No. 4401** (both Chasselas Rose × Rupestris).

Pée-Laby says that they are two different vines although they are frequently confused. They arose from the same cross. Both have dark green leaves with the same number of sinuses. The young shoots of 4101 usually remain green, while those of 4401 assume a violaceous colour. 4101 nearly always develops two shoots from the same eye, but 4401 only one. 4101 is more productive and has 3-4 bunches to every main shoot and 2-3 bunches to the shoot from the side bud of the same eye, thus making 5-7 bunches per eye (node) of the bearer, while 4401 usually has 2-3 bunches per eye of the bearer. Fertilisation takes place much quicker with 4101 than with 4401, hence the latter suffers more easily from coulure than 4101. Both grow with the same vigour, whether grafted or ungrafted, while both stand a good deal of lime in the soil. Further, both are equally resistant to downy mildew. One of their great differences is the great sensitiveness of 4401 to oidium, while 4101 is hardly ever touched by it. 4101 yields a heavier crop than 4401. Both ripen at the same time: mid-season. The grapes of 4101 have a good taste and become properly sweet. Both give a very dark wine, but that of 4101 is the finer and stronger, and is considered to be very good. Alcohol up to 16 vol. per cent. 4101 can be recommended for its robust health and vigour, together with a good yield and good quality.

Couderc 7120 (Rup.-Linc. × Aramon).

It is a vigorous grower in good soil; elsewhere it shows only moderate vigour. It resists phylloxera well and is not very sensitive to lime. For American stocks it has a good affinity. Its resistance to fungoid diseases is good. Its yield is moderate. In order to be a success it requires a warm climate, and it is grown most largely in the warm south and south-east of France. Its wine is not very dark but has a good taste. Alcohol 8-10 vol. per cent.

Gaillard Girerd No. 157 (Triumph × Eumélan × Seibel No. 1).

This is the most widely grown white direct producer of all the collections. It has been known for a long time and is grown on account of its regular, heavy yields. It grows and lasts well on its own roots and always gives a heavy yield with short pruning, whether grafted or not. In bad years it should be treated at least twice against both powdery and downy mildew. Its resistance against these two diseases is insufficient to render these treatments unnecessary, but it is considerably greater than that of the European varieties. Its bunches are often above medium size, berries medium. Its grapes ripen early mid-season. When fully ripe they have a pleasant taste and can be eaten. Its white grapes sometimes assume a reddish tinge at full maturity. Its wine is fairly strong (for French notions of a natural dry wine), having 9-11 vol. per cent alcohol, with a pleasant, typical after-taste, which some connoisseurs regard as an attenuated foxy taste, but which in any case is pleasant. The grower who treats it against diseases will be assured of a good crop every year (as contrasted with his European or Vinifera varieties), and it is not likely that this variety will soon be excelled for growing on a large scale.

CHAPTER VI

PROPAGATION OF THE VINE

UNTIL the phylloxera came to Europe, the vine was propagated there and elsewhere by means of cuttings, which were planted out directly or after being rooted in a nursery. A vacancy in the vineyard was usually filled by layering. Ripe canes were generally used as layers. After about two years the layers were severed from the mother vine. Propagation by means of seed and through grafting was seldom resorted to. Since the arrival of the phylloxera in the European vineyards these two methods of propagating the vine acquired a new significance. Propagation from seed is now used to produce new varieties that can serve as stocks or as direct producers or for improving existing varieties which are grown for their grapes and are grafted on resistant American stocks.

The bulk of the vineyards now consist of grafted vines, namely, Vinifera varieties grafted on American stocks. Propagation by means of cuttings growing on their own roots is now limited almost exclusively to American stocks and direct producers. European varieties now are practically never propagated by layering, as these layers would be growing on their European roots and hence be destroyed by the phylloxera. This method of propagation is now adopted almost exclusively for rapidly propagating a valuable new variety or a stock which roots very badly when its plain cuttings are planted out. Layering has thus lost nearly all its former importance, and therefore I shall not discuss it at greater length. Also a discussion of propagation from seed can be omitted here, as it has been treated in sufficient detail at the beginning of Chapter III. In the present chapter I shall limit the discussion to propagation by means of cuttings and through grafting.

A. PROPAGATION BY MEANS OF CUTTING.

1. **The Formation of Roots.**

In Chapter II. we have seen that the roots developing on cuttings of canes are called *adventitious* roots, and that they are formed mainly on wounds and at the nodes. Ravaz[1] has shown that every adventitious root is formed in a medullary ray in the cambium-ring, and that it penetrates the bark through the latter forming numerous young cells at this spot which are pressed outwards and resorbed as the root develops (quoted from Foëx[2]).

The different varieties and species do not develop roots from cuttings in the same manner or with equal facility. This has already been pointed out in Chapter IV. We shall now proceed to discuss different aspects of this problem which are of practical value.

2. **Selection, storing, and transportation of Cuttings.**

The best cuttings are those of medium thickness, with fairly short internodes, and such as are properly ripe and sound. Cuttings that are planted out as direct producers or are used for obtaining scions must be selected only from vines that bear well and regularly. The best plan is to mark such vines in the vineyard just when the grapes are ripe. Vines so selected must be true to type both as regards their general character and as regards their grapes. In case of varieties like Bicane, Gros noir des Beni-Abès, Malbec, Muscat of Alexandria, etc., that are very susceptible to coulure, those vines must be marked that have produced a good crop of well-filled bunches.

It is of the utmost importance to select cuttings for scions with the greatest care and to prevent cuttings of different varieties from being mixed up. It causes unpleasantness and financial loss if a grafted vineyard afterwards contains wrong varieties and vines that bear badly.

On the whole it is best to leave the canes on the vines until they are required. For practical reasons, however, this is not usually done. Cuttings for stocks are generally cut as

---

[1] Ravaz, L., "Recherches sur le bouturage de la vigne," *Comptes rendus,* 15 Sept. 1890.

[2] Foëx, G., *Cours complet de viticulture,* 4th ed., 1895, p. 287.

362

soon as the leaves have dropped. At the Cape this takes place during May (after the 15th) and June, sometimes as late as July. Cuttings for planting out ungrafted or for scions are cut when required—at the Cape usually in July or sooner, and in any case before the buds have swollen. When cuttings must be stored for some time, care must be taken that they retain their original moisture content as far as possible. This is best attained by storing them in *moist, not wet,* sand in a cool building. If stored in the open, it is best to cover them up in deep soil where they will get neither too dry nor too wet. Lay them flat in trenches and in layers that are not too thick, spread some soil between them, and cover them up with a ridge and higher than the surrounding ground to prevent water from standing on them. Select a dry site for this purpose, preferably a slope. Avoid a clay soil. A sandy soil or a light loam is the best for this purpose. Canes lying flat will not commence budding so readily as those standing more or less straight up.

For long *transportation* cuttings should be well packed. If their journey is to last a week or longer, it would be best to pack the cuttings in moist sand in a strong box lined with oil paper. This is usually not done, as sand is so heavy. Instead of sand, half-rotted leaves, moss, sawdust, or some similar material is used for packing after being well moistened. If transportation does not last longer than a week, the oil paper can be omitted. If it lasts only one or two days and takes place during cold, rainy weather, the cuttings can be packed in ordinary bags. On arrival such cuttings should be packed in moist sand or steeped in cold water for one or two days. Do not throw away cuttings that look dry on arrival, keep them in fresh water for 24-48 hours. If after this soaking they sink in the water and show a bright green colour on a freshly cut wound, they can safely be used.

3. **Length and thickness of Cuttings.**

The shortest cutting, that is, a single-eye cutting, produces the most vigorous rootling (shoots and roots), but then the soil's condition as regards moisture, etc., must be ideal. In practice longer cuttings are used, their length depending upon the nature of the soil. On a deep soil that becomes fairly dry near the surface and that cannot be irrigated, we plant a long cutting to have its lower portion in the moist soil and thus to give it a better chance of striking root. On soils with 10-12 in. of loose soil on a compact sub-soil we use much shorter cuttings. If the cutting is too long, it forms numerous roots which grow much more vigorously near its middle than near its bottom end. Sometimes no roots at all are formed at its bottom end, which then rots away. This is dangerous for the vine's future existence. Fig. 62 gives us a fine illustration of the influence of the length of the cutting

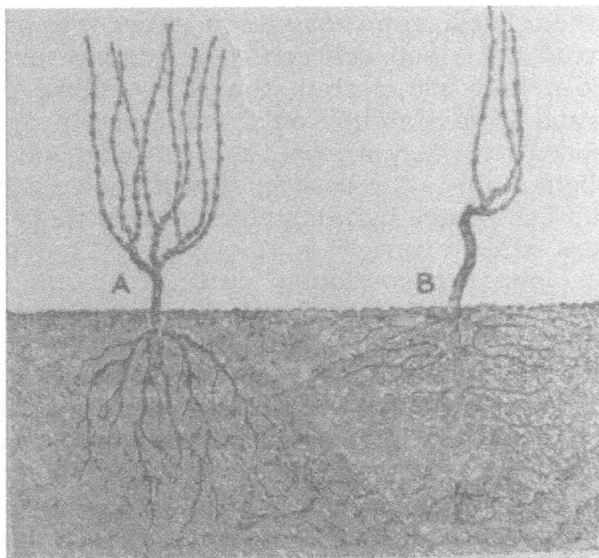

FIG. 62.—A. Vine formed from a short cutting.
B. Vine formed from too long a cutting.
A and B from Foex, *Cours complet de Viticulture*, 1895.
Coulet et fils, Montpellier. G. Masson, Paris.

upon its future development. When we consider that most of the available plant-food is usually stored in the first 8-10 in. of the soil, we readily see that the cutting's bottom end should not be much deeper either. Hence cuttings for planting out are usually cut in lengths of 10-14 in. or 25-35 cm. (see Foëx, *l.c.* p. 294). At the Cape they are usually cut 14 in. long, which is sufficient for all our soils.

The longer the cutting is, the more stored food it will contain, which can be utilised for the development of its first rootlets and very young shoots. We have, however, seen already that it is a mistake to make the cutting too long and to plant it too deep in the soil. Therefore do not make your cuttings too long, and use only well-ripened material, which will be well provided with stored food. In Europe cuttings are sometimes made as long as 80 cm. or 32 in. for specially dry soils.

The *thickness* of the cuttings is also of importance. For scions and for cuttings to be planted out directly, we discard the rather thick cuttings. For scions we select those that are as thick as or thinner than the stocks, but not thicker. For bench-grafting we use the good, medium thick cuttings. Those that are too thick should be discarded as they strike root badly and knit badly. Those that are too thin can likewise be discarded, or they can be planted out in the nursery to be grafted on the spot during the following season. The best average thickness is about ⅓ in. or 8.5 mm. Cuttings that are thinner than ¼ in. or 6 mm. and thicker than $2/5$ in. or 10 mm., cannot be recommended for bench-grafting. According to Mader (see Babo u. Mach, *l.c.* p. 195), cuttings of 6-8 mm. thick and coming from near the base of the canes are the best for bench-grafting and are therefore regarded as of the highest quality. As all shoots start ripening from the base, the bottom portions of the canes ripen best and are therefore also most suitable for bench-grafting or for planting out directly.

4. **Treatment of Cuttings to improve their rooting qualities.**

We have previously seen that air, heat, and moisture are necessary for the formation of roots. Our nursery soils will contain sufficient air. As soon as the temperature is high enough, the buds of the cuttings start developing. This usually happens before roots have yet been formed. If roots are not formed in time, the young shoots ("bourgeonnement") will wilt and die for lack of water. This explains why cuttings of Jacquez and V. Berlandieri develop their buds up to a point and then frequently die. In order to promote rooting we should try to hasten the formation of roots and to retard the development of the buds, or at least to oppose their wilting and death until roots have been formed.

In order to promote the formation of roots we can *remove some bark* on the cuttings, which, according to Ravaz,[1]

---

[1] Ravaz, L., "Reoherohes sur le bouturage," etc,

facilitates the absorption of water from the soil and hence promotes the formation of roots, or we can adopt some method or other of *stratification.* By the latter it is here meant that the cuttings are covered with soil. Thus we may tie the cuttings in bundles and put these upside down in holes made in loose soil on a warm site, cover them with a layer of moist moss over which is thrown a layer of moist sand. If this is done on a sunny spot, callus (new growth covering up a wound) is soon formed at the ends of the cuttings. Before the roots have appeared, the cuttings should be planted out. In case roots have already been formed, planting out should be delayed until the roots show a brown layer of cork near the cutting. With careful treatment they can then be planted out without appreciable damage being done to the roots, and such cuttings will soon grow well. When thus treated, the cuttings will form roots before the buds begin to develop.

We can also put the cuttings in a horizontal position in trenches along a slope, putting the bottom ends outward, so that they will be heated more than the top ends, and roots can thus be formed before the buds begin to develop. The cuttings are put in layers a couple of inches thick and loose soil strewn between the layers. They are covered with sufficient soil on top to prevent them from drying out too much.

In order to prevent the drying up of the young shoots, the nursery soil can be irrigated. In all nursery work it is always desirable that the soil should have the best moisture content for the special operation. It should be slightly too moist rather than too dry when the roots are being formed. In beds or small areas in a nursery, temporary shade can help a great deal to retard the development of the buds and oppose wilting of the very young shoots until roots have been formed and sufficient water can be absorbed from the soil. Another well-known and commonly adopted practice is to cover the cuttings (and scions) completely with soil, which retards the development of the buds and gives the roots a better chance of being formed in time.

When a cutting is bench-grafted and then planted out in the nursery, the buds of the scion do not develop as soon as those of the cutting itself would have done. The consequence of this is that the cutting (stock) now gets a better chance of forming roots. It is, indeed, well known that bench-grafted Jacquez cuttings give a much higher percentage of well-knitted grafted vines than plain Jacquez cuttings give rootlings when planted out directly in the nursery.

Viala and Ravaz (quoted from Foëx, *l.c.* p. 297) have shown that cuttings of varieties that strike root with great difficulty, do so much more easily if the canes are cut from the mother-vines when they have already developed young shoots 10-12 cm. or 4 to nearly 5 in. long. The shoots are broken off and the cuttings are then planted out. They then have ample time to form roots before those buds can develop that have not already done so. Viala and Ravaz applied this method with success to Berlandieri cuttings which otherwise root very badly. In the long run such a practice is bound to weaken the mother-vines very considerably.

In Viala and Ravaz,[1] p. 75, there is a footnote by P. Viala, 1900, in which he states that millions of Berlandieri cuttings have been planted out in the Eastern French Pyrenees and have rooted as well as Riparia (60-80 per cent). In order to obtain this favourable result, only fully ripe canes from at least 4- to 5-year-old mother vines should be used for cuttings, and these should be cut *immediately after the leaves have dropped* (italics, A.I.P.) and should be planted out immediately in the nursery, *i.e.* in autumn and not in spring, as is the usual practice. They are therefore planted very early. The nursery must be properly irrigated, and in other ways receive the necessary attention.

This hint of Viala is worth remembering for obtaining rootlings of varieties and species that strike root with great difficulty.

5. **The planting out of the Cuttings.**

Although cuttings can be planted directly in the vineyard (whether they are subsequently to be grafted over or not), particularly in case of varieties that root easily, and especially if two cuttings are planted at the same spot, it is nevertheless

---

[1] P. Viala and L. Ravaz, *American Vines: Adaptation, etc.* Translated by Dubois and Wilkinson, Melbourne, 1901, from *Les Vignes américaines,* Montpellier, 1892.

much better to let them strike root in the nursery first. This applies with even greater force to bench-grafted cuttings. The practice of planting these out directly in the vineyard is fortunately of rare occurrence, and cannot be too strongly deprecated, as it results in an uneven stand which is difficult to rectify afterwards. In the nursery these cuttings are concentrated on a small area in good soil that has been well prepared, hence the formation of roots and the subsequent growth of the rootlings takes place under more favourable conditions than usually obtain in the vineyard, and these conditions can easily be controlled on account of the limited area occupied in the nursery. I therefore strongly recommend that all cuttings (grafted or ungrafted) be rooted in a nursery before being planted out in the vineyard.

As regards the best time for planting out the cuttings, it can, in a general way, be pointed out that this should be done at a time when the nursery soil is warm and dry enough to enable the cuttings to start growing almost at once. At the Cape this is usually the case in September. In any case the cuttings should be planted out before their buds have appreciably developed, and preferably after having been stratified in some manner. For varieties that root very badly, we can adopt the hints given by Viala and Ravaz.

The cuttings are planted in a vertical opening made by means of a spade or in trenches made with a plough and a little slanting. In the rows they are set 3-4 in. apart and so deep that 2-3 eyes (about three inches) show above ground. The soil is pressed firmly against the cuttings, particularly near their base, then they are earthed up till they are almost or just covered with soil. The rows are 30-36 in. apart in order to allow the cultivator to pass between them.

6. The production of Canes: American Mother-plantations.

Canes for scions we obtain from the ordinary vineyards where they are selected, as was previously indicated. Canes of varieties that serve as stocks in grafting are specially grown in mother-plantations of American vines.

Soil for a Mother-plantation.—As it is our object to produce a large number of canes of medium thickness that are well ripe, we must select a plot of good soil for the mother-plantation. The soil should be deep, sufficiently moist, warm, and well drained—*i.e.* not too wet—and in addition fairly fertile. The last-named quality is not absolutely essential, as we can improve the soil's fertility by suitable manuring. Look out for a warm, sunny site where the canes will ripen well without being sunburnt. On the whole it is best not to irrigate a mother-plantation, as we might thereby increase the production of cuttings whilst lowering their quality. Where irrigation, however, is necessary, it can be given without harming the quality of the canes so long as it is judiciously applied. A cool, deep soil that does not require irrigation is to be preferred. Generally speaking, a sandy, or in any case a light, loose soil is much to be preferred to a heavy, clayey soil. In selecting a site for a mother-plantation, the variety or varieties to be grown on it must be borne in mind. It might be situated right in the vineyard where it could later on be grafted over with a vinifera variety when no longer required as a mother-plantation, but it will generally be found most advantageous to select a plot of ground which is specially suitable for a mother-plantation.

The soil must be properly cultivated and freed from weeds before the young shoots have reached much length, as cultivation by machinery is no longer possible once the young shoots cover the ground, at least not without considerable damage to the young growth.

Establishing a Mother-plantation.—After the soil has been well prepared we can proceed with planting. For this we prefer rootlings, although cuttings also can be used. They must in any case be sound, pure, and typical of the variety or varieties to be planted. The greatest care should be taken to keep every variety absolutely free from other varieties. If, subsequently, vines are noticed in a mother-plantation which are not of the correct variety, they *must immediately be uprooted,* as the purity of the cuttings can otherwise not be guaranteed. If the nurseryman himself does not possess a sufficient knowledge of the different varieties of stocks used, he should get an expert to inspect his mother-plantation very carefully, and point out all vines that are not of the variety grown in the particular plot. It always pays best to purchase the rootlings or cuttings for establishing a mother-plantation from some one who can absolutely guarantee their purity and identity.

Plant the vines about 6 × 6 ft. square and leave an opening of at least 12 ft between different varieties; their shoots must not be allowed to intergrow. This is to prevent their canes from being mixed up when they are harvested. Planting is done in the usual manner (see later).

Manuring.—This will largely depend upon the soil. To an average soil we shall annually apply a dressing of four tons well rotted farmyard manure per acre. It is the best fertiliser for the purpose. We can substitute other fertilisers for it, but the chemical composition of the soil must be taken into account. The nitrogen dressing must not be excessive, as it will promote rank growth, with the result that the canes will ripen badly and hence be of poor quality. We have to regulate the nitrogen dressing according to the vigour of the mother-vines. To an average soil we could apply 16-32 lbs. of nitrogen (*i.e.* 80-160 lbs. ammonium sulphate or 100-200 lbs. nitrate of soda), 100-200 lbs. of superphosphate or basic slag (about 16 per cent citric soluble P_2O_5), and 25-50 lbs. of muriate or sulphate of potash (or 150-300 lbs. of karroo sheep manure ash) per acre. It is desirable to replace the fertiliser dressing every second or third year by the dressing of farmyard manure, as green manuring is in this case usually out of the question. The reason for this is that the green manure crop (say peas) is usually not ready to be ploughed under when the young shoots begin to grow, and these will be much damaged if ploughing under takes place later on, especially where the shoots are allowed to grow on the ground. Where green manuring can be applied successfully, it should be, and then the mother-vines should be planted in rectangles, say 5 × 8 ft.

Cultivation.—This consists of ploughing in the fertiliser dressing in winter after the canes have been harvested, digging the strips of soil that remain in the rows of vines, freeing the soil from weeds and keeping it loose on the surface by suitable cultivation until the young shoots begin takeover the ground so much that further cultivation will damage them. Where the growing shoots are trained on wires or poles (see Fig. 63), cultivation can and should be practised in the free rows during summer to keep down weeds and to preserve a soil mulch, especially if irrigation is not practised. The Planet Jr. (see Fig. 89) is very suitable for this purpose.

Treatment of Mother-vines.—The first point to be considered here is the *pruning* of the mother-vines. This is very simple, as the canes are cut close to the stump, so-called, to one eye. The stump therefore does not develop arms. At the Cape, *training* the shoots is unnecessary and not practised. In the colder wine countries the shoots are sometimes trained on vertical poles in order to promote their growth and better ripening (see Fig. 63).

FIG. 63.—High training of Riparia gloire [near Geisenheim a Rh.]. Original.

Sometimes a pole is planted between four mother-vines, from each of which a wire runs to the top of the pole (up to 20 ft. high). In this case the growing shoots of each vine are attached to the wire running from it to the top of the pole. Where the shoots can grow and ripen well on the ground they should not be specially trained, as it is more economic and better not to. Shoots growing upwards tend to form very long internodes, which is not desirable in cuttings for any purpose.

When the young shoots have reached a length of about one inch, all those that are not wanted should be removed. Every vine can keep 6-20 and even more shoots, according to its age and vigour. It is a mistake to leave too many shoots, as they will then remain too weak and produce inferior cuttings for grafting purposes. To leave too few is also bad, as the canes will then become too thick. Experience will teach one how many shoots one's particular conditions require.

Where shoots are inclined to form many laterals, these latter should soon be topped to one leaf and should be completely removed a little later, in order to prevent the vine from wasting its energy on a number of worthless laterals at the expense of the good shoots. As soon as the shoots commence to turn brown near the base, they should be topped in order to put a stop to any further growth in length (which will be worthless) and to promote better ripening. Too early topping, when the shoots are still growing vigorously, is a mistake, as it will promote the growth of laterals. At the Cape little or no attention is paid to the points raised in this paragraph. It is explained by our favourable climate and by the indifference of our nurserymen and wine-farmers about the quality of the cuttings they use. It is, however, very desirable that first-class cuttings should be produced in order to be able to produce first-class grafted vines.

Harvesting of the Canes.—Once the leaves have been shed the canes are usually harvested, especially where large quantities have to be dealt with. They are immediately cut up in lengths of 10-14 in. Sometimes they are cut in lengths of 40 in., which are later on cut into three equal parts. *Cuttings for grafting* are cut just below the bottom node, and at the upper end just below the node which will form the bottom part of the next cutting. In this way no material is lost. *Cuttings for planting out* are cut just below the bottom node and just above the top node, as such a cutting must be terminated at each end by a node. So here we lose nearly a whole internode for every cutting that is obtained.

The cuttings thus obtained are tied with raffia in bundles of 100, care being taken to have all their bottom ends together. They are then put in the ground and covered up as was previously explained. The soil should be loose and moist but not too wet.

When cutting the canes only perfectly ripe and sound material is made use of, and only such cuttings as have a diameter of 6-10 mm. or $1/4$-$2/5$ in. are selected for grafting purposes: The thicker parts of the canes are discarded. Cuttings of $^1/_6$ - $^1/_4$ in. diameter may be planted out in the nursery to be grafted the next year, while the thinner ones are discarded.

B. PROPAGATION BY GRAFTING

The theory and influence of grafting will be discussed in a special chapter. Here the practical side of grafting will be dealt with. We are here concerned with the production of grafted vines, which will include re-grafting of a grafted vine with a different variety or with a better scion of the same variety.

Where conditions are suitable, it will pay the farmer best to produce his own grafted vines, as it then lies within his power to use only first-class material. Should local conditions, however, be unfavourable, he will be better advised to buy his grafted vines from a reliable nurseryman, whom he might possibly supply with the necessary cuttings for scions. In this case it is best to place the order a year ahead, as the farmer will then know that he will have the necessary variety on the desired stock when he wants to start planting. Whoever produces grafted vines should have his own mother-plantation of American vines in order to be able to guarantee his stocks. He should also preferably grow the varieties himself which he wants to use as scions, unless he can select material in his neighbourhood, in order to make sure that only good and unmixed material is used for scions.

1. **The Nursery Soil and its treatment.**

The soil for the nursery must be chosen very carefully, as the nurseryman's success will to a large extent depend upon it. Look out for a piece of land which is not wet, but still is naturally moist or can easily be irrigated. Unless the soil is deep, it should be well drained. Further, it should be loose, *e.g.* a sandy loam or a sandy soil, but under no circumstances a clay soil. A dark (rich in humus) sandy loam is very suitable. It will be fairly fertile and well aerated, will retain moisture well, will not be cold, and will thus promote the growth of the grafted vines, and is easily kept in good tilth.

As it will nearly always be necessary to irrigate the nursery, the land should be properly levelled when the soil is being prepared for a nursery, if it has not yet been levelled. Where necessary, it should, at this stage, be efficiently drained, as a wet nursery produces bad and sick vines, which are particularly subject to eel worms, which cause large swellings on the roots, and to wound bacteria which cause crown-gall at the joint. Hence the nursery should be irrigated very carefully, and the soil should not become too wet, especially if it is not very deep. In this connection proper provision must be made for fresh water. Brackish water is quite unsuitable for this purpose, and should therefore on no account be used.

Rotation of crops is necessary in a nursery and helps *to keep the soil healthy*. I suggest that the whole nursery block be divided into three plots: A, B, C, of equal size. When A is planted with grafted vines, B has a green manure crop *(e.g.* field peas) which is ploughed under, and C is put under a crop like potatoes, sweet potatoes, turnips, carrots, etc. The next year the grafted vines are grown on B, which has been kept a bare fallow with its surface soil loose and free from weeds after the green manure crop had been ploughed under, C gets a green manure crop, whilst A is put under the vegetable crop. The process is continued in this way.

The plot on which the grafted cuttings are to be planted out can be *manured* as was previously recommended for the mother-plantation, especially if it has been given a green manuring the year before, in which case the nitrogen dressing can usually be reduced. The soil must in any case be well supplied with nitrogen if we wish to produce fairly large and strong-grafted vines. If a green manure crop has not been ploughed under during the previous year, the fertiliser dressing can be almost doubled, especially where the nursery will repeatedly be irrigated. Well-rotted farmyard manure is also in this case the best fertilising material to apply, especially on sandy soil, where green manuring is also particularly needed.

Trench the soil by hand to a depth of about 24 in. (the nature of the soil will decide to what depth it ought to be trenched), but be careful under no circumstances to bring a raw subsoil to the surface. Keep it down and loosen the soil in the bottom furrow. Spread the farmyard manure or the fertilisers over the soil and have it ploughed or dug in. The soil is trenched before the winter rains are over (at the Cape in July); the fertilising is done after the soil has had a good deal of rain and has therefore settled sufficiently; the grafted cuttings are planted out in it about a month later, when the cold weather and most of the rains are over and the soil is consequently warm enough to enable the grafted cuttings to commence growing immediately.

I once more wish to emphasise the fact that the land for the grafted vines must be kept free from all crops and weeds after the green manure crop has been ploughed under. Where no green manuring is given, the land should be kept a bare fallow and should repeatedly be cultivated to keep it loose and free from all weeds during the previous summer. This will effectively get rid of the trouble of *cut-worms,* which may easily become very troublesome and cause great damage in the nursery. If the land is covered with vegetation, the moths lay the eggs, out of which the cut-worms subsequently develop, on such vegetation, and these eggs are introduced into the soil with the vegetation, where they afterwards hatch and attack the developing buds of the scions of the grafted cuttings. Hence the whole trouble is avoided by keeping the land quite bare during the previous summer.

2. **The scions and the American Cuttings.**

Under A 2 of this chapter enough has been said about the selection and storing of the scions and the American cuttings. If canes for scions are so stored that their ends show above ground, their buds will frequently have begun to develop when the scions are required. For scions only those parts of such canes should be used of which the buds are still perfectly dormant, *i.e.* have not yet begun to swell. For successful grafting the development of the stock must be in advance of that of the scion. Otherwise the buds of the scion will generally develop too far before the joint has been callused and the cutting has formed proper roots for feeding the young shoots, when they perish for lack of moisture and nourishment. The cuttings for scions and stocks are taken from the soil as they are required for grafting. In order to free them from soil particles they are steeped in water. Before being used they should be more or less air dry. The cut scions can be prevented from drying out too much by keeping them in a bucket with moist moss and covering them with a moist cloth, if need be without the moss. This is particularly necessary when standing vines are grafted in the vineyard, as this is done in the sun when the cold weather and the rains are over. Failures with grafting can very soon result from scions partially dried out.

3. **The necessary instruments.**

These comprise a *pruning saw,* a *secateur,* a *pruning bill,* and a *grafting knife.* The saw is used for decapitating old vines that must be grafted over. The pruning bill is used for splitting a thick stem or for giving it a slanting cut in order to execute the graft shown in Fig. 66. The secateur is used for cutting the scions and making the cuttings (stocks), and for decapitating medium thick vines when grafting them on the spot. The best secateur is that of Rieser of Neuchatel, Switzerland. It is and remains sharp with proper treatment, and is almost indestructible. Although it is the most expensive secateur known to me, it is the cheapest in the long run and does the best work of all. I therefore recommend it most strongly in preference to every other make. A good grafting knife should be made of good steel and must be sharp. Near the end its back should be thin without being sharp enough to make a cut. It is used in inserting a bud on fruit trees. Some grafting knives have a flat piece of ivory at the bottom end of the handle, which is used for the same purpose.

Take good care of your grafting instruments by cleaning them from time to time; keep them *sharp*; use only instruments made of good steel; rub them clean and dry after use, in order to prevent rust. A little oil or vaseline will be useful here.

4. **Binding material and grafting wax.**

The most commonly used binding material is raffia, which is obtained from the Raphia palm *(Sagus Raphia),* which is indigenous to Madagascar. By first moistening the raffia it becomes tougher and better for binding. Strips that are too broad or thick can be torn in the middle.

Grafting wax is seldom used in grafting vines, and is therefore not so important here as for fruit trees. It is used only when an old vine is to be grafted so high above the ground that the joint cannot be covered with earth in the usual way. In this case the wound at the top of the scion is also covered with grafting wax, and it should have two eyes. According to Foëx[1] this method of grafting is not very successful. Last year I applied it to nine-year-old vines with disappointing results. The scions all developed their buds to a point, then stopped, and finally died in practically every instance. The grafting wax evidently hindered the formation of callus too much. The nature of the grafting wax will no doubt influence the result.

Some kinds of grafting wax are applied cold and others arm. The cold grafting wax Lhomme-Lefort gives good results, according to Foëx, *l.c.* p. 347. In his well-known treatise, *L'Art de Greffer,* Ch. Baltet, p. 31, gives the following recipe for a grafting wax which closely corresponds with the grafting wax Lhomme-Lefort:

Black pitch	2.5 kg.
Rosin	2.5 "
Finely ground white Spanish earth	1.2 "
Methylated spirits	0.9 "
Turpentine	0.6 "
Yellow wax (beeswax)	0.075-0.1 kg.

Note.—The same proportions are maintained by using respectively 25, 25, 12, 9, 6, ¾-1 ozs. of the above-mentioned ingredients in the order given above.

First melt the pitch in an open pot over a small fire whilst stirring with a stick. Now pour in the rosin and continue stirring I then add the beeswax. After everything is molten and well stirred, the pot is taken off the fire and placed a good distance away. Now one person slowly pours in the methylated spirits and turpentine from two bottles, held one in each hand, whilst a second person keeps stirring the mixture. After everything has been well mixed, the Spanish earth is added little by little while the mixture is being stirred, until everything is well mixed. Now allow it to cool,

---

[1] *l.c.* pp. 319, 347.

and apply when cold. Baltet is of opinion that Lhomme-Lefort is still the best grafting wax obtainable.

Hedrick[1] quotes Bioletti (Cal. Exp. Sta., Bull. 180, 108-112) as follows: "A good wax for this purpose [namely, to wax string for grafting. A.I.P.] is made by melting together one part of tallow, two parts of beeswax, and three parts of rosin".

5. Systems of Grafting.

Baltet[2] divided the different systems of grafting into the following three groups:

I. Grafting by approach or Siamese grafting (after the Siamese twins).

II. Grafting by means of a detached branch or scion.

III. Grafting by means of a detached bud with a piece of bark, *i.e.* budding.

This classification will be adopted by me.

Group I. Grafting by approach is the oldest way of grafting and is Nature's only way of making a graft. This consists in two branches that press against each other, growing together whilst each remains attached to its own root system. This is of no importance in grafting vines, and will therefore not be discussed at greater length.

Group II. Grafting by means of a scion. —This is the group which mainly interests us here. We can distinguish between cleft, crown, and tongue grafting.

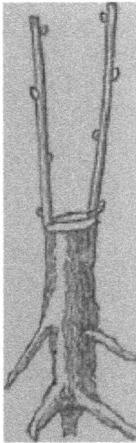

FIG. 64.—Cleft-grafting with two scions. From Foex Cours complet de viticulture, 1895. Coulet et fils. Montpellier. G Masson, Paris

Cleft - grafting is applied mainly to grafting on the spot, and then usually to thick stems. The cleft runs right through the stem if two scions are to be inserted, as is clearly seen on Fig. 64. When only one scion is to be inserted in the cleft, it does not run across the

[1] Hedrick, U. P., *Manual of American Grape-growing,* p. 54.

[2] Baltet, Ch., *L'Art de greffer,* 8th ed., 1907; Masson et Cie, Paris.

whole stem. Cleft-grafting; can also be applied where scion and stock have the same diameter, both with bench-grafting and with grafting on the spot. It is, however, not as good as tongue - grafting, which gives a better joint and not such a swelling just above the joint. It can also be applied to green grafting.

Crown-grafting is not applied to vines, as far as I know, but it is applied to fruit trees. The vine is decapitated and the scion, after being given a slanting cut on the one side, is inserted between the bark and the wood with the cut surface to the inside, thus without a cleft in the stem. After being tied with raffia, the wound is covered with grafting wax.

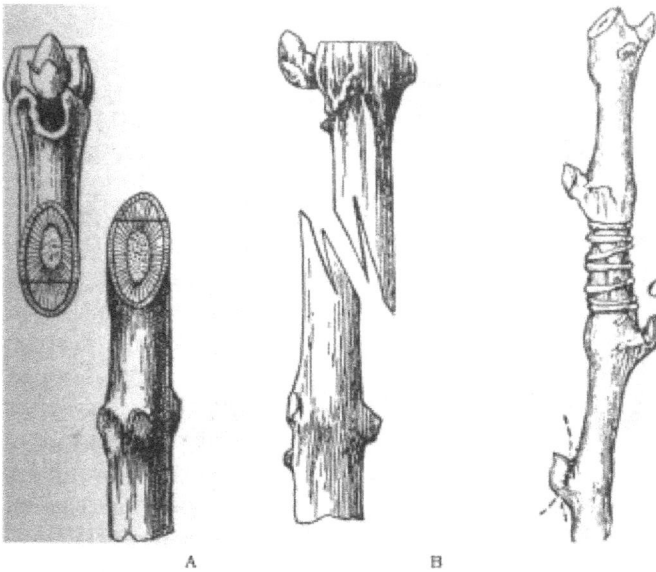

FIG. 65.—A. Tongue- or whip-grafting. From Babo u. Mach, *Handbuch des Weinbaues*, 1923. Paul Parey, Berlin. B. Tongue- or whip-grafting after Trapet From Baltet, *L'Art de greffer*, 1907. Masson et Cie., Paris.

Tongue-grafting or *whip-grafting* is the graft most generally adopted with vines, especially in bench-grafting, although it can be and is also frequently applied in case of grafting on the spot. Both scion and stock are given an equally slanting cut, and on each cut surface a cut is made as is shown in Fig. 65A. In withdrawing the knife it is pressed slightly outwards to open the tongue just formed.

Now insert the scion into the stock by slipping its tongue behind that of the latter, as is seen in Fig. 65B.

Although formerly not much attention was paid to this point, Louis Trapet, a wine-farmer of Burgundy, found that he got his best results in grafting when the slanting cut on the scion begins just slightly below the bottom eye of the scion, and that of the stock just above the uppermost eye of the stock, as is clearly shown in Fig. 65. At the same time he cuts out all the eyes on the stock excepting the top one, which is kept as a "point d'appel". The shoot developing from the latter is topped at an early stage at a length of 4 in., and is completely removed as soon as the scion is growing well, so that the stock can now no longer develop wild shoots. If the scion were to die, the wild shoot is left, and the rootling can again be grafted the following year. Trapet's improved method of grafting is now generally applied by everybody who knows about it.

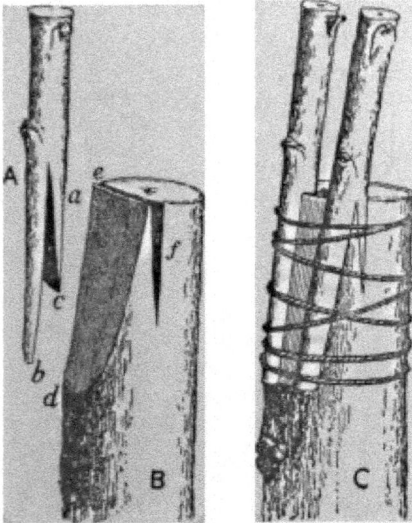

Fig. 66.—Tongue- or whip-grafting with two scions on a thick stock. From Baltet. From L'Art de greffer, 1907. Masson et Cie., Paris

In bench - grafting this method of tongue-grafting is usually applied. In grafting on the spot it is also applied, and is to be preferred for stocks with thin stems. With stocks with thick stems it can be used. After the vine has been decapitated, we make a slanting cut on the stem with a pruning bill and a more or less vertical cut to obtain a tongue on the stock (see Fig. 66B). Now make a tongue on the scion (see Fig. 66A) and insert it in the stock as on Fig. 66C. If the cut on the stock has been made sufficiently broad, two scions are inserted, as in Fig. 66C. Care must simply be taken that the bark of stock and scion correspond on one side. Although Fig. 66 is not an illustration of grafting a vine, it can very well be applied to the vine, where it usually gives better results than cleft-grafting.

Group III. Grafting by means of a Bud or Budding.—Here we can use a bud with a small piece of bark attached to it or with the whole bark cylinder (flute-budding) about an inch long and with the bud in the middle. With vines, budding is used only in green-grafting, which will now be briefly discussed.

Green-Grafting.—In Hungary this method of grafting has in the past been used to a considerable extent. Prof. Horvath especially has applied it there with great success. I It can be carried out as soon as the eye is fully developed and the wood and diaphragm have a whitish appearance on a cross-section, and while the pith is still green and full of water and the bark is easily detached. Under these conditions the eye develops a shoot during the same summer, and can bear fruit in the following year. If the operation is carried out later, when the bud is resting, it will not develop until the following spring.

According to Horvath, the best results are obtained by making a horizontal cut half way round the shoot and about half an inch above and half an inch below an eye, through which a vertical cut is made to unite the two horizontal cuts after the leaf has been cut off. The bud plus bark is also an inch long and should exactly cover the opening made. It is inserted below the bark, which is then tied over it with a woollen thread.

The bud is cut beforehand from the basal part of a shoot after the eye is completely developed. It has a length of about one inch with the eye in the middle. It is cut in such a way that there is a layer of cell tissue about $1/16$ in. thick below the bark. It helps to prevent the drying out of the bud. Also, one-third to one-half of the petiole is left with the bud. A number of buds can thus be prepared beforehand, provided they are kept in moist moss and in water. According to Foëx, *l.c.* p. 332, Horvath emphasises the importance of placing the eye of the bud with bark on the spot where the stock had an eye, in order to obtain the best results.

Green grafts can also be made by cleft- and by couple-grafting. I shall not say any more about green-grafting, as it is not of much practical value in the grafting of vines. Green-grafting might render valuable service when an overhead trellis is to be partly grafted over to another variety.

6. Practical execution of grafting and the formation of Callus (stratification).

Bench-grafting and grafting on the spot will be discussed separately. *Bench-grafting,* which is generally a tongue graft, is usually done indoors or at least in a shed. The place should be well lighted. Where large quantities of cuttings are to be grafted, it is desirable to divide up the work. One or more persons should cut the scions just above the upper eye.

FIG. 67.—Grafting room of Mr. F. Richter, Montpellier, with work in full swing on 29th March 1909. Original.

Each scion has one or two eyes (most frequently only one in bench-grafting). The scions are cut into baskets and then placed on the tables in front of the grafters. The same is done with the American cuttings after all their eyes excepting the top one have been cut out. As has already been pointed out, the canes for scions and the American cuttings are taken out of the ground as they are required. They are first washed in water to remove earth and sand, which would otherwise make the grafting knives blunt. They can be used when more or less air dry.

The grafters sit at the grafting tables and merely graft. Fig. 67 shows us the grafting-room of Mr. F. Richter in Montpellier, South of France, with the work in full swing as I saw it during my visit on the 29th March 1909. Mr. Richter then annually grafted about seven million cuttings and employed over sixty women as grafters, who on an average made 1500 grafts per day. I quote the following from my diary: "The American canes from the mother-plantation are cut in lengths of one metre (about 40 in.) and tied in bundles.

FIG. 68.—Bench-grafting without tying, at Mr. Richter's in Montpellier. On the left the late Mr. J. Leenhardt-Pommier. Original.

Then they are covered with earth and kept thus till required for grafting, when they are taken out, washed, and cut into three parts, always just below a node. The cuttings of proper length are then removed to the grafting-room. Here the same grafting knives are used as we use at the Cape. The slanting section is made very short—more or less at an angle of 45° with the axis of the cutting; the cut for the tongue is made pretty deep, begins in the wood behind the pith, and continues steeply forwards" (as is well shown on Fig. 65A).

This *short* cut gives better knitted joints than a long, more slanting cut. Richter's grafters have in front of them wooden trays with five partitions, each to take twenty grafts. This is well seen on Fig. 68. Into each partition twenty grafts are put. The tray with its 100 grafts, *which here are not tied,* is taken to the box into which they are packed as shown in Fig. 69. In the tray the grafts lie with their scions against the back board nearest the packer, hence they come on the open side of the box when packed into it.

Fig. 69.—Packing grafts into box for callusing, at Mr. Richter's, Montpellier. Original.

The box is lined with sea-grass or moss on its bottom and sides, and stands on one of its short sides when being filled with grafts. When full, it is put on its bottom and a layer of moss or sea-grass is put over the scions of the grafts which face the open side. Richter uses sea-grass. Now the box is dipped in water, or water is poured over it to wet its contents thoroughly. The joints are open, and in the bottom extra holes are bored if necessary, in order to allow the water to drain off. Fig. 70 shows us such a box filled with grafts and ready to go into the *callusing room.*

With this method of procedure it is not necessary to tie the joints of the grafts, although this is frequently done. Grafts that are immediately to be planted out in the nursery or are stratified in the ground, should be tied with raffia. The *callus* is a jelly-like, later on fairly firm mass of fresh wound tissue that grows out of the cambium ring in order to cover the wounds at the joints and at the top and bottom of the graft, and thus also unites scion and stock.

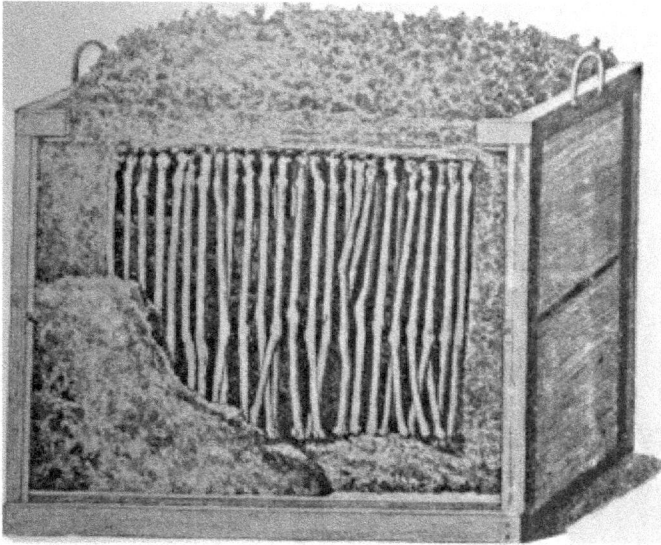

Fig. 70.—Box with moss and grafts ready to go into the callusing room. From Babo u. Mach, *Handbuch des Weinbaues*, 1923. Paul Parey, Berlin.

Under favourable conditions it is formed first at the lower end of the stock and on the scion at the joint. Here the callus is formed a little later on the stock, as callus forms most readily at the bottom wound of a cutting.

The callusing room is kept at a temperature of 25-30° C. or 77-86° F. It should be well ventilated to allow fresh air to come in from time to time. The air inside the room must be neither too dry nor too moist. If it becomes too dry, water must be sprayed on the floor. If it is too moist, moulds start growing. After having been in the callusing room for about twenty days the joint between scion and stock will have been well cemented by callus. The grafts are then put in a room at 15-16° C. or 59-60.8° F. for a few days in order to be hardened before being planted out in the nursery. Richter keeps his callusing room at about 25° C. or 77° F., and after about twenty days the grafts are planted out *directly* in the nursery, when good callus has been formed. Some grafts now already have some roots, and the buds of the scions will also have begun to develop. This is seen on

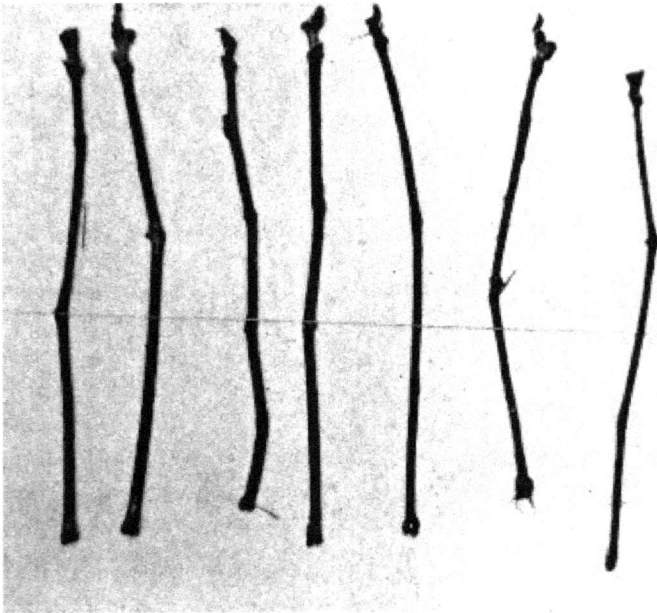

Fig. 71.—Grafts coming out of the callusing room and ready to be planted out, at Mr. Richter's, Montpellier. Original.

Fig. 71. Richter tested the strength of such a callused joint by tying a weight of 5 kg. or 11 lbs. to the stock and suspending the whole by twine tied to the scion. Thus the joint stood a tensile force of 11 lbs. after coming out of the callusing room. Richter plants out his grafts from the callusing room directly in moist nursery soil, leaving the callused joints exposed to the sun. He thus prevents the formation of adventitious roots on the scion. Although he obtains good results (about 30 per cent good, grafted vines) with this method, it seems to me rather dangerous, particularly if dry and warm weather follows immediately after the grafts have been planted out in the nursery.

Callusing in the ground can be done very satisfactorily in warm countries. The grafts are tied with raffia and stored in moist sand in a room or in deep, well-drained soil in the open. Make a trench about 12 in. deep and about 3 in. wider than the length of the grafts. Put a layer of grafts 1-2 in. thick in the trench, scatter a thin layer of loose sandy soil over these (the soil should be dry enough to sift in between the grafts), pack another layer of grafts, scatter soil and continue until the pile is 6-12 in. higher than the surrounding ground. Now cover well with soil and make a shallow ditch around the pile to carry off rain water. If the soil is sufficiently moist it is best to protect the callusing bed against rain. Otherwise it should be exposed to the sun's rays to get the desirable temperature (about 75° F.). By using some cover or other *(e.g.* a waterproof wagon-cover) the heating can be regulated fairly well. In about a month's time (the length of time depends upon local conditions) the grafts will be ready for planting out in the nursery. It is a greater mistake to plant them out too soon than a little too late. The development of the buds of the scions and root-formation on the stocks are apt to outstrip the formation of callus. This is improved by keeping the moisture content of the callusing bed somewhat low, as callusing requires less moisture than budding and root-formation. The grafts should be properly callused when planted out.

Bench-grafting is done in the latter half of winter; at the Cape usually in July and August. The grafts take about a month for callusing. They are planted out about September (at the Cape). If the nursery soil is then not too wet and the days are no longer cold, the callused grafts can start growing immediately and success is assured.

Grafting on the spot is applied where an existing vineyard is grafted over to another Variety, or where American vines have been planted out in the vineyard to be grafted on the spot, or where American cuttings have been planted out in the nursery to root and to be grafted there on the spot in the following year. The last-named method is good under such favourable climatic conditions as are experienced at the Cape, and usually gives strong, well-knitted grafted vines and a *high* percentage of successful grafts. In this case tongue-grafting is best, but here, too, the slanting cut should be short, and the stock should be carefully cut for the tongue— not too deep. The grafts can be tied with raffia if this is carefully done. Otherwise moist soil is carefully pressed against the joint without shifting the scion. The grafts are now covered with loose earth until the heads of the scions are *just covered,* but they must under no circumstances be covered too deep. Where the soil is not loose but somewhat clayey and apt to cake, the scions should be just covered with a layer of sand.

Where vines with thick stems are grafted over, cleft- or tongue-grafting is applied, as has been previously stated. Such vines are decapitated slightly above the surface of the ground or a little lower, and in any case at a height where the stem is smooth and not knotty, after having been opened up somewhat with a spade. If the vines have a copious flow of sap when decapitated, it is best to *delay the actual grafting for a day or two* until the flowing of sap has more or less stopped. When cleft-grafting thick stems it is usually necessary to insert a wooden wedge of some description in the middle of the cleft to keep it sufficiently open for the scions. The latter are given a slanting cut through the pith only on the one side. On the other side they are cut to expose the cambium but *without touching the pith.* With cleft-grafting this is the best way of cutting the scions. If both cuts penetrate the pith the wedge-shaped end of the scion is easily damaged when it is inserted in the cleft, owing to the softness of the scion's pith. Take care that the bark of the scion on its outside corresponds with that of the stock. After the scions have been inserted, raffia is sometimes tied around the joint, and moist clayey soil is firmly pressed around the stem and scions without disturbing the latter. Then they are covered with soil until the tops of the scions are just covered. Make the mounds of soil at least 15 in. wide and not higher than just stated. If the soil is clayey or caking, cover the tops of the scions with sand to prevent a hard crust from forming over the scion. If this happens, the young shoots of the scion frequently cannot penetrate the crust, and first grow in a whorl before being able to burst the crust; this is most undesirable. Also, under such conditions, the graft gets too little air to develop properly.

Instead of cleft-grafting we can apply tongue-grafting here, as has been pointed out already. The scion and stock are cut in the usual manner for tongue-grafting. The best time for grafting vines on the spot is when the buds of the stock are near bursting. When the buds have actually burst it will be a little too late, as the vines then bleed very heavily when decapitated, and the formation of callus is hindered.

The temperature of the soil has a great influence on the success obtained with grafting on the spot. If the soil has by then dried off sufficiently and is fairly warm, success is generally assured. Where the soil is too cold and wet, grafting should be delayed till the soil is drier and warmer, otherwise the results will be disappointing. Also, it is of the utmost importance that *bright, warm weather* should prevail during and shortly after grafting, as callus is then rapidly formed. That is why we delay grafting on the spot as much as possible.

Keep the scions in a bucket with moist moss or a moist cloth to prevent them from drying out too much, press moist earth against the grafts as soon as they have been made, and cover with soil as previously indicated. If the canes for scions look somewhat dry, they should first be soaked in fresh water for twenty-four hours before the scions are cut. Such canes should be kept in a cool, shaded place to delay the development of the buds.

7. **Planting out of Grafts in the Nursery.**

One of the most important points in this connection is the time of planting. It is impossible to fix any date as the *best time* for this work, as it is fixed by the ruling conditions of climate. I strongly advise not planting out grafts directly in the nursery prior to callusing, as this leads to bad results. Where stocks and scions have been kept dormant till very late and are grafted when fairly warm weather has set in, the grafts planted out directly may succeed fairly well. On the whole, however, they should first be callused.

Stretch a garden-line where the first row of grafts is to be planted and make a trench with spades (or plough) which will be deep enough to have the joints of the grafts just on the surface of the ground. As it is most desirable to have the tops of the grafts all at the same height, most nurserymen grafting large numbers of cuttings prefer to use scions with one eye only, as in this case it is possible to have all scions very nearly of the same length, and hence their heads at the same height when the joints of the grafts are all at the same height. Where scions with two eyes are used, the unequal lengths of the different internodes make this impossible, hence the preference for scions with one eye. Where cut-worms are not to be feared, there is no need for scions to have more than one eye.

The grafts are planted 2-3 in. apart and the rows 30-36 in. apart to allow for proper cultivation, etc. As the grafts are put in, the trench is half-filled with soil, which is trodden firmly against the grafts to give them proper contact with the soil and so as not to have too much air around them. Now some well-rotted farmyard manure is sometimes spread in the furrow. This is unnecessary if the nursery soil has already been well fertilised. The trench is now completely filled with soil. The line is now put in position for the next row, and the planted grafts are covered with a ridge of loose soil so as just to cover the tops of the scions. By putting the grafts 2-3 in. apart and the rows 30-36 in. wide, we can plant 58,080-104,544 grafts per acre. If the soil is clayey or of a caking nature, the tops of the scions should be covered with sand.

8. Treatment of Grafted Vines during Summer.

As soon as the grafts have been planted, the ground between the rows is dug over with a spade without disturbing the ridges around the grafts. In this way the soil is loosened and air introduced, which favours the formation of roots on the stocks. From now onwards the nursery soil is kept free from weeds and loose to minimise the loss of water by evaporation. The nursery is irrigated as often as may be necessary, and after every irrigation follows a surface cultivation by means of a cultivator or by hand. Although it is very important to keep the soil sufficiently moist during the first month, it must be pointed out that irrigation cools the soil and thus retards growth. Hence *irrigation should not be practised when unnecessary.* Nurseries that have been repeatedly irrigated in summer may give big vines, but their wood will be soft and spongy, and their roots will develop in such a way that they will give unsatisfactory results when they are planted out later on in moderately dry vineyard soil.

Keep the soil between the rows loose by frequent cultivation, and do the same with the ridges. Keep these also loose and free from weeds. When the shoots of the scion are about 6 in. long, the soil is removed to just below the joints of the grafts and the top eye of the stock, which had in the first instance been left, is cut away. At the same time any roots that might have been formed on the scion are carefully cut off. The grafts are then again covered with soil to a little above the joints. After that the roots that develop on the scions are removed twice more during summer. When this is done for the last time towards the end of summer (in March at the Cape), the joints are left exposed to the air. By then the greatest heat is over. The removal of these roots of the scion is very important. If this is neglected the scion will develop stronger roots than the stock, and the result will be a poor grafted vine with a thick scion-portion and a thin stock-portion. This applies to all cases of grafting where the scion comes in contact with the soil.

In addition to the foregoing, the leaves and shoots of the growing scions should be kept free from disease in order to be able to obtain first-class grafted vines.

9. Digging, Sorting, Storing, and Transportation of Grafted Vines.

The grafted vines are dug as soon as their leaves have fallen. In large nurseries they are sometimes taken out with a plough. They are then immediately sorted. All those that are badly cemented, *i.e.* whose joints have not been properly overgrown with callus, are put aside. Take the vine in your hands and bend it backwards and forwards at the joint to test the quality of the latter. The well-cemented ones are divided into two classes, namely, those that have grown well both above ground and in the roots, and those that have made insufficient growth. Only the former class constitutes the first-class grafted vines. The latter class and those that have not been well cemented can again be planted out in the nursery, when a considerable percentage of them will come up to the first-class standard after another year's growth in the nursery.

Plant only first-class grafted vines. The badly cemented vine usually breaks at the joint after having grown in the vineyard for a few years, or it becomes diseased and dies sooner or later. This causes vacancies in the vineyard which are difficult to fill. Again, only *healthy* vines should be planted. Here I think particularly of phylloxera, eelworms, and crown-gall. In a phylloxera region we need not worry about phylloxera. But if grafted vines are to be sent to a district free from phylloxera out of a phylloxera-infested district, they should be disinfected before being despatched. This is easily done, as we shall see later on. The other diseases are much more troublesome. Eelworms and crown-gall occur most frequently, if not exclusively, in rather damp nurseries. By applying the system of rotation which I have recommended for the nursery and by thorough drainage, the nursery can be kept fairly free from these troubles.

Once grafted vines are infected with these diseases, the infected parts should be cut away as far as possible and the wounds should be disinfected (see later under Vine Diseases), while the parts cut away should be burned there and then. If such grafted vines, treated as above, are planted on fairly dry soil, they will usually be cured after a few years. In moist soils they should not be planted.

As soon as they are sorted, the grafted vines are heeled-in in a cool, moist place which will not become too wet, and well covered with soil. At this stage they still have their roots and canes *unpruned.* This is not done until they are to be planted out. Take care to heel-in the vines in such a way that loose soil gets in between them and that no appreciable air spaces are left. Hence press down the soil firmly on the vines when the trenches are half-filled. Large air spaces will favour drying out of the vines and the development of moulds on them. If the vines are to be planted out late, the canes must be half-covered with soil to keep the bottom eyes dormant till planting time.

When grafted vines are to be transported for some distance, they must be packed in the way previously described for cuttings in this chapter. If transportation lasts a week, they can be packed in cases with well-moistened leaf-mould.

Vines that have grown very much can be reduced in size by cutting back their roots and canes to about 8 in. Sometimes drafted vines are packed with moist leaf-mould in hessian transportation lasts under a week and takes place in cold weather.   On arrival they must immediately be taken out and heeled-in in moist soil.   Vines that look rather dry on arrival should be kept in fresh water for about twelve hours prior to being heeled-in.

CHAPTER VII

THE THEORY OF GRAFTING

HERE we shall briefly discuss the union of the scion or bud with the stock, the influence of grafting on the life of the grafted vine and the formation of new varieties through grafting (the so-called graft-hybrids).

A. THE UNION OF THE SCION OR BUD WITH THE STOCK

Except in green grafting, where other tissues besides cambium can take part in cementing scion (bud) and stock together, it is exclusively in the cambium-ring of the ripe canes that new cells are formed. This takes place on the wounds made by the cuts on both scion (bud) and stock, when bench-grafting as well as when grafting established vines on the spot. This young meristematic tissue is composed of parenchymatous cells rich in protoplasm, which grow over the wound and become visible from the outside. It forms the preliminary union between stock and scion, and is known as *callus* or wound-tissue. This formation of tissue takes place, as we have already seen, in presence of enough heat, moisture, air, and the necessary foodstuffs in the uniting parts.

In this new tissue a differentiation of tissues gradually takes place until it consists of cortex with a cork layer, meristem (cambium), parenchyma and xylem, which unite the corresponding tissues of stock and scion. According to Professor L. Daniel,[1] one notices on thin sections of the joints ("bourrelet") when well cemented:

1. "That they present a considerable complexity, with tissues very irregularly intergrown, so that the normal symmetry of the organ no longer exists at this level.

---

[1] Daniel, Lucien, *La Théorie des capacites fonctionnelles et ses consequences en agriculture,* Rennes, 1902, pp. 165-166.

2. "That the trachea (xylem vessels) are less numerous, of irregular shape, of variable and generally smaller diameter.

3. "That where normally conducting tissues ought to occur, one sees parenchyma as little islands or more or less irregular bridges, which also ensure the communications between the stock and the scion."

This parenchyma allows the conduction of plant food only by osmosis, which is a different and slower process than when it takes place by capillary conduction through the long vessels.

Sometimes scion and stock are everywhere well cemented, and then we get a good joint. Frequently, however, we find that the joint is properly cemented on the one side only, or that there are at least one or more places where the two have not been perfectly cemented. This naturally hinders the transportation of food still more than where a perfect union has been established, is frequently an entrance for organisms causing disease, and also constitutes a spot of mechanical weakness where such a vine may snap later on. The joint remains a spot of greater or lesser hindrance to the transportation of food in both directions, and therefore is a disadvantage to the grafted vine, although this is inevitable.

At the joint we also find, side by side, cells that originally belonged to the scion and stock respectively, and that have grown on to each other by grafting. According to Daniel[1] it is thinkable that two or more such cells can unite to form together a mixed bud, which will contain parts of both plants. In this way, as a consequence of grafting, *asexual or graft-hybrids* could be formed. Under C in this chapter we shall see that this has actually occurred, though it happens only very rarely.

B. THE INFLUENCE OF GRAFTING ON THE LIFE OF THE GRAFTED VINE

I. **General Considerations.**

In order to be in a position to deal properly with the influence of grafting, we shall first have to consider briefly

---

[1] Daniel, Lucien, *La Question phylloxérique, le greffage et la crise viticole,* Bordeaux, 1908-1911 (3rd Fascicule, 1919), p. 35.

the life of the ungrafted or autonomous plant.   Then we shall consider in how far alterations occur with grafted plants, and finally deal with the question of graft-hybrids under C. Here I shall make extensive use of the two excellent works by Prof. Daniel that I have just quoted. They contain a store of most valuable information and are fruitful in suggesting further research on the part of the reader. I would therefore recommend their serious study to all those interested and able to read French.

It is a pity that, in applying the results of his investigations, Daniel went further, in my opinion, than these results warranted him to go. In so doing he raised all those who support grafting as the best means of overcoming the phylloxera evil, in fierce opposition against him, especially the well-known Professors Viala and Ravaz in France. In this struggle the scientific problems were obscured by economic considerations. Daniel, however, rendered our science great services and opened new vistas for which all of us ought to be thankful.

1. **The Life of the Ungrafted or Autonomous Vine.**

Where a plant grows in nature without its development being influenced by any artificial means, it leads its normal life, in which the roots take up the mineral food ("sève brute") from the soil (the leaves can also absorb a little water from moist air and ammonia), whilst the green parts take charge of the carbon-assimilation or photosynthesis and of respiration. The result of this is the formation of the elaborated food ("sève elaborée") which, together with the mineral food, reaches the places in the plant where food is consumed as it is required. Daniel[1] calls the places where food is consumed "points d'appel".

These he divides into two groups:

1. Those serving for the absorption of food from the soil.

2. Those serving transpiration, photosynthesis and the consumption of food, hence those serving to elaborate the absorbed mineral food and to consume the elaborated food.

---

[1] Daniel, L., *La Théorie,* etc., *l.c.* p. 3.

The total value of the first group, *i.e.* the total value of the absorption during time *t*, Daniel calls the "*Capacité fonctionnelle d'absorption*," or the "*Functional capacity of absorption*", and indicates it by Ca.

The total value of the second group, *i.e.* the total value of the consumption during the same time *t*, he calls the "*Capacité fonctionnelle de consummation*" or "*Functional capacity of consumption*", and indicates it by Cv. (He chose *v* because the evaporation of water plays such an important part in the life of the plant.) See Daniel, *l.c.* (*La Théorie*, etc.), p. 8.

During the active period of growth of a plant living normally, absorption and consumption balance for a sufficiently long time that we may say that the plant then lives in a state where $Cv = Ca$. Here, therefore, we find a complete equilibrium in the general nutrition of the plant. By suddenly pruning the plant's roots or by cutting off one or more close to the stump, we reduce Ca and get the condition where $Cv > Ca$, which will be seen, for instance, in the plant suffering from drought or at least growing less vigorously. If we do the reverse, *i.e.* prune its shoots or branches (topping or summer-pruning), Cv is reduced, and hence the plant's condition is now expressed by $Cv < Ca$, and the plant grows more vigorously, or can at least resist drought better. This knowledge will be valuable to us when discussing the summer treatment of the vine.

It is, however, also applied when winter pruning. If we allow the vine to retain all the shoots formed during the previous summer, we shall soon reach a condition, during the early part of the vine's summer growth, which will be expressed by $Cv > Ca$, with the natural consequence that the further development of the individual shoots and bunches suffers. By strongly reducing the number of shoots and by shortening even further those that are retained, the vine will for a long time during its coming summer growth be in the position $Cv < Ca$. Hence its shoots will grow vigorously, especially if, by topping and by breaking out shoots and suckers, we help to maintain the condition $Cv < Ca$ for some considerable time.

By making $Cv > Ca$ or $Cv < Ca$ we can at pleasure make the vine live as if we had put it in a drier or in a damper soil with the further consequences this entails. In damp soil (which is not too wet) we know that the vine will grow more vigorously, will produce softer and poorer quality grapes and shoots, and will be more subject to disease than where the soil contains just enough moisture. In too dry soil the vine will make poor growth and even suffer from drought. Its grapes will more easily suffer from sunburning, their juice and wine will contain more acid than under normal conditions, as was shown by Fonzes Diacon[1] with regard to the French red wines of 1922, which were made in parts where the whole summer and autumn up to the vintage were unusually hot and dry.

In this connection it is interesting to note how much water the vine requires in the soil at the different stages during its period of growth. E. Gain, in *Recherches sur rôle physiologique de l'eau dans la végetation* (Paris, 1895), quoted from Daniel,[2] has found that when 100 expresses the quantity of water which will completely saturate the soil, *i.e.* which the soil can take up, the most favourable moisture content of the soil during the growing period of the plant (lupins, sunflower, etc.) is as follows:

1. Germination, 25-30 per cent.
2. Fixation of the root, 15 per cent.
3. First leaf-development, 40-45 per cent;
   Second „ „   20-25 per cent;
   Flowering, 45 per cent;
   Ripening of fruit, 10 per cent.

We can apply this to the vine, and conclude that the soil should be fairly moist, but not too wet, for the germination of grape-seed and for the formation of roots on plane cuttings planted out, that the soil should be well supplied with moisture during the early growth in summer, particularly during flowering—hence irrigation and short topping at this stage frequently promote better setting and development of the berries; and finally, that the soil should contain much less moisture during the ripening period, and hence irrigation at this stage is frequently very undesirable., Where irrigation is required it should be given before ripening has properly begun.

---

[1] Fonzes Diacon, in *Revue de Viticulture,* vol. lix. p. 231 (1923).
[2] Daniel, L., *La Théorie des capacités . . . l.c.* p. 14.

2. The Life of the Grafted Vine.

Here we distinguish between two possibilities; stock and scion may be of the same variety or they may be different, either different varieties of the same species or varieties of different species. The first combination we find where a variety of grape is grown autonomous or ungrafted but is unproductive owing to bad selection of the canes planted out, and is subsequently grafted over with properly selected scions of the same variety. Here the aerial portion of the vine is served by its own kind of roots, and differs from the autonomous vine only by possessing a joint which has been cemented by new wound-tissue. On account of the identical nature of the stock and scion, we should expect a minimum interference with the flow of sap from and to the roots in this case. This indeed we find. The effect of grafting as such is here seen at its minimum. In viticultural practice we very rarely meet with examples of this kind of grafting.

Where two different varieties or species are grafted on each other, say B on A, and we indicate the functional capacities of A by Cv and Ca, and those of B by $C'v$ and $C'a$, we obtain, after the joint has been properly cemented, a vegetative system which is expressed by the relation $\frac{C'v}{Ca}$, and with which one of two things occurs: either $C'v = Ca$, i.e. the functional capacities of scion and stock are equal, or $C'v \gtrless Ca$,[1] i.e. the functional capacities are unequal, and the functional capacity of consumption of the scion is greater or smaller than the functional capacity of absorption of the stock, with all the consequences of this inequality of the functional capacities (see Daniel, *La Théorie des capacites* . . . *l.c.* pp. 183-184).

The relation expressed by $C'v = Ca$, according to Daniel, never occurs, and theoretical considerations lead us to believe that it probably stands no greater chance of being found than the philosopher's stone. In grafting practice we have therefore always to do with an *unequal* relation between the functional capacities of scion and stock. In addition to this we have the obstruction caused by the joint ("bourrelet") which can support or oppose the unequal relation between

---

[1] 2012 Editor's Note: $\gtrless$ is printed in the original with the *greater than* symbol placed on top of the *less than* symbol between $C'v$ and Ca

the functional capacities. We shall discuss these two cases separately.

1. *Grafted vines with the relation C'v > Ca.*

We generally make use of stocks of which the functional capacity of absorption is greater than the functional capacity of consumption of the scions, *i.e.* $C'v <$ Ca. Where, however, a somewhat weak stock is grafted on the spot, especially when it is done for the second time and takes place on a shoot out of the main stem, or where the joint has not been well cemented or has been injured during the vine's period of intensive growth, I have more than once observed the scion to grow vigorously in the early part of summer and to wither and die suddenly or gradually towards the end of summer. At first the soil contained a lot of moisture (our rains at the Cape fall mainly in winter!) and the scion's development was still moderate, so that it was in the position $C'v < Ca$. Then the soil moisture became less and the scion grew bigger till it reached the position where $C'v = Ca$. Later on consumption (particularly of water) exceeded absorption, *i.e.* $C'v > Ca$, and this became continually worse, until the scion died when its limit L to drought resistance was exceeded. Its position was then expressed by $\frac{C'v}{Ca} > L$. Here the difference between the water evaporated by the scion and that received by it from the stock had become too great. The same result can be observed where moles cut off sufficient large roots of trees. In this way I lost several large prune trees (on peach stock) on my farm.

In the position $C'v > Ca$ the joint acts in the same sense as the difference between the functional capacities of scion and stock as it also helps to diminish Ca and thus helps to bring about the position $C'v > Ca$.

Where the $C'v > Ca$ as it were places the grafted vine in a *drier spot,* $C'v < Ca$ as it were places it in a *moister locality.* The same position arises when we graft pear on quince and pear on wild pear. Seeing that the wild pear possesses a greater functional capacity of absorption than the quince, pear grafted on quince should, according to Daniel's theory, be in the position expressed by $C'v > Ca$, whereas pear on wild pear is in a position where $C'v < Ca$ or where the relation $\frac{C'v}{Ca}$ is at least smaller (because Ca is larger with the wild pear than with the quince stock) than where pear is grafted on quince. Hence pear on quince will grow as if it stood in *drier* soil than pear on wild pear in the same soil, and will thus form a smaller tree which will sooner bear fruit and be more productive, whilst at the same time producing better flavoured and sweeter fruit than pear on wild pear. This is confirmed in practice and was also clear from an investigation by G. Kiviere et Bailhache, "Contribution à la physiologie de la greffe" (*C. R. de l'Acad, des Sc.,* 1897), conducted with the pear Triomphe de Jodoigne grafted on quince and wild pear, according to which the pears grown on quince were more aromatic and sweeter than those grown on the wild pear stock (see Daniel, *La Théorie des capacites . . . l.c.* p. 197).

2. *Grafted vines with the relation $C'v < Ca$.*

As has been previously stated, this is usually the case with grafted vines, as we here select stocks that will produce large and vigorous grafted vines. Where the autonomous (un-grafted) plant lives during the greater part of its life very nearly in the condition $C'v = Ca$, and has a shorter or longer life according to its nature, *the life of the grafted plant, and hence also of the grafted vine, is shortened* accordingly as we deviate more from this condition and make $C'v < Ca$ or $C'v > Ca$. We have already seen that the joint hinders the upward and downward flow of the nutritive sap and hence tends to make $C'v > Ca$. Where scion and stock come from the same plant, the effect of the joint will naturally be least, and it will be less according as the joint is more perfect.

Therefore where we have a grafted vine with the relation $C'v < Ca$, the effect of the joint will be to make the difference between $C'v$ and Ca smaller. By regrafting the scion we get two joints on the same original stock. Thus I grafted Greengrape on Aramon × Rup. Ganzin Nos. 1 and 2, on which Muscat of Alexandria had died after three years for want of affinity, and the following year grafted Muscat of Alexandria on the Greengrape scion. This is a case of an *intermediate graft.* As Muscat of Alexandria has a good affinity for Greengrape, and as the latter has a good affinity for the Aramon stocks used, we can expect two good joints in this case. Although we have two joints, they may be so much better than where Muscat of Alexandria is grafted directly upon the Aramon stock, that this intermediate graft may succeed much better than the direct grafting of Muscat of Alexandria on Aramon.

The nature of the joint and the cementation of the tissues of scion and stock at the joint, together with the difference between the functional capacities of consumption and absorption of the scion and stock respectively, may yet prove to be the cause of good or bad *affinity* between scion and stock.

The perfectness of the joints of grafted vines, all of the same variety and on the same stock, can differ very much although their outward appearance may not show it. This will often, to a large extent, explain the marked differences in the behaviour of such grafted vines. At the same time it should be borne in mind that the adaptation of the scion, and particularly of the stock to soil and climate, in a large measure influence their functional capacities and hence their success. The system of trellising, winter and summer pruning, the crop which the vine has to ripen, etc., will also very largely influence its functional capacity of consumption. By giving a greater development to the vine, we shall make $Ca - C'v$ smaller and thus place the vine, in a seemingly drier locality, which will ensure a more abundant crop. This applies particularly to a variety like Ohanez (the Almeria grape).

We have already seen that the grafted vine gradually passes from the position $C'v < Ca$ to $C'v = Ca$, later on to come in the position $C'v > Ca$ because the old roots no longer grow well in the soil. As a result of this the vines begin to show signs of old age sooner than ungrafted vines; they have less vigour and produce less wood, till they become insufficiently productive, when they have to be uprooted, and the land is again put under vines after having been under cereals or having been in bare fallow for a couple of years. This is illustrated particularly well by most vines grafted on Constantia Rup. Metallica or Cape Metallica.

In all cases of grafting, therefore, we should endeavour to keep the grafted plant (vine) as nearly as possible in the position $C'v = Ca$. The position $C'v > Ca$ will, where the difference is moderate, produce fruit of better quality than $C'v < Ca$ but $C'v - Ca$ must be kept as small as possible. By improving the relation $C'v \gtrless Ca$, *i.e.* by bringing it nearer to $C'v = Ca$, we prolong the life of the grafted vine.

We can improve $C'v > Ca$ by allowing the scion (European vine) partly to grow on its own roots. This, Daniel calls *mixed grafting*. This might be allowed where vigorous, six-to-ten-year-old vines are grafted over. The danger here of course is that the scion will grow more and more on its own roots and less on the American roots, which will ultimately die, leaving the European vine on its own roots only. In soils where phylloxera is dangerous, such vines will gradually succumb to its attacks. With young grafted vines we must keep on suppressing any roots emanating from the European scion. By pruning the vines severely, we can make $C'v$ smaller and thus improve the relation $C'v > Ca$.

In order to improve $C'v < Ca$ we can allow wild shoots to develop on the stock, which will be an advantage so long as $C'v + Ca$ is not $> Ca$. Here it will be a good thing at first not to remove the wild shoots where strong stocks are grafted on the spot—this is apart from the other advantage that cut-worms will then feed upon these wild shoots and pay less attention to the developing buds of the scions—whilst this will also improve the chances of the graft being successful. After the scion has made some growth, the wild shoots should be topped, and they should be completely removed once the shoots of the scion have grown to some length. By leaving the wild shoots to develop freely during summer, we shall greatly retard the development of the scion and it will consequently remain small. In this instance Cv has become relatively much greater than $C'v,$ which, has remained too small.

If a vine grows too vigorously and bears too little, we can improve matters by giving it more bearers, and especially long bearers, by which means we shall increase its development, so that $C'v - Ca$ will become smaller, and the vine will consequently produce more and better grapes. This is precisely what is done with success in actual practice.

Daniel[1] quotes an interesting case in point which he observed on a graft of cherry on cherry-laurel. One year he allowed the stock to grow too vigorously, with the result that the cherry grew poorly and became infested with aphids. The following year he pruned back the branches of the stock very severely and kept them fairly short in summer, with the

---

[1] Daniel, L., *La Théorie des capacités* . . . l.c, pp. 243-244.

result that the cherry grew vigorously and freed itself from the aphids.

The *longevity* of a grafted vine thus depends upon the stock, scion, affinity, perfectness of the joint, soil, climate, and treatment (training, pruning, summer treatment, manuring, etc.). Under unfavourable conditions it may last for a few years only, but under favourable conditions this may be much longer. At the Cape we have grafted vines thirty years old that still do well. We can, however, assume that the average life of most grafted vines is not much longer than twenty years. As we get to know more about grafted vines, it will no doubt be possible to obtain grafted vines which will reach a much older age, say fifty years. At present we have not yet reached this stage, at least proofs to this effect are still wanting. It seems to me extremely improbable that the grafted vines (European upon American) will ever reach an age of 100 years and more, as the ungrafted European vines did repeatedly before the phylloxera came on the scene. We can be quite satisfied if they last for thirty years and remain sufficiently productive.

II. **Practical Considerations and Results**

It is much more difficult to farm successfully with grafted vines than formerly with ungrafted vines. Enough has already been said to explain why this is so. Here we have to do not merely with the symbiosis between scion and stock, but also with the phylloxera which occurs in smaller or larger numbers on the roots of all American stocks. Such a thing as a stock which is absolutely immune to phylloxera we simply do not possess. Hence the life of the grafted vine is continually in an *unstable state of equilibrium,* which is influenced by the treatment of the vine, the soil, and the climate. Hence every vine should be treated on its own merits when pruned in winter. We should perform this operation with still greater care when dealing with grafted vines than is necessary with ungrafted vines.

After we have found that a certain stock does well in a certain piece of ground when ungrafted, it does not necessarily follow that it will do well in this soil when grafted or that we can successfully graft any variety of grape on it. Thus, during my visit on 2nd September 1908 to the

Viticultural Experiment Station at Freyburg a.d.U., the most northerly viticultural area in Germany and Europe (over 51° north), I found that Riparia gloire and Riparia Geisenheim grew very well in the local highly calcareous soil, which contained many stones of limestone and up to 36 per cent $CaCO_3$, but *only when ungrafted.* Once they are grafted, the grafted vines suffer from lime chlorosis and soon perish. Hence they are useless as stocks in this soil. Therefore we must graft stocks with the desired varieties after they have been planted out in experimental plots to test their suitability. The example quoted above illustrates the great influence which the scion can exert on the life of the stock.

For the sake of clearness, I shall now discuss the principal effects of grafting on the life of the grafted vine, under the following seven headings.

1. **Non-setting and Millerandage.**

Where vines absorb a great deal of water and mineral food, especially nitrates, during the period of flowering and setting, as frequently happens with grafted vines during the first four or five years, the berries often set badly and numerous small seedless berries are formed which ripen (millerandage). This we find especially with stocks like 1202, Rup. du Lot, Aramon, and sometimes even Rip. gloire, where they grow very vigorously. Less vigorous stocks will cause better setting of the fruit of varieties grafted on them. The trouble caused by over-vigorous stocks will become less as these become older and less vigorous. The bad fertilisation is the result of too dilute food brought about by the absorption of too much water, and of too rank growth, with consequently too high a consumption of food by the young shoots and leaves, which results in the flowering clusters receiving too little nourishment. This might happen also with ungrafted vines, but a vigorous stock greatly strengthens any tendency in this direction.

Here I wish to refer to the peculiar behaviour of Muscat Hamburgh in the ungrafted state; its pedicels dry up when the berries turn colour. In England it is known as *shanking.*

This was the reason why the cultivation of this fine grape was almost abandoned in England until some one discovered that the difficulty was completely overcome by grafting it on another grape. Dr. Hogg[1] mentions Black Hamburgh, and especially Muscat of Alexandria, as good stocks for it. This would, of course, only apply in the absence of phylloxera. At the Cape it is successfully grown on a number of American stocks, and its bunches are well filled and ripen well.

2. **Lime Chlorosis.**

It is well known that most Vinifera varieties are more highly resistant to lime chlorosis than most American and Americo-vinifera vines used as stocks. It is also known that the Vinifera varieties differ considerably in their resistance to lime chlorosis. Chancrin[2] classifies these as follows:

1. *Varieties that are highly resistant to lime chlorosis:* Aligoté, Carignan, Clairette, Cabernet franc, Castet, César, Chenin blanc, Etraire de le Dui, Gamay, Malbec, Merlot, Morrastel, Muscadelle, Poulsard, Koussanne, Syrah, Sauvignon.

2. *Varieties that are fairly resistant to lime chlorosis:* Aramon, Cinsaut (Hermitage of the Cape), Chardonnay, Durif, Enfariné, Marsanne, Pinot de Pernaud, Pinot Meunier, Sémillon, Terret.

3. *Varieties not resistant to lime chlorosis:* Alicante Bouschet, Corbeau, Chasselas, Folle blanche, Gros-lot de Cinq Mars, Grappu, Jurancon blanc, Melon, Meslier, Mondeuse, Mourvèdre, Picquepoul, Pinot gris, Pinot Giboudot, Pinot Renevey, Teinturier du Cher (Pontac), Viognier.

By grafting these European varieties on American stocks we can obtain grafted vines that are highly or only slightly resistant to lime chlorosis. This will depend upon the resistance of both the stock and the scion. Here the scion can influence the resistance of the stock (see Chancrin, *l.c.* pp. 105-106). Daniel[3] says that Viala and Ravaz quote the following examples of this: *(a)* Herbemont, which soon becomes chlorotic in limestone soils, remains green if Clairette (highly resistant) is grafted on it; *(b)* Merlot (highly resistant) remains green and vigorous in the limestone soils of the Vendee when grafted on Vialla, whilst this stock soon becomes chlorotic and perishes when growing ungrafted in this soil. Here, grafting with a highly resistant scion has *increased* the resistance of the stock.

---

[1] Dr. Hogg, *The Fruit Manual,* 5th ed., London, 1884.

[2] Chancrin, E., *Viticulture moderne,* Paris, 1908.

[3] Daniel, L., *La Question phylloxérique . . . l.c.* pp. 122-123.

Further, Daniel quotes Couderc as follows: "Grafting in a large measure *diminishes* the adaptation. The ungrafted Riparias, for instance, often remain fairly green and in the end grow well in most limestone soils in which they at first turned yellow. It is only after being grafted that they become so chlorotic as to die. The chlorosis is stronger or weaker according to the variety used as scion." My previously quoted observation made at Freyburg a.d.U. completely corroborates this opinion of Couderc regarding Riparia as a stock in limestone soils. His assertion that grafting has a detrimental effect cannot be applied generally, for, as we have already seen, grafting can also have the effect of increasing the stock's resistance to lime chlorosis, and thus extending its area of adaptation.

3. **Vines die suddenly.**

It sometimes happens that vines which have been growing very vigorously, suddenly die. We find this particularly with grafted vines if the early part of summer has been rainy, and warm and dry weather suddenly sets in. The grafted vines are more sensitive to drought and humidity than the ungrafted vines have been. We suddenly notice the shoots on one arm of a vine or on a whole vine wilting and later on drying up. In France this is known as *folletage* and also as *apoplexie, i.e.* apoplexy, because it can kill the vine so suddenly. This depends upon an excessive development of the scion which begins with $C'v < Ca$ and is thereby encouraged to grow very vigorously, but during warm dry weather, changes to $C'v > Ca$, until $C'v - Ca$ becomes so great that the vine dies from drought.

According to Gouy,[1] vines grafted on 1202 frequently die suddenly like this in Italy during the first and second year. At the Cape I have repeatedly found the same thing happening with Muscat of Alexandria grafted on 1202. In this connection it is necessary to study carefully all the factors exercising an influence on the variable relations $C'v \gtrless Ca$[2] with grafted vines.

---

[1] Gouy, *Revue des hybrides,* 1905 (quoted from Daniel, *La Question phylloxérique . . .* p. 71).

[2] 2012 Editor's Note: >< is printed in the original book with the *greater than* symbol placed on top of the *less than* symbol.

Sometimes grafted vines die slowly or fairly soon without any apparent cause. The stock remains healthy, forms vigorous wild shoots and shows perfectly sound roots, but the scion dies. Sometimes the bottom leaves begin to dry up and fall prematurely, while the grapes still ripen well and the canes dry up only afterwards, and are found to be dry and dead during winter pruning. Such cases I have come across with Hermitage on Aramon in Bottelary and with Stein on Aramon near Sir Lowry's Pass, both in the Stellenbosch district. The vines in question were at the time four to six years old, and later on the trouble grew less. Here we have to do with a disturbance in the food, and particularly the water supply of the vine. This is frequently due to *thylosis,* where the unthickened parts of the walls of the vessels are pressed into these by strongly growing parenchymatous cells, thus partly or completely blocking such vessels. This is the result of a higher osmotic pressure in the surrounding parenchymatous cells than in the vessels, and of a diminution of pressure in the vessels. After a good rain or irrigation, water can pass relatively quicker by osmosis at the joint into the parenchymatous cells of the scion than into its vessels, and thus causes *thyloses* to be formed. Further, the hindrance caused by the joint can bring about an increased osmotic pressure in the cells of the scion that contain and conduct organic food, and can thus also favour the formation of thyloses.

This thylosis or internal blocking of the vessels naturally hinders the absorption of food, and especially of water, from the soil, and will, according to its intensity, cause the sudden or slow death of the vine or of a part of it only. Thylosis is frequently accompanied by the *formation of gum.* This we notice also with vigorously growing, grafted, young peach, apricot, etc. trees, if the soil contains plenty of moisture and the trees have been heavily pruned during the preceding winter, whereby $C'v$ was made much smaller than Ca. We practically never meet with this trouble among ungrafted vines, whereas we do come across it among grafted vines.

4. **Susceptibility to fungoid diseases and insect pests.**

Daniel maintains that, through grafting on American stocks the European vines have become much more susceptible to diseases and pests, and that grafting therefore is the cause of the severe fight the European grape growers have to put up to-day against the enemies of the vine. Where the relation $C'v < Ca$ exists with a grafted vine and the difference between Ca and $C'v$ is great, the vine, as it were, lives in moister surroundings, and we should therefore expect it to be more susceptible to diseases and pests. Where the relation $C'v > Ca$ obtains (this as a rule is not the case with grafted vines), we should expect the opposite result.

Daniel[1] says that he saw grafted vines on Chateau Margaux with M. Mouneyres of which the grapes were strongly attacked by downy mildew, whilst those of the ungrafted vines were sound. He adds that M. Seibel wrote to him that the direct producer, Seibel 1, when grafted on Jacquez, is much more susceptible to anthracnose than the ungrafted vines of the same direct producer, and that also other numbers, *e.g.* 2 and 47, are much more susceptible to fungoid diseases when not growing on their own roots. Daniel, *l.c.* pp. 216-217, quotes the following communications of M. Delafosse to him: "Othello, when grafted, he writes to me, does not only grow more vigorously, but escapes the brown rot, which is the great enemy of its ungrafted vines.

"In two rows of ten-year-old Jacquez vines there was one vine of Cinerea-Rupestris de Grasset. They were all grafted on the same day with 156. The one on Cinerea became a big, very productive vine. The 156 is fairly sensitive towards sulphur. Owing to the severe attacks by oidium, all these vines were sulphured last year (1907) by accident. All the vines grafted on Jacquez were more or less affected and lost some of their leaves. The one grafted on Cinerea was not harmed at all and remained brilliant green." Daniel further states that Delafosse has found at the School of Agriculture at Montpellier, that grafted 132-11 Couderc, without any treatment, was much damaged by oidium, whilst its un-grafted vines remained vigorous and healthy.

Daniel, *l.c.* pp. 217-218, quotes the following communication from M. Basille, Administrator of the Prince of Paterno in Sicily, regarding his local experience with downy mildew in 1900: "In those parts of a vineyard in the

---

[1] Daniel, L., *La Question phylloxérique . . . l.c.* p. 215.

region of Melilli which had been renewed with Nero d'Alova[1] grafted on 101-14, the infection with plasmopara viticola was so bad in 1900, that not a single vine escaped untouched, that the grapes were totally destroyed, and that the vines suffered very much in their growth. It was different with the vines grafted on Riparia, which were only slightly attacked by the disease, and then (2nd August) promised a fair crop. On the same piece of ground there was also a block of old, ungrafted vines of Nero d'Alova which was infected with phylloxera.

"The vines grafted on American stocks received the proper treatments both in the form of liquids and of powders at the proper times; the ungrafted vines were dusted only once with sulphur and copper sulphate. Yet the vines grafted on 101-14 were *totally damaged* by the plasmopara viticola, those grafted on Riparia were *only slightly attacked,* and the European vines *remained sound.* In all three cases, whether grafted or not, the European vine was the same variety."

Daniel goes on to quote a large number of other examples from viticultural practice to prove his assertion. The case of Othello quoted above, however, shows that grafting can also help a vine against disease. Hence grafting does not necessarily always weaken a vine's resistance to diseases, though this might often be the case. In any case, it is certain that grafting, owing to its great influence on the whole life of the grafted vine, makes the latter more or less (not often) susceptible to fungoid diseases.

Ravaz[2] discusses the influence of the stock on the susceptibility of the scion to downy mildew, and comes to the conclusion, which will no doubt be generally accepted, namely, that the vigour of the grafted vine determines its susceptibility to downy mildew with one and the same variety, and that this is due to the larger or smaller amount of water present in the leaves. It is, however, useless for Ravaz to add that the nature of the stock bears no relation to the susceptibility of the scion to downy mildew, as, apart from the treatment meted out to the grafted vine, the stock is most intimately connected with the vigour of the scion, hence with the amount of water in its leaves, and therefore also with its susceptibility to fungoid diseases.

---

[1] 2012 Editor's Note: This is almost certainly a misspelling of Avola
[2] Ravaz, L., *Le Mildiou,* 1014, *l.c.* pp. 102-105.

For the same reason grafted vines are often more susceptible to insect pests. It has frequently struck me that the Mealy Bug (*Pseudococcus capensis,* Brain) is worse in very vigorous vines with large, compact bunches hanging in the shade than in the less vigorous vines with smaller and looser bunches that are more exposed to sun, wind, and weather generally. And, as we know by this time, the scion has a strong influence on the vigour of the grafted vine.

Inversely, the scion can influence the stock. This we have already seen in discussing the influence of stock and scion on chlorosis. The stock's resistance to phylloxera is also influenced by grafting. If we allow the scion to overbear, or if we allow it to evaporate so much water that the grafted vine suffers from drought, we lower the vitality of the stock and therefore also its resistance to phylloxera.

In the different countries the resistance of the various stocks seems to vary a good deal with the local climatic and soil conditions. Thus Jacquez, at the Cape, is a better stock than in France, whereas Aramon is sometimes less resistant to phylloxera than Jacquez. During the last five or six years about 100,000 vines grafted on Aramon (*i.e.* A.R.G. Nos. 1 and 2) have been destroyed by phylloxera at the Cape. Either Aramon has here lost in resistance to phylloxera or has been destroyed by a new biological race of phylloxera. In any case there is no such thing as a fixed resistance of a stock to phylloxera. As soon as it is grafted, its resistance to phylloxera depends largely upon the development of the scion, and generally upon the treatment given to the grafted vine.

5. **Influence upon size of Crop.**

By selecting vigorous stocks that suit the particular soil and scion, we can obtain a vineyard which will produce heavier crops than ungrafted vines would have produced under the same conditions. According to their nature and suitability under local conditions, the different stocks, when grafted with one and the same European variety, will produce crops varying a good deal in quantity. On the whole, grafted vines come into bearing sooner than ungrafted vines. European vines grafted on Constantia metallica grow very rapidly and soon produce heavy crops, but, in consequence, the grafted vines on most soils soon exhaust themselves and show declining vigour. Such vines are badly infested with phylloxera on their roots and, after eight to ten years, they either die or do not pay.

It is well known that vines grafted on Riparia, as a rule bear heavily and regularly, but where this stock does not answer well and the vines grafted on it consequently remain small, vines grafted on Jacquez and on many other stocks will give heavier crops than when grafted on Riparia.

6. Influence on quality of Crop.

The different stocks do not absorb precisely the same mineral substances from the soil, neither do they absorb them in the same proportions or in the same absolute quantities. Hence the grafted vines are not equally vigorous and productive on different stocks. Further, the same variety, when grafted, will usually not show the same vigour and productiveness as when growing on its own roots. Hence grafting affects also the quality of the crop. A heavier crop usually means poorer quality, and *vice versa*. As the stock can also cause the grapes of the variety grafted on it to ripen later or earlier, this can also cause a difference in the quality of the grapes.

A stock which terminates its summer growth early in a certain locality will ripen its grapes earlier than one which continues its summer growth to a later date. Thus I have frequently observed at the Cape that vines grafted on Jacquez will ripen their grapes earlier than when grafted on Aramon. Muscat of Alexandria grafted on Riparia in Robertson (Cape) frequently ripens its grapes eight days earlier than when grafted on Jacquez. Local conditions may sometimes reverse the order of ripening. The point always simply is, which stock continues growing the latest, as is shown by the growth of the scion.

Daniel maintains that the quality of the grapes and their wine has, on the whole, deteriorated a great deal in France on account of grafting the French varieties on American stocks, and that quality has been sacrificed for quantity in this reconstitution of the French vineyards. At the Cape this unfortunately holds good to a large extent. But it need not necessarily be so. By choosing stocks that are sufficiently resistant to phylloxera without causing the vines grafted on them to grow too vigorously, and especially by grafting them with varieties that produce grapes of high quality and by limiting the size of the crop, we can obtain excellent quality in the grapes and wine produced by grafted vines. Only, reconstitution by grafting gave the grape grower the chance of obtaining heavy yields, and in reconstituting his vineyards he usually did not worry much about quality. It would appear to be an established fact that the wines from grafted vines mature more rapidly and reach their highest development sooner than wines from ungrafted vines, without ever reaching that height of perfection which is attainable in wine from ungrafted vines.

According to the stock on which the grapes have been grown, we find appreciable variations in the chemical composition of their musts. With the assistance of one of my students, Mr. C. J. Theron, I tested the musts of some varieties in my ampelographic collection in 1924 for sugar and total acidity. The grapes were picked when ripe. All the grapes of the variety were gathered separately for every stock in order to get the must for analysis. There were twenty vines of every variety, and they were grafted on the spot in 1920 on the following ten stocks (i.e. two vines on every stock): Jacquez, Rip. gloire, Aramon × Rup. Ganzin No. 1, Aramon × Rup. Ganzin No. 2, 1202, 101-14, 106-8, Rup. du Lot, 420A, 333. The numbers 1-10 in the tables overleaf refer to the above-mentioned stocks in the same order, beginning with Jacquez as No. 1. The sugar is given in degrees on the Marloth saccharometer, which gives approximately the percentage of sugar in the must by weight, and according to the readings of the mustimètre of Salleron-Dujardin. The total acidity of the must is expressed as grams tartaric acid per litre must, and in titrating litmus paper was used as indicator.

This investigation will be continued. Meanwhile we can merely conclude from these data that grafting on different stocks causes a considerable amount of variation in the sugar content and total acidity of the grapes. The same stock does not always produce the sweetest must, and the highest total acidity does not always occur with the lowest sugar content. By a judicious choice of our stock we can thus grow grapes with more or less sugar and acidity.

413

1. MUSCAT OF ALEXANDRIA	1	2	3	4	5	6	7	8	9	10	Date, 7.3.24 Remark*.
Degrees Marloth	20.1	20.7	20.9	19.8	20.7	22.4	20.5	21.5	17.7	19.9	Fully ripe excepting which was very full and not yet fully ripe.
Mustimetre	1093	1095	1096	1091	1095	1104	1094	1099	1080	1092 6	
Total acidity	5.3	6.2	5.2	6.5	5.9	5.3	5.6	5.6	7.1	6.5	

2. CLAIRETTE EGRENEUSE											Date, 10.3.24
Degrees Marloth	14.9	15.5	18.7	16.7	18.2	18.1	19.4	16.7	16.6	17.7	Just well ripe; 1 very full.
Mustimetre	1066	1069	1086	1075	1083	1082	1090	1076	1075	1080	
Total acidity	8.6	7.8	7.0	7.8	5.7	5.7	6.6	6.9	7.5	7.6	

3. MOLAR											Date, 10.3.24
Degrees Marloth	19.6	240	23.7	22.3	22.7	24.2	22.0	24.4	22.3		Somewhat overripe and pitch black.
Mustimetre	1089	1112	1110	1103	1105	1114	1102	1115	1103		
Total acidity	7.3	6.4	6.8	7.15	6.4	6.3	7.1	6.8	7.9	—	

4. UGNI BLANC — MACCABEO											Date, 7.3.24
Degrees Marloth	18.3	19.2	.18.7	18.9		21.6	20.9	19.6	20.9	17.3	Just fully ripe, even.
Mustimetre	1083	1087	1085	1086		1100	1096	1089	1096	1077	
Total acidity	4.4	4 1	4.9	4.7	—	4.7	4.8	3.8	4.8	4.6	

Grafting may have its disadvantages, but it can also have its advantages.

7. **Specific variations.**

By specific variations I mean variations in the characters of the grape varieties of which we make use in describing them precisely with a view to their differentiation and classification. I shall now briefly discuss some of these variations which will at once show that the grafted vine does not continue its normal life as though it had never been grafted, but that scion and stock influence each other mutually.

That the *habit of growth* of the scion can be strongly influenced by the stock was shown by Perold and Tribolet,[1] who found that Muscat of Alexandria grafted on Herbemont and two years old in a black sandy peaty soil showed a *creeping* habit of growth, *i.e.* the shoots grew flat on the ground, and the bunches were roundish and smaller than usual, whereas Muscat of Alexandria of the same age and in the same soil, but grafted on Jacquez, showed the usual more or less erect growth of the shoots and large, long bunches of the usual size and shape.

---

[1] Perold, Dr. A. I., and Tribolet, I., *American Stocks for Cape Vineyards*. Pretoria, 1912, p. 15.

The *angle of geotropism* of the stock's roots can be altered by the scion, so that its roots become more plunging (smaller angle of geotropism) or more spreading. Thus M. Baco, according to Daniel,[1] has found that the angles of geotropism of the roots of 1202 and 3309, which were respectively 55° and 50° when ungrafted, after being grafted with Baroque were in both cases increased to 70°. Similarly grafting with Baroque has caused the roots of Rip. gloire, 420A, 33A, 3306, 1616, 157", Rup. du Lot and Noah to spread out more. Inversely it caused the roots of 41B and 101-14 to plunge more by diminishing their angles of geotropism from 50° to 40° and from 65° to 53° respectively. By grafting 1202 with Tannat its angle of geotropism sank from 55° to 30°, whereas it rose from 55° to 70° when grafted with Baroque.

M. Jurie, according to Daniel,[2] has grafted Semillon on Rup. du Lot and has grown both ungrafted next to it in order to study the *influence of grafting on the leaves.* Where the leaf of Semillon has an almost closed petiolar sinus and is fairly woolly on the lower surface, that of Rup. du Lot has a very open petiolar sinus $a + \beta = 71°$, whereas with Semillon it is 110°) and is glabrous on both surfaces. On comparing leaves of ungrafted Semillon with comparable leaves of Semillon grafted on Rup. du Lot, the latter have shown numerous intermediate types between those of the ungrafted Semillon and of the stock; they were less woolly and possessed a more open petiolar sinus which was about intermediate between that of the stock and that of the ungrafted Semillon.

The following most interesting results obtained by M. Jurie by grafting show that the *taste* and time of ripening of a grape can be altered thereby, and that the scion can acquire the susceptibility of the stock to lime chlorosis and its *resistance to phylloxera.* M. Jurie says (quoted from Daniel[3]): "As subject of study I chose one of my hybrids, the 340A: it is an (Othello × Mondeuse) × (Rupestris × Monticola) with little resistance to phylloxera. Its grape is *late* and *foxy,* berries compact, do not crack and do not rot. In 1898 I grafted this hybrid on ten vines of Cordifolia × Rupestris de Grasset, which I had available at the time. Now this latter

[1] Daniel, L., *La Question phylloxérique . . . l.c.* p. 682.
[2] *Ibid.,* pp. 516
[3] *Ibid.,* pp. 520-521

plant produces an *early grape;* it is *highly resistant to phylloxera* and *sensitive to lime.* In 1899 I was surprised to find all my grapes *golden* and absolutely *free from a foxy taste,* ripe on 15th August, when those of the mother-vine were still green. The ten vines all showed the same variation, which was decidedly the result of grafting; one cannot indeed here ascribe the result to bud-variation, as we have here to do with ten grafts that have all varied in the same sense and as the parents of the hybrid scion thus modified are all late.

During the spring of 1900 I planted thirty cuttings from these ten grafted vines affected by the variation. All these new rootlings have preserved *in toto* their character of earliness, obtained as a result of grafting, and this year, on the 15th August 1901, I was able to show to the members of the Commission of Enquiry of the Agricultural Society of France on direct producers, these rootlings having in their second leaf golden ripe grapes absolutely *clean in taste ('droits de gout')* and in every respect similar to those of the vines from which the cuttings had been taken.

"These same rootlings have revealed to me another interesting fact. This year, in spring, they have been attacked by *chlorosis.* According to Millardet's studies the hybrid 340A is most highly resistant to this disease and possesses the same resistance as Rup. du Lot. The lowering of its resistance on being grafted on a stock sensitive to lime is thus another character of the stock transmitted to the scion. Finally, I have this year also noticed on this same 340A a new phenomenon concerning its resistance to phylloxera.

"Being desirous to find out whether the stock has transmitted its resistance to phylloxera to the scion, I conducted the following experiments. In two fairly large pots I planted together cuttings of 340A from the mother-vine, and from the 340A grafted on the Cordifolia × Rupestris. Later on I placed between them roots covered with phylloxera.

On the 8th November I dug up the rootlings from one pot the rootling coming from the mother-vine had ten nodosities caused by the phylloxera, whereas the one from the grafted vine had none. These rootlings have been sent to M. Millardet, who verified the fact.

"On the 14th November 1901 rootlings were dug up in the presence of M. Daniel, who was on his way to the Congress of Lyons. The result has been as conclusive: only the rootling from the grafted vine was sound ('indemne'); the other had numerous nodosities. These four rootlings were shown at the Congress of Hybridisation at Lyons on the 16th November last, and every one was able to convince himself of the reality of the fact."

Here I still wish to mention Strasburger's classical experiment which he made in 1884 by grafting Datura stramonium on a potato plant, and in which he proved that atropine passed from the Datura (the scion) into the potatoes. Here we thus have the case of a foreign substance passing from the scion into the stock. Daniel, and later on Vöchting,[1] have proved that inuline does not pass from the stock into the scion, for instance, when *Helianthus iuberosus* (Jerusalem artichoke) is grafted on *Helianthus annuus* (sunflower).

C. GRAFT HYBRIDS

These are not hybrids in the ordinary sense of the word, where cross-fertilisation gives rise to the formation of a true hybrid, which is propagated from the seed formed as a result of the fusion of the sexual cells of the plants which have been crossed with each other. In any case, as a result of grafting, new varieties of plants have already been formed with new properties which have been acquired by grafting, and remain more or less constant when such plants are subsequently propagated. The changes just quoted that M. Jurie's 340A underwent after having been grafted on Cordifolia × Rupestris de Grasset, exemplify this.

In 1825 the gardener Adam of Vitry, near Paris, noticed an adventitious shoot arising in the grafting zone (zone of the joint) of *Cytisus purpureus* and *G. laburnum.*

---

[1] See Daniel, L., *La Question phylloxérique... l.c.* p. 129.

Whereas the latter has sulphur-yellow flowers and the former has purple-red flowers, the adventitious shoot bore flowers with an intermediate colour, dirty yellowish-red. It got the name *Cytisus Adami*. When propagated it sometimes produced branches that were branches of pure *C. laburnum* or pure *C. purpureus*. This is an example of a graft hybrid. (According to Molisch.[1])

Crataegomespilus is an intermediate form between the medlar, *mespilus germanica,* and the hawthorn, *Crataegus monogyna,* which arose in 1900 in the grafting zone of a 100-year-old Medlar grafted on hawthorn in Dardai's garden in Bronvaux near Metz. It also arose as an adventitious shoot where the tissues of the medlar and hawthorn meet (see Molisch, *l.c.* p. 253). Molisch says that three intermediate types are in existence. Daniel[2] mentions two intermediate types, namely, Crataegomespilus Dardari and Crataego-mespilus Jules d'Asnières, which he has studied, and of which the former is closer to the medlar and the latter is closer to the hawthorn. Whereas the hawthorn's branches have thorns and its leaves are glabrous and deeply serrated, and the medlar has no thorns and its leaves are woolly and entire *(i.e.* without incisions), the intermediate form which is nearer the hawthorn resembles it very much, but its leaves are broader, less serrated, and woolly, while the form which is nearer to the medlar also possesses its appearance, but here and there has thorns, together with other variations which Daniel describes fully. He regards these intermediate forms as graft hybrids.

The *Bizzarria,* according to Molisch, *l.c.* p. 253, has arisen in a garden in Florence in 1640 from a graft of orange on lemon, and this as an adventitious shoot from the swelling of a part where a bud had been inserted long ago. The intermediate forms are to be seen in the leaves, flowers, and fruits. The hybrid fruits appear as lemons in orange peels or *vice versa,* or with divisions of lemon and orange side by side.

Shortly after 1900 Daniel discovered in a garden in Rennes a graft hybrid, Pear-Quince, on a 60-year-old Williams pear tree which had been sawn off at 1.50 metres or

---

[1] Molisch, Hans, *Pflanzenphysiologie als Theorie der Gärtnerei,* 5e Aufl.; G. Fischer, Jena, 1922.

[2] Daniel, L., *La Question phylloxérique . . . l.c.* p. 297.

5 ft. above the ground about the year 1900, and had been grafted sixty years earlier on quince. It was in the grafting zone of one out of a hundred pear trees which had thus been sawn off, that Daniel discovered the graft hybrid. In this case, too, the origin was an adventitious shoot from the zone where pear and quince tissues met. He noticed three such shoots. The leaves were intermediate types between those of pear and quince. He continues their study (see Daniel, *La question . . . l.c.* pp. 300-302).

Daniel *(l.c* p. 308) also quotes the following experience of Luther Burbank: *"There exists"* he says, *"a close and remarkable analogy between hybridisation and grafting.*

"When I brought with me from France a variety of plum *(Prunus myrobolana,* var. Pissardi), of which there was not a second specimen in America, I grafted it on the Kelsey plum, a variety of *Prunus triflora.* The scion itself did not flower, but the presence of the scion caused a *hybridisation* (crossing) of the two species in the stock. This is the only case, to my knowledge where the scion has affected the propagative system of the plant, and thus gave rise to a cross between two forms which had hitherto never yet been crossed. Several hundreds of descendants of this cross are now growing" (according to the *Tribune horticole,* 1907).

The problem of the graft hybrids entered a new phase when Hans Winkler[1] obtained his graft hybrids. He set out with the idea that, in order to bring about the fusion of two plants in an asexual manner, one should attempt to obtain adventitious shoots out of the cementing tissue at the joint, which are built up out of parts of both parents. For his purpose he made use of the property of certain Solanaceae to form adventitious shoots out of the callus of the stem. If a seedling of *Solanum lycopersicum* (tomato) is topped or decapitated and the laterals then developing are all removed, adventitious shoots arise out of the callus which is formed on the cut wound of the stem. If, immediately after the tomato stem has been decapitated, we graft it with *Solanum nigrum* (nightshade), they will make a successful joint after a while. Now make a slanting cut through the joint, so that part of the cut surface will consist of tissue of the stock and part of tissue of the scion, and remove all laterals that are

---

[1] Hans Winkler, "Über Pfropfbastarde und pflanzliche Chimaren," *Ber. d. D. Bot. Ges.,* 1907, p. 595.

formed, when new adventitious shoots will form out of the cut wound. This Winkler did, and by repeating the experiment a couple of hundred times, he frequently found that next to many ordinary shoots in the border zone of scion and stock, a shoot was formed of which the right side consisted of nightshade tissue and the left side of tomato tissue. This double being he called a *"Chimdre"*.

By repeating these tests and using nightshade as stock and tomato as scion, Winkler obtained a large number of intermediate forms between the two parents. Amongst these there was one which looked more or less like a sexual hybrid between tomato and nightshade would look if some one should succeed in bringing about this cross. He called it *Solanum tubingense.* This plant combines the characters of the parents not only in the leaf but also in the flower and fruit. (The above is an extract from Molisch, *l.c.* pp. 251-252.)

This discovery of Winkler induced other investigators to carry out researches on this subject. It had been known for a long time that a type *Pelargonium zonale* is grown in the gardens which has leaves with white borders or of which one-half or the whole is white, but in 1909 Baur (see Molisch, *l.c.* p. 254) found that these pelargoniums possessed growing points of which one-half is green and the other half white, and thus represent "Chimaren" like those of Winkler. Leaves originating on the green side are green, those originating on the white side are white, whilst those originating on the border between white and green are whitish - green. Baur calls them *"Sektorialchimären"* because the tissues here lie side by side. They can, however, also lie over each other as a glove over the hand, when he calls them *"Periklinal-chimären"*. Both kinds occur with the *Pelargonium zonale.* The latter case, where we can cover the organs of one plant with the epidermis of another, may in future still give valuable results in producing crop plants that will be more highly resistant to diseases and pests than the present varieties. This is a field of research that has many possibilities in store.

CHAPTER VIII

VINE DISEASES

What do we mean by Disease?

ALTHOUGH many persons might consider it unnecessary to discuss this question, it will soon be evident that it is desirable to state clearly what is meant by disease. Disease is not an absolute term. Thus we consider *couture* (non-setting of berries) a disease and treat it here under the non-parasitic or physiological diseases, but it is only a *relative* disease. If Muscat of Alexandria develops just sufficient coulure to produce a sufficiently filled but loose bunch of ripe grapes, then it is a disease in the sense that a certain number of flowers did not form fruit and seed; but from the table grape-grower's point of view it was not only no disease, but of direct economical advantage as it renders the pre-thinning of the bunches practically unnecessary. If the berries, however, set so badly that too few normally developed berries remain to produce well-filled bunches, then it is only harmful and in every respect a disease. And even in this case the disease is relative, for, although it be true that it hinders and damages the propagation of the vine through seed and the object for which it is grown, namely, the production of good grapes, it nevertheless remains a fact that, as a direct result of intense coulure, the vine will grow with extra vigour during the rest of the growing period upon which it has entered. Hence intense coulure will be advantageous to the existence of the particular vine itself. If, owing to this, the vine grows very vigorously, it might happen that its tissues become so much softer and more juicy, that it is now more exposed to the attacks of parasites (fungi and insects, etc.).

This example will no doubt show clearly how elastic the idea of disease can sometimes be, and that one and the same phenomenon can be regarded as a disease or as no disease, according to the point of view from which we regard it or the extent to which it occurs.

If a cabbage or beetroot plant runs into seed before having formed the desired cabbage or beetroot, we may call it a disease from the point of view of our object in cultivating it, but it is no disease from the point of view of the plant itself, as it ensures its propagation in good time. In this connection it is therefore necessary to distinguish between our object in growing a plant and the object of the plant itself, namely, to propagate itself. Sorauer[1] calls the diseases that threaten the first object, *relative diseases,* and those that threaten the latter object he calls *absolute diseases.*

We can define "*disease*" by saying that it is a *deviation from the plants normal life which is harmful either to the object of the plant itself or to the object for which it is grown.* According to this definition we must therefore include all such deviations under diseases, irrespective of the causes that produced them. When we come to consider the causes, we can divide them into *parasitic* and *non-parasitic* causes. The first group includes *animals* and *plants* that live on other plants and animals as parasites and cause a disease in them, while the latter arise from unfavourable conditions of heat, light, air, moisture, foodstuffs, and the presence of poisons in the air or in the soil that have not been produced by plants or animals.

The factors just mentioned are subject to much variation in nature, whereby they can so influence the life of the plant (or animal) that it is put into a state in which some parasite or other can attack it and make it sick. For every disease there are, with regard to the different factors that influence it, *optimum* or most favourable conditions under which the disease can develop best and do most harm. As the conditions depart from this optimum, the chances become less favourable to the disease and more favourable to the plant or animal. Thus we see that even the parasitic diseases require, for their development, not merely the presence of the parasite, but also a certain state of the

---

[1] Sorauer, P., *Handbuch der Pflanzeitkrankheiten,* 3ᵉ Aufl., 1909, vol. i. p. 3.

attacked plant or animal, which state in turn can be brought about by a number of different factors.

In this chapter we shall frequently have an opportunity of seeing diseases depend upon certain states of the weather (cold, heat, moist air, rain, etc.) and upon a certain stage of development of the vine. Thus, for instance, oidium andanthracnose do not develop on ripe grapes, whereas Grey rot or Botrytis does.

In this connection Sorauer says: "The healthy organism possesses a natural immunity, and a disturbance of it constitutes the condition for the parasitic attack" *(l.c.* p. 15). "The parasites have very definite, sometimes narrow limits. Such a limit which the parasite cannot exceed under normal conditions is that state of a living being which we are accustomed to describe as *'healthy',* without being thus far able to fix it more precisely" *(l.c.* p. 15). "Experience has taught us that we have been protected against epidemics, not by an ideal complete destruction or exclusion of such parasites, but because the parasites concerned did not find the climatic conditions which favour their development"*(l.c* p. 16).

Here we shall briefly discuss the principal vine diseases as well as their treatment, and shall divide them into the following three groups: *Non-parasitic* or *Physiological Diseases, Diseases caused by Fungi,* and *Diseases caused by Animals.*

A. NON-PARASITIC OR PHYSIOLOGICAL DISEASES

1. FROST IN SPRING AND AUTUMN

Why does Frost kill a Plant or a part of it?

The different authorities are not yet unanimous in replying to this question. On the whole there are two theories regarding this. According to the one the plant's death is caused mainly by *drying out,* while the other maintains that each plant has a *specific minimum temperature* which it can stand, and that it is killed by frost when its temperature sinks below this minimum.

Müller-Thurgau and Molisch[1] are amongst the supporters of the desiccation theory. They point out that when a plant cools down very considerably, undercooling takes place until the formation of ice suddenly begins and proceeds rapidly. The formation of ice takes place at first and mainly m the intercellular spaces. Later on it can take place inside the cells also. Whilst this happens, it may sometimes happen that the tissues are torn, but the main point is that the cells lose so much water through the formation of ice that the protoplasm coagulates and shrinks owing to lack of water, and if it proceeds far enough, this causes such an alteration in it that it dies and cannot recover later on when the ice-crystals melt. The chances of recovery are greater when these crystals melt slowly than when this takes place rapidly. The colloidal protoplasm has then lost the power of taking up water and returning to its original state.

Sorauer and Mez are the principal supporters of the second theory. Thus Sorauer, *l.c.* vol. i. p. 508, says: ". . . death by freezing is no specific process of desiccation, but must be attributed to a molecular *irreparable destruction of the structure of the protoplasm*. . . . This specific minimum (temperature. A.I.P.) is no fixed quantity but rises with the quantity of cellsap, *i.e.* death from freezing takes place at a higher temperature, and inversely the loss of water will increase the resistance against all factors and thus death from freezing will take place at a lower temperature". Mez (quoted from Sorauer, *l.c.* vol. i. p. 509) says: "A plant which can endure at all the formation of ice in its tissues does not die owing to desiccation of its protoplasts, but as a result of cooling below the specific minimum".

However that might be, it is well known that unripe shoots are killed by winter frosts and turn black. This must be attributed partly to the high-water content and low content of stored food in such shoots. Further, leaves are killed by frost first along their nerves, which are the channels conducting the water through the leaves. A heavy dressing of nitrogen produces vigorous growth and the absorption of much water into the body of the plant; this also causes it to be killed sooner by frost. All plants and all varieties of grapes are not equally sensitive to great cold and frost.

---

[1] Hans Molisch, *l.c.* pp. 208-210.

Factors influencing the danger of Death from Frost.

1. *Cold Winds.*—Where winds blow across ice-fields in spring and are therefore cold and dry, they frequently cause rain in warmer parts, which causes the air to become drier and colder. If the wind is cold enough and keeps blowing long enough, it might become so cold that the young shoots of the vines are frozen to death. Against this we are powerless, but winds which are not too cold are a protection against frost.

2. *Calm and Cloudless Sky.*—If during cold weather, often with rain and hail or snow, the wind goes down near sunset and the clouds disappear, leaving a clear sky over night, there may easily be a frost. Under these circumstances the soil and vines already cooled lose so much heat by radiation into space, that they may easily become frost-bitten. If the sky remains overcast during the night, the clouds reflect the greater part of the radiated heat and radiate another part back to the earth. Here the clouds act as a kind of blanket and the vines will not be frost-bitten.

3. *Nature and State of Cultivation of the Soil.*— Compact clay soils conduct heat better than coarse sandy soils which contain more air, and retain water better than the latter. The result is that vines on clay soils are less subject to frost than those on sandy soils.

Recently cultivated soil does not become so warm during the day as uncultivated soil owing to the amount of air between the loose particles of the cultivated soil, which greatly diminishes the thermal conductivity of the loose layer of soil. The increased surface of the loose soil increases the loss of heat by radiation at night. Therefore vines will sooner be frost-bitten on cultivated than on uncultivated ground. In the vineyard this can often be seen sharply at the row where cultivation has stopped. In places where frost is to be dreaded in spring, cultivation should begin only after the danger of frost is over. But where this is undesirable, the cultivation should be followed by an irrigation. In this way the soil settles again and the drawbacks of loose soil are removed.

4. *Soil covered with Grass.*—This prevents the soil from being properly heated during the daytime, and the grass further cools the soil by evaporating a lot of water. Vineyards that are covered with grass will therefore be more

susceptible to frost than such as are free from it. The grass can be cut at the surface and removed from the vineyard, or it can be ploughed under in winter in time to allow the soil to have a firm surface and little or no grass in spring.

5.   *Dry Straw or Manure on the Ground.*—This acts in the same manner as loosening the top soil by cultivation.

6.   *Irrigation of Soil.*—Under (3) we have already seen how irrigation can sometimes help as a preventative against frost in spring. Apart from this case, however, it is a sure preventative against spring frost if the soil is properly soaked. The explanation is that the specific heat of water is about four times as high as that of the soil, which means that, for equal weights, water will absorb about four times as much heat as soil for a rise of one degree in temperature. Water also conducts heat better than soil. Hence wet soil will absorb more heat during the day than dry soil, although its temperature, for the reasons just mentioned, remains lower. For the same loss of heat by radiation at night, the wet soil will sink much less in temperature than the dry soil. The day and night temperatures of wet soil therefore differ much less than those of dry soil.

7. *Large Masses of Water.*—For the reasons just mentioned large masses of water contain a large quantity of heat, which is partly transmitted to their surroundings, and thus helps to protect vineyards in their neighbourhood against spring frosts. They further help by making the air in their neighbourhood moister and by causing fogs, both of which help against frost.

8. *Time and Method of Winter Pruning and Trellising.*— Generally speaking, we can say that vines which are pruned very late will start budding late.   Hence we can often protect the vines against spring frost by pruning them very late, so that they start budding when the danger of frost is over. Chauzit recommended washing the wounds made in winter-pruning, shortly before the buds begin to open, with a 30 per cent solution of copperas (ferrous sulphate), which, in his experience, delays budding for seven to fifteen days.   I have no personal experience of this.   It is also a good thing to give the vines a preliminary pruning early in winter, leaving only those canes that are to be used for bearers, and to prune these when their buds are about to swell.   Where long bearers are given, these are not bent down until they have budded and the worst danger of frost is over.

Vines on high trellises have much less to fear from spring frosts than those on low trellises or those that are not trellised at all. This explains why we find the high pergolas in Tyrol and the training of vines on trees in Northern Italy.

9. *Site.*—Vineyards on low-lying land and in depressions are more exposed to frost than those planted along slopes. The reason for this is that the cold air from the heights descends to these low-lying sites, which consequently become the coldest.

10. *Variety of Grape.*—All varieties are not equally susceptible to frost, one reason being that they do not all start budding at the same time. In parts where spring frost is much to be feared, late budding varieties should be grown or varieties like Gamay, which will still give a fair crop although the first shoots and inflorescences may have been killed by frost.

Frost Forecasts.

With a view to protecting the vines against frost, as will be discussed later on, it is of the utmost importance to be able to know beforehand whether there is danger of frost in the coming night. Fortunately this can be done with fair accuracy. The best method for practical use is the *dew-point method.* If the dew-point[1] of the air lies above 0° C. (the freezing-point of water) the water vapour in the air will be precipitated as dew, but if the dew-point lies below 0° C. it will be precipitated as frost. All that is required is to determine the hygrometric state of the air just after sunset. This is most easily done by means of a hygrometer which consists of two thermometers; the mercury bulb of one must be kept dry and that of the other kept wet, and both must hang side by side. This instrument is known as the Wet and Dry bulb hygrometer.

Fig. 72 gives us Lang's night frost curve. Supposing we read the thermometers shortly after sunset and find the dry thermometer showing 10° and the wet one 6°, then the hygrometric difference is 10° – 6° = 4°. On Fig. 72 we now find the vertical dotted line through 10 and the horizontal line through 4 meeting each other before either of them has reached the night frost curve. In this case frost need not be

---

[1] The *dew-point* is the temperature to which the air must be cooled in order to form dew.

feared that night. Had the readings, however, been respectively 10° and 5°, the hygrometric difference would have been 5°, and the dotted lines through 10 and 5 both cut the night frost curve before meeting. In this case it is practically certain that there will be frost that night. The greater the psychrometric difference the drier is the air and hence the lower the dew-point, and therefore the greater the danger of night frosts.

Based on Lang's figures, Maresch and Kappellerof of Vienna constructed a so-called frost-defence thermometer (Manufacturer: Heinrich Kappeller, Vienna V, 1, Franzensgasse 13). In Babo u. Mach[1] this thermometer is described. It gives a reliable indication.

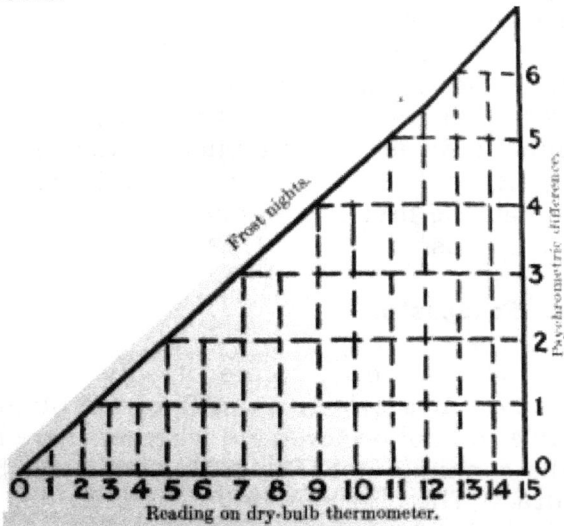

Fig. 72.—Night frost curve after Lang.

Automatic instruments with an electric bell have also been constructed, which are set up outside in the vineyard and ring as soon as the temperature of the air has sunk to the temperature at which the instrument had been set, say, for instance, +1° C. In this way the farmer is warned inside his house that frost may soon form. Amongst others, J. Richard of Paris has constructed such an automatic instrument.

Means of protecting Vineyards against Frost in Spring.

(a) Smoke or Artificial Clouds.—Even in 250 B.C. the grape-growers of Carthage deposited heaps of manure in their vineyards and set them on fire when they feared frost

[1] Babo u. Mach, Handbuch des Weinbaues, Zweiter Halbband, 4. Aufl. 1924, S. 421, Paul Parey, Berlin.

during the night (see Portes et Ruyssen,[1] vol. iii. pp. 290-291). The Romans also knew of the favourable effect of artificial clouds or smoke in preventing frost. In Peru the Incas burnt manure to protect their maize plants against frost if the evening was cold and the sky clear. So this is a very old remedy which is still used to-day very effectively.

For this purpose the farmers in a certain area are organised, and if on any evening there is danger of frost that night, all light their fires simultaneously. Coal-tar is frequently burnt in special receptacles for this purpose and forms excellent smoke. These receptacles stand 11-22 yards apart in rows that are 220-440 yards apart. Each receptacle burns about 33-44 lbs. of tar in four to five hours. It is necessary that smoke should be generated everywhere in such an area to have it thick enough over the vineyards. The smoke should also be made in good time. It is best to make it early in the evening, before too much heat has been lost by radiation, for the smoke has a protective effect like natural clouds in reducing the loss of heat by radiation to a minimum. It is also evident that this method can be applied only on fairly level ground and not on hillsides.

(b) Some Covering or other.—This method is used particularly in districts where frost is to be feared every year and for some considerable time. In the Champagne *straw mats* are fixed in rows some 6 ft. above the ground with openings of about 1 metre (39.37 in.) between the rows. Also the sides are covered with mats, which answer the purpose very well. *Muslin* is suspended in strips over the vineyard along the Moselle and its tributaries, *e.g.* the Ruwer, so as to cover the whole vineyard. This is done when the vines start budding in cold weather, and they sometimes remain covered for eight days continuously. In favourable weather the muslin is pulled to the one side of the vineyard at about an hour after sunrise and spread over again about an hour before sunset in order to protect the vines against frost in the night. This answers admirably. As these vineyards are planted on steep slopes, smoke is out of the question. It has cost the owner of six acres of vines on a steep slope along the Ruwer £225 to cover his vines with muslin, and it takes seventeen men three hours per day to uncover and cover

---

[1] Portes et Ruyssen, *Traité de la vigne et de sea produits,* 3 vols. 1886-1889. Ootave Doin, Paris.

these vines. It is an expensive remedy, but it ensures the safety of the crop.

(c) *Other means* are irrigation, keeping the soil compact and free from grass, late pruning, high trellising, etc., which have already been discussed.

Treatment of Vines Killed by Frost.

As will appear from what follows, the treatment depends upon the stage of development of the vine.

(a)   *The Shoots only* 1½-2 *in. long.*—In this case the shoots killed by the frost are broken off, when some dormant buds will develop and produce grapes.

(b)   *The Shoots* 6-10 *in. long.*—If only the tips of the shoots have been killed and not the inflorescences, we merely cut off the dead tips of the shoots. If the inflorescences have also been killed, we cut back all such shoots with a sharp knife to about $1/5$ in. from the old wood.

(c)   *The Shoots* 14-24 *in. long.*—If merely the tips have been killed, nothing need be done.

Frost in Autumn.

At Wiltingen along the River Saar in Germany I saw a Riesling vineyard on the morning of the 23rd October 1908, which had been badly frost-bitten the previous night. The leaves had nearly all dropped, the stalks of the bunches were as brittle as glass, so that they dropped when touched, and the berries were reddish-brown. This gives them a cooked taste, so that they cannot produce a wine of high quality. If a severe frost comes before the canes are properly ripe, it might happen that not merely the grapes suffer, but also the whole vine, should it, in consequence of the frost, shed its leaves too early.

2. FREEZING TO DEATH IN WINTER

It has already been pointed out that those parts of the shoots that did not ripen well, easily die and turn black in winter. Very intense cold can, however, cause the death of well-ripened wood also and even of the entire vine. If the ground is covered with snow it protects the underground parts of the vine from becoming frozen. All varieties of vines are not equally subject to freezing. In viticultural areas where very severe cold is to be feared in winter, very long cuttings are planted, and frequently, as I saw near Valladolid in Spain in 1909, the vines are covered with soil in winter and opened again in spring. A layer of straw or grass covering the soil helps very considerably to protect the underground parts of the vine against winter cold. Where vines have been frozen to death, they are decapitated above the ground, when they will usually form new shoots which can serve to build up a new vine.

3. HAIL

Hail storms usually occur in summer, when they can destroy not merely the whole crop, but can also damage the shoots to such an extent that even the following crop is affected. It is well known that certain regions are more in the track of hail storms than others. One should, as far as possible, refrain from planting vineyards in such places. Apart from the direct damage, the wounds caused by the hail constitute splendid points of entrance for fungi and other enemies of the vine.

Remedies against Hail.
(a) *Shooting with Hail Cannons.*—They have funnel-shaped barrels and are not charged with a ball. The idea is to shake the clouds and to cause rain to fall instead of hail. In France (especially the Cote d'Or), Northern Italy, Styria, etc., this remedy is applied with uncertain results.

(b) *Paragrèles.*—These are high electric towers which are built in rows on high elevations to protect entire districts. They send out powerful electric waves and seem to be fairly successful in the South of France, although their success is doubted by some.

Treatment of Vines Damaged by Hail.
If vines are damaged by hail early (June in Europe and before the end of the year at the Cape), the vines should be cut back severely in order to develop new shoots that can provide bearers for the next year. Should it happen later in the year, it is best to do nothing until the next winter pruning, and thus keep the resting eyes dormant.

4. DROUGHT

Like other plants the vine requires a certain amount of moisture in the soil for its normal development. If the soil is too shallow, the lateral drainage too severe, or the air too dry, vines easily suffer from drought in a dry summer and autumn. Vines suffering from drought have a dull colour, and later on the leaves partly dry up and the grapes wilt or burn and ripen badly. The adult leaves are seen to grow smaller and the berries remain small under the influence of drought. When they are finally ripe they give a must with a high acidity and frequently contain a great deal of free tartaric acid,[1] owing to the strongly reduced absorption of mineral salts from the soil. According to the intensity of the drought, the sugar content can be poor or good. The crop will naturally be small. The grower of table grapes should see to it that his vines are well supplied with moisture until the crop has been harvested, as the value of his grapes mainly depends upon the size of their berries. Later on I shall revert to this point.

Remedies against Drought.
Loosen the soil to a good depth before planting the vines, irrigate when desirable and possible; keep the top soil loose and free from weeds, keep the vines topped fairly short, use American stocks that are fairly drought resistant (like Rip. gloire, 106-8, Rup. du Lot, 1202, etc.) and suitable in the particular soil, and plant the vines far apart, say 6 ft. × 6 ft. square or 36 sq. ft. per vine.

5. SUNBURN

Vines suffering from drought are more susceptible to sunburn than those that have enough water at their disposal. Grapes will also burn more easily if the early summer has been wet and rainy and very hot weather suddenly sets in, since the berries are then tender and have thin skins. The bloom on the berries is a natural protection against sunburn; therefore where it is rubbed off by prethinning the bunches (table grapes) too late or by strong

---

[1] See Fonzes Diacon, *Revue de Viticulture,* lix. p. 231 (1923).

winds, such berries will burn much more easily than those that have their bloom intact. If vines have been growing vigorously and the bunches have been hanging in the shade all the time, the grapes can easily burn if they are suddenly exposed to the direct rays of the sun by removing shoots or leaves. Where the moisture content of the soil is the same, grapes will sooner be sunburnt in a dip or valley than on a slope where there is usually a breeze of some sort and the air does not get quite so hot.

Grapes usually burn with calm dry hot weather, but can also burn if a dry hot wind (temperature of air say 104° F. or 40° C.) blows for twenty-four hours or longer. The most dangerous period for sunburn is from when the berries are nearly full grown till when they begin to ripen. In the Western Province of the Cape this is between 15th December and the end of January, 30th December to 3rd January being usually the worst. Once the grapes are fairly ripe, the danger of sunburn is over.

Sometimes only odd berries get burnt and dry up, but sometimes the stalk may get burnt, and the whole bunch perishes. If great heat follows shortly after the vines have been sulphured, spots are easily burnt on the berries, which spoil them completely for sale as table grapes. Grapes on high trellises are much less exposed to sunburn than those near the ground. The most dangerous part of the day is from 11 A.M. till 4 P.M. As the air does not get equally hot everywhere, and as the moisture content of the soil varies a good deal on different spots, sunburn is not equally bad everywhere. Again, the different varieties of grapes are not equally sensitive to sunburn. The most sensitive are Gros Maroc and Flaming Tokay; less though still fairly sensitive are Muscat of Alexandria, Gros Colman, Henab Turki, and Hermitage; Muscadel, Greengrape, and White French (Palomino) are considerably less sensitive, whilst Tribodo bianco and Rosaki hardly ever become sunburnt. Sunburn can cause the greatest damage to table grapes.

Remedies against Sunburn.
1. See that the soil contains enough moisture,
2. Do not damage the bloom.

433

3. Do not sulphur the vines during the dangerous period if hot weather prevails then.

4. Do not grow varieties that burn very easily in your locality.

5. Harden the grapes by exposing them to direct sunlight from the beginning, by removing some of the bottom leaves, but at the same time try to have them in the shade from 12 noon–3 P.M.

6. Let the rows of trellised vines, wherever feasible, run east and west, as the sun will then remain on the rows and the grapes will be more in the shade. Try to have the bunches on the southern side of the row (in the southern hemisphere), as it will be the best shaded, unless the shoots are tied to side wires, when the bunches will generally hang in the shade.

7. Train very sensitive varieties on high pergolas or trellises, when the foliage will protect the grapes against the rays of the sun.

8. Use vigorous stocks that are drought resistant.

9. Start topping early to get strong shoots with sufficient laterals to provide shade for the grapes.

6. CHLOROSIS *(Icterus)*

Chlorosis is recognised by the yellowish-green to yellow colour of the leaves, which is frequently seen before flowering, when the berries set very badly. If it is bad, the leaves later on turn brown and drop. It is then usually worse in the second year and may cause the death of the vine if there is no early improvement. Often the leaves turn green again later in summer and the disease gets less and less in succeeding years. The internodes of the young sick shoots are shorter than is normally the case.

Causes.—Different causes can bring about this disease. It is, however, always due to an *unhealthy state of nutrition* of the vine. Hence it is a very old vine disease, which, however, is more troublesome among the grafted vines than it was formerly among the ungrafted European vines. In this case, the unhealthy state of nutrition can be due to:

(a) Wet, badly ventilated soil, which can be put right by good drainage;

(b) lack of potash, which can be rectified by an appropriate potash dressing;

(c) lack of iron in the leaves in a *soluble* state, which is usually associated with the presence of *too much lime* in the soil;

(d) too dry soil, which causes chlorosis through under-feeding the vines;

(e) too cold soil and air;

(f) hereditary chlorosis through propagation from cuttings or scions taken from chlorotic vines;

(g) Chlorotic symptoms which are the result of certain definite parasitic diseases will be discussed under these.

As by far the largest number of cases of chlorosis in vines are caused by the presence of too much lime in the soil, we can suitably speak here of

Lime Chlorosis.

In 1907 Dr. Emil Molz[1] published an important study of this disease. In discussing the favourable effect which an application of ferrous sulphate to the soil or washing the wounds made by pruning with a 35-40 per cent solution of ferrous sulphate according to Rassiguier has on lime chlorosis, he denies that the iron as such brings this about. He says *(l.c.* p. 14): "The ferrous sulphate applied to the soil thus acts partly by destroying the calcium bicarbonate, and also directly and indirectly by making the food ingredients in the soil soluble".

In order to prove that this chlorosis is not due to lack of iron, he quotes *(l.c.* p. 10) amongst others the following ash determinations by Schulze:

Fe_2O_3 in healthy leaves 1.26 per cent; in sick leaves 1.58 per cent.
Fe_2O_3 in healthy wood 0.53 per cent; in sick wood 1.24 per cent.
K_2O in healthy leaves 13.02 per cent; in sick leaves 5.29 per cent.
K_2O in healthy wood 32.20 per cent; in sick wood 16.42 per cent.

From these analyses it appears that the sick leaves and wood contained only about half as much potash as the healthy leaves and wood, but *more iron.* Hence Molz con-

---

[1] Molz, Dr. Emil, "Untersuchungen über die Chlorose der Reben," *Centralblatt für Bakt.,* II. Abt., Bd. xix., 1907.

cludes that the chlorosis cannot be attributed to lack of iron. He is of opinion that plenty of lime in the soil can cause an alkaline soil reaction, which easily causes the fine roots to rot and the leaves to obtain so much calcium bicarbonate from the soil that they show a neutral or even faintly alkaline reaction. This plant sap is the reason why even the still healthy roots can absorb the foodstuffs, especially potash, from the soil only with difficulty. The lack of potash hinders the carbon assimilation (photosynthesis), to the detriment of the whole plant. He further surmises that the yellow colour of the leaf is due to the action of the alkaline plant sap upon the chlorophyll pigment.

Viala and Ravaz[1] *(l.c.* p. 33) say: "Whatever the case may be, although the question has not been sufficiently elucidated, carbonate of lime is the true cause of chlorosis".

It is generally admitted that excessive quantities of carbonate of lime in a soil can cause chlorosis, but opinions differ as to the way in which it acts. Contrary to the opinions expressed by Dr. Molz, we must assume that lime chlorosis is due to lack of iron, but, be it noted, lack of *iron in solution.* What happens is more or less what Dr. Molz has stated, but to this we should add that the iron becomes insoluble in the neutral or faintly alkaline plant sap in the leaf, which causes a lack of dissolved iron (though there be plenty of undissolved iron present), which in turn hinders the formation of chlorophyll and thus causes the yellow colour of the leaf.

In the presence of large quantities of calcium bicarbonate in solution, the dissolved iron in the leaf will be or will pass into the form of ferrous hydrocarbonate which, through the loss of carbonic acid, changes into the carbonate which is insoluble in a neutral or alkaline medium:

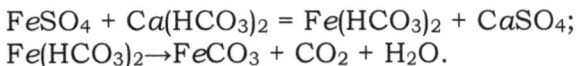

$$FeSO_4 + Ca(HCO_3)_2 = Fe(HCO_3)_2 + CaSO_4;$$
$$Fe(HCO_3)_2 \rightarrow FeCO_3 + CO_2 + H_2O.$$

This explains both the chlorosive effect of lime in the soil and the temporary healing effect of spraying chlorotic leaves with a 1 per cent ferrous sulphate solution, and should therefore be accepted as being the most satisfactory

---

[1] P. Viala and L. Ravaz, *American Vines, Adaptation, etc.* Transl. by Dubois and Wilkinson, Melbourne, 1901, from *Les Vignes américaines,* Montpellier, 1892.

explanation. Here the excessive amount of lime in the soil still remains the primary cause. Whereas lime is usually present in the soil as carbonate and hydrocarbonate (or bi-carbonate), it can occur as calcium hydrate just after the soil has been given a very heavy dressing of quick or slaked lime, which can for some time cause a marked alkaline soil reaction. This hydrate will cause the soluble iron in the soil solution to separate out as insoluble hydrate or basic carbonate, and thus hinder its absorption by the roots.

Thus the *chlorosive power* of a limestone soil depends mainly upon the quantity of calcium hydrocarbonate which is present in the soil solution, and this again depends upon the subsoil's content of water, humus (as potential source of carbon dioxide), and calcium carbonate; further, it will depend upon the soil's temperature and upon its state of drainage. Soft, porous, fine limestone, such as chalk, will naturally dissolve more easily than a hard limestone or, still worse, a dolomitic limestone. If the lime is mixed with clay it will naturally dissolve more slowly than when mixed with sand. The clay will therefore diminish the chlorosive power of the soil. A good irrigation or good rains in spring or summer will therefore also make the lime chlorosis worse, whilst drought will have the opposite effect. This is precisely what European experience has taught us.

Fertilisers like nitrate of soda with a physiologically alkaline reaction will make chlorosis worse.

In Chapter IV. we have seen that all the American stocks are not equally sensitive to lime chlorosis. The reader will find there a classification of the better known stocks with regard to their sensitiveness to lime in the soil, and should bear in mind what has been said above concerning the chlorosive power of lime under different conditions.

It will now be clear that we ought to know how much lime a soil contains before we can with fair safety select a suitable stock for it. For this reason soil samples have systematically been taken in Europe in districts where the soil is rich in lime, and their lime content has been determined; a *lime chart* has been made for every district, and could serve as a guide in choosing suitable stocks for it. At the Cape I have done this work for all the vineyard soils of Montagu and for a number of the vineyard soils of Robertson (see Chapter IV.). The result is that very few of these soils will cause lime chlorosis.

For this purpose the lime is determined by means of a calcimetre, for instance that of Bernard (see Guillon, *l.c.* pp. 372-374). The registering Calcimètre Houdaille can be used to determine the rate at which the lime dissolves. The *rate of solution* or "vitesse d'attaque" is expressed by the number of mg. $CaCO_3$ dissolved per second from 1 gr. of limestone soil by a 25 per cent solution of tartaric acid. This rate of solution × per cent $CaCO_3$ in the soil very nearly gives us the chlorosive power of the soil (see Guillon, *l.c.* pp. 374-380).

In Chapter VII. we have seen that the different European varieties differ in their sensitiveness to lime chlorosis, and how grafting affects it. Chlorosis is worse with grafting on the spot than when grafted vines are planted out. Where the stock is grafted on the spot, it is weakened in its second year, when chlorosis usually is most to be feared. If cuttings instead of rooted vines are planted out directly in the vineyards, as has happened at Cognac, and if they are longer and are therefore planted deeper than rooted vines, they suffer more from chlorosis, as their bottom roots must grow in soil richer in lime.

Vines suffering from attacks by phylloxera will suffer from chlorosis sooner than healthy vines, as they are then generally weakened.

Treatment of Chlorosis.

This must be according to the circumstances, and can be briefly summed up as follows:

1. If wet, badly ventilated soil is the cause, proper drainage should be applied.

2. If lack of certain food ingredients, *e.g.* potash, is the cause, the necessary fertilisers should be given.

3. If the soil reaction is neutral or faintly alkaline, apply a heavy dressing of superphosphate; where this occurs in soil very rich in lime, use stocks with a high lime resistance. The safest procedure is first to lay out experimental plots with different stocks and different scion varieties, and to apply the experience thus gained in selecting the stocks.

7. COULURE AND MILLERANDAGE

These are related phenomena. In case of coulure or non-setting, the pedicels and pistils drop after flowering, so that nothing of the flower remains behind. It sometimes happens that the pistil does not drop, but does not develop into a normal berry either. The little berry remains green and quite small when the normal berries are ripe, or it grows larger but still remains much smaller than the normal berries. It is *seedless,* always *round,* even when the normal berries are oval or elongated *(e.g.* Muscat of Alexandria, Rosaki, etc.), and ripens earlier than the normal berries. In this case the French speak of "millerandage" (see Fig. 73).

Müller–Thurgau[1] has carefully studied the question of seedless grape berries, and came to the following conclusion: "1. When fertilised by their own pollen and when foreign pollen is excluded, the ordinary grape varieties form berries with seeds. 2. If every influence of pollen is excluded *(i.e.* if the flowers are emasculated and foreign pollen is kept away), some varieties can form bunches with seedless or parthenocarpic berries, others not. 3. In the partheno-carpically formed berries (as under 2) the seeds can also grow to a certain size, which depends partly upon the variety and partly upon the supply of organic food. Such seeds which have been formed without

FIG. 73. — Millerandage. From Foëx, *Cours complet de viticulture,* 1895. Coulet et fils, Montpellier. G. Masson, Paris.

pollination are, however, always empty, containing neither embryo nor endosperm. 4. It is of no advantage in our viticulture to give preference to varieties where parthenocarpy[2] can take place. 5. In planting large areas, one variety can safely be planted, as our cultivated varieties are independent of foreign pollination (cross-pollination), and bear just as surely and heavily, and form just as large

---

[1] H. Müller-Thurgau, "Kernlose Traubenbeeren und Obstfrüchte," *Landw. Jahrb. d. Schweiz,* 1908.

[2] *Parthenocarpy* means virgin fruit formation, *i.e.* formation of fruit without fertilisation.

berries with seeds, as when fertilised by other varieties.   6. The parthenocarpy of the vine is influenced by the nutrition of the flowers, *i.e.* by the more or less copious supply of sugar to them. 7. By ringbarking the fruit-bearing shoots, they become enriched in sugar and starch, which favours parthenocarpy. The inflorescences then become bunches of grapes with seedless berries, even with varieties where, without ringbarking, the berries drop without developing any further. 8. It must still be inquired into in how far penetrating pollen-tubes can participate in the formation of seedless berries. ..."

From this we see that seedless berries arise through not being fertilised, and that it can happen with some varieties without any irritation by a pollen-tube being required, whereas such irritation is absolutely necessary with other varieties, as the pistils would otherwise simply drop. With coulure also no proper fertilisation has taken place.

The question now arises, *why fertilisation does not take place at all or does not take place well.* The reasons are:

(a) **Abnormal Structure of the Flower.**

Male flowers naturally cannot form any fruit. Female flowers ordinarily cannot form fruit without cross-pollination, as their pollen is usually sterile. The normal structure of the flower with which we are dealing here refers to a normal hermaphrodite flower, *i.e.* one with well-developed male and female organs. Several grape varieties that are very susceptible to coulure and the formation of seedless berries (millerandage) possess very short and sometimes somewhat bent stamens, although their pollen may not be sterile, *e.g.* Ohanez, Bicane, etc.   Others again sometimes possess only four stamens, *e.g.* Pinot Chardonnay, Sabalkanskoi, Sultana (which always forms seedless berries). Double flowers *(Chloranthy)* we find with Clairette, Malbec, etc., which are very susceptible to coulure. The double flowers naturally are not fertile (see Foëx, *l.c.* p. 495).   Sometimes the fault lies with the pistil. This is particularly so with varieties like Sultana, Currant, etc., the berries of which are always seedless.

(b) **Abnormal Vigour.**

It may be that the vine grows too vigorously on account of too high a nitrogen dressing, or because it is pruned too short and requires a larger development. But it might also be that the vine is too feeble, owing to over-production or under-feeding or drought or a diseased root-system (phylloxera, too wet soil, etc.).

(c) Inclement weather during Flowering.

Rain or very strong winds hinder the transport of pollen to the stigma both directly and by hindering the insects in their work that visit the flowers. Cold weather during flowering and setting hinders the development of the pollen-tubes and the real process of fertilisation.

Remedies against Coulure and Millerandage.

(a) If the trouble is due to an *abnormal structure* of the stamens, whilst the pistils are normal, artificial pollination by carefully passing over the flowering bunches a couple of times with a soft woollen brush helps if the pollen of that variety is not sterile (this is done with Ohanez in Almeria); otherwise cross-pollination must be resorted to. Further, *only properly ripe canes that have themselves borne good bunches of grapes should be used for planting out or as scions.* They should be marked in good time during summer.

(b) If *abnormal vigour* is the cause, the treatment should fit the special case. An excessive nitrogenous nutrition is best corrected by a heavy dressing of phosphates and potash without nitrogen, *i.e.* superphosphate or basic slag and sulphate or muriate of potash.   Very vigorous growth and coulure which are due to too short pruning can best be corrected by giving the vines *long* and short bearers at the winter pruning, say four long bearers (of 12 in.–18 in. each) and four short ones per vine, or more if each vine has more than 30 sq. ft. of soil to itself.   In addition, it is good to prune such vines very late—when they are near budding—which will cause a heavy loss of sap by bleeding and later budding and flowering, thus giving the vines probably more favourable weather for flowering. Topping before and during flowering sometimes also helps.

If there was lack of vigour because the soil was too wet and cold, drainage will be effective. If the soil was not too wet but too poorly supplied with plant food, or if the roots were attacked by phylloxera or another disease, a complete dressing of fertiliser rich in nitrogen will help, unless the vines are already too weak and had better be uprooted. If the low vigour was caused by over-production, short and less bearers should be left and heavy manuring given, and, in addition, all the inflorescences should be removed during the first year of the treatment in order to enable the vines to recover. If it is a case of chlorosis, apply the remedies already prescribed.

Some vines grow too vigorously, because their berries set very badly, so that they produce hardly any grapes, and because they themselves were propagated from precisely similar vines. If long pruning does not help here, they should be grafted over with scions from vines that bear well. Select the canes for scions very carefully and do not take any from bad vines.

(c) If *unfavourable weather* during flowering is the cause, we are almost powerless. By pruning the vines very late we can delay the date of flowering, and thus sometimes obtain more favourable weather during flowering. By planting suitable windbreaks and bushes, a good deal of protection against cold and strong winds can be supplied. If the weather has been very dry from the time the vines budded and the soil was too dry during flowering, irrigation will help if given shortly before flowering. Such a case was brought to my notice at Robertson. Irrigation helps here, as the vine then procures more nourishment from the soil. Do not plough or cultivate the vineyard during flowering if there is danger of coulure, as this will make it worse.

It is good to sulphur vines when flowering. The sulphur may possibly have a useful irritating effect on the development of the process of fertilisation, or it may help merely mechanically in bringing the pollen on to the stigma.

Finally, we should give preference to varieties which give well-filled bunches under local conditions and with proper treatment. Hence we should consider carefully beforehand the choice of varieties to be grown. In this connection the stock should not be forgotten. A very vigorous stock like 1202 or Rup. du Lot frequently induces too great vigour and too much coulure in the vine grafted on it. Varieties that are very subject to coulure should be grafted on stocks with more moderate vigour.

Ringbarking is a sure remedy against coulure in certain respects. It is discussed in Chapter XII.

8. BRUNISSURE OR BROWN LEAF DISEASE

Ravaz[1] has made a thorough study of this disease. He shows also that Viala,[2] *l.c.* p. 403, was wrong in attributing it to a myxomycete or slime fungus. He states that the disease usually shows itself late in the season, September to October in France. If this happens before the grapes turn colour, it is very bad. It never appears after the vintage. On white varieties it shows as yellowish-brown patches on the upper surface of the leaves, and as brown, nearly black matches with red and black varieties. The patches are small but numerous, and soon merge into each other, when they form yellow or dark brown fields which may extend over the whole leaf. The lower surface of the leaf remains green for a long time, but finally the discoloration penetrates the whole leaf. This happens particularly where the disease has shown itself early. In this case the leaves drop before or after the vintage, which is bad for the quality and quantity of the crop and for the vine's future prosperity.

Where the discoloration does not penetrate right through the leaf, it often drops at the same time as the healthy leaves in autumn. Only the leaves are directly affected, and at first only the lower ones, the upper ones being, affected in serious cases only. Ravaz comes to the following conclusions *(l.c.* p. 181):

"1. Brunissure is not a parasitic disease. ...
"2. The state of the weather naturally controls the development and life of the plant, but in general cannot make normally constituted leaves turn brown. ...
"3. Brunissure is the result of over-production, which causes the impoverishment and exhaustion of the tissues.
"4. Hence causes that lead to the impoverishment of the tissues (downy and powdery mildew, etc.) or that will retard the nutrition of the plant, will favour brunissure. Inversely, all factors favouring the plant's nutrition will also assist the vine against brunissure. It is therefore an easy matter to prevent this disease,

---

[1] Ravaz, L., *La Brunissure de la vigne,* 1904; Coulet et fils, Montpellier, and Masson et Cie, Paris.
[2] Viala, P., *Les Maladies de la vigne,* 3ᵉ ed., 1893.

which, by the way, becomes rarer and rarer as the vines grow older or bigger. Brunissure is a disease of young vines."

I have not yet observed this disease in the Cape vineyards.

9. APOPLEXY OR STROKE

As the name suggests, this disease appears suddenly through the end of a shoot and its leaves suddenly wilting. Sometimes this happens only to a few and sometimes to all the shoots on the same arm or to most of the shoots on a vine. Within a few days such shoots and the whole vine may be dead. Sometimes only one arm dies. It is best to dig up such vines and replace them with others.

The disease is caused through the vine or shoot evaporating more water than is taken up in the same time from the soil. Thus it occurs when dry hot weather suddenly follows upon a cool, rainy, early summer. It is of most common occurrence where the subsoil is wet. It is always only a few vigorous vines that are thus affected here and there in the vineyard. According to Viala it occurs everywhere, but is most common in cool, wet, deep soils, *e.g.* wet sandy soils and fertile alluvial river-soils. The hot and dry Siroco (wind) sometimes causes it in Algeria.

A grafted vine with a poor joint may easily develop this disease if it has grown vigorously in early summer. Where the subsoil is too wet, drainage will help. I have repeatedly come across cases of apoplexy in Cape vineyards, but not to any extent. It has always been limited to an odd vine here and there.

What has thus far been described is a kind of apoplexy also known as "folletage" in France. There is a totally different vine disease also called apoplexy in France, which Viala named "maladie de l'esca", and which is caused by a fungus, *Stereum necator.* In California (see *California Grape Grower,* April 1926) it is known as *Black Measles.* This disease is most prevalent amongst the older vines (15 years and older). P. Viala, who studied it for many years, recently published a memoire on it in the *Annates des épiphyties.* In discussing the treatment against this disease, Viala[1] recommends spraying the vines, after they have been winter pruned, with an arsenical liquid that can be made up as follows:

Arsenious Acid (White Arsenic)	20 lb.
Sodium carbonate	15 "
Soap (preferably black)	18 "
Water	10 gallons.

This spray is diluted 15–20 times with water before being used. It is very poisonous, and should preferably be bought from some manufacturer, ready for use after being adequately diluted. Pyralion and Pyrafollol are examples of such preparations.

In *Le Progrès agricole et viticole* of the 7th March 1926, A. Bachala recommends that the spraying of the vines with an arsenical solution against apoplexy should take place at least 10 days after winter pruning, and, in order to avoid burning, before the buds begin to swell. Treat two years in succession and then skip two years, when treatment is resumed.

10. RONCET

Viala, *l.c.* pp. 422-423, considers the *roncet* of Burgundy and the *court noué* of the South of France to be one and the same non-parasitic disease. The shoots of the sick vines are thin, have short internodes, and are much branched; the leaves are small and fingered like the parsley leaf. The bunches set their berries very badly, suffering from both coulure and millerandage. The leaves look thicker than usual but remain green. The vines steadily become weaker until they die. Later on we shall meet with another kind of court noué.

Ravaz says that roncet was known in France prior to the outbreak of phylloxera, but that it became worse in the grafted vineyards. According to French publications, roncet became more serious after 1890. Where diseased vines have been uprooted and replaced by healthy ones, these also

---

[1] P. Viala, "Traitements de l'esca," *Revue de Viticulture*, Tome LXIV., 18th March 1926.

became sick. Rup. du Lot and Riparia are amongst the most sensitive to it.

Pantanelli[1] has made a very thorough study of this disease in Sicily and published his results in 1911. He distinguishes between three stages of this disease with Rupestris, which in his opinion shows the disease in its most typical form.

First stage: the leaves become fingered like a parsley leaf (long teeth with deep incisions) and remain small but green; second stage: the shoots have short internodes; third stage: the leaves develop light patches—Pantanelli calls this "mosaico della vite", or mosaic disease of the vine.

By planting cuttings that showed the different stages, he succeeded in propagating the three stages on different vines. With Berlandieri the fingered leaves are wanting, and are replaced by an irregular, bullate leaf; already in the first stage we find short internodes and mosaic patches. With the Viniferas, those varieties which normally possess incised leaves show the fingeredness and short internodes, whilst those that normally are more or less entire show the symptoms observed with Berlandieri with very short internodes. The mosaic patches occur more often with the second than with the first group.

We further notice a branching of the shoots, which, however, remain green, and the inflorescences that appear set no fruit. With the exception of the fingeredness of the leaves, the symptoms are observed also with other diseases. Hence Pantanelli used the *fingeredness* as the characteristic symptom of the disease.

Causes of the Disease.

According to Pantanelli it is not a parasitic disease. Diseased cuttings again give sick vines but cannot infect healthy cuttings when stratified together. He comes to the conclusion that one should not look for the cause of the disease in the portion of the vine above ground, but in its root system.

---

[1] Pantanelli, E., Roncet in *La Viticoltura moderna,* Anno XVII. (1911), Nos. 10 and 11

The formation of new hair-roots usually begins at least a month before the vines start budding. With the sick vines this takes place later or to a lesser extent, and may even not take place at all if the vines are already very sick. In Sicily the formation of hair-roots stops in July (in non-irrigated vineyards), and begins again with the first rains in autumn. The recovery of sick shoots, which occurs in all cases of initial roncet after the sickly development of the first month, follows upon the formation of new hair-roots. Also cuttings that have been planted out can recover—either immediately or as they grow, and will do so sooner if they form roots readily in a certain soil. The thin canes recover more easily than the thick ones, as they contain less of the harmful substances and sooner develop sufficient roots. Treating diseased canes with warm water or different solutions just when they are planted, helps a good deal, as this promotes the formation of roots.

"With the formation of the roots and the recovery of the young vines, the normal relations between the ash constituents are also restored; the formation of albuminoids is intensified, the phosphoric acid and magnesia are increased, the lime and potash diminish" (Pantanelli, *l.c.* p. 19). This makes us think that the roncet of the portion of the vine above ground is connected with the inadequacy of the absorption system, which is also shown by the characteristic *spreading of the disease according to the soil conditions.* The chemical analysis of the soil of patches of vines suffering from roncet has shown nothing abnormal. The physical analysis of patches of soil where healthy cuttings have developed roncet has brought out characteristics which can be summed up as follows: "These consist of most minute particles, amongst which those predominate that can easily be washed away, and are poor in quartz pebbles, sand and coarse limestone. Here the top soil is not very deep; the subsoil is compact and rests upon a hard or impenetrable pan of marl or of a cemented clayey-sandy mud, or of a spongy and wet rock, formed into a depression, in any case with bad or no drainage. The result of this is the compactness, bad ventilation and difficult drainage of the layer of soil in which the deepest roots grow" (Pantanelli, *l.c.* p. 19). This confirms the results of previous investigators.

By bringing soil from sick vines directly into boxes without breaking the sod or removing the rootlets and then planting healthy cuttings (especially from susceptible varieties) in it, they will, according to Pantanelli, develop roncet in the first or second leaf (year). Thus roncet can be reproduced experimentally. Sterilisation of the diseased soil with dry heat causes the vines to strike root less well, but is sufficient to remove the toxic effect of the soil. When diseased soil, freed from roots, dries out in direct sunlight, it loses its toxicity. The same is done by 2% lysol, but 2‰ lysol does not do it.

Pantanelli is of opinion that the toxic substance belongs to the *enzymatic toxines* and, according to him, its origin should be looked for in the process of the slow death to which the bits of the roots spread in the soil have been subject. Clay soil absorbs the poison and holds it.

He recommends that vines be planted on healthy soils which, through their structure and site, do not lend themselves to an accumulation of the toxic substances; by previously draining and ploughing deeply soils on which diseased vines have been uprooted, and especially by first growing cereals on such land for some years before again planting vines on them, and then planting varieties that are only slightly susceptible to the disease, we can ensure success.

He classifies the American stocks as follows, beginning with the most susceptible ones:

I. Rup. du Lot.
II. 3306, Berl. Resseguier No. 1, Rip. tomentosa.
III. Aramon × Rup. Ganzin No. 1, 3309; Rip. Grand glabre, 420A.
IV. Berl. Résséguier No. 2; Rip. gloire; Rip. × Cordif.—Rup. 106-8.
V. Rup. Metallica, Rup. Martin, Rup. Ganzin, Solonis, Rip. × Solonis 1616.
VI. Cabernet × Rup. 33A, Mourv. × Rup. 1202, Auxerrois Rup.
VII. 157-11, 34 E.M., 101-14.
VIII. Rup. × Berl. 1737, 220A, 301A.

At the Cape this disease has not yet been specially studied as it is of no importance, though it seems to occur occasionally.

11. THE CALIFORNIA VINE DISEASE OR ANAHEIM DISEASE

This disease broke out in California in 1884, and was very serious in 1886, especially in Anaheim, where it destroyed thousands of acres of vines. In 1892 N. B. Pierce[1] published the results of his investigations about this disease, in which he left the cause of the disease an open question. Viala, *l.c.* p. 412, attributes it to a myxomycete, *Plasniodiophora californica,* but Ravaz[2] came to the conclusion that it is a physiological disease, and that it is caused by over-production, as in the case of brunissure.

The sick vines grow badly, ripen their wood badly, and show yellow, later on red or reddish-brown to blackish-brown patches between the nerves of the leaves and along the borders of the leaves. "The unaltered nerves are always surrounded by a green border. The leaves are decidedly variegated, and dry up and curl round along their edges. They frequently drop in spring or in summer. The young leaves developing on laterals are also changed. The inflorescences dry up on the vines and drop" (Viala, *l.c.* p. 411). We find practically only the symptoms of over-production with the subsequent exhaustion of the vine.

12. THE RESULTS OF OVER-PRODUCTION

Ravaz[2] has made a careful study of over-production and has found it to be the cause of several physiological diseases of the vine, such as brunissure, the California vine disease, etc.

Some years ago I investigated a typical case of over-production near Durbanville in a cool, deep, well-trenched gravelly soil. It was Muscadel grafted on Jacquez (Lenoir) and trellised on wire. In its fourth leaf this vineyard showed a patch of vines which looked very poor and gave the impression that these particular vines were suffering from phylloxera. On examining the roots I found them to be badly infested with phylloxera, which had already damaged them

---

[1] Pierce, N. B., *The California Vine Disease,* U.S. Dept. of Agric, Div. of Veget. Pathol., Bull. No. 2, 1892.

[2] Ravaz, L., *Influence de la surproduction sur la végétation de la vigne,* Montpellier, 1906, Coulet et fils.

very considerably. As Jacquez should have answered well in this soil, the attacks by phylloxera could be attributed only to a weakening of the vines, which in turn should have been caused by over-production. These conclusions were perfectly correct. The owner afterwards told me that these vines were trellised in their second leaf, when they were long pruned and bore a fairly heavy crop. In their third leaf they were also long pruned and bore a *heavy* crop. In their fourth leaf, when I examined them, the result of this over-production was clearly visible and serious. On my advice all the grapes were at once taken off the weak vines, which were pruned short and heavily manured the next year. During last year the weak vines, according to their vigour, were not allowed to bear any grapes or only a few. Now they have recovered, but had they not been properly treated in time, they would by now have been dead.

On poor soils, as has often happened with Stein and Hermitage on Aramon in Helderberg (Cape), and in an unfavourable season (especially one of drought), over-production can give the phylloxera such a hold on grafted vines that some are weakened very much and die. Where the stock is not very highly resistant to phylloxera, over-production can soon weaken the vine sufficiently to let it succumb to phylloxera if it is not aided in time by a drastic reduction of the crop and by extra manuring.

Ravaz, *l.c.* (just quoted), pp. 13-15, says that numerous cases occurred in the South of France in 1905 which made the wine farmers fear that a new disease had broken out or that the phylloxera had again attacked the grafted vines. There were also patches in the vineyards where the vines grafted on Jacquez, 3306, 3309, Riparia, Rup. du Lot, Aramon × Rup. Ganzin No. 1 were very poor. This happened after there had been over-production in 1904 in many instances. He further mentions numerous instances of vines in rich, deep, moist soils which had been most vigorous and had borne very heavy crops in 1904, and had been much weaker in 1905. Thus on p. 18 he mentions the following case: "Near Montagnac, again, in a deep, fertile, moist soil which forms depressions in some parts, M. Dessalles planted a block of Carignan four years ago. In 1904 the vines were very vigorous and bore a heavy crop, which ripened none too well, as the leaves turned brown and dried up towards the time for gathering the crop. In the spring of 1905 many of the vines, especially in the depressions, budded badly and formed only very thin shoots. One finds here again the same symptoms as in the previous cases: bearers, arms, vines on the whole live, but they are *empty* [of stored food, A.I.P.], *exhausted;* the roots are more or less rotten; their living tissues likewise are empty."

In Tunis some cases of deterioration of vines in 1905 caused the "Direction" of Agriculture to ask Prof. Ravaz to come over for investigating these cases. Here also he came to the conclusion that the trouble was not caused by some new or old disease, but that it was simply the effect of over-production by *young* vines. The microscopic examination of still living tissues of the weak vines has shown that their cells contained no starch and that their nuclei were either very small or altogether wanting. In 1904 the vineyards in France, Algeria and Tunis bore very heavy crops!

The investigations of Bioletti and Twight (quoted from Ravaz, *l.c.* p. 44) concerning the deterioration of certain vineyards in the Santa Clara Valley, brought them to the conclusion that this deterioration was caused by *over-production* and *drought.* This led Ravaz to think that the California vine disease of Anaheim might be due to the same causes or to over-production alone. Bioletti and Twight pointed out that the varieties most affected are those that bear most heavily, and that the poor croppers are the ones to escape. The deterioration took place in 1898-1900 (many vines then died), and the vineyards produced very heavy crops in 1896 and 1897, whilst 1898-1901 were four very dry years. They were of opinion that the California vine disease of Anaheim was something different. After having carefully studied their publication, Ravaz was of opinion that both cases correspond with the numerous cases of exhaustion through over-production which he had studied in France, Tunis and Algeria.

Whilst Pierce considers over-production as the first sign of the disease (as when ungrafted Viniferas are just being attacked by phylloxera), this, according to Ravaz, is the cause of the California vine disease. At the end of his study on *Influence de la surproduction sur la végétation de la vigne,* pp. 73-79, Ravaz makes the following

"1. All the cases of deterioration in Tunis, Algeria, France, Austria, and California which we have studied, came about suddenly: everywhere they developed with great rapidity. The previous year, from spring to autumn, nothing made the *grape growers* foresee them. The vegetation was fine, and, what was a more reassuring sign, the vines were *laden with fruit.* It was during winter that the vine began to dry out, and it was in spring that it was observed that it did not grow or grew badly." He then proceeds to point out that it was the large number of bunches that exhausted the vines.

"2. All vines can be exhausted by their grapes. Even the least productive (varieties, A.I.P.) sometimes, under specially favourable conditions, bear more bunches than they can nourish properly. Consequently we should not be surprised to find all the productive grape varieties offering cases of deterioration such as we have studied. But it goes without saying that the most productive (varieties, A.I.P.) are also most subject to this trouble."

At the Cape such cases of deterioration were limited mainly to Hermitage and Stein, both of which are very productive. Ravaz quotes the case of Alicante Bouschet, which produces up to 250 hectolitres wine per hectare (about 18 leaguers or 2225 gallons per acre) in Algeria. Near Durbanville (Cape) on the farm Evertsdal, trellised Hermitage grafted on a Rupestris seedling has for several years been giving 42 leaguers of wine per Cape Morgen *(i.e.* 291⅔ hl. per ha., or about 4590 gallons per acre) on a reddish loam with a gravelly subsoil and without irrigation.

In Franschhoek (Cape) a block of 1.1 Cape morgen of Muscat of Alexandria, trellised and grafted on Herbemont and Jacquez, has for many years been giving 55-57 leaguers of wine, thus fully 50 leaguers per morgen of 346 hl. per ha. or 3080 gallons per acre, *which is very nearly a world record.* The soil is a broken mountain soil with a clay subsoil. Both vineyards are over fifteen years old and still in splendid vigour, but they are heavily fertilised and never suffer from drought, though neither of them is irrigated. Such cases are very great exceptions.

"3. It is in the depressions in the soil where the deteriorations are generally most serious. Now, it is precisely in these places that the vine grows least well." He observes that there are exceptions, but that the soil usually gets wettest in such depressions, is badly ventilated, and frequently gets driest in summer. Consequently the vine here has the least chance of again forming new young roots. (Compare the corresponding case with Roncet.)

"4. As the exhaustion ultimately is the result of insufficient nutrition, the deteriorations must be more or less subordinate to the nature of the soil. They can certainly occur everywhere, but it seems to me that they are of more especially frequent occurrence in certain soils." The nature, etc., of these soils still requires further investigation.

"5. Drought can undoubtedly cause the death of the vine; but this does not happen except in certain special soils. When not very intense, it accentuates the importance of the exhaustion in hindering the nutrition of the plant. But when more intense, it can, *for certain of its values,* actually prevent exhaustion taking place, for then it opposes the migration to the bunches of the substances contained in the body of the vine. It is like overheated wheat. But its role after all is very secondary."

"6. I have already indicated what should be done to avoid and attenuate these deteriorations.

"To avoid them one should—

"i. Watch the fructification as closely as the vegetation.

"ii. Apply a reduced pruning *(i.e.* fewer and shorter bearers. A.I.P.).

"iii. Remove the surplus bunches.

"iv. *And particularly not demand a heavy crop from young vines.*

"v. Further the plant's nutrition by manuring and irrigation if possible, and by good cultivation. Like drought, irrigation can be useful or harmful. It is useful if given before the vine stops growing, whether before or after flowering. If given late, when growth has stopped, it can favour the migration (of stored food. A.I.P.) and hence the exhaustion of the plant. This is the effect produced by late rains.

"To attenuate them, one should restore to the vine the roots it has lost. This one attains by heavy manuring and very reduced pruning, and by an early removal of the bunches if present. It is hard, I admit, to sacrifice a crop, even if small, or a portion of a crop. Nevertheless it is the most efficient means of rapidly restoring its original vigour to the vine."

As I have already pointed out, one has to be doubly careful with grafted vines, as they can easily fall a prey to phylloxera, organisms causing putrefaction (on the roots), etc., when in a weakened state, whether caused by overproduction, drought, or something else. *The normal, healthy life of the plant depends upon a state of equilibrium which must be maintained as far as possible. A disturbance of it, for instance, by overproduction, endangers the plant's whole life.* It is in the weakened state that the vine is particularly exposed to the parasitic diseases which we shall now discuss.

B. FUNGOID DISEASES

1. OIDIUM OR POWDERY MILDEW

Historical.

This disease is indigenous to North America. It came to Europe in the first half of the nineteenth century. In England Tucker discovered it in 1845 in a grapery of Margate. Berkeley found that *Oidium Tuckeri* is the cause of the disease, and gave this name to the fungus. In France it was first discovered in the grapery of Baron Rotschild at Suresnes. Within a couple of years it had spread over all the European and North American vineyards. The French wine crop, which in 1850 had been 45.3 million hl., dropped in 1854 to 10.8 million hl., or less than a quarter, and this was mainly due to oidium—this was before the days of phylloxera, plasmopara, and black rot. This caused great misery.

Since the discovery of this disease in England, sulphur has been used there to combat it. In 1853 Messrs. de la Vergne and H. Mares published an article on the sulphuring of vines against oidium which led to its universal use as a means of combating this disease. Mares prosecuted this study, and his splendid publication, *La Maladie de la vigne,* of 1856, led to the general adoption of sulphuring.

At the Cape this disease was probably introduced shortly after the middle of the nineteenth century. It is our most common vine disease as well as of all grape-growing regions.

Note.—It seems to be certain that the phylloxera was introduced into France on American grape varieties that were imported on account of their high resistance to oidium.

External Characteristics of the Disease.

The fungus causing this disease grows on the surface of all the green parts of the vine.   It forms light grey patches covered by an ash-coloured powder consisting of the *conidia*[1] of the fungus.   On the shoots it later on forms brown spots or patches.   Unlike downy mildew, it appears on both surfaces of the leaves.   If it attacks the vine early, it can hinder the development of the shoots and leaves.   The infected berries show patches with whitish-grey powder, later on speckled with brown spots, and often burst on one side when the seeds become exposed.   It is to the grapes that the fungus usually causes the greatest damage.   If you put your nose to a vine and notice a mouldy smell, you can be sure that oidium is present.   Once the grapes have turned colour, they are seldom attacked by the fungus.

Factors that influence the Development of the Disease.

(a) The Variety of Grape.—Under identical conditions the different varieties of grapes differ greatly in their susceptibility to this disease. At the Cape, Stein, Cabernet Sauvignon, and Gros Maroc are amongst the most susceptible varieties; Muscat of Alexandria, Muscadel, Pontac, and Gros Colman are less but still fairly susceptible; Greengrape, White French, and Hermitage are much less susceptible.

Viala, *l.c.* p. 12, gives the following grouping:

Varieties that are *strongly attached:* Muscats, Chasselas, Frankenthal, Malvoisies, Teinturier (Pontac), Folle blanche, Clairettes, Piquepouls, Gamays, Cabernet-Sauvignon, Syrah (Shiraz), Roussanne, Carignan, Riesling, Sicilien précoce, Ugni blanc, Cinsaut (Hermitage), Alvarelhão, Nebbiolo, Trebbiano, etc.

---

[1] *Conidia* are small cells, summer spores, formed by the fungus for its immediate propagation.

Varieties that are *slightly attacked*: Aramon, Sauvignon, Marsanne, Dolcetto, Grenache, Morrastel, Petit-Bouschet, Alicante-Bouschet, Pinots, Merlot, etc.

Varieties that are *attacked very little*: Cots (Malbec), Duriff, Verdese, Catawba, Isabella, York-Madeira, and most American varieties: *V. riparia, V. rupestris,* etc.

(b) Heat.—This is the principal factor. Hence oidium is a more serious disease in warm countries than in those with a cooler climate. As long as the weather remains cold or cool, there is no danger of the disease developing to any appreciable extent. Thus in 1920, with the continuous rains and cool weather until in November in the Western Province of the Cape, oidium only began to develop here in November, when warm weather set in. It was then unnecessary to sulphur the vines any earlier. According to Viala, the fungus can develop rapidly at an average temperature of 20° C. (68° F.) with a maximum of 25-30° C. Its optimum lies between 25° and 30° C. Between 35° and 40° C. it grows very little, and at 45° C. it dies (see Viala, *l.c.* pp. 13-14).

(c) Humidity.—Contrary to anthracnose and plasmopara, which require liquid water, oidium can develop and propagate itself if the air only possesses enough moisture, without liquid water being required. As soon as the air has become warm enough in a certain place, the disease will begin to develop if the air contains enough moisture. This explains why a hothouse or a depression in the vineyard or densely grown vines in a fertile patch of soil or on a pergola constitute such suitable places for the development of this disease. Here the movement of air is at its minimum, so that the vine easily gets enough moisture for the rapid development of the fungus. This also explains why the disease is worse in Constantia and Helderberg (near the Ocean, False Bay) than in Paarl (further inland).

In Dal Josaphat (Paarl district, Cape) vines are seldom sulphured. There the air is fairly dry and, in the summer, quite warm. At Prince Alfred's Hamlet (Ceres district, Cape) the vines are usually not sulphured and yet remain free from oidium. There the air is usually very dry in summer, which is further evident from the fact that Red Spider is very troublesome in the prune trees. In the irrigated vineyards of Worcester, Robertson and Montagu oidium can cause a lot of damage, as the humidity of the air underneath and inside the usually vigorously growing vine is much increased by the successive irrigations, notwithstanding the fact that the open air in these parts is also very dry in summer.

Botanical Study of the Fungus.

The cause of oidium is a fungus called *Uncinula necator (Schwein.)*, Burr. Earlier names for it were *Oidium Tuckeri*, Berk., and *Uncinula spiralis*, Berk, et Curt.

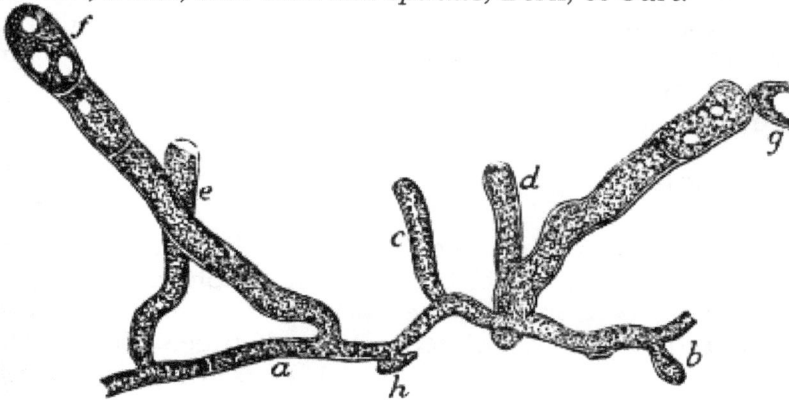

FIG. 74.—The organs of Oidium Tuckeri. *a*, mycelium; *b*, *c*, *d*, *e*, de conidiophores; *f*, conidiophore fully developed; *g*, conidium nearly separated; *h*, s 400. From Viala, *Les Maladies de la vigne*, 1893. Masson et die, Paris.

The *mycelium*[1] threads grow on the surface of the green parts of the vine and draw sap from the epidermis cells by means of suckers or *haustoria* which grow into these cells and thus keep the fungus attached to the plant. They are projections of the mycelium threads on their side next to the plant. Out of the mycelium threads the more or less erect *conidiophores* (see Fig. 74, *c-f*) arise, at the ends of which the conidia (see Fig. 74, *g*) are formed which can immediately propagate the disease.　They are on an average 16μ long and 8μ broad (μ = $^{1}/_{1000}$ mm. = $^{1}/_{25,000}$ in.)

During summer the disease is propagated by these conidia and by the mycelium filaments.　It passes the winter in the form of altered and more resistant mycelium and of

---

[1] This is the body or thalius of the fungus, which consists of thin white threads or filaments that arise out of the conidia when these germinate; these threads can branch, and the different organs for propagating the fungus are formed out of it.

special fruit-bodies called *perithecia* (really *cleistoihecia*). The latter contain 4-8, usually 6, asci, each of which contains 4 or 6 *ascopores* (they are winter spores) which, on germinating in the following spring or summer, can again propagate the disease. These perithecia possess a number of appendages, *i.e.* threads more or less spirally curved near their ends and with cross-walls. They are not formed until late in autumn (October-November in Germany). In warm climates they are rarely found, and then the disease is no doubt propagated from year to year by means of altered mycelium which successfully passes through winter. I am not aware that any one has yet found the perithecia of this fungus in South Africa.

Control of Oidium.

This disease is most successfully combated by dusting the green parts of the vine with sulphur. Marès (see Viala, *l.c.* p. 39) attributes the favourable action of sulphur on the fungus to the direct contact between them at a temperature of at least 25° C, which causes the mycelium, conidiophores, and conidia to wilt, shrivel up, and die. It is well known that sulphur is much more effective against the disease when it is hot than when cool weather prevails. According to E. Mach (quoted from Viala, *l.c.* pp. 42-43) it is particularly the sulphur dioxide formed from the sulphur in hot air that kills the fungus. He found that the hotter the air in a heavily sulphured vineyard, the more sulphur dioxide it contained.

Poliacci (quoted from Sorauer, *l.c.* ii. p. 197) has shown that sulphuretted hydrogen is formed where sulphur is in contact with the fungus and with the green parts of the vine, and this gas is known to be a strong fungicide. In this case the contact mentioned by Marès is also necessary. In any case it is certain that the sulphur should be dusted as far as possible over all the green parts of the vine.

Instead of sulphur an aqueous solution of lime-sulphur is sometimes sprayed on the vines, which also liberates some sulphuretted hydrogen. If the solution is too strong or contains too much calcium mono-sulphide, it can easily burn the grapes, shoots, and leaves. It is used much less than pure sulphur mixed with dry slaked lime, which diminishes the danger of sunburn and makes the sulphur dust better.

If valuable table grapes are attacked by oidium late in the season (January at the Cape), when it would be dangerous to sulphur, we can effectively combat the disease by spraying with *hot water* at 70–75° C. or 158–167° F. between 11 A.M. and 3 P.M., *i.e.* the hottest part of the day. Keep the nozzle of the pump close to the parts to be treated to prevent too much cooling, and spray copiously. Such water is obtained by mixing two parts of boiling water with one part of water at the ordinary temperature of the air. I have found that water at 80° C. easily burns the grapes.

The Effect of Sulphur on the Vine.—Marès maintains that regular sulphuring makes the vines more vigorous and more productive; the leaves get a darker green colour and the grapes ripen more simultaneously and earlier (about ten days)—see Viala, *l.c.* pp. 44. Viala proceeds to say that black grapes develop more colour when sulphured and that fertilisation is much favoured thereby, so that better-set bunches are obtained by sulphuring during the full bloom. He therefore recommends that vines be sulphured when in full bloom, especially if the weather is not favourable for a good fertilisation. One should not sulphur wine grapes later than a month before the vintage, as some of the sulphur might still be on the grapes during the vintage and thus get into the must. Should this happen, the wine will have a smell of rotten eggs due to certain organic sulphur compounds (mercaptans) formed by the yeast cells, etc., out of the sulphur and alcohol. Another danger of *late* sulphuring (after 15th December at the Cape) is sunburn. When grapes with sulphur on them are exposed to direct sunlight on a hot day (35–38° C. or 95–100.4° F. in the shade), they easily get sunburnt. Although they may still be good for wine-making purposes, they will certainly be spoilt for sale as table grapes.

When should we Sulphur?—In parts where oidium is very bad, it will be safest to sulphur for the first time when the young shoots are about 6 in. long. Then sulphur heavily. This at the same time combats Erinosis. In some cases this one sulphuring will suffice. The second sulphuring is given when the vines are in full bloom. We have already seen that sulphuring also favours fertilisation. In many cases no further sulphuring will be required, especially if it was done thoroughly on the first two occasions. Sometimes, however, it will be desirable to sulphur a third time, which will be about a month after the second sulphuring or a week before the grapes turn colour. In the Western Province of the Cape this will have to be between the end of November and the middle of December.

When necessary, and especially with such susceptible varieties as Stein and Cabernet Sauvignon in Constantia, two to three additional sulphurings should be given between the normal sulphurings already indicated. Inversely, sulphuring may sometimes safely be delayed until the disease has just appeared on the vine or spot in the vineyard where it always shows itself first. But then such a vine or spot should be carefully watched to detect the disease right at the beginning. Whoever is not prepared to do this, should rather sulphur at the stated times. Where a vineyard has developed oidium very badly late in the season, it will pay to sulphur the vines very thoroughly immediately the crop is off. This will reduce the intensity of the attack in the following year very much indeed. In the Stellenbosch district oidium frequently becomes bad late in December, and for this reason it is desirable to sulphur well about the 15th of December.

Sulphur as early in the morning as possible, but only after the dew is off. Dew causes the sulphur to collect in clumps, thus preventing its uniform distribution over all the green parts of the vine, and it tends to block the sulphur dusters. We sulphur early in the morning in order to have the sulphur on the vines as long as possible during the hotter part of the day, and because winds frequently spring up about 11 A.M., when sulphuring has to be stopped. In the absence of any wind, we can continue till midday. On the whole it is best not to sulphur in the afternoon.

Sulphuring Appliances.—We can use little bags made of hessian for hand dusting. This is very satisfactory for the first sulphuring (shoots 6 in. long). Then it does not matter if the sulphur falls a little thick here and there. Sulphur-bellows are also used, and are very suitable. The best of all is a sulphur knapsack pump like that of Vermorel. It costs about £5, but lasts for a long time, saves a great deal of sulphur (about two-thirds), and the work can be done much more rapidly (fully three times) and thoroughly than with little bags. When everything is taken into consideration, we must certainly give preference to such a machine. It is most invaluable when table grapes have to be sulphured late in the season and there is some danger of sunburn. As it spreads the sulphur so evenly and finely, there is even then little danger of sunburn. Larger power sprayers might be useful where very extensive vineyards have to be sulphured, but in most cases the portable type referred to above will be the most economical, even though several of these have to be bought for working simultaneously.

The Sulphur itself.—We have to consider here flowers of sulphur and ground sulphur. Both are good. Their value and suitability depend upon their *purity* and *fineness.* Good sulphur should contain at least 99 per cent sulphur, burn without leaving hardly any residue, and contain at most 0.2 per cent impurities. The fineness of sulphur is determined by means of Chancel's sulphurimeter. According to Viala, *l.c.* p. 53, good flowers of sulphur should show a fineness of 50-70 degrees Chancel, and the superior quality 75-90 degrees Chancel. According to Lüstner[1] ground sulphur goes up to 85° Chancel, and a fineness of at least 70° Chancel should be insisted upon. By blowing the ground sulphur through silk sieves a fineness of 100° Chancel can be obtained, but such a fine sulphur is too expensive. Fineness is the most important quality of sulphur for vines. The finer the sulphur is, the better we can get the sulphur on all the green parts of the vine and the further we shall get with a pound of sulphur.

The sulphur should naturally be thoroughly *dry* to dust well. If necessary, it should first be spread out on a tarpaulin exposed to the sun and stirred from time to time to get thoroughly dry. By mixing a little slaked lime with the sulphur it will dust better, although this should be not necessary. Slaked lime and sulphur are sometimes mixed to combat anthracnose at the same time. It is supposed to help. Where sulphuring with little sacks has to be done late, there will be less danger of sunburn if the sulphur is mixed with slaked lime. On the whole, pure, dry, fine sulphur used at the proper time and in the right manner is the best and most

---

[1] In Babo u. Mach, *Handbuch des Weinbaues,* Zweiter Halbband, 4. Aufl., 1924, p. 370. Paul Parey, Berlin.

461

effective means of combating oidium where vines are sprayed with Bordeaux mixture for other diseases (downy mildew, black-rot, etc.) it will also combat oidium, but is more expensive and not so effective as sulphur.

2. ANTHRACNOSE

Historical.

This is one of the oldest European vine diseases. It is indigenous to Europe and Northern Africa, and occurs in all the grape-growing countries. Fabre and Dunal gave the name "Anthracnose" to this disease in 1853. Meyen in 1841 expressed the opinion that the disease is caused by a parasite. In 1873 de Bary recognised a fungus as the cause of the disease, and described it under the name *Sphaceloma ampelinum*. In 1877 Saccardo gave it its present name, *Gloeosporium ampelophagum* (Pass.) Sacc.

External Characteristics.

This disease attacks all the green parts of the vine during its whole period of growth, but most of the damage is done to the young parts in spring. On the whole it is not a very dangerous disease unless climatic conditions favour its development very much, in which case it can become very destructive on susceptible varieties of grapes. On the young shoots and leaves it at first causes light red spots with brown margins which later turn brown and black all over. In the leaves it ultimately causes holes and on the shoots and berries sunken spots. On the full-grown berries it forms hard black scales or crusts which sometimes unite and may cover the greater part of the berry, on which, by the way, the disease is most characteristic. Such berries are worthless as table grapes, although it will not cause them to rot. Where young shoots are seriously attacked, they can be completely destroyed, will not produce any grapes, and will hardly give suitable bearers for the next crop. Thus this disease can sometimes work terrible destruction in a vineyard. It is most to be feared by the table grape-grower, as all the affected berries have to be removed, which might result in completely spoiling the bunch.

Factors that Influence the Development of the Disease.

(a) *Climatic Conditions.*—If rainy weather prevails during the first month after the vines have started budding, anthracnose is usually very bad, even though the air be then rather cool. It is indeed only during a wet spring or summer that this disease assumes dangerous proportions. The conidia of the fungus germinate only on the surface of liquid water. Moist air, therefore, does not suffice, and this explains why the disease is not noticed in a dry summer. In parts with summer rainfall it is naturally dangerous, like most fungoid diseases, for although it can develop fairly well in rather cool weather, it develops more rapidly if warmer weather prevails and rains fall at the same time. Whilst, at the Cape, it sometimes does great damage in the Cape, Stellenbosch, and Paarl districts, it is practically unknown in the drier districts (Ceres, Worcester, Robertson, and Montagu), even on the very susceptible Sultana vine.

(b) *Variety of Grape.*—The different varieties differ greatly in their susceptibility to this disease. *Very susceptible are* Sultana (Thompson's seedless), Muscat of Alexandria (Hanepoot), White French, Muscadel, Crystal, etc. *Fairly susceptible* are Rosaki, Molinera Gorda, and a number of other table varieties. *Almost* or *practically immune* are Greengrape, Stein, Cinsaut or Hermitage (Viala calls it susceptible!), Gros Colman, Barlinka, etc.

(c) *Trellising.*—Vines grown on high trellises are less susceptible to anthracnose than those growing near the ground; the bottom leaves of these vines are the first to show this disease.

Botanical Study of the Fungus.

The fungus causing this disease grows inside the tissues of the affected green parts of the vine. At certain places below the cuticle the mycelium threads form a fairly dense kind of fungus tissue out of which a dense mass of parallel thin rods grow at right angles to the surface of the leaf, shoot, or berry, and at the ends of which very minute conidia (1-$2\mu \times 3$-6μ) are formed, which rupture the cuticle here and there and serve to propagate the fungus in summer. We find them in the brown and black spots caused by the disease. Each conidium usually has near each end a very small drop of oil; sometimes the one is wanting. They germinate only in liquid water, *i.e.* rain or dewdrops. It is therefore dangerous to walk through a vineyard which has this disease, as the conidia might in this way be spread to uninfected vines.

The fungus passes winter in the diseased tissues in the form of mycelium and of special fruit-bodies called *pycnidia,* which open during spring in water and liberate large numbers of spores which germinate in water, forming mycelium threads and thus propagating the disease. These mycelium filaments can penetrate the healthy epidermis and thus enter the plant. The hibernated mycelium can also propagate the fungus in the next spring and summer. The pycnidia are formed at the surface of the anthracnose wounds towards autumn.

Control of the Disease.

The most effective way of combating this disease is the following late winter treatment. Immediately after the winter pruning, the whole of the vine showing above ground, including the wounds made by pruning and the eyes, is washed with dilute sulphuric acid. Take one bottle of strong, crude sulphuric acid and pour it slowly into four gallons of water. This gives a so-called 4 per cent solution. Although we can use a paraffin tin (not rusty) for this purpose—it will just hold the $1 + 4 \times 6 = 25$ bottles of liquid—it is better to use a wooden tub and wooden buckets. Take a stick, tie some sacking around it, dip this into the dilute acid, and carefully swab the vines. Instead of swabbing we can use an acid-proof spray pump and spray the vines. Cut one hole in the bottom and two holes in the sides of an ordinary old sack and let every workman slip one over his clothes, as the acid will burn holes in them unless thus protected. Do the work thoroughly and in bright weather, so that the acid can do its work properly before it can be washed away by rain. Pour the acid slowly into the water and stir; *do not pour the water on the acid!*

Viala and Vermorel have recommended the following formula:

> 50 lbs. ferrous sulphate (Copperas).
> 1 lb. *(i.e.* ⅓ bottle) com. sulphuric acid.
> 10 gals, hot water.

Put the copperas in a wooden tub, pour the sulphuric acid on to it, then add ten gallons of hot water and stir till the crystals are dissolved. This solution must be used at once, as the copperas will partly crystallise out when the liquid cools. It is more troublesome to prepare, and the dilute sulphuric acid by itself is just as effective. Vermorel (see Babo u. Mach, l.c. ii. p. 399) has found a dilute sulphuric acid of 10 per cent to be equally effective. At the Cape we have found a 4 per cent acid to be sufficient, if applied in fine weather. Much depends upon the weather following the treatment. If no rain falls for, say, a week after treating the vines, the acid gets concentrated on the vine, and a fairly weak solution will be as effective as a stronger solution if rain fell soon after treating with the stronger solution.

In cases where the disease had been very bad in the previous season, it is best to repeat the treatment about a fortnight before the vines start budding. At the Cape this would be late in July or early in August.

Diseased wood of the previous season and the first sick leaves and shoots of the new growth should be burnt.

Dusting with lime and sulphur early in the season is believed to have a good effect, and spraying with Bordeaux mixture as a preventative measure in spring and summer is fairly effective, but becomes unnecessary if the winter treatment has been properly applied. The latter is the most satisfactory and effective means of controlling anthracnose.

3. DOWNY MILDEW OR PLASMOPARA

Names—Downy Mildew, Plasmopara, Falscher Mehltau, Blattfallkrankheit, Lederbeerenkrankheit, Mildiou, etc.

Historical.

This disease is indigenous to North America, and it was known to exist there long before it was introduced into Europe. In 1863 de Bary carefully described this disease and the fungus causing it, which he named *Peronospora viticola.* In 1888 Berlese and de Titoni called it *Plasmopara viticola.*

After 1872 Cornu pointed out the great danger to the French vineyards if this disease were introduced into France. He considered the importation of American vines into France very dangerous. In 1878 the disease was discovered in

France almost simultaneously by Planchon and Millardet on leaves of Jacquez and of various European varieties. From then it spread with great rapidity over all the European vineyards. In South Africa it is known in Natal and the Eastern Province of the Cape. It has never yet been found in the Western Province of the Cape, but would probably not do much damage in our normally dry climate (in summer). It is, however, safest to keep it away as long as possible.

External Characteristics of the Disease.

The fungus attacks all the green parts of the vine—leaves, shoots, bunch, flower, fruit. Once the tissues are ripe, they are no longer attacked by this disease. On all the green parts it causes characteristic alterations.

(a) *Leaves.*—They are attacked in all stages. One first notices light patches on the leaves which the French call "taches d'huile" *i.e.* oil patches or spots, where a white downy mass is formed on the lower surface of the leaf. This consists of the conidiophores and conidia or summer spores of the fungus. It is characteristic, and cannot be confused with any other disease. This stage marks an intensive development of the fungus, which now kills the cells of the leaf. In the centre of the light-coloured patch they begin to turn brown. This extends to the borders, when the leaf develops a hole at that spot and tears. At the same time the conidiophores disappear on these patches. With favourable weather the patch may in one to two days' time have a diameter of an inch. In serious cases the leaves drop. This hinders the further growth of the vine, as we see by the bad maturation of its grapes and canes. Usually the leaves suffer first and most from this disease.

(b) *Shoots.*—On these it causes dark brown or nearly black patches on or near the nodes and seldom on the internodes. It is very seldom that conidiophores develop on the shoots. The fungus grows mainly in the bark (bast) and seldom in the wood and pith. It first makes the cells grow quicker before killing them. This causes a curving of the shoots, with the diseased part on the outside of the curve. When this is repeated an S-shape is produced. Here it causes little damage.

(c) *Bunch.*—The young bunch is attacked on the stalk, pedicels, flowers, and later the little berries. In very moist air the small berries can also be covered with white masses of conidiophores and conidia, when the disease is called *grey rot*.

The mycelium filaments of the germinating spores enter the berries through the stomata, which are fairly numerous on the bourrelet (thickened portion of the pedicel near where it enters the berry). Under the influence of the fungus the berries turn brown; this is why the disease in this case is called *brown rot* ("Lederbeerenkrankheit"). The diseased berries shrivel up and readily drop.

Sometimes this disease can destroy nearly the whole crop and weaken the vine very much through the loss of leaves, which hinders the photosynthesis and causes the grapes and wood to ripen badly. Such grapes possess little sugar when ripe, and produce a poor wine. In many grape-growing areas plasmopara is the most important fungoid disease of the vine. Once the grapes have turned colour, the danger of plasmopara is over. Hence it is not a disease of ripe grapes.

Factors that Influence the Development of the Disease.

A very wet, rainy summer with sudden changes of temperature and little sunshine is very favourable for the development of this disease, whilst dry weather brings it to a standstill. See Lüstner in Babo u. Mach, *l.c* ii. p. 344, and Ravaz,[1] *l.c.* pp. 73-116. The germination of its conidia takes place only in liquid water. This explains why vines that are growing under a roof (*i.e.* in graperies) and that are therefore not wetted by dew, fogs, or rain, remain free from this disease. Vines growing in the open are, however, exposed to the weather, and can thus be wetted by rain, fogs, and dew. According to Ravaz, *l.c.,* just quoted, p. 77, the conidia retain their germinating power for at the most five days in very moist air. In less moist air they lose it very soon; this is probably due to desiccation. This is why dry winds and dry weather check this disease so effectively.

---

[1] Ravaz, L., *Traité général de viticulture,* IIIme Partie—tome III. *Le Mildiou,* 1914, Coulet et fils, Montpellier, and Masson et Cie, Paris.

According to Ravaz, *l.c.* p. 76, its conidia begin to germinate at 6-5° C, and at 10° C. or 50° F. it proceeds fairly rapidly. As the temperature rises the speed of germination increases, until the maximum speed, is reached at 27° C. or 80.6° F. Then it proceeds more slowly, and comes to a dead stop at 35° C. or 95° F. The conidia which did not germinate because the temperature was too high, seem thereby to have lost for ever their power of germinating. Hence very hot weather, like dry weather, stops the disease, whereas cool weather still suits it quite well. This explains why it is so dangerous in rainy, foggy weather. After the fungus has entered the green parts of the vine, its further development still depends upon the temperature and humidity of the air. Dry hot weather will oppose it, whilst damp, moderately warm weather will favour it. Hence closely grown, low vines that are irrigated are more open to attack by plasmopara than open, trellised vines that are not irrigated. The humidity of the soil naturally influences that of the air in and between the vines.

The various varieties differ considerably in their susceptibility to this disease, and their susceptibility is strongly influenced by the stock upon which they are grafted. The same variety will remain fairly healthy or become very sick according as it shows moderate or great vigour, and this is largely determined by the stock. Thus Ravaz has found that 1202, 3306, 3309, Aramon R.G. Nos. 1 and 2, Rip. gloire, Jacquez, etc., make the disease attack the vines grafted on them more intensely where they grow very vigorously, whilst the Rip. × Berl. hybrids like 420A, etc., cause a less exuberant growth, with the result that the vines grafted on them are less attacked. We have previously seen that exuberant growth exposes the vine more to all sorts of diseases and pests than a moderate, healthy growth. Therefore we also find that a heavy dressing of nitrogen, which urges the vines to tremendous vigour, makes them more susceptible to disease.

Apart from all that has just been said, we find that the various species and varieties of vines, under the same circumstances, differ in their susceptibility to this disease. According to Rayaz, *l.c.* p. 112, not a single species of the genus Vitis is quite immune to it. Some, like V. *riparia,* V. *cordifolia, V. rupestris,* are affected only under extraordinarily favourable circumstances, and then only slightly. Others are less resistant, but are usually only slightly damaged, *e.g. V. Lincecumii, V. Berlandieri, V. labrusca,* etc. Others, however, are so susceptible to the disease that they can lose all their leaves, *e.g. V. vinifera, V. Californica, V. Arizonica.*

We do not yet know to what we have to ascribe these differences in susceptibility. It is possible that the chemical composition of the cell content might play a dominant or decisive role in this connection.

But within the same species the different varieties are not equally susceptible. This is also the case with the European varieties, all of which belong to the species *V. vinifera.* Ravaz, *l.c.* p. 116, says that, generally speaking, varieties with fleshy thick, soft leaves, like Grenache, Clairette, Mourvèdre, Terret, Carignan, Morrastel-Bouschet, Malbec, etc., are attacked more intensely than varieties with thin or dry leaves, like Aramon, Mondeuse, Durif. The constitution of the leaves is of course also influenced by local conditions and cultural operations, and not necessarily always in the same way; hence the differences recorded in the relative susceptibility of different varieties to plasmopara (see Babo u. Mach, *l.c.* ii. p. 345).

Botanical Study of the Fungus.

The fungus causing this disease is now called *Plasmopara viticola,* Berlese et de Titoni (its former name was *Peronospora viticola,* de Bary).

Its mycelium filaments grow without cross-walls in the leaf (and other green parts of the vine), and there between the cell-walls, which the fungus does not penetrate, but frequently presses inwards by means of round suckers through which it absorbs food from the cells, whereby these latter are ultimately killed. The mycelium is frequently branched. On Fig. 75 this is indicated by *a.*

Conidiophores.—After a sufficient development, and if the air is sufficiently moist and warm (according to Ravaz, *l.c.* pp. 96 and 100, an air temperature of 20° C. or 68° F. seems to be the most favourable, provided the air is then saturated with water vapour to 97–100 per cent), the conidiophores emerge through the stomata or any casual wound in the epidermis and rapidly develop.   As they emerge through the stomata, we usually find them on the lower surface of the leaves, where the stomata mainly occur. Sometimes we also find them along the nerves on the upper surface of the leaves, where a few stomata are known to occur.

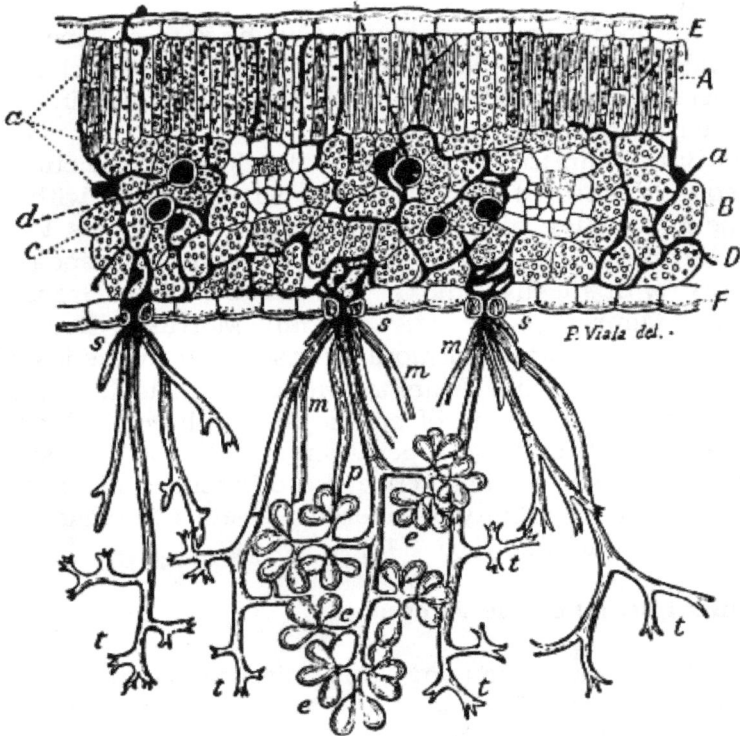

FIG. 75.—Theoretical cross-section through a leaf infested with Plasmopara viticola. A, upper side of leaf with palissade tissue. B, lower side of leaf with mesophyll tissue. D, vascular bundle of nerf. E, upper and F, lower epidermis, a, mycelium of fungus growing between the cells; c, suckers of the mycelium; b, antheridium and oogonium uniting; d, winter spore or egg ; s, s, s, stomata through which the bundles of conidio¬phores emerge; p, a conidiophore with the summer spores or c onidia; e, e, e, conidia attached to the extremities of the branches; m,.m, lower p art of conidiophores of which the upper part is not shown; t, t, conidiophores with sterigmas that have borne conidia; t (to the right) a conidiophore with special b ranching. From Viala, Les Maladies de la vigne9 1893.
Masson et Cie, Paris, × about 200.

It is only when the conidiophores are thus formed that the leaf begins to suffer seriously. These conidiophores can emerge in numbers of 1–10 through one stoma (see Ravaz, l.c. p. 45). (Compare Fig. 75.)   Their branching, and especially the formation of *sterigmata,*[1] is quite typical of this vine disease and immediately enables us to recognise it. (See Fig. 76,which is particularly characteristic.)   Here the conidia have already dropped from the sterigmata, as is the case with the sterigma *kf.*

Conidia.—They arise on the sterigmata as round points which, however, soon assume an egg shape, as can be well seen on Fig. 76, *k.* They average about 10μ by 16μ. According to Ravaz they are bigger according as there are fewer of them on the same conidiophore, hence according as it is less branched.

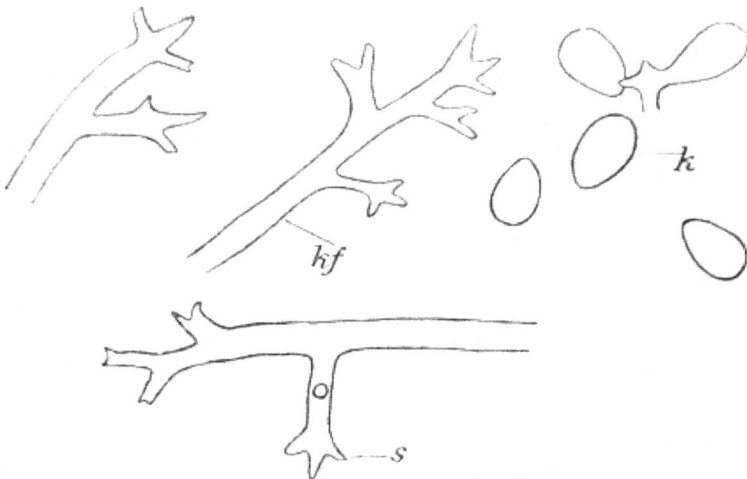

FIG. 76.—Conidiophores and conidia of Plasmopara viticola. kf, a conidiophore whose conidia have dropped ; k, a conidium ; above it is a conidium that just became detached from the sterigma, and to the right of it is one still attached to the sterigma ; s, a typical sterigma group. Original,   x 660.

The conidia are rich in protoplasm and possess a thin cell-wall. As soon as they get into a drop of water (rain or dew) their content undergoes a change and they become zoosporangia. Within two hours the cell-wall bursts, and averagely 5–6 (sometimes 3–17) zoospores emerge from one original conidium. The zoospores are naked masses of protoplasm with cilia, and they swarm for a quarter of an hour.

---

[1] The sterigmata are the sharp points at which the conidia are formed

They then become round masses enclosed by a thin wall of callose, which begin to germinate within the next quarter of an hour if at a temperature of about 28° C, The mycelium filament thus formed is very thin, about 1μ diameter, and it enters the plant through its stomata or through casual wounds. Thus *infection* takes place. Should the drop of water evaporate before the mycelium has entered the leaf, the zoospore with its mycelium filament will die. The conidia are scattered by the wind as they become detached from the sterigmata, but, as we have already seen, soon die from desiccation.

Hibernation.—Towards the end of summer and in autumn this fungus forms its *winter-spores* or *oospores,* which are the product of the fusion or sexual fertilisation between two differentiated, swollen ends of mycelium filaments (see Fig. 75). These spores are round bodies and have a diameter of $25\text{-}30\mu$. They are dark brown, have a thick membrane, and sometimes occur in large numbers in the leaves. They are liberated when the leaves rot in the soil. According to Babo u. Mach, *l.c.* ii. p. 343, they are formed on the shoots also. Ravaz, *l.c.* pp. 60-68, describes the germination of the winter spores. He was one of the first to observe this, which he did in 1911. The germinating winter spore gives rise to 1-3 conidia, which agree with the summer conidia, except that they are usually larger. Therefore Ravaz calls them *Macroconidia* (on an average $23\mu \times 35\mu$). Under favourable conditions they also become zoosporangia, burst, and allow a large number (up to 40) of zoospores to escape, just as we have seen happening with the summer conidia. These zoospores are identical with those of the summer conidia and their further development also. The winter spores can remain alive for some years in succession, and are thus an efficient means of propagating the fungus from year to year.

Also this fungus can pass winter as *mycelium.* According to Cuboni the mycelium sometimes passes winter beneath the outer scales of the eyes, from where it emerges in spring together with the young leaves, and causes a fresh infection. Istvanffi has found it also in hibernating berries and shoots (quoted from Babo u. Mach, *l.c.* ii. p. 343).

Control of the Disease.

Copper in some combination or other is generally used in combating this disease. In the laboratory the fungus is killed by the following concentrations of the substances mentioned: Slaked lime 1 : 10,000, ferrous sulphate 1 : 100,000 (Fe in solution), copper sulphate 3 : 1,000,000 (Cu in solution). It is therefore clear that an extremely small amount of copper in solution suffices to kill the fungus. From the preceding it is clear that the poison must be on the plant (leaf) when the conidiophores emerge, in order to be able to kill the conidia and the mycelium formed on germination of the zoospores, before it has penetrated the leaf. Hence the poison must be spread over all the green parts of the vine, and should remain there for some time in order to be able to act when required. Therefore mixtures are used in which the copper is presented in a very slightly soluble combination.

Gayon, quoted from Ravaz, *l.c.* p. 123, has proved that copper hydrate dissolves slowly in water containing carbon dioxide (as hydrocarbonate, A.I.P.), and this at the rate of 40 mg. per lit. at 15° C, thus fully eight times as much as is required to kill the fungus. Water which is exposed to the open air naturally always contains carbon dioxide and a little ammonia in solution. In this way are dissolved the slightly soluble copper compounds which are present in the various mixtures used in controlling this disease. They are used as liquids or as powders. Of the former there exist a large number, of which the most important will now be discussed.

Bordeaux Mixture.—It is commonly used for controlling this disease, the potato disease caused by *Phytophthora infestans,* black rot, and many other fungoid diseases on our cultivated slants. Prof. Millardet of Bordeaux gave the first formula for preparing this mixture, hence the name. To make this mixture we use copper sulphate, freshly slaked lime, and water. The most commonly used formula for vines is 10 : 5 : 50, *i.e.* 10 lbs. copper sulphate, 5 lbs. quicklime, and 50 gallons water, but 5 : 5 : 50 and 4: 4: 50 are also used. It is prepared in a number of different ways, but usually the copper sulphate is dissolved in water, and the lime, after slaking with water, is made into a thin paste by adding more water and stirring, before they are mixed. Sometimes the two are mixed as a fine dry powder which need simply be added to the necessary quantity of water and stirred to prepare the Bordeaux mixture. The following rule is very important: *prepare the mixture freshly every day.*

Bordeaux mixture is prepared in the following ways:

A.   Galloway or American Method.—Dissolve the copper sulphate in half the total quantity of water required for the mixture (do this in a wooden tub or cask and not in an iron or zinc vessel!), stir the lime well with the remaining water, now pour both *simultaneously* into some suitable vessel, stirring well all the time. In this way we obtain a very light and voluminous precipitate which remains suspended for a long time (see Ravaz, *l.c.* p. 130). This is the method most generally adopted for preparing Bordeaux mixture.

B.   Supposing we use 100 gallons water, then the copper sulphate is dissolved in 90 gallons water by suspending it overnight in a muslin bag; the lime is stirred into 10 gallons water and *slowly* added to the copper sulphate solution, stirring vigorously all the time.

C.   Lüstner recommends the following in Babo u. Mach, *l.c.* ii. p. 348: Dissolve 1 lb. copper sulphate in 5 gals, water, slake ½lb. freshly burnt lime with water and make up to 5 gals., now pour the copper sulphate solution slowly into the lime milk *(i.e.* just the reverse of what is done in B), and keep stirring all the time.

Burgundy mixture is made by substituting soda for lime in the Bordeaux mixture. It is made alkaline, neutral, or acid.

In order to make the mixture keep longer, sugar is added to it. Kelhofer (quoted from Babo u. Mach, *l.c.* ii. p. 349) maintains that the mixture keeps well for a year if we add 1 gr. sugar to every lit. of the mixture, *i.e.* 1 lb. sugar per 100 gals, mixture. In order to increase the "wetting" or spreading and adhering power of the mixture, hard soap, resin, casein, etc., are added to it. Casein is used in alkaline mixtures *(e.g.* the Bordeaux mixture above recommended). It is most conveniently dissolved by adding ½ oz. caustic soda and 5 ozs. casein to ½ gal. hot water. This quantity is sufficient for 50 gallons of spray mixture.[1]

---

[1] See F. de Castella and C. C. Brittlebank, *Downy Mildew of the Vine,* Dept. of Agric, Victoria, Australia, Bull. No. 49 (new series), p. 45.

Neutral Copper Acetate ("Verdet neuter"), Cu (CH_3COO)$_2$ + H_2O, contains 31 per cent Cu and is generally-used in a 1 per cent solution in water. After having been on the leaves for some time, it passes into a basic acetate (not easily soluble) which adheres well to the leaves.   It is effective.

Green of Montpellier ("Verdet gris") consists mainly of $Cu(CH_8COO)_2$. $Cu(OH)_2$, which is partly soluble and partly dissolves with difficulty.   It acts well.

Powders.—Sulfosteatite is an example of these. It is prepared by making a saturated solution of copper sulphate into a paste with talc or steatite, $H_2Mg_3(SiO_3)_4$, which is then dried and ground to a fine powder. It is used especially for the grapes as a supplementary treatment between the sprays. As was previously stated, the Bordeaux mixture is sometimes sold as a powder.

*Time of Treatment.—*It is best to spray just before the conidiophores appear. Where the disease is dangerous, spray as follows:

First spray as soon as the young inflorescences (bunches) are clearly visible;

Second spray two weeks later;

Third spray shortly before flowering;

Fourth spray when flowering is nearly over;

Fifth spray three weeks later;

Sixth spray when the vineyard ceases growing; with powder between the 1st and 2nd, 2nd and 3rd, and 3rd and 4th sprays (according to Ravaz, *l.c.* p. 185).

Under more favourable circumstances three sprays will suffice, namely, the first, second, and fourth mentioned above.

*Spraying Outfit.—*Use a power sprayer or a knapsack pump like that of Vermorel.

4. BLACK ROT

Hedrick[1] says of this disease: "This is the most widely distributed and the most destructive fungous disease of the grape in the regions east of the Rocky Mountains. For-tunately, it is unknown on the Pacific coast." It is principally

---

[1] Hedrick, U. P., *Manual of American Grape Growing*; New York, The Macmillan Co., 1919, p. 219.

due to this disease that the European grape varieties can hardly be grown in the States east of the Rocky Mountains, and that the varieties grown in these States are mainly varieties of Labrusca and other indigenous species which are more or less resistant to this disease. In Europe it occurs mainly in S.W. and S.E. France. It is particularly dreaded in the Gironde, where it sometimes causes greater damage than downy mildew and other diseases. It occurs in Natal and the Eastern Province of the Cape, but is unknown in the Cape viticultural districts (Western Province, etc.), where the dry, warm summer climate is against it. Practically the same applies to California and Australia.

Historical.

According to Viala, *l.c.* p. 156, Batheam and Longworth were the first, and that in 1848, to make mention of the destruction wrought by this disease in Ohio and elsewhere. In 1885 Viala and Ravaz discovered it for the first time in France. This was on grapes from Ganges (Hérault). It was most fortunate that the disease broke out in the French vineyards only after these had already been sprayed against downy mildew with sprays containing copper, else it would have caused much greater damage. Viala and Ravaz have studied this disease, and gave its present name, *Guignardia Bidwellii* (Ell.) Viala et Rav., in 1892 to the fungus causing this disease.

External Characteristics of the Disease.

It attacks principally the grapes, where it does the great damage. It also attacks the other green parts of the vine, but here the damage done is much less. The berries are usually attacked when they are almost turning colour. From a small round spot it can give the whole berry a dead reddish-brown colour within 24-48 hours. The berry soon begins to shrivel and the skin to wrinkle, and assumes a darker colour, beginning from the spot where discoloration began. After three to four, sometimes two, days, it is quite dry and pitch black. The skin remains intact. When the berry turns from reddish-brown to a darker colour, we notice the formation of small black dots or pustules. They are smaller than a pin's head, but very numerous. All this happens within three to four days, and is so characteristic that we can thereby easily recognise the disease. Later the berry drops; sometimes part of a bunch or a whole bunch drops. All the bunches on a vine and all the berries on a bunch are not attacked simultaneously. Berries that are still sound at the veraison (turning colour) are not subsequently attacked by this disease. Its development is prevented by the sugar which is then present.

The disease usually begins on the young leaves, and about three to four weeks before it attacks the berries. On the leaves it forms dry, reddish-brown patches, on which we also notice the black pustules. Also on the young shoots a discoloration with the formation of pustules takes place. A severe attack can destroy practically the whole crop.

Factors that Influence the Development of the Disease.

For the development of black rot the air has to have a high temperature and a high moisture content. Minimum temperatures of 15-20° C. with maximum temperatures of 35-37° C. are most favourable for its development if rain falls then, and in any case if the air then is very moist. Liquid water favours its spreading, and would seem to be necessary for the germination of its spores. Heavy dew falls and foggy weather favour this disease, especially if the air is then warm. It is particularly in deep and fairly enclosed valleys where warm, moist air frequently occurs, that black rot becomes very bad. The different species and varieties of the same species are not all equally susceptible to the disease.

Botanical Study of the Fungus.

Viala, *l.c.* pp. 174-191, gives a splendid description of *Guignardia Bidwellii,* the fungus causing this disease. Its mycelium possesses cross-walls, and grows between and through the cells of the leaves and other organs of the vine. The black pustules on the berries, leaves and shoots consist of *pycnidia* and *spermogonia,* in which conidia and spermatia are respectively formed. Under suitable conditions the conidia germinate and can again cause an infection.

In autumn we find also *sclerotia,* which are formed out of the mycelium and the empty pycnidia. During the following spring they develop conidiophores on which conidia are formed.   In North America and in France the fungus forms in May and June, *i.e.* early summer, also *perithecia,* in which asci with ascospores are formed. They are also formed out of the mycelium and pycnidia. In addition the mycelium sometimes also forms *chlamydospores.* Hence this fungus shows a great variety of reproductive organs. For propagating it from year to year, Viala attaches most importance to the pycnidia, either directly or through the formation of perithecia in the following spring. The sclerotia, however, are of importance, as conidiophores or pycnidia are formed out of them.

Control of the Disease.

Grapes grown under glass do not get this disease, even if European varieties are thus grown in Eastern America. The reason is the absence of liquid water on the vines. In the same way all bunches are protected against black rot if they are covered with paper bags when the berries have reached the size of peas. According to Viala, *l.c.* p. 197, this was done in the Atlantic coast states of the United States. This disease, like plasmopara, is effectually controlled by spraying with Bordeaux mixture. Hedrick, *l.c.* p. 220, recommends a Bordeaux mixture of 4 : 4 : 50, though in France 10 : 5 : 50 is more commonly used. Also here the treatment must be a preventative. Vines that are sprayed for plasmopara need not be sprayed specially against black rot, except that a thorough spray is needed about two months after blossoming.

5. WHITE ROT OR ROT BLANC

This disease is caused by the fungus *Charrinia diplodiella* (speg.) Viala et Ravaz, formerly called *Conithyrium diplodiella* (speg.) Sacc. It is principally a grape disease, although it does occur also on the leaves and shoots of the vine. It came from America (Missouri), and was observed for the first time in Europe by Cataneo near Florence in 1876. Now it is known in most European wine countries, North Africa, Asia Minor, Caucasus, and North America. At the Cape it is not yet known. It develops only with *great heat* and in *very moist air,* and then usually only when the grapes have begun to ripen. In some very exceptional seasons it might do damage at the Cape.

The fungus enters the plant through wounds. It attacks principally the bunches and causes brownish-yellow patches on the affected parts. The berries turn whitish-grey and shrivel up, whilst their surface gets covered with the black dots of its pycnidia. At last the dried-up bunch drops. The diseased stalks assume a rust-brown colour and dry up. Where the fungus is already present, it can cause great damage to the crop within a very short time if weather conditions favour its development. This actually happened in Northern Italy and in Hungary. Its fruit-bodies are principally *pycnidia,* the macroconidia of which pass winter as *sclerotia. Perithecia* have been seen only once. Frequently *Botrytis cinerea* appears together with white rot and completes the damage done by the latter.

Control.

It is difficult to kill its spores. A 3-4 per cent Bordeaux mixture will kill the mycelium filaments formed when these spores germinate. It is best to spray with this mixture just when the pycnidia are bursting through the epidermis, as the fungicide is then absorbed there, and prevent the germination of the spores. After the winter pruning and after a hailstorm the vines should be sprayed immediately. It is also important to keep the vineyard free from weeds. De Istvanffi,[1] quoted from Babo u. Mach, *l.c.* ii. p. 392, recommends spraying with a 2½ per cent solution of potassium meta-bisulphite and sulphurous acid, and a 3 per cent solution of magnesium sulphite, which kill the coniothyrium spores in twenty-four hours.

6. GREY ROT *(Botrytis cinerea)*

Names: Grey Rot, Noble Rot, Pourriture grise, Pourriture noble, Graufaule, Edelfaule, Vaalvrot, Edelvrot, etc.

According to its degree of development the fungus, *Sclerotinia Fuckeliana,* Fuckel *(Peziza Fuckeliana,* De Bary) = *Botrytis cinerea,* Persoon, can cause the grey or the noble

---

[1] G. de Istvanffi, "Études sur le rot livide do la vigne (Coniothyriutn diplodiella)". *Ann. de l'Institut centr. ampélog. royal hongrois,* tome ii., 1002.

rot. If it rains every now and then when the grapes begin to ripen or are already ripe, as during the vintage, this fungus causes the grapes to rot, which may result in great loss. The berries then become soft and rot, and we see a grey powder on them. With black grapes the fungus destroys a lot of pigment. In any case wine made from such grapes tends to become turbid and to assume a dark colour when exposed to air. This is caused by an enzyme, an oxydase, which the fungus produces and brings into the must. It also consumes a lot of sugar. The rotten berries drop very easily, and hence are easily lost. The best remedy is, as soon as one notices that rotting has begun, to remove a number of leaves from the inside of the vine in order to expose the grapes directly to the rays of the sun. The disease will come to a standstill if the grapes can get properly dry, and as soon as dry weather sets in. In normal summers it does no damage in the Western Province of the Cape.

In Sauternes, the Palatinate, the Rheingau, and along the Moselle the wine farmers purposely leave their grapes on the vines till fairly late in order to develop the noble rot. In this case the weather should be moderately warm with a fair amount of moisture in the air. Then the disease develops just so far that the berries turn brown, and later become a kind of moist raisin which is covered with the grey powder, the conidia of the fungus. The fungus causes little cracks in the skin of the berry through which a good deal of water evaporates, thus gradually converting the berry into a sort of noble rot raisin. Such berries are gathered one by one, and produce a thick, brown must containing 35-40 per cent sugar when the must of the ordinary sound berries will have only about 20 per cent sugar. In this case the Germans speak of "Edelfaule" and the French of "Pourriture noble". Such musts produce the delicious and highly aromatic, natural sweet wines of the Rhine, Moselle, and Sauternes.

Botanical Study of the Fungus.

If we examine the grey powder on the berries under the microscope, we notice that it consists of the oval conidia of the fungus. These are formed on conidiophores that are branched very much like the stem of the grape bunch, hence the conidiophore with its conidia closely resembles a bunch of grapes in the grouping of its parts. (See Fig. 77.) These conidiophores are so typical in appearance that we can easily recognise the fungus thereby. The conidia can be transported to sound berries by wind and insects. They germinate only in moist weather. In dry air they are dead in a couple of hours. We usually find the fungus in this conidium form on the grapes, and it is this form which is called *Botrytis cinerea*.

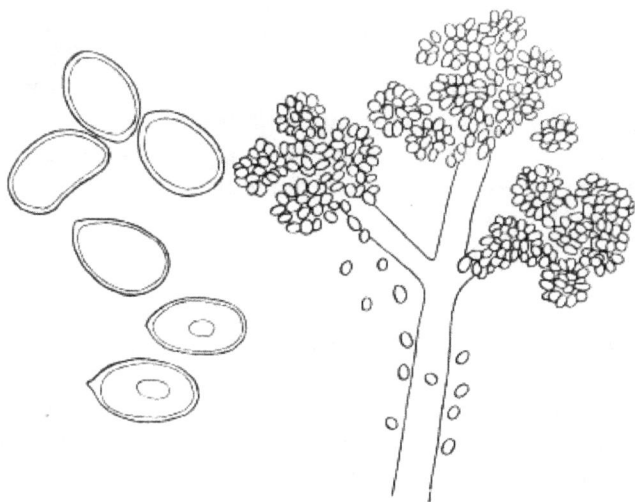

FIG. 77.—Botrytis cinerea. Conidiophore with conidia. x 135. To the left conidia.   × 1088. Original.

On germination of the conidia the mycelium then formed enters the berries through wounds or through the stomata on the thickened part of the pedicel (the "bourrelet"), or through the lenticels on the berry or through the soft skin of the ripe berry. Although it is pre-eminently a disease of the berries, the fungus can also attack the young shoots and the leaves if damp weather prevails. It can also attack the stalk, when the whole bunch may drop.

It passes winter as mycelium and as *sclerotia,* which are black bodies of mycelium filaments. Out of these the typical conidiophores and conidia (Fig. 77) again arise in spring or later; these propagate the disease in summer. Out of the sclerotia may also be formed a cup-shaped fructification, an *apoihecium,* with asci and ascospores which form mycelium filaments on germination, out of which the summer sporesare again formed.

When cuttings are bench-grafted and then moistened (especially in moist sand) and kept in a fairly warm room for the formation of callus, this fungus easily develops and hinders the formation of callus, and may cause great damage. Under these circumstances the black sclerotia are readily formed.

Control.

The removal of leaves has already been mentioned. De Istvanffi (quoted from Babo u. Mach, *l.c.* ii. p. 388) further recommends spraying with 1 per cent Bordeaux mixture (5 : 5 : 50) and dusting with a powder consisting of 10 per cent sodium bisulphite and 90 per cent clay. The latter could be used on ripe grapes, but not the former.

7. ROOT ROT

Names: Root rot, Pourridié de la vigne, Wurzelfäule, Wurzelschimmel, etc.

This disease can develop wherever the soil conditions are favourable, *i.e.* where the soil is too wet, especially with stagnant water in the subsoil. The fungus causing this disease was first carefully studied by R. Hartig, and called *Dematophora necatrix,* Hart., by him. It is now known as *Rosellinia necatrix* (R. Hart.), Berl. It lives in the soil and goes on to the roots of the vine and fruit trees if conditions favour this.

Where it attacks vines, we look for afflicted patches as in the case of phylloxera. The vines grow badly, and perish after a year or two. At first the smaller roots rot and then the thicker ones. The patches extend concentrically if the soil allows this. In dry, permeable soils it cannot exist. In nurseries it can do great damage, as the vines are here very near each other.

Botanical Study of the Fungus.

It can be both a parasite and a saprophyte. Where it is present as a parasite, we find only its mycelium. It shows its fructifications only when acting as a saprophyte, hence on dead parts of the vine, also when killed by itself. The *mycelium* propagates the fungus and spreads it in the soil. It occurs as a woolly, white and brown mass, also as white threads or strings, a true pseudo-parenchyma, on the living roots that are being attacked. The brown mycelium threads regularly show *pear-shaped swellings* at their cross-walls, which are absolutely characteristic of this fungus. The mycelium grows on the outside of the roots, in the cambium, in the wood, and in all the tissues, and also penetrates their cell-walls.

Apart from its propagation by mycelium, the fungus propagates itself by means of chlamydo spores, conidiophores with conidia, sclerotia, pycnidia, and perithecia. The chlamydo spores are usually egg-shaped (rarely spherical), thick-walled resting spores, which are formed out of the mycelium at its pear-shaped swellings.

Control.

The only effective preventative is *energetic drainage*. As soon as sick patches are noticed in the vineyards, the sick vines should be carefully dug up, roots and all, and burnt on the spot. Do not allow the fungus to form its fructifications on the dead wood. Dig up also all vines for 6-9 ft. around the sick patch. For two to three years nothing should be grown on such patches excepting cereals. Finally, the soil should be worked deeply, and should be injected with carbon bisulphide at the rate of 33-41 grammes (about 1⅓ ounces) per square yard. Repeat this for a couple of years. Where this fungus has become troublesome in a nursery the above treatment should be applied, and later a system of rotation should be followed, as I have already recommended for nurseries.

8. "ROTER BRENNER"

This disease has for the first time been carefully studied and described by Müller-Thurgau[1] in 1903. It causes intensely red to purple-red patches with yellow or light green borders on the leaves of black grape varieties. On those of white varieties the patches first become yellow, sometimes nearly white, and only when the tissue in the patch dies, does the colour become a light reddish-brown.

---

[1] H. Müller-Thurgau, "Der rote Brenner des Weinstocks," *Centralblatt für Baku,* Abt, II., Bd, X., 1903, pp. 1-38.

This disease should not be confused with other discolorations of leaves caused by physiological diseases. It occurs in Switzerland, Germany, etc. In the northern part of the canton of Zurich it was particularly bad in 1900.

The lowest two to three leaves of the shoot are attacked worst of all and drop early. The disease hinders the ripening of the grapes and canes, weakens the vine, and causes a rapid decline in its productiveness. Müller-Thurgau is of opinion that the same fungus may possibly destroy the young bunches. The disease is caused by a fungus that lives in the leaf nerves, and which Muller-Thurgau has named *Pseudopeziza tracheiphilus* (Müller-Thurgau). It seems that the fungus is restricted to the leaf nerves so long as the leaf is alive. Towards autumn the mycelium filaments penetrate the dead leaf tissue, where conidiophores and conidia make their appearance on the lower surface of the leaf. On old leaves that had been lying on the ground for a long time he found numerous *apothecia,* up to over 100 per square cm. They contain asci with ascospores. One apothecium can form up to 800 ascospores. The ascospores are liberated in spring and propagate the disease. The infection takes place on the lower surface of the nerves. Müller-Thurgau has made the disease develop on healthy leaves by means of an infection with these ascospores.

Control.

The development of the fungus depends upon the weather and soil conditions. This explains why year after year it occurs only on certain places and vines in a vineyard. The vines are especially attacked on soils that become very dry in summer *(e.g.* sand, stony soils and loams, and shallow soils). Hence it is much worse on hot, dry, steep slopes than on moist, deep, fairly level-lying soils that are rich in humus. A copious dressing of farmyard manure makes the vines much more resistant to this disease. It is a good plan early to plough under the leaves that have dropped, and thus to keep the spores in the soil as far as possible. Further, the vines should be sprayed *early* (end of May—beginning of June in Switzerland) with Bordeaux mixture, and as far as possible on the lower surface of the leaves. The above was extracted from Müller-Thurgau's publication on this disease.

9. BACTERIAL DISEASES

Sorauer, *l.c.* ii. pp. 56-60, mentions a number of bacterial diseases of the vine which have in most cases, however, not yet been sufficiently studied. Hence it is often doubtful whether we are dealing with the primary cause or with a secondary phenomenon of some of these diseases, *e.g.* Mal nero, Gommose bacillaire, etc. The *bacteriosis of grapes* has been studied by Cugni and Macchiati, who attribute it to a bacillus, *Bae. Uvae*. It colours the young berries and pedicels brown, and they dry up to a brittle mass. According to Macchiati the disease has been successfully propagated by infection with pure cultures of the bacillus.

E. Prilleux has studied a bacterial disease on grapes in hothouses, which causes light brown patches on the berries that penetrate deeper into the berry and expose the seeds. Such berries dry up. It can destroy all the grapes if it comes early. In the cells he found a swarming bacillus of $1.25\mu \times 0.75\mu$.

In Italy the disease *"Rogna delta vite"* occurs on the shoots, on which soft swellings are formed which later become hard and woody. This is attributed to *Bacillus ampelopsorae Trevisan,* by Cuboni. Cavara regards the *tuberculosis of the vine,* which occasionally occurs in Italy, as identical with the foregoing disease, because he was successful in causing it by an infection with a pure culture of *Bac. ampelopsorae*. The above has been extracted from Sorauer.

In South Africa, America, and elsewhere we have to do with a bacterial disease of vines and fruit trees which Sorauer does not mention. It is the *crown-gall* or *cancer* of the vine, and is the same as "Rogna della vite" and tuberculosis of the vine mentioned above. It enters the plant through wounds. Hence we sometimes see the cancerous growth caused by this disease on the joints of grafted vines. However, they occur on the roots also, where they should not be confused with the swellings caused by root-eels and phylloxera. The disease is particularly bad in damp soils (nurseries). The only remedy is to drain the soil, not to grow vines or fruit trees on it for some years, to follow a system of rotation in the nursery, and to use healthy grafting material. Cut out the cancer deeply till healthy wood is reached, and wash the wounds with a 5 per cent solution of copper sulphate, where it occurs on grafted vines. If such vines are planted in fairly deep and fertile soils, they will usually either rid themselves of the disease or not suffer appreciably from it. The disease is caused by *Bacillus tumefaciens.*

C. DISEASES CAUSED BY ANIMALS

1. PHYLLOXERA, *Peritymbia vitifolii,* Fitch *(Phylloxera vastatrix,* Planchon)

A. **Historical.**
The insect was discovered for the first time in 1854 by Asa Fitch in the State of New York. It was the gall or leaf form, and he called it *Pemphigus vitifolii.* In 1867 Dr. H. Shiner found and described also the winged insect which he called *Dactylospora vitifolii.* Meanwhile Westwood had discovered it in hothouses in Hammersmith near London in 1863, first in the leaf galls (see Fig. 49a), but later also on the roots. In 1867 he gave it the name *Peritymbia vitisana.*

Near Pujault (Gard) sick patches were observed in the vineyards in 1863. In 1867 it was so bad in other places too that a special Commission consisting of Messrs. Sahut, Bazille, and Planchon investigated the trouble in 1868, and found the insects on the roots of the vines. Planchon is usually mentioned as the discoverer of phylloxera in France, though others attribute the discovery to Sahut.

On the 28th August 1868 Planchon saw the winged insect formed out of the root form, and then called it *Phylloxera vastatrix.* In 1869 he found galls on vine leaves that were analogous to those of Fitch's *pemphigus.* He then got the idea that it must be the same insect as the root form. With the help of his brother-in-law, Lichtenstein, he succeeded in making the leaf form of the insect live on the roots. Riley came over specially from America to check the identity of the different forms, and corroborated it.

The phylloxera then rapidly spread over the whole of France and the wine countries of the world. At the Cape it was discovered in patches of sick vines in Constantia and Moddergat (now Helderberg) in 1886, and spread very rapidly. Baron Carl von Babo[1] mentions the discovery of

---

[1] Von Babo, Baron Carl, *Reports on Viticulture in the Cape Colony,* 1885 and 1886.

phylloxera at the Cape in his 1886 report, whilst in his 1885 report it was stated that phylloxera had not yet been found at the Cape.

Phylloxera is indigenous in the eastern states of the United States, and was probably brought from there to England and France between 1854 and 1860 on the numerous rooted American vines that were then imported because they were more resistant than the European varieties to oidium, which was then causing great damage in the European vineyards.

B. **Biology of Phylloxera.**

This insect lives only on the vine (vitis), and occurs both on the leaves and the roots. The leaf form has been called *Phylloxera vastatrix gallicola* and the root form *Phylloxera vastatrix radicicola.* The leaf form does no direct harm, but helps to propagate the insect. On the whole the galls (see Fig. 49a) occur very seldom on European vines; on the American species and varieties, on the contrary, they occur fairly often and are sometimes very numerous. I believe the galls have never yet been observed in South Africa, either on American or European vines. Notwithstanding this absence of the gall form, the phylloxera does not seem to have been weakened during its forty years' existence at the Cape. On the contrary, vines grafted on Aramon R.G. Nos. 1 and 2 are now killed that were formerly not damaged. It would seem as if a more dangerous or virulent biological race of phylloxera has been evolved at the Cape.

In this connection it is interesting to mention the conclusion to which Börner[1] came as a result of his studies of the biology of phylloxera. According to Babo u. Mach, *l.c.* ii. pp. 297-299, Börner concluded that there are two races of phylloxera, one occurring in the south of France and the other in Lorraine. The latter he called *Peritymbia vitifolii pervastatrix,* whilst the former retains the name *vastatrix.* Whereas both attack and damage the European vines equally, the pervastatrix race avoids certain American vines

---

[1] C. Börner, "Exptr. Nachweis einer biol. Rassendifferenz zw. Rebläusen aits Lothringen u. Südfrankreich. *Peritymbia (Phylloxera) vitifolii pervastatrix.*" 1910. *Zeitschrift für angewandte Entomologie,* vol. i. p. 59, 1914.

or their hybrids completely or more or less. According to Börner, the following are *immune* from the pervastatrix race: Rip. gloire de Montpellier, Rip. × Rup. 3306 and 3309, Cab. x Rup. 33A1, Rip. Geisenheim 1, Rip. × Rup. Geis. 13 and 107, Cordif × Rup. Geis. 19 and 20, Solonis × Rip. 1616A, etc. On these immune stocks the pervastatrix race cannot propagate itself and perishes. Hence these stocks remain sound. He says that thus far only the pervastatrix race has been found in Germany, and warns against the introduction of the vastatrix race from France. He further recommends that only the stocks that are immune against the pervastatrix race should be used in Germany, when the phylloxera will in course of time die out in Germany. This can certainly not happen as long as there are ungrafted European vines in Germany. The following stocks Börner considers *half immune* to the pervastatrix race: Aramon R.G. Nos. 1 and 2, and Mourv. × Rup. 1202. Grassi and Topi in Italy and Schneider-Orelli in Switzerland have checked Börner's investigations and confirmed his results. It is desirable that similar investigations should be undertaken in other wine countries. The two races differ in that some American stocks are attacked by the one and not by the other.

Viala, *l.c.* p. 500, classifies the different stages in the development of the insect as follows: 1, The sexual form; 2, the leaf form; 3, the root form; 4, the winged form (see Fig. 78).[1]

1. *The sexual form.*—It arises out of the eggs laid by the winged insect on the lower surface of the young leaves between the nerves (seldom on the bark of the shoots). The eggs are of two sizes, 0.40 × 0.20 mm. and 0.30 × 0.15 mm. Out of the small eggs males hatch and out of the large ones females. Both sexes are devoid of sucking and digestive organs. After they have mated the male dies. The female lays just one large egg—the winter egg—in the crevices of the bark of the two-year-old or older wood.

2. *The leaf form.*—In spring one insect hatches out of the winter egg. It crawls up the young shoots and begins to suck on the lower surface of a young leaf, when a gall is formed at that spot. In this gall the insect lays a number of eggs out of which insects are again hatched. This propagation is parthenogenetic.

---

[1] *2012 Editor's Note:* Fig 78 is here reproduced on page xiv. (PFM)

3.   *The root form.*—These are the insects that are hatched in the leaf galls, whence they emerge especially in autumn in order to enter the soil and get on the roots, which they puncture to suck sap for nourishment.   It is here that they do great damage. These insects can propagate themselves directly on the roots in several generations during the year and from year to year without passing into other forms. This is what mainly happens at the Cape. These insects have a yellow to brown colour and are very small, but can be seen with the naked eye.   In the warm wine countries with their long summers the root insects can pass through more generations in one year than in the colder countries, and are therefore more dangerous and can destroy the vines more rapidly.   They lay eggs on the roots out of which the same kind of insects hatch.   Here some hibernate under the bark of the thick roots, whilst others crawl out of the ground in the warm summer months (July in Europe and January to February at the Cape), moult several times on the ground or on the stem of the vine, and pass into

4.   *The winged form,* which now begins to fly about and lays the two kinds of eggs that were mentioned under the sexual form.

M. Max Cornu[1] has made a splendid and detailed study of phylloxera to which I refer the reader for further details. Fig. 77 is a $1/6$th reduced reproduction of PL XX in his work.

C. **The damage the Phylloxera does.**

In Chapter IV, *(h),* "Resistance to Phylloxera", the formation of nodosities and tuberosities on the roots, and the rotting of the young and older roots, have already been discussed. After the phylloxera has killed nearly all the young rootlets and some of the old roots, the vine lives on its stored food for some time and then dies from exhaustion, as it can no longer take up sufficient food from the soil. As soon as the vines have been weakened so much that they can no longer form new roots, the phylloxera leaves them and goes to more healthy ones. Hence when looking for phylloxera in a

---

[1] Max Cornu, "Études sur le Phylloxera vastatrix,' *Mémoires présentés par divers savants à l'Academie des Sciences de l'Institut National de France,* tome xxvi. No. 1, pp. 1-367, MDCCCLXXXVIII. Imprimerie nationale, Paris

phylloxera patch, one should not examine the roots of very weak vines, but those of the ones nearby that are still looking fairly healthy.

The phylloxera thus attacks and damages the roots, but one can notice the effect of this on the parts of the vine above ground. The very first effect on a vineyard attacked by phylloxera is to make it unusually productive. This was noticed also with the arrival of phylloxera in our Cape vineyards, and was a sure sign that such vines would be much worse the following year. As the insect can fly in summer, it is quite clear why it attacks a vineyard in patches and not steadily from one side.

A *phylloxera patch* is easily detected in a vineyard by the experienced person. During the growing period the patch shows a round depression, the vines in the middle being usually the smallest and weakest. At a more advanced stage one or more vines in the patch will be dead and the roots so rotten that the vine can be pulled out by hand. The seriously affected vines form only poor, short shoots, and thus a phylloxera patch on a slope looks like a bare patch from a distance. The very first sign is that the affected vines stop their summer growth sooner than the healthy ones. Their leaves at first have a dull green colour; they turn yellowish later in summer and drop early. Grapes that may still be found on such sick vines ripen badly.

The following year we notice a number of new patches, until after three years a large vineyard on clay soil will have only odd patches of vines that are still alive, and these will have to be uprooted, as it will not pay the farmer to keep them. This is precisely what happened at the Cape. The wine farmer's income thus dwindles away to almost nothing in a few years' time; the phylloxera has caused terrible losses in all the grape-growing areas, and many growers were ruined financially. According to Viala, *l.c.* p. 500, the phylloxera has caused a loss of at least £400,000,000 to France. Vineyards in full bearing which before the arrival of the phylloxera were worth £130 to £260 per acre, were subsequently sold for grazing or grain lands, *i.e.* at £5 and £10 to £17 per acre, and France then had about six million acres under vines. The day labourers earned only half as much as previously in wages, and one-fourth to one-third of the population emigrated from the viticultural areas.

At the Cape, too, we had a hard time, but by immediately planting fruit trees in the doomed vineyards, often before they showed signs of the attack, and by growing vegetable crops and living as cheaply as possible, our wine farmers struggled through until they could again produce wine from the grafted vines. It is a notable achievement which should be recorded in honour of these men, that so few at the time became insolvent.

The phylloxera brought also certain advantages. Thus the vines were planted further apart (4'6" × 4'6" and 5' × 5' square) in the new vineyards than they had been in the old ones, because grafted vines were scarce and expensive. This made possible the cultivation of the vineyards by machinery —the Planet Jr. was introduced about this time—and thus enabled our wine farmers to adopt the system of dry land farming long before it was ever mentioned in South Africa. This enabled them to grow vines on slopes more than before, and practically put a stop to the irrigation of vineyards in the Western Province (Cape). It should also be recorded, as was previously mentioned, that the phylloxera gave a great impetus to the development of the fruit industry at the Cape.

D. **The Spreading of the Phylloxera.**

This happens in a natural way or by the agency of man. The insects crawl slowly in and on the ground from vine to vine. Although the winged insect cannot fly far on a calm day, it can soon cover a good distance when there is a wind. When infected vines are carried away by floods and landed at a great distance away, the insect might soon spread over a great distance. It is possibly also spread by other flying insects and birds. Man can spread it by means of his boots and clothes when going from an infected into a non-infected vineyard, and also by sending infected leaves, canes, and especially rooted vines, into clean areas. This latter danger we can reduce to a minimum by first thoroughly disinfecting all parts of the vine that go from an infected into a clean area. The safest seems to be the treatment with water of 60° C, which soon kills the eggs and the insects. Faes and Staehelin[1] dip the whole vine or cutting for twenty-four hours in a 3 per cent solution of K_2CS_3 (pot. thio-carbonate) and 1 per cent black soap, which effects a complete disinfection.

As the phylloxera lives exclusively on the vine, it can be present on other plants only by chance and only for a while, so that these could only very rarely assist the spreading of the insect.

E. **Control of Phylloxera.**

Here we must distinguish between treatments aiming at the complete destruction of the insect and those with a more limited aim, where the phylloxera is to be kept in check to such an extent that viticulture can be practised with success in its presence. We therefore distinguish between methods of extinction and cultural methods.

(a) Methods of Extinction.

They are applied when phylloxera first makes its appearance in a country, and the object is to destroy the insect completely. It has been shown that the progress of the insect can thus be retarded, but in not a single country where the insect has once taken a footing has it been possible to get completely rid of it. Here the vines are killed together with the insect, which naturally is not the case with the cultural treatments.

Where total extinction is aimed at, the vines on the whole phylloxera patch, together with a considerable zone of healthy vines around, are uprooted, deposited in the middle of the patch, sprayed with paraffin oil, and burnt on the spot. Holes 1 ft. deep and 20 in. apart are made in the ground with crowbars, 150-200 grammes of carbon bisulphide is poured into every hole, which is immediately closed and the soil rammed down, after which some water is poured over the place where the hole was made. Finally paraffin oil is poured over the whole bare patch at the rate of 2 litres per square metre. All tools, etc., are thoroughly disinfected. Such patches are regularly examined for several years, and for a number of years they are not replanted with vines.

---

[1] Faes and Staehelin in *Revue de Vit.,* tome lx., 1924, p. 284

(b) Cultural Methods.

1.   *Treatment with carbon bisulphide,* CS_2.—This is a most inflammable liquid and therefore dangerous to handle unless it is treated with great care. It should be injected once a year in dry soil by means of a special injector at the rate of 15–25 grammes per square metre. Instead of carbon bisulphide its aqueous solution, containing 0.5 gramme per litre, can be used. This is no longer inflammable. Vines thus treated have a darker green and healthier colour, and grow more vigorously than untreated vines. It seems as if the carbon bisulphide in some way increases the vine's vigour. It no doubt has an influence, seemingly favourable (from the vine's point of view), on the micro-flora and -fauna of the soil.

2.   *Treatment with potassium sulfo-carbonate,* K_2CS_3.—This substance decomposes in the soil into potassium carbonate, carbon bisulphide and sulphuretted hydrogen, according to the equation $K_2CS_3 + CO_2 + H_2O = K_2CO_3 + CS_2 + H_2S$. Both these sulphides are powerful insecticides. This treatment is much more expensive than the first one, and requires a lot of water. It does not answer in clay soils. Other soils are suitable and are treated in winter. Hollows are made around the vines and the solution is poured into these. Apply 40-50 grammes pot., sulfocarbonate dissolved in 10-15 litres water per square metre of soil, *i.e.* 360-450 lbs. dissolved in 9000-13,500 gallons water per acre. This amounts to an irrigation with half an inch of water. At the same time the soil gets a heavy dressing of potash, amounting to 267-334 lbs. pot. carbonate per acre.

3.   *Submersion in Winter.*—This has been done in certain parts of France. It naturally requires a level vineyard with a wall around it, readily available water, and a sufficiently long and cold winter. The vines must be absolutely dormant when this is done. All varieties do not stand it equally well. In 1905 I saw such a vineyard near Narbonne in the South of France. It took three weeks to cover the soil with 6 in. water, which remained standing for five weeks before it was again run off. This requires about 1–2½ million gallons water per acre according to the nature of the soil. (See Foëx, *l.c.* p. 717.) - I doubt if this is now still practised. Such vines grow very vigorously and are very productive, but must be heavily manured. At the Cape it

could not be applied, as the vines are really dormant for too short a period owing to our mild winters. With this treatment the insects and a number of eggs die, but some eggs escape. Hence the treatment must be repeated every winter.

4. *Planting Vines in Sand.*—From the beginning it was noticed that the vines in certain sandy soils, especially those along the coast of the Mediterranean, were hardly damaged by the phylloxera, whilst vines on clay soils perished very soon. Even now this cannot be explained. The fine sand seems to hinder the movements of the insect, and the finest splinters might kill them when attempting to move. Such soils should contain about 70 per cent and more sand, also fine sand, and should not be loamy. Foëx says that they must contain at least 60 per cent quartz.

In France numbers of vines have been planted in sand along the coast of the Mediterranean Ocean. Cinsaut (Cape Hermitage) answered best for this purpose. Such vineyards should be heavily fertilised, and preferably with artificial fertilisers, as continuous applications of farmyard manure will make the soil less unsuitable for phylloxera.

At the Cape, Stein and White French do fairly well in fine, moist, sandy soils. In moist, fertile, fine, sandy silt in Montagu and Robertson (Cape) Sultana and Muscadel do well ungrafted, and produce good crops for six to eight years and often longer periods.

Heavy dressings of Government guano (Peru type), potash and phosphates can keep ungrafted vines in moist, sandy soil for many years in a satisfactory state of vigour and productiveness although attacked by phylloxera. I have seen cases of ungrafted Green grape in such soil, where the vines had already been much weakened by phylloxera, which were restored satisfactorily by heavy fertilising, especially with Government guano, and now after nine years are still in a fair condition.

5. *Grafting on American Stocks.*—This has so far been the most generally applied and the most satisfactory cultural method of solving the phylloxera problem. It has already been fully discussed.

6. *Direct Producers.*—These have also been discussed (Chap. V.). It would be an ideal solution if direct producers could be obtained which are sufficiently resistant to phylloxera and at the same time produce grapes of high quality.

2. ROOT-EEL WORMS OR NEMATODES, *Heterodera radicicola,* Greef

Although Lüstner (Babo u. Mach, *l.c.* ii. p. 204) maintains that root-eelworms do no damage to vines, my experience at the Cape is different. In moist and in wet sandy soils (Stellenbosch, Retreat, Prince Alfred's Hamlet, etc.) they have killed grafted vines where phylloxera was not dangerous. This was where 1202, Rup. Constantia metallica or Cape metallica, and Aramon R.G. Nos. 1 and 2 had been used as stocks. On the Stellenbosch University Farm I noticed that the one-year-old roots of the Vinifera scions were sound whilst the Aramon roots were badly damaged by the nematodes. The two- to three-year-old grafted vines were sometimes damaged by eelworms on the root-stump of the American stock and showed slimy, rotten patches.

The nematodes penetrate the young roots and cause swellings in them something like the nodosities caused by phylloxera. On the older roots they also cause swellings. The swellings caused by nematodes are usually much larger and softer than those caused by phylloxera; this enables one to see at once with fair certainty whether they are the work of nematodes. Absolute certainty we get by examining under the microscope thin sections made through such a swelling, when the presence of cysts containing eggs or young nematodes will definitely prove that nematodes were the cause of the swellings. Such a cyst represents the swollen females, which now consist merely of a skin with eggs or young eels. After a while the swellings rot and the eelworms are liberated, when they again move about in the soil. Rodrigues Moraes has observed this in Portugal, and also distinguished between males and females. Mating probably takes place when the eelworms are liberated. In Portugal they have done a lot of damage (quoted from Portes et Ruyssen, *l.c.* iii. p. 783).

Nematodes cause patches to be formed in the vineyard like those formed by phylloxera. The eelworms themselves are at most only ¼ mm. or $^{1}/_{100}$ in. long and about $^{1}/_{1500}$ in. broad. Hence they cannot be seen with the naked eye. They attack wheat, sugar-beet, potatoes, tomatoes, tobacco, fig trees, peach trees, and many other plants, and can kill them if the soil conditions favour their development. An excess of water in the soil is a primary condition, although the soil need not always be *very* wet. Further, sandy soils suffer most from eelworms.

This eelworm is also known as *Anguillula radicicola,* Greef. The eelworms belong to the Nematodes. They are difficult to combat once they have established themselves in a soil. Sometimes energetic drainage will help. Ottavi[1] further recommends that the soil should be given an injection of 15-20 grammes carbon bisulphide per square metre (about ½ oz. per square yard or 151 lbs. per acre) when the eelworms are moving about in the soil. According to Moraes this is the case in Portugal during May or about one month before the vines flower. At the Cape this would be about the end of September. This will not kill the eelworms that are in the roots. It will probably be best to use stocks that are resistant to the eelworms, and at the same time to drain the soil well. Thus far it seems to me as if Jacquez and Rip. gloire are highly resistant to eelworms.

3. MEALY BUG, *Pseudococcus capensis,* Brain

One usually sees individual ones crawling about from berry to berry when the grapes are about to turn colour. Later on one notices the white masses, hence the name *mealy* bug, in the bunches. They remain in one place. It would be better not to have this bug on wine grapes. It completely spoils table grapes, and this is where it does most damage. It seems to be getting gradually worse in our Cape vineyards. According to my own observations on the Paarl Experiment Station, it seems to be worst on vines that grow vigorously and form big, compact bunches like Grenache noir. When ants run about on a vine, it is usually a sure indication of the presence of the mealy bug.

Control.

Dr. C. Mally, formerly chief entomologist for the Cape Province, has worked out a method of controlling this pest by fumigation with hydrocyanic acid. Suitable gas-tight tarpaulins are put over the vines and covered along the edges with soil in order to obtain an enclosed space. Underneath

---

[1] O. Ottavi, *Viticoltura teorico-pratica,* 4th ed., 1922; Fratelli Ottavi, Casalmon-ferrato.

these a small glass tube containing the required quantity of anhydrous hydrocyanic acid is broken, when the acid gradually evaporates and fills the enclosed space. It kills both the insects and their eggs. The treatment takes place in winter whilst the vines are dormant, but is troublesome to execute. If one notices the development of the bugs in time, they can be killed by spraying with hot water about 70° C. or 158° F. when still crawling about as isolated individuals. This treatment will have to be repeated. Where the pest is bad, clean cultivation the whole year round is essential. From the time the bugs appear on the bunches till the grapes turn colour, spraying with 4 per cent Clensel may be tried (according to Prof. Dr. C. K. Brain).

4. CALANDRA, *Phlyctinus callosus*, Bohem

This insect is indigenous to the Cape. Lounsbury[1] gives a brief description of the insect. On Fig. 79 we see the larval, pupal, and beetle stages of the insect. The worm or larval stage the insect passes in the ground, where it lives on the roots of the vine without doing appreciable damage. It becomes a pupa in the soil, usually early in spring, and the beetles emerge from the soil during the latter half of October and the first half of November. In Constantia, where the pest is worst, they are usually expected about the 5th November. There is only one generation of beetles a year. The young ones are lighter coloured and softer than the old ones, which are ash grey and fairly hard. Some of them remain alive right through winter and continue their work of destruction at the beginning of the following spring.

---

[1] Lounsbury, C. P., "The Calandra of the Vine *(Phlyctinus callosus,* Bohem)," *Agric. Journ., Cape of G.H.,* vol. xxxvii., Oct. 1910, pp. 448-450.

The calandra attacks a number of plants, both wild and cultivated. Here it interests us as a vine pest. It is usually worst in clay and peaty soils, but can do a lot of damage also in other soils. It is usually worst during the first year after the soil has been trenched or deeply ploughed and planted with vines. Also, it is worse in some years than in others. It eats leaves, shoots, bunches (stalks, pedicels, and berries), and generally all the green parts of the vine, where it is very bad. It usually does the greatest damage to the grapes. Sometimes it eats the stalk so much that the whole bunch dries up and is lost. Sometimes it eats the pedicels, when the berries may dry up. Sometimes, and usually, it eats the berries. If damaged while small, the whole berry is usually lost. Once the berries are nearly full grown, it merely eats one or more holes in them, which means total destruction of such berries with table grapes, as they have to be removed. In this way a whole bunch might be spoilt, and therefore the calandra usually causes greater damage to table grapes than to wine grapes, although it may be serious enough with the latter.

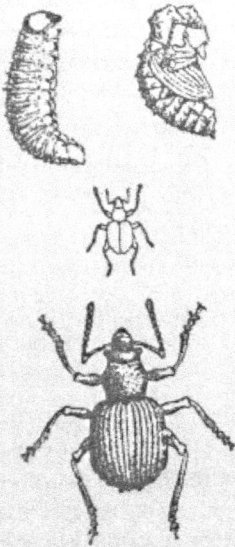

Fig. 79.—Calandra. From Lounsbury, "The Calandra of the Vine," *Agric. Journ.*, *C.G.H.*, vol. xxxvii. (Oct. 1910).

The beetles can continue eating until the grapes ripen, when they decrease in numbers and gradually disappear, until we notice them again on the vines in the next spring. Most people notice only the beetles, as the larvae and pupae live in the soil. The beetles eat mainly at night. During the day they hide in the soil under clods of earth, beneath the old bark, and even under dense foliage and in compact bunches. If one touches them or shakes the vines, many of them drop to the ground and feign to be dead. As their colour often corresponds with that of the soil, and as they often crawl into the soil very quickly, one cannot easily detect them on the ground.

Control.

Spraying with lead arsenate is very effective in a mother plantation of American stocks, and generally in a nursery or with young vines where no crop is expected. According to Lounsbury, *l.c.,* one should take 3 lbs. lead arsenate paste to 50 gallons water. Spray well when the young shoots have reached a length of 6 in., or as soon as the first beetles are noticed; then again after a rain, and after the summer growth of the vine is well developed. Bearing vines can be thus sprayed until the young berries have a diameter of about 1 mm. or $^1/_{25}$ in. Do not spray later with lead arsenate, as the poison remains on the ripe grapes and renders them unfit for human consumption.

Calandra can be successfully controlled by tying strong paper around the stems of the vines and around the standards and droppers where the vines are trellised, and covering the paper with some sticky material like tangle-foot. At the same time the vines should be kept well topped to prevent any shoots from touching the ground, and thus enabling the beetles to get on to the vines.

Another method is to place bundles of crumpled vine leaves or old pine cones in the trunks of the vines, when large numbers of calandra will hide there during the day. After a couple of days the leaves will be dry, and are collected in a good bag soon after mid-day and then burnt. By repeating this a couple of times, most of the beetles can be destroyed. Some farmers have the soil around the trunks of the vines removed by spades early in winter (say beginning of May at the Cape) and scattered among the rows of vines. This is supposed to help a good deal. Others, again, throw slaked lime on and around the vines early in winter, and believe that it helps. We are not by any means so sure of the effectiveness of the last two treatments as of the others, which enable us to effectively control the calandra.

5. ERINOSIS, *Eriophyes (Phytoptus) vitis* (Land.), Nal

This disease is caused by a microscopically small mite. It punctures the leaf, usually on the lower surface, when it gradually bulges out on the upper surface and forms a sort of blister which, when small, has a reddish colour, but gradually turns yellowish to greenish as it grows larger.

Underneath these blisters we notice a white felty mass of hairs that have developed out of the lower epidermis cells where the leaf was punctured. Later, these hair's take a reddish-brown colour. The leaf usually shows a number of such blisters (see Fig. 80). The mites can be seen amongst the hairs by means of a microscope. A magnification of 50 diam. suffices.

The mite has a length of 140μ (males) to 160μ (females), thus about $1/170$ in., and is about one-fourth as broad. The females are usually much more numerous than the males. They lay fairly large eggs amongst the hairs, where these hatch into new mites. The mite hibernates behind the scales of the eyes on the canes of the vine, from where they emerge when the eyes start budding during the next spring, crawl up the young shoots, and puncture the young leaves on their lower surface. The full-grown leaves are hardly ever attacked.

FIG. 80.—Erinosis of the vine. A, leaf from above; B, leaf from below. From Babo u. Mach, *Handbuch des Weinbaues*, ii., 1924. Paul Parey, Berlin.

Certain varieties, *e.g.* Muscat of Alexandria, are attacked much more than others. Some American stocks are also attacked, *e.g.* 1202, 333, whilst I have never yet observed it on 420A. In moist spots and in hollows it is much worse than on warm, dry sites. Where the disease is very bad, one sometimes sees the white felt-like masses also along the nerves on the upper surface of the leaves. In such cases the leaves remain small and the internodes short, which naturally means that the vine is going to suffer a great deal. Ordinarily it does not really damage the vine. It can be spread by means of infected cuttings

Control.

If the vines are regularly sulphured early in the season, the disease will soon disappear almost completely. Although the efficacy of sulphur against erinosis has sometimes been doubted, it has certainly proved very effective at the Cape.

Note.—There is a slight resemblance between the white patches caused on the lower surface of the leaf by plasmopara and by erinosis, although they ought never to be confused. Only erinosis causes the blisters, and it shows a mass of felt-like hairs, whereas plasmopara shows a soft white mass of conidiophores and conidia.

6. FRUIT FLY, *Ceratitis capitata,* Wied

This pest occurs in the countries bordering the Mediterranean Ocean (hence also called the Mediterranean Fruit Fly), the Canary Islands, the Azores, South and West Africa, Madagascar, Mauritius, the West Indies, the Bermudas. According to Sorauer, *l.c.* iii. pp. 416-417, it occurs also in England, but is so rare as not to cause any damage. It attacks many kinds of fruit (and also other plants) just when the fruits begin to ripen, but no longer once they are ripe (see Sorauer, *l.c.* iii. p. 417). As soon as the fruit begins to ripen and thus also becomes *soft*—this happens when the grapes turn colour—the fly punctures its skin and deposits an egg inside the fruit. This egg develops into a maggot which eats the fruit inside and causes it to rot. The full grown maggot becomes a puparium, out of which a fruit fly is once more formed.

As far as I know, I was the first to point out that the fruit fly can cause great damage in grapes. In 1915 I noticed numerous half-ripe berries in the Almeria (Ohanez grape) pergola on the Paarl Viticultural Experiment Station with brown stripes under the skin, which were found to contain very active maggots on being opened. At my request Dr. F. W. Pettey, entomologist at the Elsenburg School of Agriculture, bred the fruit fly out of this material, and thus proved that it was a fruit-fly infection. That year about 90 per cent of the Almeria grape crop was destroyed by fruit fly. The following year the loss was under 1 per cent, but then I controlled the pest, as will be stated below.

It is only the *late* grape varieties that are troubled by this pest. As the Almeria grape, really Ohanez, is the last to ripen, it is no wonder that it is so badly attacked. The infestation will be all the worse if there are late peaches and other fruits in the neighbourhood which are infested with fruit fly. On the whole it is only very late varieties of grapes that are threatened by this pest.

Control.

The most effective and at the same time a very easy treatment is Dr. Mally's "Fruit Fly Bait", which is made up as follows:

>   3 lbs. brown sugar,
>   3 oz. lead arsenate (50 per cent paste),
>   4 gallons water.

Stir well until all the sugar is dissolved and spray the vines with a hand syringe every eight to ten days from the moment the berries begin to turn colour or become soft (thus about four weeks before the grapes will be properly ripe) until about eight days before the grapes are picked. There is no danger of being poisoned by such grapes. Let the loose drops fall just here and there on the leaves, and try to miss the grapes as much as possible. In addition the fruit fly should be properly controlled in adjacent orchards. The fly is attracted by the drops of poisoned bait, eats it, and dies. This treatment is cheap and quickly applied.

7. OENOPHTHIRA PILLERIANA, Schiff (Pyralis vitana, Audouin; Tortrix pilleriana, Schiff).

In France it is known as "pyrale" and in Germany as "Springwurmwickler". At the Cape it is fortunately not known. Where it occurs in large numbers it is one of the most dangerous pests of the vine. It is particularly bad in the

South of Hungary and in France. The moth flies at night. In July (in Europe) the female moth lays 20-60 eggs on the leaves of the vine. These eggs are oval in shape and about 1 mm. or $1/25$ in. long. In nine to ten days they are hatched and little worms (caterpillars) (2 mm. or $1/12$ in. long) emerge from them, which suspend themselves by means of a thread and land on the trunk of the vine, where they crawl underneath the old bark, spin themselves in, and hibernate. The following spring they crawl out again, spin threads over the opening buds or around the young leaves, and eat the leaves. Where they are very bad they can eat up all the leaf-blades, leaving only the leaf-stalks. Sometimes they destroy the clusters also. When one wants to catch them they drop to the ground and crawl into the soil. When full grown they have a length of 2.5–3 cm. or fully an inch. In Europe they eat the leaves towards the end of June and spin in the drying leaf, in which they become a pupa (10–12 mm. or nearly ½ in. long), out of which in ten to fourteen days' time (usually towards the middle of July) a moth emerges, which again lays eggs as we have seen at the beginning. Naturally it is the worms that do the harm.

During the day the moth sits on the under-side of the leaves, and in any case inside the vine in the shade. This insect can live also on other plants (e.g. lucerne, strawberry, etc.) and lay its eggs on them.

Control.

Of the numerous control measures against this pest, the following are the most important:

1. *Gather Leaves with Eggs.*—The little heaps of eggs can easily be seen on the leaves. By employing women and children for gathering the leaves on which eggs have been laid, the work can be done at a cost of about £2, 5s. per acre. As the moths will again lay eggs, this work must be repeated a couple of times in order to be effective. This is a simple and effective means of control.

2. *Fumigation with Sulphur Dioxide.*—This is done in winter after the vines have been pruned but long before they start budding. The soil should be dry and the vines covered with metal covers. Under each cover burn 20-25 grammes or almost 1 oz. sulphur per vine. Let it act for ten minutes.

3. *Hot-water Treatment.*—In Europe the vines are sprayed with hot water in March, *i.e.* before the buds burst. The whole trunk above ground is thoroughly sprayed up to where the bearers are attached to the old wood. Do not spray the one-year-old wood, in order not to damage the eyes of the bearers; it is unnecessary, since the worms are only under the old bark. In Burgundy and down to Lyons this treatment is commonly applied. Every vine requires 1–3 lit. or about ½ gallon water. In France it costs about £1, 3s. 6d. per acre of 4000 vines. A portable boiler for making hot water is used in this work.

4. *Rub off Old Bark*—This is done with Sabaté's chain glove in winter to remove the worms. The rubbed-off bark is carefully collected and burnt.

5. By collecting and burning the leaves with the pupae.

6. By *spraying with lead arsenate* or some other stomach poison in spring to kill the worms that have hibernated.

8. CLYSIA (CONCHYLIS, TORTRIX) AMBIGUELLA, Hub.

This is one of the oldest-known pests of the vine. In 1713 it caused great damage on the Island Reichenau near Constance. It is very difficult to combat, and can sometimes cause great damage. In Germany it is called "einbindiger Traubenwickler" and also "Heu und Sauerwurm" etc.: in France it is known as Cochylis.

The moth is light grey; the front wings are yellow with a bluish-brown stripe across. There are two generations per annum. In Germany the moths of the first generation appear from about the middle of May to some time in June. During the day they sit on the leaves or in the vine, and they fly early in the morning or just after sunset. Now they fly around the vines, and the females, after mating, try to alight on a blossom cluster there to lay their eggs. These eggs are oval, ½ mm. or $1/50$ in. long, and look like pearl glands. Ten to fourteen days after the egg has been laid the worm appears. It immediately begins to eat and destroy the blossoms (which have not yet opened) until it is full grown. The dried-up blossoms are now spun together so that dried up, hay-like masses of blossoms remain in the cluster. Sometimes the peduncle is eaten, when the whole cluster dries up. If they are not yet full grown when the berries have set, they eat these too, and even the shoots. The spinning in affords excellent protection to the worms and makes their destruction very difficult. If fine, warm dry weather prevails during flowering, and it is consequently soon over, much less damage is done than if cool, moist weather prevails at this stage. In the latter case the worms can do great damage. After four to five weeks the worm stops eating, and towards the end of June to the beginning of July it gets ready to pupate. Now it usually leaves the bunch and hides underneath the bark or in a crevice in the stakes, where it pupates. Sometimes it remains in the bunch, spins itself in, and pupates there. Out of these pupae arise the moths of the second generation from the end of July to the beginning of August. They deposit their eggs on the berries, pedicels and stalks, sometimes also on the leaves. These eggs hatch soon as the air is now warm, and after eight to ten days (middle of August) the first worms of the second generation appear. They immediately begin to eat the berries and bore into them. In dry weather such berries drop, in wet weather they rot. The worm spins the berries together, and one rotten berry can infect the others so that a whole bunch can soon perish. When the grapes are ripe, end of September to beginning of October, the worms are full grown and leave the berries to pupate in the same places as those of the first generation. The pupae pass winter and give rise to the moths that appear the following May (middle) to the beginning of June.

All grape varieties are not equally damaged by this pest. Besides the vine, lilac *(Syringa persica),* ivy *(Hedera helix), Ampelopsis hederacea,* etc., serve as host plants to this insect.

Sometimes it destroys ½ to ¾ and sometimes the whole crop. It is calculated that in 1897 it caused a loss of £2,000,000-£3,000,000 to the wine crop along the Moselle. It is now a permanent enemy of the grape-grower in Germany, France, etc. Fortunately it is unknown at the Cape.

Control.

As for Polychrosis, see below.

9. POLYCHROSIS (EUDEMIS) BOTRANA, Schiffm.

The destruction described under 8 is also caused by this insect. In Germany it is called "bekreuzter oder gesprenkelter Tranbenwickler", and in France "eudemis". It likes heat and is a native of the South European countries, probably Italy. It has now spread fairly generally over all the European vineyards, and occurs even in North America. Its life-history corresponds largely with that of Conchylis. Its moths start to fly between 4 and 5 P.M. and stop flying as soon as it is dark. In the morning they fly from the time it gets light until about 8 o'clock. Contrary to Conchylis, it has three generations every year. The first and second are at the same time as those of Conchylis, but the moths of the third generation only appear in autumn, lay their eggs on the ripe grapes in September, out of which worms arise on the ripe grapes. In the Palatinate even a fourth generation has been observed in warm autumns.

Its moth is smaller than that of Conchylis. Its worm has a dirty yellowish-green colour with a light brown nearly yellowish head, whilst that of Conchylis is reddish-brown with a dark brown head.

Control of Conchylis and Polychrosis.

(a) *Control of the Moths.*—They can be caught on some sticky surface as they fly about in the morning and evening. Also a large number of moths can be caught by means of lamps in calm weather and if there is no moonshine. The moths that will be caught are mainly those of Pyrale and Conchylis, as those of Polychrosis do not fly about much in the dark. The light is put on a water and oil surface in which the moths perish.

(b) *Control of Worms:*

1. *Catch by Hand.*—Use small pincers or bits of wire to remove the worms from the clusters.

2. *Contact Poisons.*—Spray under high pressure with 3 per cent soft soap. This is very effective.

3. *Stomach Poisons.*—The most effective are lead arsenate and nicotine, but with these we can spray only for the worms of the first generation, as if used later we run the risk of having poison on the ripe grapes, and poison and the smell of nicotine in the wine. The later worms are gathered

by hand together with the affected berries and destroyed. This is done twice.

Note.—In discussing pests No. 7, 8, 9, I have obtained my information mainly from Lüstner in Babo u. Mach, *l.c.* ii. pp. 216-241.

10. MINOR PESTS

Caterpillars sometimes do great damage to vines and fruit trees. They destroy mainly the young clusters. They are most to be feared in rainy springs. As soon as the caterpillers are noticed, the vines should be thoroughly sprayed with lead arsenate (2 ozs. powder to 4 gals, water) and a little sugar. They can also be caught by hand. Do not top the vines when they have caterpillars on them, as they will then mostly remain in the growing tips.

"Krompokkel" is a local name given to the young stages of long-horned grasshoppers (American Katydids) occurring in the Cape vineyards. As a rule they are not very troublesome, but can sometimes eat the leaves sufficiently to damage them appreciably. Spraying with lead arsenate can be applied.

The *large green Caterpillar* of *Theretra capensis*, Linn., eats the leaves so that only the leaf-stalks remain. It sometimes does a lot of damage in nurseries. Its excrements betray its presence. In the vineyards it usually eats the leaves after the vintage, thus hindering the proper maturing of the wood. It should be caught by hand and killed. The difficulty is that it is not readily noticed on the vines, owing to its green colour.

Red Spider, Tetranychus tellarius (?), can sometimes cause the leaves to take a red colour. Dry air favours its development. At the Cape the young leaves near the ends of the growing shoots of Clairette blanche and White French late in the season sometimes curve in to the lower surface and remain rather small, whilst the shoots form short internodes. Such leaves are covered with thin threads below. This is probably due to red spider.

Court noué.

Müller-Thurgau describes under this name a disease caused by a mite, *Phyllocoptes vitis*, Nal., which shows a close agreement with the *Eriophyes vitis* causing erinosis. The symptoms in this case, however, differ from those of erinosis. Here the young shoots are hindered in their development. They remain small, the internodes are short, and the leaves remain small and sickly. The blossom clusters remain small and soon perish. Out of adventitious buds shoots arise that grow normally and can serve as bearers in the following season. In order to control the disease the dwarfed shoots should be broken off and burnt, and the vines should be sulphured early as with erinosis. In winter the vines can be treated with hot water or 4 per cent lysol. I have not yet noticed this disease at the Cape.

Magarodes capensis, Giard, is a native of the Cape and occurs on vine roots, but has thus far not done any appreciable harm.

CHAPTER IX

THE NUTRITION OF THE VINE

LIKE all other green plants with a root-system the vine obtains its nourishment out of the air and from the soil. The nourishment out of the air is brought about by the carbon-assimilation or photosynthesis, and that from the soil by absorption through the roots of the soil solution which contains the mineral food ingredients required by the vine in a very dilute aqueous solution. The food formed from the air is also known as the organic food, and that obtained from the soil as the mineral food. By means of chemical analyses of plants, including their ash, and by nutrition experiments it was found that the following elements, in suitable combinations, are **indispensable** for the plant's life: *carbon, hydrogen, nitrogen, oxygen, phosphorus, sulphur, potassium, calcium, magnesium, iron.* As **dispensable but useful** must be considered: *chlorine, silicon, sodium,* when present in suitable combinations (see Schneidewind,[1] *l.c.* pp. 97-118). All this naturally applies also to the vine. We shall first discuss the nutrition out of the air and then that from the soil.

A. NUTRITION OUT OF THE AIR

Although it has been proved by experiments that a plant can, to a very limited extent, take up its food in aqueous solution through the leaves, this practically never takes place in Nature. The nutrition out of the air is therefore

---

[1] Schneidewind, Prof. Dr. W., *Die Ernährung der landwirtschaftlichen Kultur-pflanzen,* 5e Aufl., 1922. Paul Parey, Berlin.

practically limited to the taking in of carbon dioxide [apart from the oxygen taken in for respiration] which the plant uses for the building up of organic food in the process known as

Carbon-Assimilation or Photosynthesis.

In it the carbon dioxide in aqueous solution (the water is absorbed by the roots) is reduced and oxygen gas is liberated at the same time. There are different theories regarding the chemical side of this process, but that of Baeyer[1] seems to be preferable to the others. According to him the immediate product of reduction of the carbonic acid is formaldehyde, of which six molecules are condensed to form sugar, dextrose. This is expressed by the following chemical equations:

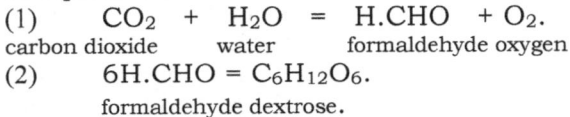

(1) CO_2 + H_2O = H.CHO + O_2.
carbon dioxide   water   formaldehyde oxygen

(2)   6H.CHO = $C_6H_{12}O_6$.
       formaldehyde dextrose.

The two processes are often expressed together by the following equation: $6\ CO_2 + 6\ H_2O = C_6H_{12}O_6 + 6\ O_2$.

This process takes place only in the green *chlorophyll* grains and under the action of *sunlight*. The presence of carbon dioxide in aqueous solution in the plant's cells is of course an absolute necessity.

Although this process takes place on a small scale also in the other green parts of the plant, the leaf is its principal seat. Indeed, the whole structure of the leaf is adapted to the necessary exchange of gases and exposure to sunlight.

The *sugar* (dextrose) which is formed in this process is soluble in water, and is partly again transformed into other substances.   The first of these is *starch*,[2] which can be observed under the microscope as small grains that are coloured blue-black by an iodine solution.   By means of certain substances (enzymes) the starch is again changed into sugar and transported in aqueous solution to the

---

[1] Von Baeyer formulated his theory in 1870. See *Handwörterbuch der Naturwissen-schaften,* vol. vii. p. 794.

[2] See Hugo de Vries in Oudemans en de Vries, *Leerboek der Plantenkunde,* 4ᵉ druk. 1906, i. pp. 209-210.

growing parts of the plant, where it serves both as food and as building material. This sugar can also be changed to *cellulose.* Both cellulose and starch are insoluble in the plant sap as such. The former serves as building material for the cell-walls, whilst the latter is stored as food which can again be changed into sugar (dextrose) when required as food for immediate consumption, *e.g.* when the plant (vine) awakens to new life after its winter rest.

The following equations express the changes mentioned above:

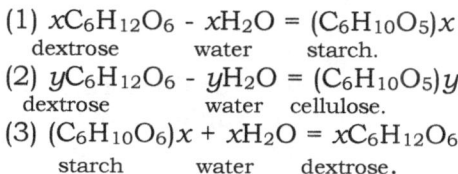

(1) $x C_6H_{12}O_6 - x H_2O = (C_6H_{10}O_5)x$
dextrose   water   starch.

(2) $y C_6H_{12}O_6 - y H_2O = (C_6H_{10}O_5)y$
dextrose   water   cellulose.

(3) $(C_6H_{10}O_6)x + x H_2O = x C_6H_{12}O_6$
starch   water   dextrose.

Sometimes the dextrose loses less water than during the formation of starch and cellulose, when cane sugar can be formed, which we use daily.

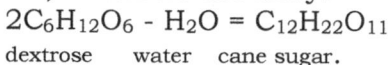

$2 C_6H_{12}O_6 - H_2O = C_{12}H_{22}O_{11}$
dextrose   water   cane sugar.

This can again be split up into dextrose and levulose by acids and enzymes present in the cell sap.

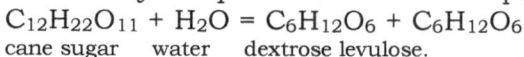

$C_{12}H_{22}O_{11} + H_2O = C_6H_{12}O_6 + C_6H_{12}O_6$
cane sugar   water   dextrose levulose.

When cane sugar is thus split up, the mixture of dextrose and levulose in equal quantities is called invert sugar. In the juice of ripe grapes the sugar consists of nearly equal quantities of dextrose and levulose. These sugars and starch and cellulose are called *carbohydrates.* In the plant we find, however, also other products of an organic nature which, in the first instance, owe their origin to the carbon-assimilation. Thus the *acids* like oxalic acid, tartaric acid, malic acid, are formed from the sugar by *oxidation.* They can therefore be considered as products of metabolism formed during the respiration of the plant, although they are merely intermediate products and not the final products, carbon dioxide and water.

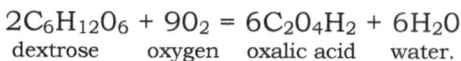

$2 C_6H_{12}O_6 + 9 O_2 = 6 C_2O_4H_2 + 6 H_2O$
dextrose   oxygen   oxalic acid   water.

$$4C_6H_{12}O_6 + 9O_2 = 6C_4O_6H_6 + 6H_2O$$
dextrose   oxygen   tartaric acid   water.

$$4C_6H_{12}O_6 + 6O_2 = 6C_4O_5H_6 + 6H_2O$$
dextrose   oxygen   malic acid   water.

$$C_6H_{12}O_6 + 6O_2 = 6CO_2 + 6H_2O$$
dextrose oxygen carbon   water.
                    dioxide

We also find *oils* and *fats* as reserve or stored food in plants, *e.g.* grape-seed oil in the seeds of ripe grapes. It consists of glycerides of palmitic, stearic, and other acids, and is formed out of sugar by *reduction*. Here, as with the carbon-assimilation, energy is required to bring about the reduction. The energy thus required for the formation of oils and fats is not given directly by the sunlight, but is supplied through the combustion or oxidation of organic material that has previously been formed.

$$17C_6H_{12}O_6 + 86H_2 = 2C_{51}H_{98}O_6 + 90H_2O$$
dextrose   hydrogen tripalmitin   water,
                             (fat)

or

$$17C_6H_{12}O_6 = 2C_{51}H_{98}O_6 + 4H_2O + 86O,$$

in which latter case the oxygen would be consumed by the previously mentioned process of combustion.

The *tannin* and the *pigments* are formed out of the carbohydrate by oxidation.

$$7C_6H_{12}O_6 + 6O_2 = 3C_{14}H_{10}O_9 + 27H_2O$$
dextrose   oxygen   tannin   water.

The *proteins* are formed by reduction out of dextrose and nitrogenous substances of which the nitrogen has been absorbed in a chemical compound dissolved in the soil solution.

$$2C_6H_{12}O_6 + 3NH_3 + H_2 = C_{12}H_{19}N_3O_4 + 8H_2O$$
dextrose   ammonia hydrogen protein   water.

Protein also contains some sulphur which is absorbed in the soil solution. Schneidewind's empirical formula for protein given above is thus not quite correct. See Schneidewind, *l.c.* p. 45. The proteins, like starch, can serve as stored food.

All these organic foods and building materials are therefore intimately connected with the process of carbon-assimilation, and hence it is of the utmost importance to the plant's life that this process should be furthered as much as possible. The sunlight consequently is the plant's great source of energy, and man and beast make use of this stored-up energy when using the plant or its products as food or as a source of mechanical power, heat, and light, in the burning of wood, coal, vegetable fats, and oils.

Requirements for Carbon-Assimilation.

1. *Presence of Carbon Dioxide.*—From the preceding it is obvious that carbon dioxide is absolutely necessary for this process to proceed. Air usually contains about 0.03 vol. per cent carbon dioxide. Although the green plants bind daily a very large quantity of this gas through the process of carbon-assimilation, very large quantities of it are continuously discharged into the atmosphere by the breathing of men, animals, and plants (especially at night when no carbon-assimilation takes place), through the various processes of combustion, volcanic activity, and the activity of microbes in the soil. In this way the carbon dioxide content of the air is kept fairly constant. Should the carbon dioxide content of the air diminish appreciably, plant growth would be considerably weakened thereby.

According to Schneidewind, *l.c.* p. 20, the optimum carbon dioxide content of the air for the building up of organic material, hence for carbon-assimilation, is about 1 vol. percent, and it is certain that an increase in the carbon dioxide content of the air can appreciably increase plant growth. By bringing a great deal of farmyard manure and green manuring into the soil, and by providing for good ventilation and drain-age at the same time, very large quantities of carbon dioxide will be formed in the soil, sufficient to increase the carbon dioxide content of the air above such soil and thus to increase the plant growth.

2. *Heat.*—According to Schneidewind, *l.c.* p. 24, carbon-assimilation begins near 0° C, reaches its maximum at about 37° C, and ceases near 45° C. According to him, the favourable influence of heat up to a certain point is associated with its influence on the life of the plant as a

whole. Kniep[1] quotes the work of Miss Matthaei (1904) on this point, and shows that she found the carbon-assimilation to be twice as strong at 20° C. as at 10° C. He attributes this principally to the greater speed of the chemical reaction at the higher temperature. Too much heat can cause the leaves to wilt, when the stomata are almost closed and carbon-assimilation is at a minimum.

3. *Light.*—This is one of the most important requirements for carbon-assimilation. Special investigation on this point has shown that it is particularly the yellow to red part of the spectrum of sunlight (hence the rays with a wave-length λ = about $590\text{-}700\mu\mu$,) which is most active in the process of carbon-assimilation; blue also helps, but green helps very little. According to Kniep, *l.c.* vii. p. 808, Engelmann put forward the theory that "the assimilatory power is a function of the absorption of light by the chlorophyll". According to this view we should expect green to be very inactive. Engelmann's curves for the carbon-assimilation and absorption of light by green cells agree very well, except that the former falls from blue to violet whereas the latter still rises. This we can explain as due to the fact that the absolute amount of energy here sinks rapidly (see Kniep, *l.c.* p. 809).

Sunlight is as indispensable for the healthy growth and development of the vine as for that of other green plants. We shall have to reckon with this in the training and general treatment of the vine in summer. It is the part which sunlight plays in the process of carbon-assimilation which makes it of such great importance to the life of green plants and hence also of the vine.

4. *The Necessary Plant Food in the Soil.*—Unless the plant has this and enough water, it cannot develop normally, and can therefore not maintain a normal carbon-assimilation. If the plant absorbs too little iron in the soil solution, it gets chlorosis. The leaves then have a whitish-yellowish colour with hardly any green, and can hardly maintain any appreciable carbon-assimilation. The result is that the plant languishes and can even die. We have already seen that drought hinders the carbon-assimilation. This is due to the diminished intake of carbon dioxide, as the

---

[1] In *Handw. der Naturw.* Vol. vii. p. 812.

stomata are more or less closed, and to the lack of water in the leaf necessary for the carbon-assimilation.

5. *The Chlorophyll Grains.*—They contain the pigments which absorb the sunlight that is necessary for carbon-assimilation. These chloroplasts contain a mixture of different colouring substances. The real chlorophyll is an organic magnesium compound. Willstätter and his school did pioneer work concerning the pigments that occur in the chlorophyll grains. We know that the presence of dissolved iron is necessary for the formation of chlorophyll, but how this precisely happens is not yet known. In any case iron is not an ingredient of chlorophyll.

Light is also necessary for the formation of chlorophyll. Plants that are kept in the dark become yellowish (etiolated), and it is only when they are again exposed to light that they reassume their green colour. Here we assume that the plant has all the necessary food-stuffs at its disposal in the soil solution, including dissolved iron.

B. NUTRITION FROM THE SOIL

We have previously seen that the root-hairs in the absorption zone of the young roots absorb the soil solution. This is a very dilute solution which contains the plant's mineral food. By the continuous evaporation of water, mainly through the leaves, there is a continuous stream of absorbed soil solution mainly to the leaves, where it takes part in the carbon-assimilation. We have already seen that iron and magnesium are necessary for this process, but water is also required for the building up of organic material. It is evident that a good deal of water is thus bound chemically. From the leaves the elaborated and built-up food goes to the centres of consumption in the plant.

The reader will by this time have realised that *water* is one of the most important foods of the vine, as of other plants. Indeed, all forms of life in the first instance depend upon a sufficient supply of water. Every farmer knows by experience how much the success of his crops primarily depends upon a favourable rainy season. The vine is no exception to this rule. Most soils contain all the food ingredients in larger quantities than are required by our cultivated plants. It will, however, largely depend upon the soil's *water content* and *ventilation* (state of drainage and cultivation) whether the mineral food that is present will be dissolved sufficiently in the soil solution. The factors mentioned influence and determine the nature and development of the soil's microflora (bacteria, etc.), which, together with the carbon dioxide and oxygen (from the air), render the plant-food in the soil soluble and thus available for the plant.

Once we have determined by chemical analyses which ingredients and how much of these the vine takes up from the soil, and what is present in the soil, it will depend upon the factors mentioned what quantity and kind of fertilisers will have to be applied to a vineyard in order to keep it healthy and in a good state of production. Whilst it is wrong to adopt a system of "Raubbau" (*i.e.* taking plant food out of the soil without replacing it by the addition of fertilisers) in agriculture, it is certainly uneconomical not to use a part of the food that is present in the soil. One can therefore adopt "Raubbau" to a certain extent so long as one does not go too far with it. Sometimes we shall, for certain purposes, add much more of certain plant foods to the soil than the vine removes from it. The first question of great importance to us here is:

What does the Vine take out of the Soil?

It should now be clear that the vine takes all its food out of the soil with the exception of carbon and oxygen which are obtained from the air. The oxygen is partly taken in through the roots, but then it also came originally from the air. Of the remaining food ingredients there is usually a sufficiently large quantity in an available form in our-cultivated soils for an almost incalculable time, with the exception of potassium, phosphorus, and nitrogen, of which we have usually to give certain quantities to the soil in manuring, and lime which we ought sometimes to give. Occasionally the soil solution might contain too little iron, but this happens only where we have given the soil too much lime, and thus made its reaction neutral or even faintly alkaline. Hence the problem of manuring most soils reduces itself to the application of certain quantities of potassium, phosphorus, and nitrogen in suitable combinations (they are also called the *three critical plant foods),* a certain quantity of lime from time to time, and a certain quantity of organic material. In answering the above question we shall therefore mainly take note of the quantities of potash (expressed as K_2O), phosphoric acid (expressed as P_2O_5), and nitrogen (N) which are annually removed from the soil. In analyses of soils and fertilisers these three food ingredients are usually expressed in the forms just given.

Dr. Paul Wagner[1] in 1907 published the results of the investigations on the manuring of vines which he had conducted for many years. As a result of these he has calculated that a vineyard with an annual maximum production of 15,000 kg. grapes, 3200 kg. wood (the weights for wood and leaves here refer to the dry substance, "Trockensubstanz"), and 2900 kg. leaves per hectare (i.e. 13,380 lbs. grapes, 2854 lbs. wood, and 2587 lbs. leaves per acre), annually removes out of the soil in round numbers 100 kg. potash, 80 kg. nitrogen, 30 kg. phosphoric acid per hectare, i.e. 89 lbs. potash, 71 lbs. nitrogen, 27 lbs. phosphoric acid per acre. These figures are based on the assumption that about one-third of the leaves are returned to the soil and are therefore not lost. Table 1 gives the average total quantities of the foods mentioned that are annually removed from the soil by the above-mentioned crop, according to Wagner, l.c. pp. 115 and 117:

TABLE 1

	Per Hectare.			Calculated per Acre.		
	$K_2O.$	N.	$P_2O_5.$	$K_2O.$	N.	$P_2O_5.$
	kg.	kg.	kg.	lbs.	lbs.	lbs.
For the production of grapes	42	25	10	37·5	22·3	8·9
,, ,, ,, wood	27	26	9	24·1	23·2	8·0
,, ,, ,, leaves	40	49	13	35·7	43·7	11·6
Total	109	100	32	97·3	89·2	28·5

It should be noted that these results were obtained in *pot experiments* and were calculated at 10,000 vines per

---

[1] Wagner, Dr. Paul, "Forschungen auf dem Gebiete der Weinbergdüngung," *Arbeiten der Deutschen Landwirtschajts-Qesellschajt,* Heft 124, February 1907

hectare. Every pot had one vine, and the crops could there-fore be weighed and analysed very accurately. Wagner has found very slight differences in the food requirements of the different grape varieties.

The average crop in his pot experiments, calculated for one hectare, was 27,100 kg. grapes, 3200 kg. wood, and 2900 kg. leaves containing on an average (in round numbers) 150 kg. potash, 120 kg. nitrogen, and 45 kg. phosphoric acid. This is a very high crop of grapes, and is equal to 25 leaguers wine per morgen or 24,173 lbs. grapes per acre.

Müntz[1] has for a number of years conducted very complete investigations into the composition of the soils and the annual crops of grapes, wine, canes, and leaves in all the principal wine districts of France, and in 1895 published the results of his investigations in *Les Vignes* just quoted. For the well-known French wine districts Midi (South of France), Médoc, Champagne, and Burgundy, Müntz, *l.c.* p. 451, gives the following mean results for the crops and the *mineral food annually removed from the soil* by the vines (I add the calculated "gallons and lbs. per acre"):

TABLE 2

District.	Yield of Wine.				K₂0		P₂0₆	
	Hl. per Ha.	Gls. per Acre.	Kg. per Ha.	Lbs. per Acre.	Kg. per Ha.	Lbs. per Acre.	Kg. per Ha.	Lbs. per Acre.
Midi	103	917	48	42-8	43	38-4	12	10 7
Medoc	28	249	43	38-4	60	53 5	14	12-5
Champagne	25	223	42	37-5	45	40 1	13	11-6
Burgundy	25	223	24	21-4	25	22-3	7	6-2

He also calculated how much *mineral food* the vine annually requires *to produce one hectolitre wine,* and on p. 456 gives the following results (see Table 3).

The lbs. N, P₂0₅, K₂0 required to produce 100 gallons wine, I have calculated from Muntz's figures for kg. per hl.

On pp. 462-463 Müntz shows how the annually absorbed plant food is apportioned between the leaves, canes, and wine. To his figures I added, in Table 4, a column for the food absorbed by the grapes; this I obtained by deducting the amounts absorbed by the leaves and canes from the total amount absorbed by the vine. The quantities

---

[1] Müntz, A., *Les Vignes*, Paris, 1895. Berget, Levrault et Cie.

are given as *kg. per hectare* vines, and are Müntz's average figures.

TABLE 3

	To produce 1 hl. Wine.			To produce 100 gls. Wine.		
	N.	P₂O₅.	K₂O.	N.	P₂O₅.	K₂O.
	kg.	kg.	kg.	lbs.	lbs.	lbs.
Midi	0·480	0·118	0·423	4·80	1·18	4·23
Médoc	1·485	0·496	2·065	14·85	4·96	20·65
Champagne	1·690	0·410	1·810	16·90	4·10	18·10
Burgundy	1·020	0·295	1·025	10·20	2·95	10·25

TABLE 4

	Vine.			Leaves.			Canes.			Wine.			Grapes.		
	N.	P₂O₅.	K₂O.	N.	P₂O₅.	K₂O.	N.	P₂O₅.	K₂O.	N.	P₂O₅.	K₂O.	N.	P₂O₅.	K₂O.
Midi (average of 8 vineyards)	53	13	44	30	4·5	14	6	2·4	10	3	2·0	12·0	17	6·1	20
				56·6 %	34·6 %	31·8 %	11·3 %	18·5 %	22·7 %	5·7 %	15·4 %	27·3 %	32·1 %	46·9 %	45·5 %
Médoc (average of 5 vineyards)	44	15	59	28	8·0	25	9	4·0	16	1	1·0	5·0	7	3	18
				63·6 %	53·3 %	42·4 %	20·5 %	26·7 %	27·1 %	2·3 %	6·7 %	8·5 %	15·9 %	20·0 %	30·5 %
Champagne (average of 7 vineyards)	47	11	51	30	5·4	26	7	2·0	12	1	1·0	3·0	10	3·8	13
				63·8 %	49·1 %	51·0 %	14·9 %	18·2 %	23·5 %	2·1 %	9·1 %	5·9 %	21·3 %	32·7 %	25·5 %
Average for all 20 vineyards	48	13	51·3	29·3	6·0	21·7	7·3	2·8	12·7	1·7	1·3	6·7	11·3	4·2	17
				61·3 %	45·7 %	41·7 %	15·6 %	21·1 %	24·4 %	3·4 %	10·4 %	13·9 %	23·1 %	33·2 %	33·8 %

The average figures for all twenty vineyards were calculated by myself from the figures given in the rest of Table 4, as were also the percentages of the plant food absorbed by the leaves, canes, wine, and grapes. The percentage absorbed by the grapes was obtained by subtracting those for the leaves and canes from 100. The plant food absorbed by the skins, seeds, and stalks we get approximately by subtracting the figures given in Table 4 for "wine" from those given for "grapes".

Müntz, *l.c.* pp. 465-466, also determined the total leaf-surface (reckoning both surfaces) of the vines per hectare and per hectolitre wine produced. The following were his average results:

TABLE 5

	Weight of Fresh Leaves per Ha.	Surface area of the Leaves.	
		Per Hectare	Per Hl. Wine made,
Midi (average)	5046 kg.	34,875 sq. metres	320 sq. metres
Champagne (average)	5570 kg.	29,432 sq. metres	1234 sq. metres

From Table 5 we see that the weight of the leaves per ha. is about 10 per cent more in the north (Champagne) than in the south, whereas the opposite holds good for the total leaf surface. We also notice that the total leaf surface required to produce 1 hl. wine in the north is four times as great as in the south—it is a case of quality versus quantity.

Table 4 clearly shows that the *leaves* alone absorb fully 60 per cent of the nitrogen and nearly half of the phosphoric acid and potash (more of the former than of the latter) taken up by the vine. The *canes* take about $1/6$ of the nitrogen, $1/5$ of the phosphoric acid, and ¼ of the potash, hence not very much. The *grapes* take about ¼ of the nitrogen and about $1/3$ of the phosphoric acid and potash, thus a fair amount. The *wine* removes on an average only 3.4 per cent of the nitrogen, 10.4 per cent of the phosphoric acid, and 13.9 per cent of the potash, hence only a small portion of the mineral food taken up by the vine from the soil. Therefore if we keep the canes and leaves in the vineyard, and return the husks, seeds, and stalks to it, the vines will need very little manuring. This usually does not happen, however, and therefore vines should be well manured to keep them in a good state of production.

From Table 3 it is clear that, for the production of one hl. wine, two to four times as much mineral food is required in the districts producing wine of high quality as is required in the south where quantity is produced. Table 2 shows that the total mineral food absorbed per acre does not differ much in the different districts except in the case of Burgundy, where it is only a little more than half as much as in the other districts. In Burgundy the vines are grown in most instances on stony, somewhat dry slopes that are not heavily fertilised and consequently do not develop great vigour, so that the production of leaves and canes in proportion to the crop of grapes is less than in the other mentioned.

Műntz summarises his results as follows (quoted from Michaut et Vermorel,[1] p. 93):

"1. In all the vineyards the absorption of nitrogen and potash is higher than that of phosphoric acid.

"2. Nitrogen is absorbed in large quantities by the vine, and, contrary to widely held convictions, must be applied; the dressings of nitrogen are indeed those that give the most apparent results.

"3. In the vineyards of the Midi more nitrogen is absorbed than potash; in those of the more northern parts it is potash which is absorbed most. In the latter parts potash is the dominant (factor) of the vine, whereas nitrogen is such in the Midi.

"4. Notwithstanding the enormous differences in yield, the vineyards in the south do not require an appreciably larger amount of food than those in the more temperate regions.

"5. The amount of food which the vine consumes in producing a hectolitre wine is three to four times as high in the more northern parts as in the Midi."

Unfortunately we do not possess such data as Műntz has worked out for the French vineyards in case of the South African vineyards, nor, so far as I know, for the Californian and Australian vineyards. The Cape soils and climate (especially) in the viticultural area agree to a considerable extent with those of the south of France, and therefore the dressings of fertilisers which I recommend, have been based on Műntz's figures for the south of France, These dressings have in the past given good results at the Cape. In California and Australia they may have to be altered to suit the local conditions.

Manuring of Vines.

Now that we have seen what mineral food the vine annually removes from the soil, we can begin to discuss the question of the manuring of vines. Under *Manuring* I understand "the addition of various substances to the soil

---

[1] Michaut, C, et Vermorel, V., *Lea Engraia de la vigne,* 3rd ed., 1905. Coulet et Fils, Montpellier and Masson et Cie, Paris.

that can maintain its fertility or even increase it". In this connection it is naturally assumed that the soil is well cultivated and does not contain either too little or too much moisture.

Usually more mineral food is added annually to the soil in manuring vineyards than is removed from the soil during one vegetative period. Thus Kroemer in Babo u. Mach, *l.c.* i. p. 401, states that in Germany the average annual dressing of fertilisers for vineyards per hectare amounts to 160 kg. potash, 130 kg. nitrogen, and 100 kg. phosphoric acid, *i.e.* respectively 143, 116, and 89 lbs. per acre.

Müntz, *l.c.,* pp. 474-485, quotes the following examples for France:

		Nitrogen.	Phos. Acid.	Potash.
Midi : Guilhermain	(a)[1]	74 kg.	47 kg.	56 kg.
	(b)[1]	74 ,,	17 ,,	56 ,,
Verchant	(a)	70 ,,	31 ,,	70 ,,
	(b)	38 ,,	10 ,,	31 ,,
Médoc : Château Latour	(a)	112 ,,	51 ,,	165 ,,
	(b)	39 ,,	12 ,,	48 ,,
Sauternes : Château Yquem . .	(a)	82 ,,	120 ,,	160 ,,
	(b)	24 ,,	7 ,,	33 ,,
Burgundy : Gevrey-Chambertin . .	(a)	55 ,,	165 ,,	55 ,,
	(b)	30 ,,	9 ,,	34 ,,
Beaune	(a)	46 ,,	23 ,,	54 ,,
	(b)	21 ,,	7 ,,	21 ,,
Champagne : Ay	(a)	59 ,,	47 ,,	147 ,,
	(b)	49 ,,	10 ,,	49 ,,
Verzenay . . .	(a)	110 ,,	50 ,,	120 ,,
	(b)	40 ,,	12 ,,	48 ,,

These figures of Müntz are very instructive. We notice from them that in the south of France, where quantity is mainly produced, the mineral dressings are not as high in the Médoc (Château Latour) and in Sauternes (Château Yquem) where wines of the highest quality are produced. This just shows how wrong it is to maintain that heavy dressings of fertilisers will necessarily lower the quality of the wine produced.

[1] (*a*) means annually "given in manuring", and (*b*) means annually "absorbed by the vine".

In considering the three ingredients, we find that about 1 to 3½ times as much nitrogen is given as is absorbed, about 1 to 5 times as much potash, and about 3 to 18 times as much phosphoric acid. Everywhere more is usually given than is absorbed. The excess of potash is usually as great as or slightly greater than that of nitrogen, but that of phosphoric acid is relatively much greater than that of nitrogen and potash. In Gevrey-Chambertin, where the finest wine of Burgundy is produced, 17 times as much phosphoric acid is given as is absorbed, whilst on Château-Yquem, which produces the finest Sauternes wine, 18 times as much phosphoric acid is given as is absorbed. This is done because, *inter alia,* phosphoric acid increases the quality of the wine.

At the end of his chapter on the manuring of vines, Műntz, *l.c.* p. 486, comes to the following "*Conclusions.*—As a practical conclusion from these numerous observations we can state that the economic conditions, more than the requirements of the vine, should decide the manuring to be given.

"In the vineyards producing wines that are high in price, it is the grower's interest to fertilise heavily, as even slight increases in the yield pay for the extra expenses thus incurred. But vineyards that produce much but cheap wine should get just the quantities of fertilisers that are required, as the increased yields might not pay for the increased expenses so incurred. ..."

From what has preceded it should be clear that there can be no such thing as a general manurial formula to be applied in all cases, unless it be regarded merely as an indication, which should be departed from as conditions make it necessary and desirable. Experience has to teach us what manuring is best and most profitable in every instance.

The Specific Action of Nitrogen, Phosphoric Acid, and Potash.

Before proceeding to putting forward definite formulae for the manuring of vines, I wish to draw the reader's attention to the specific action which nitrogen, phosphoric acid, and potash have on the development of the vine as can be observed where one of them is almost wanting or is present in a large excess.

Nitrogen promotes particularly the vine's vigour. A great development of shoots and leaves with a dark green colour must be attributed especially to nitrogen. Weakened vines should therefore get a preponderating nitrogen dressing, whilst very vigorous vines that are not sufficiently productive *(inter alia,* also setting their berries badly) require little or no nitrogen, but a heavy phosphate and potash dressing to render them productive.

An excessive nitrogenous nutrition results in grapes that are rich in nitrogen, fairly soft and susceptible to disease, with low-keeping qualities. Wine made from such grapes is rich in nitrogen, takes a long time to become bright, and is specially subject to bacterial diseases. In the production of table grapes, heavy nitrogen dressings are excellent for producing large berries and bunches, provided this treatment is not abused at the cost of the quality of the grapes in other respects, *e.g.* colour, taste, firmness, keeping qualities. Vines that grow too vigorously owing to excessive and one-sided dressings of nitrogen, are specially susceptible to all kinds of diseases and pests. Such vines frequently cannot ripen their grapes properly owing to a shortage of water in the soil, which is caused by excessive evaporation of soil moisture through the abnormally developed foliage of such vines. The maturation of the grapes and wood of such vines is retarded and made worse.

In this connection I wish to point out that we indirectly increase the nitrogen dressing very considerably if we introduce organic matter into the soil, *i.e.* by means of farmyard manure or by green manuring, in the presence of sufficient soil moisture, and at the same time reduce the loss of soil moisture through evaporation to a minimum by intensive surface cultivation, which also promotes the introduction of air into the soil. In this way we favour the soil bacteria that can fix atmospheric nitrogen and make it available as food for the vine. In calculating the nitrogen dressing this should be borne in mind. In warm wine countries a great deal of nitrogen can be thus fixed.

Phosphoric Acid.—It helps to neutralise the evil effects of an excessive nitrogen dressing. It promotes the production of grapes and the better setting of the berries; it promotes the better ripening of the grapes and wood, which takes place sooner and more completely, so that the grapes at maturity will often be sweeter than they would otherwise have been; and it increases the phosphoric acid content of the must, which greatly favours the alcoholic fermentation. Hence we can say that it raises the quality of the grapes and wine. This accounts for the very heavy dressings of phosphate in the French districts that produce wine of very high quality.

According to Michaut et Vermorel, *l.c.* p. 114, Dr. Paul Wagner of Darmstadt maintains, with regard to the action of phosphoric acid: "It is the phosphoric acid which forms the good wines, regulates production by preventing coulure, makes the vine more resistant to diseases, regulates the good constitution and the perfect ripening of the grapes".

According to Lyon, Fippin, Buckman,[1] *l.c.* p. 550, and Russell,[2] *l.c.* pp. 38-39, phosphoric acid especially increases root development, this being particularly valuable in clay soils, and hastens the ripening process.

According to Schneidewind, *l.c.* p. 104, a phosphatic dressing does not increase the sugar content of sugar-beet or the starch content of potatoes if both are harvested when fully ripe. If harvested earlier or if complete ripeness is not reached, the crops that had received dressings of phosphate will be richest in sugar and starch, because phosphoric acid hastens the ripening process. For the same reason phosphate dressings can raise the sugar content of the must, especially in the colder wine districts, not directly but indirectly through the earlier and therefore better ripening of the grapes. Too much phosphate has no detrimental effect.

According to Paturel,[3] quoted from Russell, *l.c* p. 42, I the best wines contain most P_2O_5 (about 0.3 gram per litre), the second and lower qualities containing successively less. Further, when the vintages for different years were arranged in order of their P_2O_5 content a list was obtained almost identical with the order assigned by the wine merchants." This confirms Wagner's statement quoted above.

Potash.—Schneidewind, *l.c.* p. 110, says: "From such experiments and from the fact that those plants that produce large quantities of carbohydrates also absorb fairly large

---

[1] Lyon, Fippin, Buckman, *Soils, their Properties and Management,* 1915. Macmillan Co., New York.

[2] Russell, Dr. E. J., *Soil Conditions and Plant Growth,* 3rd ed., 1917. Longmans, Green & Co., London.

[3] Paturel, *Bull. Soc. Nat. Agric,* 1911, p. 977

quantities of potash, we conclude that potash plays an important part in the formation, translocation, and deposition of the carbohydrates. Hence potash stands in the same relation to the carbohydrates as phosphoric acid stands to the proteins."

According to Russell, *l.c.* p. 43, "The most striking effect, however, is the loss of efficiency in making starch, pointed out long ago by Nobbe; either photosynthesis or translocation—it is not yet clear which—is so dependent on potassium salts that the whole process comes abruptly to an end without them". Potash starvation is evident from the poor, dull colour of the leaves and the lack of turgor in the shoots. Such plants are much less resistant to diseases and pests than plants with normal potash nutrition.

As we have previously seen the vine requires much potash. It helps *inter alia* partly to neutralise the organic acid in the juice of the grape, and so to prevent the must being too acid. Hence the must contains more free acid in a very dry summer and autumn. The drought is the cause of too little potash, etc., being taken up from the soil.

In France, Zacharewicz[1] of Vaucluse has for a number of years carefully conducted experiments on the manuring of vines. At the end of his publication quoted below, he says the following about potash (*l.c.* p. 114): "Potassium carbonate gives the highest yields, but, owing to its price, potassium sulphate gives the greatest profits, and at the same time the sweetest musts and therefore the heaviest wines".

Since we know (according to Ravaz) that the good quality of the wine and also the size of the crop *inter alia* depend upon the food which the vine has stored up during the previous summer and autumn, it follows that a liberal phosphate and potash dressing which promote the ripening of the wood and carbon-assimilation by the leaves, should exercise a good influence upon the quality and quantity of the wine. This is, of course, especially to be expected where the soil is fairly well supplied with nitrogen.

---

[1] Zacharewicz, Ed., *Expériences sur les engrais appliqués a la culture de la vigne,* 1900.

Formulae for the Manuring of Vines.

In drawing up such formulae one should think of the composition of the soil, yield of the vines, nature of soil and climate, especially the rainfall and its distribution over the year. The formulae which I offer here should therefore be regarded as average values, which should be departed from where necessary. When the vine is planted it is usually best not to give it any fertiliser. During the second and third years half the quantities here recommended for vines in full bearing might be given.

If vines are manured when they are planted, the young roots can easily be burnt by a concentrated soil solution around them, whereby these vines may be badly damaged and even killed. In this case use some bone-meal or farmyard manure, but not without mixing it very thoroughly with the soil in the holes in which the vines are to be planted. Such manuring might hinder the even distribution of the roots in the soil. For these reasons it will on the whole be best not to manure vines when they are being planted on a large scale.

As regards the absolute quantities of nitrogen, phosphoric acid, and potash to be given *annually per acre* of bearing vines, I would recommend the following as mean values: *30-40 lbs. nitrogen, 40-60 lbs. phosphoric acid,* and *40-50 lbs. potash.* As soon as the vines become too vigorous, less nitrogen should be given. It may sometimes, as in growing table grapes, be advisable to fertilise more heavily (especially with nitrogen).

Suitable Combinations in which to give these Foods

1. *All three* (N, P_2O_5, K_2O) we give best in the form of well-rotted farmyard manure. It always still remains the best fertilising agent. Unfortunately the farmer has not always got enough of it. It contains on an average 0.4–0.6 per cent nitrogen, 0.2–0.3 per cent phosphoric acid, and 0.5–0.7 per cent potash, hence respectively 8–12, 4–6, and 10–14 lbs. per ton of 2000 lbs. In 4½ tons of farmyard manure we would thus be giving about 36–54 lbs. N., 18–27 lbs. P_2O_5, and 45-63 lbs. K_2O. Hence 4½ tons per acre will be sufficient, excepting as regards the phosphoric acid. This

would mean that, if the vines are planted 5×5 ft. square, every eight vines will get a basket of manure weighing 40 lbs.

2. *Nitrogen* alone can be given in the form of: chile saltpetre or nitrate of soda, $NaNO_3$, which contains about 15.5 per cent N. in the trade; or sulphate of ammonia $(NH_4)_2SO_4$, containing 20–21 per cent N. in the trade; or calcium nitrate with about 12 per cent N. in the commercial article; or calcium cyanamide, CN. NCa, with 15–22 per cent N. in the trade. (See Schneidewind, *l.c.* p. 299.)

The nitrate nitrogen is directly absorbed by the vines, the ammoniacal nitrogen is partly absorbed as such, but mainly, after having undergone nitrification, as nitrate nitrogen; the calcium cyanamide must first be decomposed in the soil before the vine can use its nitrogen, which is first changed into ammoniacal nitrogen, which is then absorbed as has just been stated. Together with phosphoric acid we give nitrogen in: bloodmeal (aver, about 11 per cent N. and 1 per cent P_2O_5), bonemeal (raw bonemeal 4 per cent N. and 21 per cent P_2O_5, degelatinised 1 per cent N. and 28 per cent P_2O_5), fish guano (about 8.5 per cent N. and 17.4 per cent P_2O_5). (See Schneidewind, *l.c.* pp. 310-311.) Government guano (Cape), *i.e.* Peru type, is highly nitrogenous (8-11 per cent N.) and fairly phosphatic (10-14 per cent P_2O_5, nearly all citric soluble) with a little potash (1-2 per cent K_2O). It is therefore a very valuable fertiliser.

3. *Phosphoric acid* we give in some form of bonemeal, or calcium superphosphate (16-20 per cent water soluble P_2O_5), or basic slag (11-23 per cent P_2O_5, most of which is soluble in 2 per cent citric acid, and about 40 per cent lime), or finely crushed raw or rock phosphate—preferably tricalcium phosphate with up to 38 per cent P_2O_5 and more (46.8 per cent P_2O_5 theoretically possible).

4. *Potash* we supply in sulphate of potash (48-51 per cent K_2O and muriate of potash (55-60 per cent K_2O). Karroo sheep manure and its ash contain respectively 3 and 8–10 per cent K_2O, but they should not be used on heavy loams or clay soils, as they spoil the physical state of these soils. On the whole the high-grade sulphate of potash is the best.

A System of Manuring.

Where sufficient farmyard manure is not available to fertilise the vines annually, we should try to introduce organic matter into the soil at least every second year. This we can do alternately by means of farmyard manure and by green manuring. In between we can apply dressings of more or less pure mineral fertilisers. This gives a four-year rotation. I wish to recommend the following *per acre* of vines in full bearing:

First year : 4½ tons farmyard manure.
Second year :

1.
{
300-400 lbs. Govt. guano
100-200 lbs. basic slag or superphosphate
80-100 lbs. sulphate of potash
}

or 2.
{
200-275 lbs. nitrate of soda
250-350 lbs. superphosphate
80-100 lbs. sulphate of potash
}

or 3.
{
150-200 lbs. sulphate of ammonia
300-400 lbs. basic slag or superphosphate
80-100 lbs. sulphate of potash
}

or 4.
{
250 lbs. Govt. guano
750 lbs. Karroo sheep manure
50-200 lbs. basic slag
25-50 lbs. sulphate of potash.
}

Third year : Green manuring with 300-400 lbs. basic slag and 80-100 lbs. sulphate of potash.
Fourth year : According to one of the formulae of the second year.

After the fourth year we repeat what was done in the preceding four years and in the same order. In practice we would divide the vineyard into four equally large blocks, A, B, C, D. Suppose we begin by giving them the treatment recommended above for the first, second, third, and fourth years respectively. The next year A would get the dressing which B had got the previous year, B that which C had got, C that which D had got, and D that which A had got, and so on in rotation. In this way all four formulae are applied every year, but by rotation on all four blocks in the order indicated above.

Whoever has enough farmyard manure to fertilise his whole vineyard with it every year, need merely add to it 50 lbs. basic slag or superphosphate and 50 lbs. Government guano (if necessary) per acre per annum. We have previously seen that farmyard manure lacks phosphoric acid for the vine's requirements, compared with its nitrogen and potash contents. If the farmyard manure suffices to manure only half the vineyard every year, give the other half a dressing of mineral fertilisers according to one of the formulae recommended above for the second year. The next year the farmyard manure is applied to the half that had received mineral fertilisers the year before, and the mineral fertilisers go on to the other half. Continue in this way.

N.B.—1. Do not mix basic slag or burnt limestone with Government guano or sulphate of ammonia or farmyard manure, nor nitrate of soda with superphosphate (for some considerable time), as this can cause a loss of nitrogen. They should be spread separately, especially in the cases first mentioned.

2. In these formulae it is assumed that the basic slag contains about 15 per cent citric soluble P_2O_5, the super-phosphate about 17 per cent water soluble P_2O_5, the nitrate of soda about 16 per cent nitrogen, the sulphate of ammonia about 20 per cent nitrogen, the sulphate of potash about 50 per cent K_2O.

Bringing in the Manure or Fertilisers into the Soil

(*a*) *How.*—The manure or fertilisers can be spread or broadcasted evenly over the ground, or can be deposited in furrows or in holes. The climate, soil, and fertilisers should be considered in this connection. In clay soils, and in case of not easily soluble fertilisers, the roots will be obliged to search for their food in the furrows or holes where the fertilisers have been deposited. In light or sandy soils the plant food will be carried by the soil solution also to the farther roots, especially where readily soluble fertilisers have been used.

Where it seems on the whole to be best to spread the fertilisers evenly over the ground, except in case of young vines, the climate, together with practical considerations, sometimes make it desirable to depart from this rule. This leads us to the question: *How deep should the fertilisers be brought in?* In warm parts, especially where the system of dry-land farming is followed, it is not desirable to have the roots growing near the surface (owing to the risk of drought), so that in this case the fertilisers (manure) should be brought in fairly deep. Where copious irrigation can be "applied or where good rains fall fairly often in summer, the fertilisers need not be brought in very deep. In clay soils and heavy loams the fertilisers should be brought in to about the depth of the (feeding) roots. In sandy soils the fertilisers (manure) are brought in shallow, especially substances like nitrate of soda, superphosphate, etc., which dissolve very readily in water and hence easily penetrate into the soil, and can easily e washed out of the soil by heavy rains. In practice the distribution of the fertilisers over the ground is decided by the depth to which it is desirable to bring them into the soil. Where they are brought in shallow, they are spread evenly over the ground and then ploughed or cultivated in to a depth of 4-5 in. Where they should be brought in deep, they can still be spread evenly over the ground if the vineyard is dug by means of spades, although this nowadays rarely happens. We would usually have to make holes between four vines or draw deep furrows between the rows of vines.

Where *holes* are dug between four vines, we should have to make 436 holes per acre if the vines are planted 5 × 5 ft. square. These holes are usually made 18 to 24 in. square and 12 to 15 in. deep. It entails a good deal of work, but has its advantages. Apart from the fertilisers (manure) put into the holes and mixed with soil, we can also deposit the canes removed in pruning into these holes, which also catch water whilst they are open. They are closed towards the end of winter (July to August at the Cape), when the vineyard is ploughed. In making these holes, as few of the thick roots as possible should be cut.

The drawing of *furrows* between the rows is more general. Use the Oliver "Z" plough, which has a double share and mouldboard, and throws the soil to both sides of the furrow. Pass with it twice in the same furrow. Let the furrow now be opened to a depth of about 12 in. by means of spades, spread the fertilisers in it, and close it by means of a vineyard plough. Now plough the whole space between the rows of vines towards the middle.

(*b*) *When.*—For green manuring we must put the seed and necessary fertilisers into the soil early in April (at the Cape), in any case after the first good rain in late autumn. Otherwise we wait till winter. On clayey soils, where the danger of leaching is small, the fertilisers (manure) can be brought in as early as May to June (at the Cape). With broken or lighter soils it is best to wait until July. It is very desirable that a couple of good rains should fall after the fertilisers have been brought in and before the vine begins its new period of growth.

Green Manuring.

Although it is every grape-grower's duty to make as much farmyard manure as possible on his farm, it will usually still be necessary or desirable to apply green manuring in order to have enough organic matter (humus) in the soil for a good bacterial flora, which is intimately connected with the fertility of the soil.

In selecting a green manure crop we give preference to a legume, as certain nodule bacteria can live in symbiosis with the legume on its roots, where they cause peculiar nodules to be formed, and fix a great deal of atmospheric nitrogen which then becomes available as plant food. When such a crop is ploughed in, we do not merely introduce organic matter in the soil, which can become humus, but also a large quantity of nitrogen food.

In our Western Province (Cape) the ordinary field pea is the best legume for a green manure crop in the vineyard. It grows well in our winter, and is far enough advanced when the time for ploughing it in has arrived. Although this legume will develop nodules on its roots in most of our vineyard soils, it will usually pay to inoculate the seed with nodule bacteria (for peas) before sowing it.

How to inoculate the Seed.—Where pure cultures of these bacteria are obtainable at reasonable prices, they can be used. Otherwise I would recommend the following method which Dr. J. S. Marais has used with great success on the Stellenbosch University Farm:

Dissolve 6 ozs. ordinary glue in a gallon of water, moisten the seed peas with this solution, spread them out in a thin layer on a clean cement floor or on a tarpaulin, then sift air-dry soil on them whilst they are being stirred. Take soil on which peas have repeatedly been sown and which has developed numerous nodules—black garden soil will usually contain numbers of these bacteria. Now let the seed dry in the air. Every pea will now have a thin coating of soil around it, and at the same time some of the required bacteria. The glue in the water makes the soil stick to the peas.

Bringing in the Seed.—Draw a furrow about 6 in. deep between every two rows of vines with the Oliver "Z" plough. Now sow the necessary fertilisers and then the inoculated peas in the furrow and on the freshly-turned soil, *not merely in a narrow strip.* Now cultivate in the fertiliser and seed superficially with the Planet Jr. As has been stated, this is done in April or May after the first good rain has fallen.

Ploughing under of the Green Manure Crop.—This is done when the peas are in full flower and hence at the top of their development.   In the Western Province (Cape) this is usually done in August or (more often) September.   In any case it should be done before the winter rains are over, and when there is still sufficient moisture in the soil to enable the green material to change into humus, and to facilitate the ploughing under.   If this operation is delayed until the soil is too dry, ploughing is made difficult and the green material will to a large extent undergo a dry rot in the soil instead of becoming humus, thus not nearly producing the desired effect. Therefore plough under the green crop before it is too late, even though the plants are not yet in flower.   Roll the crop or tramp it down by foot in order to be able to plough it under properly.   Further, make use of a chain and a roll-coulter, and plough towards the middle of the row.   Have the remaining strips of ground in the rows of vines dug over by hand, instructing the workmen at the same time to cover with soil any green material showing above ground, as it would otherwise not humify.   Leave for three to four weeks before again cultivating superficially in the rows.

Quantity of Seed.—This depends upon the size of the peas and upon their germinating power, but usually 40 to 50 lbs. per acre of vines will be sufficient.

Some quantitative data on Green Manuring.

In an article on "Western Province Soils: Humus Deficiencies; Green Manuring; Some Quantitative Data", which my colleague, Prof. Dr. I. de V. Malherbe, published in *The Farmer's Weekly* (Bloemfontein) on the 15th April 1925, he supplied some quantitative data on the green manuring of vineyards and orchards in the Western Province (Cape).

A green manure crop of peas in the vineyard on the Stellenbosch University Farm gave the following yields per acre:

	Fresh Vegetable Matter.	Air-Dry.	Oven-Dry.
Tops	4472 lbs.	754 lbs.	655 lbs.
Roots	298 "	60 "	53 "
Total	4770 "	814 "	708 "

This green manure crop contained the following amounts of plant food in pounds per acre:

	Nitrogen.	Phos. Oxide.	Potash.	Lime.	Magnesia.
Tops	29.57	5.01	19.77	8.85	3.07
Roots	2.32	0.35	1.17	0.69	0.24
Total	31.89	5.36	20.94	9.54	3.31

In the orchards the quantities of fresh vegetable matter and plant food contained in the green manure crop of peas were three to four times as high as those given above for the vineyard. This is due to the relatively small percentage of the area of the vineyard soil on which the peas can be grown between the rows of vines.

In view of the low yields of fresh vegetable matter and nitrogen, Prof. Malherbe considers the advantage gained by sowing peas in vineyards for green manuring as rather problematical.

He also investigated the value of non-leguminous cover crops, of which gousblom (Cryptostemma calendulaceum) may be very valuable in the vineyard. It grows during winter and is ready to be ploughed, in during August, thus suiting the grape grower at the Cape very well. He obtained the following results with a good crop of gousblom in the orchards of the Stellenbosch University Farm, growing on a sandy loam and harvested on 9th August 1923 (see next page):

A good cover crop of gousblom will therefore introduce more than four times as much vegetable matter (considering the oven-dry material) into the vineyard soil as a crop of peas, and it will very efficiently prevent loss of plant food from the soil during winter, when the vines are dormant and most of

	Fresh Vegetable Matter.	Air-Dry.	Oven-Dry.
Tops	Per acre 39,640 lbs.	3348 lbs.	2911 lbs.
Roots	850 "	135 "	119 "
Total	40,490 "	3483 "	3030 "

Amount of Plant Food per acre yield (in pounds)

	Nitrogen.	Phos. Oxide.	Potash.	Lime.
Tops	80.55	19.73	157.92	46.84
Roots	1.37	0.62	1.91	1.51
Total	81.92	20.35	159.83	48.35

the rains fall. It should therefore be encouraged to grow, and should be sown where it is not yet established. When ploughing under the crop, leave some plants to produce seed for the next year's crop. Where gousblom will not grow but peas will, the latter should be grown as a green manure crop. In some cases peas may be too late, when some other suitable green manure crop *(e.g.* rye or barley) should be grown.

In view of the very large amount of potash present in a good gousblom crop, it will be obvious that a liberal dressing of potash early in winter will largely favour the growth of this cover crop. This explains why the early application of Karroo sheep manure favours the growth of gousblom.

Lime.

It has already been pointed out that most soils contain enough lime to provide the vine's food requirements in this respect. As most vineyards in the Western Province (Cape) stand on soils poor in lime, we should apply lime from time to time. As a green manure crop brings more acid into the soil, I recommend a dressing of lime when the fertilisers for the green manure crop are given. Use ground limestone for this purpose. In this form the lime is present as calcium carbonate, $CaCO_3$, which is not washed out of the soil as readily as quick or slaked lime. The main point, however, is that it will not destroy humus in the soil, which quick or slaked lime does. It will therefore help to keep the soil reaction right without unnecessarily diminishing its humus content.

The practice of some farmers of throwing some slaked lime on the old vines towards the end of winter will not appreciably affect the soil reaction, and thus do no harm, whilst it will generally be advantageous to the vine's health, though this advantage will be but slight. I consider it unnecessary. Rather give the vineyard a lime dressing, and control the diseases and pests as was recommended in the chapter dealing with these.

According to the scheme recommended above, the vines would get a lime dressing every fourth year. I recommend ton or 1500 lbs. ground limestone per acre of vines. Take a limestone which is rich in calcium carbonate. Should it at the same time contain some phosphoric acid, this would be an advantage that is not to be despised.

CHAPTER X

ESTABLISHMENT AND SOIL CULTIVATION OF A VINEYARD

1. SELECTION OF SITE

THE vineyard should naturally be planted where all the conditions are most favourable. These have previously been discussed, and need not be repeated here. We have seen how important a suitable climate is, and that parts where hail falls in summer, where frost is to be dreaded in spring, and strong winds prevail in spring and summer, and where most rains fall in summer, should be avoided as much as possible.

The site for the vineyard should be situated as favourably as possible with regard to the dangers just mentioned. The soil should be fairly deep, and should otherwise fulfil the conditions mentioned in Chapter I. The site should make the subsequent cultivation of the vineyard easy and should be easily accessible. Very steep slopes should be avoided, unless they are to be planted with varieties producing very valuable grapes, as they would have to be *terraced,* and as this is a costly undertaking. Along the Rhine, just below Rüdesheim, we see such terraces (see Fig. 81). These were made possible by the high quality and price of the wine produced by these vines.

The locality should be favourably situated for obtaining the necessary labour and fertilisers, and for selling the grapes or their products. We should preferably select soil which is naturally well drained and free from brack, and which will not cost too much to get ready for planting with vines.

2. PREPARATION OF THE SOIL

Here we must distinguish between land that is still virgin soil and land that has already been under cultivation. Virgin soil has first to be cleared of bush and large stones that may be present. Agricultural dynamite can render valuable assistance in removing large stumps and rocks. Hollows and sluits should be filled up and the land levelled as much as possible with a view to subsequent irrigation and washing.

FIG. 81.—Vine terraces along the right (northern) bank of the Rhine below Rüdesheim. Original.

The levelling must be carefully done where the vineyard is to be irrigated. On a slope terraces will have to be made.

Terracing is not merely required where a slope is to be irrigated. A long, steep slope should be terraced for better cultivation and removal of rain-water. Below Rüdesheim the terraces are very narrow (see Fig. 81), and each one has only a few rows of vines. Sometimes it is sufficient if a stone wall is built here and there, as can be seen on Fig. 82. his vineyard produces the famous Swiss wine called Delazey, which is one of the best dry white wines of Switzerland, and closely resembles the Cape Drakenstein. The grape grown here is the Fendant vert and roux *(i.e.* Chasselas doré). The fall is very great, being about 1 in 2½-2 or 23-30°.[1]   As this vineyard lies just above the Lake of Geneva and faces the south, it gets much sun and heat—also that reflected from the lake: this explains the good quality of its wine.

FIG. 82.—The sloping vineyards of Dezaley along the northern shore of the Lake of Geneva and east of Lausanne. Notice the two stone walls to break the steep fall. To the right a boat is visible on the lake. The fall here is about 1 in 2-2½ and 23-30°. Original (1908).

Where extensive drainage is required, it should be done before the vines are planted. A good stone or French drain is excellent where lots of water have to be drained off, provided it is laid deep enough and the stones are well packed to pro-vide good drainage and so as not to be blocked easily. For details about drainage I refer the reader to works dealing with that subject.

When the ground has been cleared, levelled, terraced, and drained if necessary, the soil can be further prepared for planting.   It is usually best to put virgin soil under cereal for one or two years, giving it liberal manurial dressings. Weeds can at the same time be suppressed. This constitutes a material advantage if the soil is merely to be ploughed deep

---

[1] *2012 Editors Note:* FIG 82 says 2-2½ which seems correct.

before receiving the vines. If the soil is to be trenched, the weeds could be suppressed in this operation. This brings us to the important question of

Trenching or Ploughing?—It is obvious that the soil can be loosened and ventilated better and deeper by hand trenching than by ploughing, but trenching costs much more than ploughing. It is therefore of the greatest importance to consider carefully under what circumstances it will pay to trench the soil. It will usually not pay to trench heavy, deep loams and clay soils, as they become compact very soon after heavy rains have fallen. A comparative test which I carried out some ten years ago at Elsenburg (Cape) on this very point, has shown that the vines grew somewhat better on the trenched than on the ploughed land during the first couple of years, but after five years no difference was noticeable. In this instance, therefore, ploughing was preferable.

On deep sandy soils a combination of ploughing and trenching can be applied. Use a big plough that can draw a furrow 15 in. deep, and post a gang of men along the furrow. As the plough passes, every man starts digging in the furrow behind the plough and takes out about 9 in. of soil. By having the loose soil scooped out, we can easily work the soil to a depth of 24 in. On the Stellenbosch University Farm I had this done with great success. The plough returns across the unploughed land and works only in one direction. It costs much less than trenching and is just as effective.

Gravelly soil and residual soils of weathered granite and shales are best prepared by hand-trenching them to a depth of 24-30 in., as the roots are thus enabled to penetrate much deeper into the soil than would have been possible with ploughing. Such soils can simply not be ploughed to such a depth. By trenching, the more or less weathered material is brought to the surface and exposed to the air, which causes it to weather rapidly and become changed into soil. We thus obtain a deeper layer of soil which will remain fairly loose for a considerable time.

In dry summers the vines will ripen their grapes much better on soil thus trenched than where it has only been ploughed to a depth of 12-20 in. The vines can also make better use of the mineral food in the soil if their roots can penetrate deeper into it. For the production of good table grapes the trenching should be deep. Where grapes are grown for wine, deep ploughing may be suitable, although grafted vines will then tend to deteriorate sooner.

Where alluvial soils with a much darker top soil than subsoil are trenched, the top soil should be kept on top. This should also be done if the subsoil is a clay and much heavier than the top soil. We can do this by scattering the top layer (depth of the spade) of the second trench over the ground and depositing the rest up to the necessary depth into the first trench, which will now obviously not be filled. The top layer of the third trench is thrown across the second to fill the first trench, the rest goes into the second trench, and so forth. In this way the top layer is kept on top.

Where the subsoil is brackish we should plough, as trenching will tend to bring the brack to the surface. As a matter of fact, all the fertile, sweet alluvial soils of Worcester, Robertson, and Montagu are ploughed, not trenched, and they are all more or less brackish in the subsoil. The Karroo soils are likewise ploughed. Their subsoil usually consists of a fairly hard layer of clay and carbonate of lime. In this soil holes are made by means of a crow-bar; the vines are inserted in them on planting, and the holes are then filled with soil and immediately irrigated. This hard subsoil is penetrated more easily by the roots, owing to the summer irrigations. Deep ploughing or trenching is not essential where vines are irrigated.

The Ground has previously been under Vines.— Where old vineyards are to be replanted it is best to grow cereals or some other crop on such land for a couple of years, as the soil may get tired of vines. The Germans call this "Rebenmüdigkeit". This depends upon various possible causes, of which I wish to mention the following: accumulation of harmful substances excreted by the vine roots in course of time, loss of plant food taken out of the soil by the vines, presence of bacteria and fungi *(e.g. Rosellinia necatrix),* which cause the roots to rot and are generally harmful to the vine. If the uprooted vines—whether grafted or ungrafted—have been infected with phylloxera, the soil will also be infected with it.

It will, in any case, be best not to replant it with vines for three to five years after the old vines have been uprooted. Then vines may again be planted on it, when it will be advisable to give the land a heavy phosphate dressing, say 500-600 lbs. basic slag per acre, before it is ploughed or trenched, as the fertiliser can then be better incorporated with the soil than when the vines have been planted.

The trenched or ploughed soil should be ploughed superficially or cultivated, harrowed, and kept loose until it is planted with vines. The most suitable time for trenching or ploughing and subsoiling the soil is after the winter rains have fallen sufficiently to moisten the soil to a great depth. At the Cape this is in June and July. After the soil has been prepared, a few good rains should fall on it to let the soil settle somewhat. If this does not happen, the soil might be too loose and contain too much air when the vines are planted, which may cause them to dry out and die. Hence have the land trenched or ploughed by the end of July or, in any case, before the winter rains are over. This is for dry-land farming conditions..

3. SPACING THE VINES

(a) **Number of Vines per Acre**.—Whoever travels through the world's wine countries will notice that the number of vines per acre grows smaller as one travels from the colder to the warmer wine regions. Thus Foëx, l.c p. 369, supplies the following data regarding France and Algeria in this respect: In Champagne there are up to 24,000 vines per acre, in Burgundy 16,000 to 20,000, in the Beaujolais—we are continually going south — 6000 to 6400, in the Ermitage (Hermitage, Rhone valley) 4000, in the Drome 2400 to 2800, in the Herault (in the south of France) 1760, in Algeria 1000.

In the south of France most vines are planted 1.50 m. × 1.50 m. square, i.e. 5×5 ft. square, as is done at the Cape. In the other European countries we find the same thing happening.

From this it would appear that the number of vines planted per acre is not arbitrarily decided. *As the deciding factor I regard the available water supply of the soil and its loss through evaporation.* Foëx, l.c. p. 369, says that where the vines stand close together and are therefore smaller, their roots grow more superficially than where they are further apart and bigger. Hence they should be planted further apart in the warm regions in order to become bigger vines that can send their roots deeper into the subsoil to absorb enough water to cover the greater loss by evaporation than in the cooler regions. The big vines have thicker roots and their root-system is more strongly developed than those of the smaller vines; this favours a greater absorption and transportation of water. To this must be added the fact that it is desirable to have the roots growing fairly near the surface in the colder regions, where the soil is not heated so intensely as in the warm regions.

Hence we should plant the vines further apart according as the soil is less deep and drier, and the climate hotter and drier. Where irrigation can be practised, we need not necessarily adhere to this rule. This is precisely what happens in Montagu (Cape), where the old vines were usually planted 3 ft. 4 in. × 3 ft. 4 in. square. This soil is usually very fertile and inclined to become brackish under irrigation, for which reason the vines are allowed to cover the soil more or less completely, and thus give the brack (salts) less chance of rising to the surface. In consequence, these vineyards have to be cultivated by hand, but it pays to do this, as the crops are heavy and of good quality—mostly sweet Muscadel wine.

Where dry-land farming is practised at the Cape, *every vine should get 20 to 86 sq. ft. of soil,* which means 2178 to 1210 vines per acre.

(*b*) **Distance between the Rows and between the Vines in the Row.**—Where vines are not irrigated and in large vineyards, even under irrigation, the rows must be wide enough to be worked by machinery in at least one direction. For this purpose such rows must be at least 4 ft. 6 in. wide. In order to be able to plough in a green manure crop, the rows should be at least 6 ft. apart, although it can be done where the rows are only 5 ft. apart, though this makes the work very difficult.

For table grapes the rows should be 8 ft. apart. In the chapter dealing especially with the production, etc., of table grapes, the reason for this width will be explained. The distance between the vines in the row will to some extent influence the distance between the rows. Thus it will he advisable to plant the Cabernet Sauvignon vines 3 ft. to 4 ft.

apart in the rows, as it should be trellised and pruned according to the double Guyot system, *i.e.* two short and two long bearers. If every vine has to get 24 sq. ft. of soil, the rows will have to be 8 ft. to 6 ft. apart. Where vines are trellised, the rows have to be 6 to 8 ft. apart. If

Where vines are not trellised, they should not be planted closer than 4 ft. 6 in. in the rows to allow cross cultivation. With trellised vines the distance between the vines in the rows depends largely upon the variety of grape. As has been stated above, vines that are pruned according to the double Guyot system should be planted 3 to 4 ft. apart in the row. Varieties requiring a large permanent stem in order to give good crops should be planted much further apart. Therefore Ohanez, Molinera gorda, Barbarossa (Danugue), Barlinka, Gros noir des Beni-Abes, etc., should be planted 6 to 12 ft. apart in the rows. The more fertile the soil, the further apart the vines can be planted. Varieties such as Muscat of Alexandria, Gros Colman, Lady Downe's Seedling, Rosaki, which bear well with short pruning, need not be planted more than 4 to 5 ft. apart.

(c) **Direction of the Rows that are Furthest Apart.**—In the colder wine districts it is desirable to select a north to south direction. This is indeed done where possible in the northerly wine districts of Europe. In this way the mid-day sun, which in these districts most of the time does not rise very high above the horizon and yet is the hottest, can strike the soil between the rows fully and thus heat the soil most; this makes the vines grow more vigorously and causes them to ripen their grapes better. This is well illustrated on Fig. 83, where the rows run approximately north and south. This is the most northerly vineyard in Europe (51° 13' N. lat.), and it is due to its special site that the grapes ripen well. Notice the stone wall which divides the slope into terraces. This vineyard is cultivated by hand.

Where vines are planted along a slope and have to be cultivated by means of machinery, the rows in which we work have to run more or less along the slope, *i.e.* more or less at right angles to the line of greatest fall. This facilitates the work in the vineyard, and enables us to check or prevent washing.

In hot countries it is better to let the rows run east to west where the ground lies more or less level or the slope allows this, as the sun then remains on top of the row of vines and there is less danger of sunburn than where the rows run north to south and the rays of the sun get a better opportunity of striking and burning the grapes, especially about 3 p.m. in December and January. Where cold and strong winds blow in spring and early summer, it is best to have the rows running across the direction of such winds. In deciding upon the direction of the rows that are farthest apart, all the above considerations should be taken into account.

FIG. 83.——The most northerly vineyard in Europe near Freyburg a.d. Unstrut (Germany), about 51° 13' north. Original.

(d) **Systems of Planting**.—We distinguish between four systems of planting: square, rectangular, diagonal, hexagonal. In every case we have first of all to *mark off a rectangular block of land*. Supposing WXYZ (Fig. 84) is the block of land we wish to plant, and that one set of rows runs parallel to WX. Put in stakes at W, X, Y, Z. Also put one at P in the line WX. From P measure PQ exactly equal to 30 ft. Hold the end of a tape measure at P and describe an arc of a circle with centre P and radius equal to 40 ft. Do the same with centre Q and radius equal to 50 ft. Let the two arcs intersect each other in R; now put a stake at R. Then the angle QPR will be a right angle, because the sides of the triangle QPR are as 30 : 40 : 50 or 3 : 4 : 5, and $3^2 + 4^2 = 5^2$.

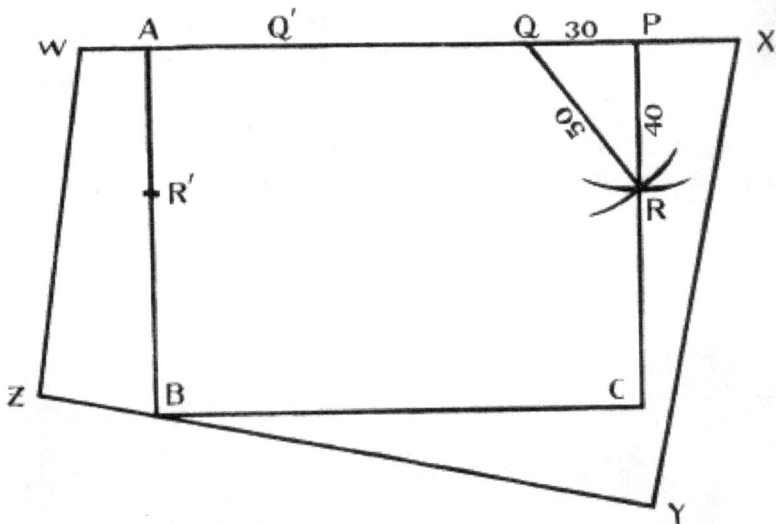

FIG. 84.—Squaring the land.   Original.

Test for Accuracy.—Measure 30 ft. from P in the direction PX and fix a stake. If the angle RPQ is truly a right angle, the distance from R to the peg in PX will be exactly 50 ft. If we find this to be so, then the angle RPQ is a right angle; if not, the operations should be repeated until such is the case. Now make a right angle RA'Q' at A in the same way (A and P are any two points on WX not too far away from W and X), and fix a stake B in the line ZY on the continuation of AR'. Then the angle BAP is of course a right angle. Measure PRC = AB. Now make a right angle at B, when the line in the direction of C will pass through it. If this is the case, our work is done, and we have the rectangle ABCP. The rest is simple.

Square System.—Supposing we wish to plant the vines 5' × 5' square.

Set a planting line or wire successively along AB, BC, and CP, and we have shallow ditches (2 to 3 in. deep) cut by means of spades to mark these lines, and continue them to the boundary lines ZY and YX. Take two straight measuring rods, each 20 ft. long, and cut a mark on them every 5 ft.; put them in ditches AB and PC with their ends exactly in the line AP. Now fix your line or wire at each mark across the ground and parallel with AP. Every time have a ditch cut across the field and along the line, which should run up to the boundary lines WZ and XY. Continue until the whole block has ditches cut parallel to AP and 5 ft. apart. Now put the measuring rods in the ditches AP and BC with their ends at A and B respectively. Plant the vines in AB, putting one at every point where a ditch crosses AB. Now set the line above and below at the first 5-ft. mark on each measuring rod, and plant a vine in every ditch where the line crosses it. The vines will then be placed as is shown

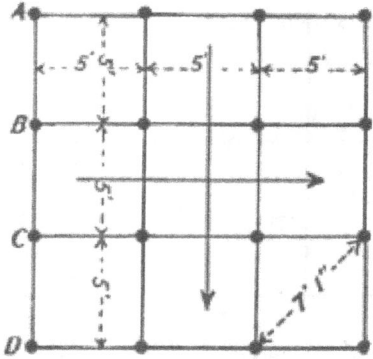

FIG. 85.—Square system of planting vines. Author's illustration (*Union of S.A. Journ. of Agric.*, Jan.-Mar. 1913).

in Fig. 85 by the black dots where the lines cross. Shift the measuring rods whilst the line is still in position. Fix the end peg of the planting line along ZY in order to be able to plant the whole block at the same time.

Here we get two rows for cultivating by means of machinery, each 5 ft. wide and at right angles to each other. The two diagonal rows are 3' 6.4" wide. This system of planting is quite good for level land and if the vines are not trellised.

An acre of land takes $\frac{4890 \times 90}{25}$ = 1742 vines when planted 5' × 5' square, and 1210 when planted 6' x 6' square.

Rectangular System.—The difference between this system and the previous one is simply that the vines are nearer together or farther apart in the rows than the distance between the rows. This applies particularly to trellised vines, where the rows should usually be 8 ft. apart and the vines are often 4 ft. apart in the rows. We proceed as before, but with this difference, that the ditches are made as far apart as the distance between the rows, *e.g.* 8 ft.; the line is set across the ditches and at right angles to them, and a vine is planted along the line in every ditch. Then, if the vines are to be 4 ft. apart in the rows, the line is again set across the ditches, but 4 ft. away from its first position, and so on. If the rows are 8 ft. apart and the vines are 4 ft. apart in the rows, an acre of land will take $\frac{4840 \times 9}{4 \times 8}$ = 1361 on this system.

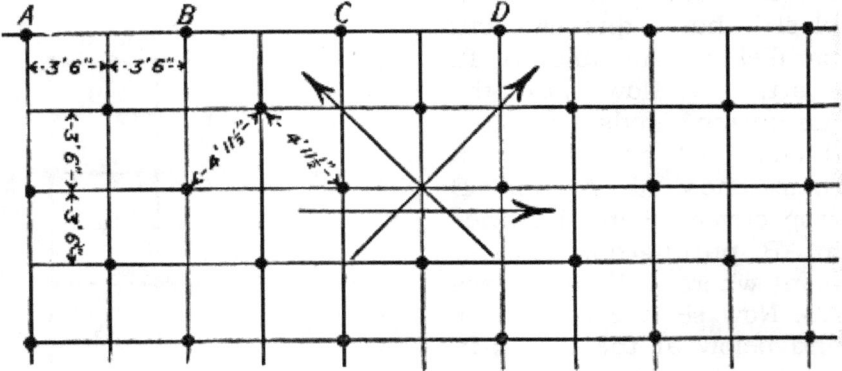

Fig. 86.—Diagonal system of planting vines.   Author's illustration (*Union of S.A. Journ. of Agric.*, Jan.-Mar. 1913).

Diagonal System.—If we wish to plant the vines 3' 6" on the diagonal system, we make the ditches 3' 6" apart and set the line at right angles across them.   Supposing we begin by planting the row ABCD . . ., then it is clear that we have planted the vines A, B, C, D in every alternate ditch. We plant A where the line cuts the ditch, skip the next ditch and plant the vine B in the following ditch, and so on.   In the second row, which is 3' 6" away from the first, we plant the vines along the line, but in the ditches that were skipped in the previous row; and so we continue till all the ground is planted.   The arrows indicate the directions in which we can plough and cultivate.   These are at right angles to each other, and the distance between the two rows of vines in one of these directions is 4' 11½ They run at an angle of 45° with the directions where the distance between the row is 3' 6". These vines can therefore be said to be planted

4' 11½" × 4' 11½" square. The 4' 11½" is obtained from $\sqrt{(3\frac{1}{2})^2 + (3\frac{1}{2})^2}$

This system is adopted where vines are planted on a slope, and it offers the advantage that, instead of having to cultivate directly uphill (if we wish to cross-cultivate), we do so at an angle of 45° with the horizontal, if AD indicates this direction. Where the *slope is fairly steep* and we still wish to cross-cultivate the vineyard, we make the ditches that run straight up the slope 5 ft. apart, set the line at right angles to them, thus along the slope in the direction of least fall, and plant a vine along the line in every alternate ditch where the line crosses it. Now set the line 3 ft. higher and still at right angles to the ditches, *i.e.* parallel with the first row of vines just planted. Now plant along the line in the ditches previously skipped. In this way the vines will be planted 5' × 3' diagonally (in Afrikaans "skuinsry-uit" *i.e.* every alternate diagonal row being omitted on a 5' × 3' rectangular system of planting). We now have two rows in which to plough and cultivate the vineyard, and which no longer are at right angles to each other, and which make an angle of 31° with the line of least fall instead of 45° as before. Here the ascent up the slope is much less steep and cultivation therefore less severe on the draught animals. The working rows are 5' 1¾" wide, but the more or less horizontal rows are only 3 ft. wide, and too narrow for passing in with a plough or cultivator.

Along very steep slopes it is advisable to plant the rows of vines that run in the direction of least fall, *i.e.* along the slope, far enough apart to facilitate ploughing and cultivation in this direction, and to give up the idea of cross-cultivation. Plant the vines here 3 to 4 ft. apart in the rows and keep the rows 6 to 8 ft. apart; hence 3' ×8' or 4' × 6' or 4' × 8'. The wider the working rows are, the less hand cultivation will be required.

If vines are planted 3' 6" diagonally, each vine gets 2' × 3½' × 3½' = 24½ sq. ft., and an acre will take $\frac{4840 \times 9}{24\frac{1}{2}}$ = 1777 vines. If planted 3' ×5' diagonally, an acre takes $\frac{4840 \times 9}{2 \times 3 \times 5}$ = 1452 vines.

Hexagonal System.—If we wish, to plant the vines 5 ft. hexagonally, we cut ditches across the field 4' 4" apart, as shown in Fig. 87. Now set the line at right angles to these ditches in the direction AH, and plant a vine along the line in every alternate ditch, *i.e.* at A, G, etc. Our measuring rods have marks 2½ ft. apart. Now set the line 2½ ft. further, and plant vines along it in the ditches previously skipped, and so on. We might equally well have cut the ditches 2½ ft. apart and set the line every 4' 4", but this would have required a greater length of ditches to be made.

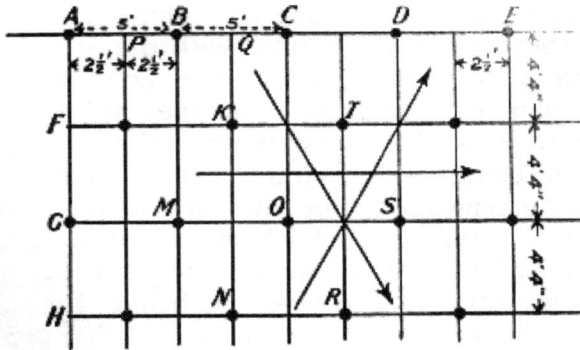

FIG. 87.—Hexagonal system of planting vines. Author's illustration (*Union of S.A. Journ. of Agric.*, Jan.-Mar. 1913).

With this system of planting we get three rows in which to work. They are all 4' 4" wide and make an angle of 60° with each other. Every three vines nearest each other form an equilateral triangle, *e.g.* the vines T, O, S on Fig. 87, where OT = TS = SO = 5 ft. In the same way four vines form a rhombus with sides 5 ft. long. Around every vine inside the block of vines we have six vines 5 ft. apart and each 5 ft. from the central vine; the six vines are at the corners of a regular hexagon with sides 5 ft. long, therefore we say that the vines are planted 5 *ft. hexagonal*. In Fig. 87 the vines K, M, N, R, S, T form such a hexagon around the central vine O.

The width of the row for working in is here $\sqrt{3}/2 \times 5' = 0.866 \times 5' = 4'\ 4"$. Where vines are planted x ft., hexagonal, this becomes $0*866 \times x$ ft., and the number of vines per acre is $\dfrac{4840 \times 9}{x \times x \times 0.866}$. If 5 ft. hexagonal, the acre will take $\dfrac{4840 \times 9}{5 \times 5 \times 0.866} = 2012$ vines, or almost 15½ per cent more vine than when planted 5′ × 5′ square. This advantage of the hexagonal over the square system is illusory. Once we have decided how many square feet every vine has to get, we shall get the same number of vines per acre no matter which system of planting we adopt. If we want a working row of 6 ft. wide, we shall have to plant the vines 7 ft. hexagonal, if this system of planting is adopted, when the acre will take only $\frac{4840 \times 9}{7 \times 7 \times 0.866} =$ 1026 vines as against 1210 when planted 6′ × 6′ square, which also gives us working rows 6 ft. wide though only two instead of three in number as in the previous case. *If the same width of working row is insisted upon,* the hexagonal system takes 15 per cent *less* vines per acre than the square system. On the whole it offers no real advantage, and therefore I do not recommend that vines should be planted according to this system.[1]

The following table taken from Perold,[2] *l.c.* p. 13, gives the distances for the square, diagonal, and hexagonal systems of planting which will take about the same number of vines per acre

Square System.		Diagonal System.		Hexagonal System.	
Distance Apart.	Number of Vines per Acre.	Distance Apart.	Number of Vines per Acre.	Distance Apart.	Number of Vines per Acre.
4 ft.	2722	2 ft. 10 in.	2722	4 ft. 3 in.	2795
4 ft. 6 in.	2151	3 ft. 2 in.	2178	4 ft. 9 in.	2220
4 ft. 9 in.	1930	3 ft. 4 in.	1960	5 ft.	2010
5 ft.	1742	3 ft. 6 in.	1778	5 ft. 6 in.	1667
5 ft. 6 in.	1440	3 ft. 10 in.	1482	5 ft. 9 in.	1521
5 ft. 9 in.	1317	4 ft.	1361	6 ft.	1397
6 ft.	1210	4 ft. 3 in.	1210	6 ft. 6 in.	1190

---

[1] It is sometimes called planting "*in quincunx*" in English and "*en quinconce*" in French, but this is bad terminology.

[2] Dr. A. I. Perold, "The Establishment and Cultivation of a Vineyard", *Agric Journ. Union of S. A.,* Jan.-Mar. 1913. Reprint No. 5 of 1913

4. PLANTING OUT THE VINES

(*a*) **Trimming and Storing.**—The vines are cut back to one bearer with two good eyes and the roots are shortened to 1 to 2 in. The thinner the root is the more it is cut back. From the time the vines are taken out of the soil they are kept in a cool place, as much as possible in the shade, and covered with wet sacks.

(*b*) **Time of Planting.**—In a region with winter rainfall it is best to plant out the vines towards the end of winter. If the soil will not become too wet, they can be planted out earlier in winter. It is best to plant them out before they have begun to form new roots where they have been heeled in; they can then begin to form their new roots after having been planted in their proper places. Where we can plant early nearly all the vines will grow. If the soil is level and likely to become too wet, planting should be delayed until the winter rains are nearly over—near the end of August at the Cape. Where frost is to be feared in spring, planting should be delayed till fairly late.

(*c*) **The Planting itself.**—Where merely a few vines are to be planted, special holes can be made. But if a large number of vines is to be planted, a spade is simply put into the ground along the line, and where it crosses a ditch, a vertical cut is made and opened sufficiently to allow the vine to be inserted. Put the vine slightly deeper than is necessary, and pull it up until it is just deep enough in the soil. Now again put the spade vertically into the soil and about 4 in. away from the first cut; by bending it backwards and forwards press the soil firmly against the vine and tramp the soil near the surface firmly down and against the vine by foot. A properly-planted vine should require the application of considerable force to pull it out of the soil. Vines that are loosely planted easily dry out and die.

The *depth* at which a vine is planted depends upon the soil in case of ungrafted vines, whereas grafted vines are slanted so that the joint is just at the surface or slightly higher. We do not want the scion to come into contact with the soil, as it will then form its own roots, which, unless cut off in time and repeatedly, will develop very strongly, and cause the roots of the stock to grow poorly until the grafted vine will be growing merely on its own roots and no longer offer the advantages of a grafted vine. On the other hand, it is inadvisable to have the joint unnecessarily high above the ground, as too much of the stem of the stock will thus be exposed to the sun, which may damage it, as it does not form as much protecting bark as the European vine. On a slope the joint should be just at the surface of the soil, and on level land one inch above it.

When the vines are planted they are usually covered more or less with fine soil. The little mounds are about 15 in. wide and high enough just or (preferably) almost to cover the head of the vine that has just been planted. Wherever the roots of the scions are not removed in time, and especially if the vine has been planted early, I prefer not to cover them with soil, but merely to make the soil level around the vine. This will prevent the formation of roots from the scions. I have done this successfully, but it requires a moist soil when the vines are planted, and a couple of good rains shortly afterwards, unless irrigation can be applied.

5. SOIL CULTIVATION OF THE VINEYARD

During the vine's period of growth, and particularly in summer, the vineyard soil should be kept *loose and free from weeds.* We are further to see to it that the fertilisers and enough water get into the soil. The soil cultivation of the vineyard aims at fulfilling the above requirements. Owing to the varying conditions of soil and climate in the different grape-growing countries and districts, we find considerable differences in the methods of cultivation adopted in the different parts.

The soil can be cultivated by hand or by means of machinery or by both. In the colder wine countries where the vines are planted fairly close, the vineyard is mainly cultivated by hand. In wine countries enjoying a warmer climate, the vines are usually planted far enough apart to make cultivation by means of machinery possible, and this is usually done, although a certain amount of cultivation by hand is also inevitable. By using suitable machinery we can cultivate the soil more cheaply and more suitably for dry-land farming than can be done by hand.

As was stated in discussing green manuring, the first cultivation after the vintage consists in bringing the fertilisers and seed for green manuring into the soil. In the Western Province of the Cape this takes place in April or May. Where calandra have been bad the previous season, the soil is removed around the stems of the vines by means of spades after the first winter rains have fallen, and scattered over the ground. This is done to kill the calandra eggs. Where the vineyard is on a slope, the necessary sluits for removing the flood water should be put in order before the rains begin to fall, and they should be given little fall. This is necessary to prevent washing as far as possible. Drain-outlets should now be cleaned to allow the drains to function normally. Open drains should also be cleaned. If the soil is not quite loose, it should be cultivated to allow the rain-water to soak in better. Otherwise we can now leave the vineyard soil alone until it is ploughed or dug over by hand near the end of winter.

During winter *grass* should be allowed to grow in the vineyard. It checks washing, and helps the rain-water to soak into the ground. In winter the grass will not evaporate much water from the soil. The grass temporarily binds some of the nitrates in the soil and prevents them from being washed out by heavy rains. The plant-food absorbed by the grass is not lost to the vine, as it is returned to the soil when the grass is ploughed in and undergoes humification in the soil. This ploughing in of grass helps to loosen heavy clay soils.

Some time before the winter rains are over and before the vines begin budding, the vineyard soil is worked once to a fair depth—say 6 to 8 in. Formerly the vineyards were dug over by means of spades, when the manure (and fertilisers) were at the same time brought into the soil. Good work can thus be done, but it is an expensive way of doing it, and is now restricted to vineyards in special localities where the vines are too closely planted to allow machine cultivation.

Most vineyards are ploughed fairly deep in winter. It is best first to draw a 6 to 8 in. deep furrow between every two rows of vines (in the middle) with the Oliver "Z" plough (see Fig. 88, A). If we wish to bring in our manure fairly deep, this furrow can be deepened to 12 in. by means of spades. The manure is usually spread in or across these furrows and the soil ploughed towards them from both sides. A smaller vineyard plough is generally used for drawing the last furrow nearest the vines. The strip of uncultivated land remaining in the row of vines is dug over by means of spades.

Instead of using the Oliver "Z" plough, we can attach a furrower (see Fig. 88, B) to the hindmost arm of the Planet Jr. cultivator (see Fig. 89). The furrower is sold in three sizes: 10, 12, and 15 in. wide. In loose soil it will draw deep enough furrows. In heavy soils the Oliver "Z" will do better work.

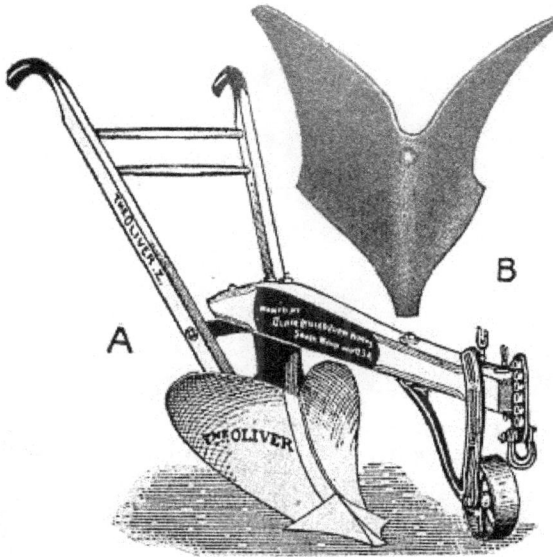

FIG. 88.—A, Oliver Ridging Plow, No. " Z ". Double mouldboard, 16 in. spread. B. Furrower of Planet Jr.

When most of the rains have fallen, *i.e.* in spring, the vineyard is cultivated with the Planet Jr. cultivator. Fig. 89, A, shows us a Planet Jr. cultivator that can be set wider and narrower by means of a lever. It has seven arms to which are attached seven cutlivator steels of $2\frac{1}{4}' \times 8'$. It can cultivate a strip of land 42 in. wide in passing once. Its wheel is spokeless, which is an advantage. By setting the wheel in front we can cultivate more or less deeply. It can be drawn by one strong horse or mule if the soil is loose and we cultivate to a depth of merely 2 in. As we usually cultivate to a depth of 3 to 4 in. immediately after winter, it will have to be drawn by two animals at this stage. It can serve a number of different purposes according to the steels we use. Thus it can be used to draw furrows, to close up furrows, to earth up soil, etc.

Fig. 89, B, shows us a Planet Jr. cultivator with only five arms; on the front two are seen with *cultivator steels,* and on the three farther back are three *sweeps.* It has a special contrivance for easily setting the wheel for deeper or shallower cultivation. One good mule or horse can easily pull it.

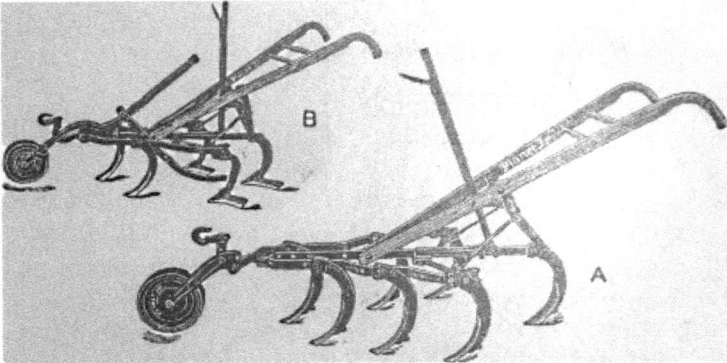

FIG. 89.—A, Planet Jr. cultivator No. 83 with 7 arms and 7 cultivator steels, with lever for setting the cultivator to the desired width, and with an adjustable spokeless steel wheel.
B. Planet Jr. cultivator with 5 arms and 3 sweeps and 2 cultivator steels, with lever for setting the cultivator to the desired' width, wheel as in Fig. 89, A, but with a special device for raising and lowering the wheel.

After winter we first cultivate in the same direction as was ploughed, making use of cultivator steels. In this way the soil is somewhat levelled. A row (really the space between two rows of vines) 4 to 5 ft. wide can be cultivated by passing once in it. Rows that are wider require to be passed twice. The same has to be done in 4 to 5 ft. rows if the first cultivation does not loosen the soil sufficiently. Where the vines are not trellised, we immediately cross-cultivate. Every time rain has fallen the vineyard is cultivated in order to keep the surface soil constantly loose. As it becomes warmer and drier (mid-November at the Cape), we use also sweeps, as can be seen in Fig. 89, B, since we now cultivate very superficially. It suffices now to *keep the soil loose to a depth of merely two inches* until the beginning of winter, but then this layer of soil must be kept loose *all the time*. Hence we must cultivate every time rain falls in summer. Deep cultivation would now be wrong, as it entails an additional loss of water by evaporation. Even though no rain has fallen, it will be advantageous to cultivate the vineyard every three weeks. Do not cultivate when the grapes turn colour. Varieties that are much subject to coulure had better not be cultivated for two to three weeks during flowering and setting, as the soil will then be warmer and the berries will set better than when the soil is constantly kept loose. After a good soaking rain we must in any case cultivate to prevent a hard crust from being formed on the surface of the soil. Wherever possible, cultivation should include cross-cultivation.

In his most interesting and valuable work, *La Goutte d'eau (i.e.* "The Drop of Water"), Mons. E. Maroger describes his system of intensive grape-growing in the South of France. His vines are planted 2.50 × 1.50 metres or 8' 2.4" × 4' 11", say 8' × 5' in round numbers, are trellised high (four to six and more wires above each other in a vertical plane), each vine has four to six long bearers, and he keeps the soil constantly loose. He thus retains much moisture in the soil and gives it plenty of air; this reacts favourably upon the soil bacteria, which can now develop in large numbers and make the soil fertile, *inter alia,* by fixing atmospheric nitrogen and making it available for the vine. He fertilises moderately.

On p. 81 he says: "We also believe that we are justified in saying *that the fertilisers occupy a second place whilst the superficial soil cultivation quite takes the first place. It has been said long before our time that two cultivations are as good as one manuring or one irrigation.*" He also uses the Planet Jr. cultivator with seven steels, and says that he passes with it once only in a 5-ft. row, and that his soil is always so loose that one horse can pull the cultivator, *but he never cultivates deeper than 5 cm. or 2 in. (l.c* pp. 260-261). He never works his vineyard deeper than 2 in., *but does it every week for nine months of the year.* Even the fertilisers are thus cultivated in. When applying farmyard manure the strip of soil remaining in the rows of vines is thrown between the rows of vines. He purposely works so as not to disturb the roots in the top soil. He works about 75 acres of vineyard, and his experience extends over a period of thirty years. His vines gave him up to 400 hl. wine per ha. or 3560 gallons per acre in 1920 (Maroger, *l.c* p. 187), and the average yield on his farm during the fifteen years from 1907 to 1921 was 127 hl. per ha. or 1130 gallons per acre *(l.c* p. 6), whilst his neighbours, who prune their vines short and do not trellise them, get only about half this yield. Still, Maroger does not manure heavily. It will be worth while to test out Maroger's system in other countries. Until it has been clearly proved that his system can be successfully applied, existing systems of soil cultivation should not be abandoned. At the Cape we plough fairly deep in winter and then bring the fertilisers into the soil. In summer we cultivate superficially (2 to 3 in. deep). It should further be remembered that high trellising and wide planting (8′ × 5′) are two important points in Maroger's system, which give him big strong vines with strong roots that penetrate deep into the soil, and can therefore resist drought well.

After the vineyard has been well cultivated a couple of times in spring, any remaining grass or weeds are suppressed by means of spades or hoes, so as to have absolutely no grass or weeds left in the vineyard during summer.

6. IRRIGATION OF VINEYARDS

For detailed information on this subject I refer the reader to special works dealing with it. Here I merely wish to draw attention to a few points in connection with the irrigation of vines.

Table grape varieties should, as far as possible, be irrigated only in winter and early summer. By irrigating repeatedly in winter we can sometimes do without any irrigation in summer, provided the soil is constantly kept loose, as has just been described. This is applicable to all varieties of grape vines. If summer irrigation has still to be applied, we should endeavour to give our last irrigation before the vintage, about a month before the grapes are picked. With table varieties it is better to stop irrigating at an earlier date—say shortly after the grapes have set. *If the vines, however, show signs of suffering from drought, we should always irrigate if water is available for this purpose.* By cultivating the soil after every irrigation as soon as it is fit to be cultivated, we can greatly reduce the loss of water by evaporation, and hence frequently reduce the number of irrigations to be given.

In irrigating, the soil should be properly wetted to a considerable depth, though not swamped. Red and black varieties sometimes colour up badly when grown on very fertile soil that is irrigated in summer, as the vines then grow very vigorously ; the soil is kept fairly cool, and the grapes hang mostly in semi-obscurity, which is unfavourable to the formation of pigment.

Before being planted with vines, the deep, fertile soils along the River Murray in Australia are usually underdrained about 5 ft. deep if they are to be irrigated. The vines are planted in rows 8 to 11 ft. apart, and one or two (where 11 ft. wide) furrows, 6 in. wide and 3 in. deep, are drawn in the openings between the rows of vines. In these furrows the water runs when the vineyard is being irrigated. The land is levelled to have only a slight fall, and the water runs for about twenty-four hours in the furrows. In this way the soil is thoroughly wetted to a depth of 4 to 5 ft. Owing to the underdrainage the vines are not harmed and brack (alkali) does not rise to the surface. When the soil in the furrows is dry enough to be cultivated, these strips are cultivated to keep a loose mulch on the surface of the ground. As the water ran only in these furrows, there is no necessity to cultivate the whole surface of the soil. Here the vineyards are ploughed 8 to 10 in. deep to keep the roots away from the surface, and thus to enable the vines to resist drought better.

CHAPTER XI

WINTER PRUNING AND TRELLISING OF THE VINE

A. D'ARMAILHACQ[1] begins his discussion of the pruning of the vine as follows: "The pruning of the vine is the principal operation and the foundation of its cultivation. If one were not to prune it, the vine would develop excessively during its first couple of years, then it would soon exhaust itself and bear no fruit, at least in poor soils; or if it should produce fruit, this would consist only of small bunches that ripen badly and of which the wine will certainly be bad."

It is indeed a fact that winter pruning is one of the most important operations the grape-grower has to perform. Hence only experienced and reliable workmen should be used for this work. This applies particularly to the pruning of young vines that have still to be formed. Therefore it is safer to pay the pruners per day and not per 1000 vines, as is frequently done at the Cape, as this may easily lead to bad pruning.

The fact that the vine possesses tendrils shows it to be a climbing plant which attaches itself by means of its tendrils to the objects it comes in contact with. We thus notice, where the vine grows wild, that it climbs up along the trees, etc., and sometimes attains a great development. But then its grapes are almost worthless, as d'Armailhacq has rightly pointed out. Only by pruning it judiciously can we obtain the desired quality and quantity of grapes.

I. AIMS OF VINE-PRUNING

1. *To regulate and ensure the production of grapes, both as regards quantity and quality.* I wish to point out that, *within certain limits,* an increase in the grape crop

---

[1] A. d'Armailhacq, *De la culture des vignes dans le Médoc,* 3rd ed., Bordeaux, 1867.

usually means a decrease in the quality. Where vines grow wild and bear too little, the quality of the grapes will improve with the increase in the crop, so long as this is not overdone. If a grape variety produces too sweet must for a *light* wine, as frequently happens in hot countries, we can obtain a less sweet must and hence a lighter wine by making the vines bear more. In this respect the quality of the crop will improve with its quantity, only to a certain extent, as the quality will soon begin to deteriorate in other respects. Vigorous and healthy vines that produce only moderate crops can, up to a certain point, usually produce heavier yields without any adverse effect on the quality of the crop. The quality will suffer only when the vines have more grapes than they can ripen perfectly every year *for our special purposes.*

2. *To obtain larger and better bunches and berries.*— Although this is true for all grape varieties and for all the purposes for which they are grown, it applies more particularly to high-class table grapes and raisin grapes.

3. *To give every vine a development above ground proportionate to its vigour.*—We have previously seen that there exists a relation between the development of the vine aboveground and that underground. If a vine therefore shows slight vigour above ground, we must give it *few and short* bearers (spurs) when pruning. If, on the other hand, it shows an exuberant growth, and especially if at the same time it bears too little, we should give it a larger development, hence *more and also long* bearers (rods) and even a longer stem or permanent stem.

4. *To give the vine the desired shape.*—This is with a view to the ripening of the grapes under the local soil and climatic conditions, the cultivation of the soil, the control of diseases, and the treatment and picking of the grapes.

5. *To provide new wood for bearers at the desired places.*— This is the reason why, in giving a vine long bearers (rods) we give it about an equal number of short bearers (spurs). If we omit to do this, the vine's permanent stem will soon become too high or too long, because the eyes near the basis of the long bearers either do not bud at all or produce only weak shoots that cannot serve as bearers.

II. PRINCIPLES APPLIED IN PRUNING

Very long experience gained in cultivating the vine has taught man how to prune and train it in order to attain the objects aimed at. This experience can be summarised in the following principles:

1. *The fertile buds occur on the one-year-old canes which are attached to two-year-old canes.* Therefore we select those canes for bearers that have grown during the previous summer and have originated on similar bearers of the year before. Canes that have developed out of older wood are called *water-sprouts,* and these, as a rule, do not bear any fruit if they are chosen as bearers. Sometimes we are nevertheless obliged to use them as bearers in order to maintain a proper shape, *i.e.* when the permanent stem becomes too high and must be brought back. The shoots that subsequently develop on them again form fruit-buds and therefore provide good bearers.

2. *The vine's productiveness is usually inversely proportional to its vigour.*—If a vine is too vigorous it bears too little fruit, and when lacking in vigour it is inclined to overbear until it dies from exhaustion. By pruning we should therefore aim at making the vine bear as much as it is capable of without injuring itself. (See what has been said on this point under Aim 3.)

3. *On one and the same cane the fertility of the eyes increases the farther they are removed from its base and thus from the old wood.* If a portion of the cane near its end has ripened badly, the eyes there will be less fertile and even altogether unfertile. Some varieties have fertile eyes at their base, *e.g.* Muscat of Alexandria, Hermitage (Cinsaut), and hence produce good yields with short pruning, *i.e.* when pruned to bearers with two well - developed eyes. Others usually have no fertile eyes at the base of their canes, *e.g.* Sultana (Thompson's Seedless), Cabernet-Sauvignon, Barlinka, Ohanez, so that we can expect good crops from them only when pruned long, *i.e.* with rods having at least six well-developed eyes.

3. *The eyes on a long bearer or leader will form more vigorous shoots as their number is smaller; they will begin budding from its end, and the shoot at its end will develop most vigour.* Hence we often notice that long bearers or leaders develop shoots only on the first two feet, whilst the

remaining eyes do not develop. This can cause bare patches on a trellis or pergola. By removing the end eye the others will develop more evenly. By experiments on the Stellenbosch University Farm I have shown that it is possible to make shoots develop wherever bearers are wanted on leaders of 6 to 10 ft. long by removing all the bottom eyes during winter pruning and leaving only one or two eyes every 12 to 15 in. where bearers are wanted. In this way the leader will retain only 10 to 16 eyes. It naturally requires a vigorously growing vine to make all these eyes develop into strong shoots. I have repeatedly done it with good success. In this way a long cordon can soon be covered. Otherwise it is not safe to make the annual extension more than 18 to 24 inches.

5. *The more erect a shoot grows the more vigour it will develop. By bending or twisting long bearers we can increase their productiveness and somewhat moderate the vigour of the shoots developing on them.* This is a result of the hindering effect the bending has on the flow of sap. Long bearers are therefore usually bent. In the Guyot system of pruning this fact is made use of in bending the long bearer and allowing the young shoots developing upon the short bearer to grow straight up.

6. *Every vine can properly nourish and ripen only a certain number of shoots and bunches, which is proportional to its vigour.* Hence every vine should be pruned upon its own merits. If the vine has borne heavily or too much during the previous season, we must prune it during the first following winter so that it will produce less in order to guard against self-exhaustion, and *vice versa.* Hence pruning will be determined by the variety of grape, the soil, the climate, the stock (with grafted vines), and the actual state of the vine when pruning. Where the soil is infested with phylloxera—and this is the case in practically all vineyards reconstituted on American stocks—we should particularly guard against over-production, which will weaken the vine and make it a prey to the phylloxera.

7. *The bunches and berries on one and the same shoot (or vine) or bunch will be larger as they are less numerous.* This should be borne in mind when the vine is summer-pruned.

Hence in this kind of pruning, especially with table grapes, we have often to remove young shoots and bunches and berries out of the bunches while still quite small. Naturally here also an extreme limit exists.

8. *The well-ripened canes give the best results and therefore, as far as possible, should be used as bearers.* The experienced pruner can immediately see whether a cane has ripened well. The more or less brown colour of the cane up to its tip is a good indication of ripeness. By cutting the cane and dipping the wound for four to five minutes in a solution of iodine in potassium iodide and water, we can judge of its degree of ripeness by the intensity of the bluish-black colour developed in the wound. Unripe canes merely assume a yellow colour with a few dark spots. If a thin section of a well-ripened cane is observed under the microscope, we notice very numerous starch grains in its cells. Unripe or badly ripened wood contains very few grains of starch in its cells.

9. *Canes with short internodes are usually riper and possess more fertile eyes than those with long internodes.* Therefore the former are chosen as bearers in preference to the latter. We further notice that varieties that bear well when pruned short, *e.g.* Muscat of Alexandria, have much shorter internodes near the base of the canes than higher up. With varieties that require long pruning, *e.g.* Barlinka, the difference is not so marked.

10. *Canes that have borne well during the previous season possess more fertile eyes than those that have borne hardly any fruit, and should therefore get preference wherever possible.* We have previously seen that they should be chosen for scions and cuttings for planting out directly. This is of special importance in short pruning. Where every vine gets a number of very long bearers, as in Maroger's system of pruning where every vine gets at least six long bearers each of 14-18 eyes, we can select the most vigorous ripe canes as long bearers. They will tend to produce vigorous canes and will still bear enough.

11. *The weather conditions that prevailed during the immediately preceding year* (reckoned from the beginning of spring until the end of- winter) *exercise a preponderating influence on the size of the coming vintage, and this should be taken into account when the vines are being pruned.*

In 1875 Arthur Marescalchi coined the word "Carpoprognosia" (*i.e.* prediction of the fruit crop) and read a paper on it, as applied to viticulture, before the Italian Oenological Congress in Verona, which he called "Primi studii di carpoprognosia applicata alla viticoltura". Marescalchi in *I Principii della viticoltura* (1909), by Ottavi-Marescalchi, discusses this subject on pp. 783-821. He points out that the formation of the eyes on the bearers, which begin to develop in spring and form shoots with or without blossoms that set their berries well or badly, has already begun in the previous spring. We have previously seen that it is decided at an early stage in their development whether they will be fertile or not. But also their further development during that year or period of growth influences their state of perfection as fertile eyes.

Marescalchi has carefully compiled the observations recorded on the weather and the size of the grape crops during the period 1855-1907 for the region around Monferrato in North-Western Italy, and concludes *that the big crop was always obtained when the preceding year had been fairly warm and dry.* A dry and mild (not cool) spring allows the young shoots and their eyes to develop well, and favours the formation of fertile eyes. A summer and autumn that are warm and dry, but not so much so as to hinder the growth of the vines, allows the fertile eyes to develop well and ripens the wood well, with the result that the following crop will be heavy if it is not destroyed by hail or diseases, or if the previous crop had not already been a very big one.

This also explains the greater fertility of varieties like Sultana, Ohanez (White Almeria), and Muscadel in Montagu, Eobertson, and Worcester (Cape) than, for instance, in Paarl and Stellenbosch, where the climate is less warm and dry. The less fertile the eyes are, the longer the bearers have to be pruned in order to obtain good crops. It thus follows that we should prune our vines more liberally (longer bearers) when the preceding spring and summer have been very cool and rainy.

According to Marescalchi we can with fair accuracy predict the size of the coming crop, even before the vines begin budding, by taking into account the weather conditions (heat and rainfall) during the preceding twelve months and the size of the preceding crop. A very big crop is never followed by an equally big crop, though it can be followed by a good crop or by a poor one.

III. THE BEST TIME FOR WINTER PRUNING

Opinions vary greatly as to which is the best time for winter pruning—owing to the differences in soil and climate, grape varieties, systems of pruning and trellising, and the objects for which the grapes are grown. In 1912 Prof. Ravaz of Montpellier published the results of his experiments on early and late pruning of the vine in a brochure called "Taille hâtive ou taille tardive?" *(i.e.* pruning early or late?). He experimented with six rows of Aramon (wine grape) grafted on Rupestris, of which the first row was pruned immediately after the vintage, the second just when the leaves had dropped, the third when the vine was fully dormant (end of December-beginning of January), the fourth when the vines began to bleed (about 20th February), the fifth when they commenced budding, and the sixth when the young shoots at the ends of the canes were 5-6 cm. or 2 in. long. Pruning was done in one operation and only short bearers of two eyes were left.

His results were interesting. The second row budded first and its grapes ripened first. The first row usually budded ten days later than the second row and about the same time as the fifth row, whilst the sixth row budded on an average 20-4 days or nearly three weeks later than the second row; the third row usually budded soon after the second, and the fourth row usually some days later than the third but before the first row. The grapes ripened in the order in which the vines had budded.

Influence on Vigour.—If we express the vine's vigour in terms of the weight of canes obtained on pruning (in winter), then the first row developed least vigour and the second row most. Ravaz, *l.c.* p. 10, gives the following average weights in grammes for the canes per vine in row one to six respectively: 478, 640, 535, 525, 516, 540. Pruning immediately after the vintage thus had a weakening effect upon the vines. This was to he expected, as the early removal of the leaves prevented sufficient food being stored up in the vine. We also notice that pruning when the leaves fall tends to make the vines more vigorous, and is therefore to be recommended for weak vines. As it also causes earlier budding, this might in some cases expose the vines to spring frost.

Influence upon the Size and Quality of the Crop.—The average *number of bunches* per vine in rows one to six was respectively: 15.33, 15.98, 16.88, 18.51, 19.45, 19.11. In this experiment late pruning has thus promoted greater fertility than early pruning. This is probably due to the fact that late pruning makes the vines bud later, so that the young eyes develop under warmer and drier conditions than when budding takes place early. Grape-growers indeed generally believe that early pruning promotes vigour and late pruning better crops.

The average *weights of grapes harvested* per vine were respectively, 2801, 3187, 2673, 3159, 3238, 3367 grammes in rows one to six. Very late priming has thus caused heavier yields than early pruning. There were no great differences in the sugar and acid contents of the musts obtained from rows one to six.

Ravaz concludes that in Montpellier late pruning is best. As the vines then bud later, they are less exposed to spring frosts, whilst the vines still remain vigorous and give the best yields. The loss of sap by bleeding when pruning late need not be taken too seriously.

Summary.

1. Varieties that are shy bearers, *e.g.* Barlinka and Ohanez, should be pruned late, say from 1st August at the Cape.

2. Varieties that are vigorous and productive, and set their berries well, can be pruned early, say from 1st June at the Cape.

3. Varieties that are fertile but set their berries badly, *e.g.* Muscat of Alexandria, Gros noir des Beni-Abès, Bicane, Rosaki, should be pruned late, say from 25th July to beginning of August at the Cape.

4. Varieties that are very vigorous but shy bearers, *e.g.* Ohanez (White Almeria), should get a first pruning (leaving only canes for bearers) immediately the grapes have been picked, and should be finally pruned late, say beginning of August at the Cape.

5. Do not commence pruning until the leaves have fallen, excepting where vines are too vigorous (as in 4).

6. Prune late where spring frosts are to be feared.

7. To have the grapes ripen early, prune early (beginning of June at the Cape). This rule can be applied to varieties like Gros Colman and Henab Turki that will set their berries well and give a good crop.

8. To have the grapes ripen late, prune very late, say about 15th August (at the Cape) or when the vines begin budding.

9. The vines can be cleaned up, *i.e.* get a first pruning when the leaves have fallen. We now remove all the canes not required as bearers. Where long bearers (rods) are required, we should leave two extra ones in case some should break when bent for tying up. With short pruning this is not necessary.

10. In the Cape viticultural area most vines are fully pruned in July. They often get a first pruning in May and June. The grower of table grapes will have to differentiate more in the dates of pruning his different varieties than the grower of wine grapes.

11. Do not prune your vines until a good rain has fallen in autumn or early winter, or (to put it differently) *do not prune if the pruning wounds remain dry.* Although the vine may not bleed, the wounds should at least become moist, as they otherwise do not heal well and constitute a port of entrance to microbes, whereby the vine's life may be threatened.

12. Although in many wine countries numerous theories exist on this point, there is no scientific proof that we should study the phases of the moon in deciding about the actual date of pruning.

IV. SYSTEMS OF PRUNING AND TRELLISING

Pruning and trellising are treated together, since a certain system of pruning usually requires a certain system of trellising.

We can distinguish between *short* and *long* pruning according as we prune to short bearers *(spurs)* or long bearers *(rods)*. Short bearers have two and at most three good eyes. Long bearers have at least six to twelve eyes. When long pruning we are obliged, in order to maintain a proper shape, to leave almost as many short bearers as long ones. Hence long pruning is really a combination of long and short pruning in almost every case. Sometimes we prune to half-long bearers (four to five eyes), as is done with Sauvignon blanc and Semillon in Sauternes, when we can speak of medium-long pruning. Both with long and short pruning we can group the bearers in different ways on the vine and make them take up different positions in space. This gives rise to the different systems of pruning and trellising. We can distinguish between three main groups:

A.   Gooseberry Bush or Goblet.
B.   Untrellised vines with stakes (échalas).
C.   Trellised vines.

A and B allow of cross-cultivation, whilst C allows only cultivation between the rows of vines.

N.B.—Sometimes the following classification is adopted:

Cordon when the stem has only one main arm that runs vertically, inclined or horizontally.

Spalier where the stem has two or more arms grouped in one plane; *vase-shaped* or *goblet* where the vine's arms are grouped around a common centre as with A above.

A. **Gooseberry Bush or Goblet**

This is the usual system adopted at the Cape as well as in the south of France and Italy, in Spain, Tunis, Algeria, California, and in most warm grape-growing regions. It does not follow that some other system of trellising may not give better results in these places, but it certainly proves that grapes can be successfully grown in this way in warm areas.

Fig. 90 shows us three vines, A, B, C, pruned according to this system. After the young vine has had one summer's growth it is pruned for the first time on the spot. It now gets *one* bearer with *two* eyes. For this purpose we select the most erect, and when possible also the most vigorous cane. Hence the cane at *a* in Fig. 90, A, has been removed and 6, that grew straight up, kept and pruned to two good eyes.

During the following winter the vine, now two years old, is pruned, as shown in Fig. 90, B.

If the vine is now sufficiently vigorous, it gets two bearers each with two eyes. At the same time the head of the previous year's bearer is cut back as far as possible, as is shown at c in Fig. 90, b.

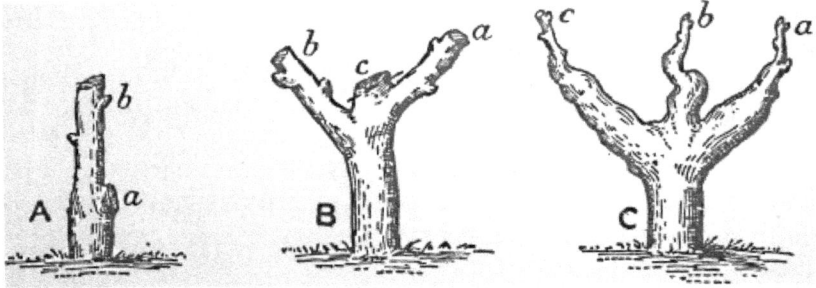

FIG. 90.—A, B, C. Unsupported vines pruned short. Author's illustration
(*Union of S.A. Journ. of Agric*, Jan.-Mar. 1913).

Vines that are now vigorous enough to be given three bearers also get only two. It is best for the sake of the shape and whole future of the vine rigidly to adhere to this rule. Those vines that have not made sufficient growth, again get only one bearer with two eyes. The third winter the vine gets three to four bearers, as is shown in Fig. 90, C. During the second winter the two bearers should be chosen in such a way as to be as nearly as possible opposite each other in the same vertical plane. The third winter we select the three to four bearers in such a way as to be evenly grouped around the vine. We do this for the even distribution of the shoots in space, whereby better lighting and ventilation are ensured, and mechanical considerations are satisfied. If the shoots and grapes are evenly distributed around the vine, their weight will also be well distributed, and the vine can maintain its upright position.

According to the nature of the soil, variety of grape, etc., we can give the vine more bearers in subsequent years. With pure short pruning we give four to six bearers, rarely more, where vines are planted 5 × 5 ft. square or closer together.

This system of pruning can also be applied to varieties like Sultana and Cabernet Sauvignon that require long pruning. We then prune the vines during the first three years as was described above.

From then onwards we begin to prune long bearers also. In the fourth winter the vine gets three to four short bearers and one to two long bearers that are bent and tied to the vine or to each other. On Fig. 91, B, we see a vine thus pruned, but with a stake to which the long bearers are tied. At the Cape the Sultana vines are usually pruned in this way, but are not staked.

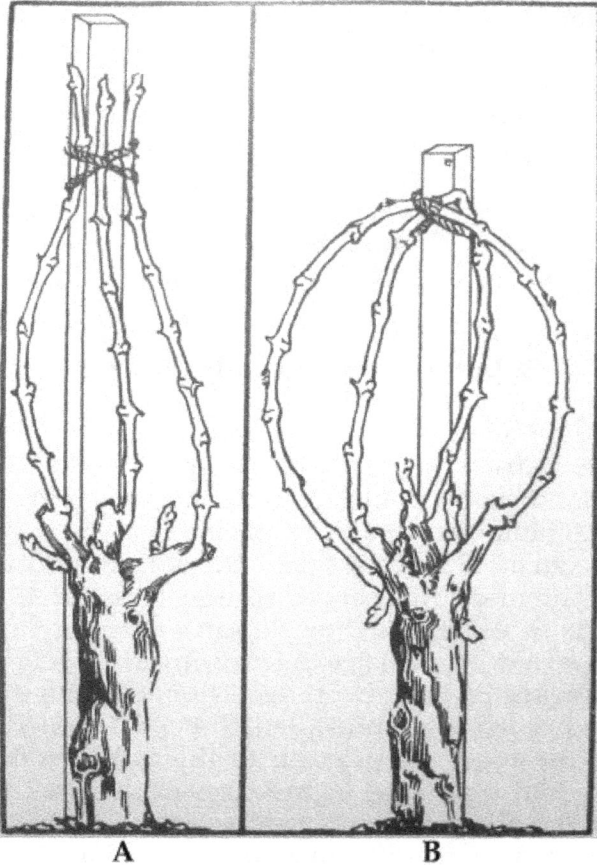

FIG. 91.—Staked vine with short and long bearers. A. Long bearers tied to stake above; B. Long bearers bent downwards. From F. T. Bioletti, *Vine Pruning in California*. Bull. No. 246 (Oct. 1914). Univ. of California Publications.

Later they get up to four long bearers that are bent and twisted around each other and/or tied. I am, however, of opinion that with Sultana and Cabernet Sauvignon trellising is to be preferred.

In winter the canes on the short bearers are used for short and long bearers, whilst the old long bearers with their canes are cut away close to the permanent stem. If the short bearers did not develop sufficient canes for bearers, we take the lowest good cane on the old long bearer as a new long bearer. Where shoots emerge on the permanent stem, we should often use them as short bearers in order to prevent the permanent stem from rising too rapidly and thus making the vine top-heavy. If only long bearers are given, the permanent stem of the vine will soon rise too high and become top-heavy. *Hence always provide some short bearers when long bearers are given.* The long bearers are there for the production of grapes and the short ones for producing canes to serve as bearers in the following season. Fig. 91 was taken from Prof. F. T. Bioletti's instructive Bulletin No. 246, being Part II. of his *Vine Pruning in California.*

B. **Untrellised Vines with Stakes**

At the Cape this system is almost non-existent. In the cooler wine districts, *e.g.* the Champagne, Moselle, Rheingau, in Switzerland, we find this system commonly adopted. Figs. 82 and 83 illustrate this. The object here is to keep the permanent stem of the vine near the ground, and to tie the shoots in summer to the stake so that both the leaves and the ground are well exposed to the sun's rays.

In warmer wine districts it is sometimes adopted with varieties requiring long pruning. Thus Sultana vines are frequently pruned and trained in this way. (See Fig. 91, A and B.) In the cooler wine countries the stake used is 18-24 in. higher than that used in the warmer wine countries, in order that the growing shoots can be tied to it. The presence of the stake makes no difference to the pruning. This is done as in A, or long bearers are also given and tied to the stake. With this system cross-cultivation is still possible.

C. **Trellised Vines**

It will suffice if the principal *types* of systems of pruning and trellising are here discussed. I shall first of all describe the Guyot system, which is the simplest, one long and one short bearer per vine; then the Medoc system with two long and two short bearers per vine, a kind of double Guyot system; then the horizontal cordon with four or more short bearers; then the horizontal cordon with an equal number of short and long bearers (four or more of each); then the short trunk with short and extra long bearers (Maroger's system); then the vertical cordon and spalier; then the high pergolas and overhead trellises of Tyrol and Almeria. In Portugal and Northern Italy the vines are sometimes trained on trees, but this is too unimportant to deserve any special discussion.

1. **The Guyot System**.—The characteristic feature about Guyot's system is that the vine always gets only one short bearer with two good eyes and one long bearer with eight to twelve good eyes. The first winter the vine gets one short bearer with two eyes. The second winter it gets one bearer with two to three eyes. The third winter it gets its short and long bearers, *a* and *bc* respectively in Fig. 92. In the second winter after pruning, the wire is put up. As is shown in Fig. 92, Guyot recommended planting by each vine a stake standing 4 ft. 6 in. above the ground, and between every two vines a stake standing 2 ft. 6 in. above the ground, whilst one wire is fixed 28 in. above the ground, the stakes being tied to this wire. It will be more practical to have two wires respectively 12-15 and 24-30 in. above the ground. Instead of stakes we can use T-section fencing standards standing 30 in. above the ground and 18 ft. apart, with two droppers also standing 30 in. above the ground between every two standards.

Guyot's long stake at each vine was to be used for tying the shoots out of the short bearer to. In this vertical position they will develop special vigour. They are topped shortly above the head of the stake. The long bearer was to be attached to the short stake, which will then later help to support the weight of the grapes. If two wires have been put up, we shall tie the long bearer to the bottom wire at *c*. Its shoots are subsequently tied to the upper wire and topped shortly above it in order to allow the shoots on the short bearer to develop extra vigour. This we see on Fig. 93, which represents the vine after its leaves have dropped in the fourth winter and before it is pruned, except that the two long canes Q, R, are not shown full length.

The fourth winter we prune as shown in Fig. 93. The old long bearer with all its canes is cut off the stem at the dotted line P. The bottom cane on the old short bearer is pruned to two eyes and the upper one is used as a long bearer and cut at the desired height (eight to twelve eyes).

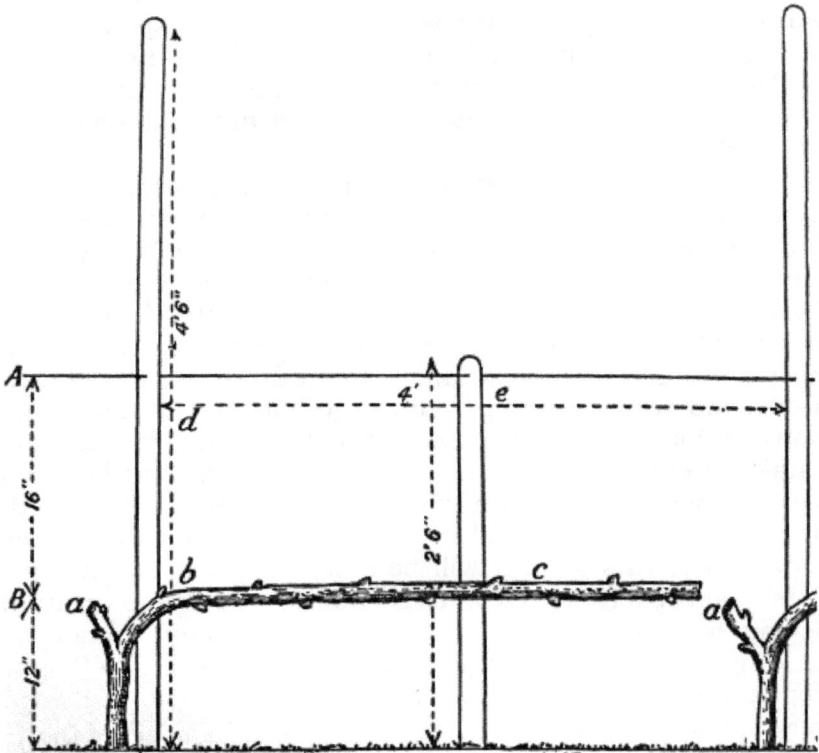

FIG. 92.—Guyot system of pruning (third year). Author's illustrations (*Union of S.A. Journ. of Agric*, Jan.-Mar. 1913).

If the old short bearer has only one good cane, it is used for the short bearer and pruned to two eyes, when the nearest good cane on the old long bearer is pruned as a new long bearer. In subsequent years we continue to prune as in the fourth winter. Where a shoot develops lower down the stem, we should use it as a short bearer when required to lower the permanent stem when it becomes too high.

2. **The Médoc System or Bordelais Spalier.**—This is a kind of double Guyot system, although it is the older of the two.

A. d'Armailhacq, *l.c.* pp. 113-169, discusses this system very fully. It is the system according to which the Cabernets are usually pruned and trained in the Médoc district, each vine getting two short and two long bearers. The length of the long bearers is varied somewhat according to the vine's vigour. If it has borne a heavy crop in the preceding season and has grown none too vigorously, the long bearers are pruned shorter than when the vine has grown vigorously and given only a moderate crop. If the shoots on the long bearer develop only moderately or even poorly, then it was pruned too long.

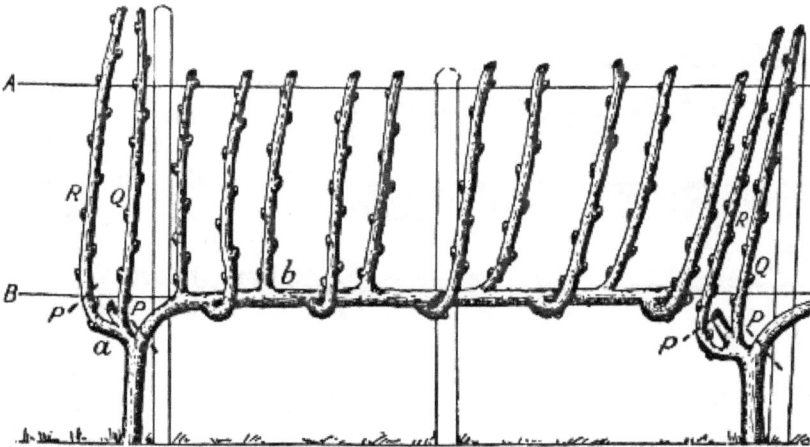

FIG. 93.—Guyot system of pruning (fourth year). Author's illustration (*Union of S.A. Journ. of Agric*, Jan.-Mar. 1913).

In the Médoc wooden stakes are fixed in the ground, one at each vine and between the vines. These stand 16 in. above the ground, and over them wooden slats are nailed or one wire is fixed. The long bearers are tied to the slats or wire. In the first winter the vine gets one bearer of two eyes. The second year it again gets one bearer of two eyes. Any vine which is now sufficiently vigorous gets two long bearers which should be well in the line of the row of vines. The third winter every vine gets two long bearers. As such, two equally vigorous shoots are selected, which leave the permanent stem to the right and left at 5-6 in. above the ground. They should not be chosen at a greater height, as the trunk of the vine would otherwise rise too fast.

575

In the second and third winters the long bearers are tied to the right and left to the horizontal slats or wire, but each retains only its two lowest good eyes, the rest being cut out.

The fourth winter every vine gets two long bearers about 15 in. in length and with seven to eight good eyes. The following winter the lowest canes on the long bearers are chosen as new long bearers. In this way the vine develops two arms that steadily grow longer, but should not go too high. In order to prevent this, two short bearers with one to two eyes are pruned as low down the permanent stem as possible, when these will provide the long bearers in the next winter or a year later, and the old arms are sawn off just above the place where the new long bearers emerge from them.

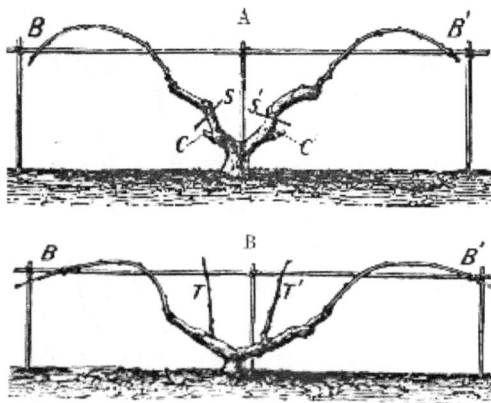

FIG. 94.—Médoc system of pruning. From Perraud, *La Taille de la vigne.* Masson et Cie., **Paris.**

Fig. 94, A, taken from Prof. J. Perraud's *La Taille de la vigne,* shows this. Here c, c' are the new short bearers, whilst s, s' indicate where the arms are sawn off. Where possible the new short bearers are selected near the bottom of the arms. They are sometimes I pruned long, as is shown by T, T' in Fig. 94, B, so that they can be tied to the slat or wire, but all the eyes with the exception of the lowest two are cut out. The next winter the arms are sawn off just above the places where T, T' come out of the trunk. The erect position of T, T' gives the shoots growing on them a better chance of becoming strong.

This is a good system for Cabernet Sauvignon. I recommend planting the vines 3-4 ft. apart in the rows and 7-8 ft. between the rows.

Trellising.—Fix a No. 8 plain galvanised iron wire about 15-16 in. above the ground, and about 11 in., higher two No. 10 plain galvanised iron wires in a horizontal plane and about 8 in. apart.

Use H-section iron standards, 5 ft. 6 in. long, with a cross-piece of the same material bolted across near the top and 9 in. long with a hole near each end, the holes to be about 8 in. apart. The two No. 10 wires run through these holes. The standards should have a hole about 11 in. lower than the top hole through which the cross-piece is bolted to it. Through this lower hole run the No. 8 bottom wire. Place the standards about 32 ft. apart, and between every two standards place one T-bulb standard 3 ft. 3 in. long (*i.e.* half of a 6 ft. 6 in. standard) with a hole about ½ in. from the top, through which the No. 8 wire runs. This is to give extra support to the bottom wire. Drive the standards into the soil so that the bottom wire will be about 15-16 in. above the ground. As corner posts use 6 ft. 6 in. T-bulb standards, weighing about 17½ lbs., and without cross-pieces. They must have an anchor or a strut.

Pruning.—Adopt a double Guyot system or the Médoc system. Tie the two long bearers to the bottom wire right and left of the vine's trunk. Where there is only one wire, the growing shoots are merely topped. Where we have the system of trellising just described, the growing shoots are pushed between the two top wires and subsequently topped. The shoots will soon attach themselves to these wires, which make it unnecessary to tie them to a wire. As the grapes will usually hang in the shade with this system of trellising, it will protect them against sunburn.

3. **Horizontal Cordon with Short Bearers.**—It is sometimes spoken of as the Royat cordon, and can be used for all varieties that bear well with short pruning. The first winter the vine is pruned to one bearer with two eyes. If the vine has made sufficient growth during its second summer, it is put on the bottom wire in the second winter, as is seen in Fig. 95, A. Usually the cane is not taken more than 2 ft. along the wire. The following winter the first cane is pruned to two eyes, and also some of the others, which are so chosen as to have the bearers 8-12 in. apart on the cordon (see Fig. 95, B). Canes not required for bearers are cut clean off against the stem. The last cane is led along the wire and tied to it. If the vines are 4 ft. apart in the row, we can cut this last cane near the following vine, as is seen in Fig. 95, B. The dotted lines on Fig. 95, B, indicate where the bearers are cut.

The following year the cordon becomes complete and then looks as is seen in Fig. 95, C. It now has six short bearers on the horizontal part of the permanent stem called cordon. It depends upon the circumstances whether four, five or six bearers are given. The best number of bearers is four to six.

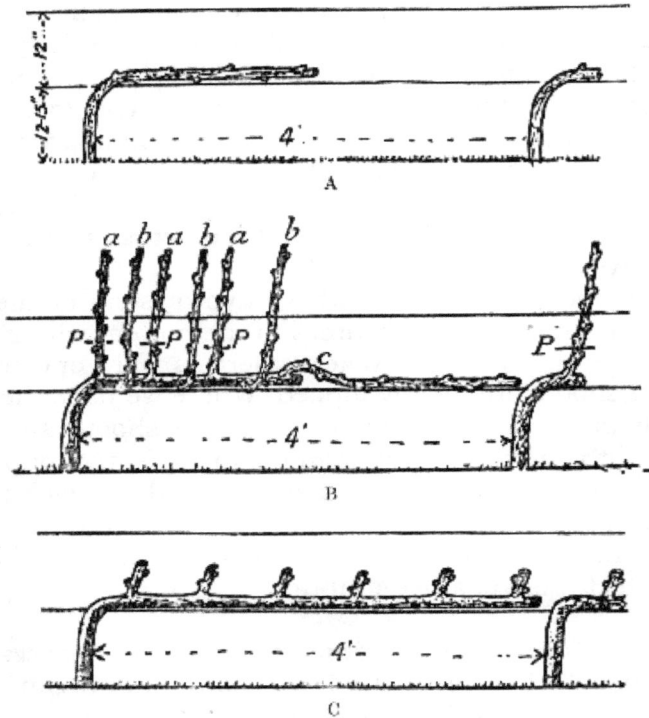

Fig. 95.—Horizontal cordon with short bearers. A. First year on wire. B. Second year on wire. C. Third year on wire, cordon complete. Author's illustrations in " Wynbou ", *Populair Wetenskaplike Leesboek*. 5. Nasionale Pers Bpk., Kaapstad.

Very vigorously growing vines, especially where well-developed vines have been grafted over, enable us to cover the 4 ft. in one year. Take the strongest shoot and guide it along the bottom wire until the next vine is reached, when it is topped. Fig. 96, A, shows my first experiment with this. By *breaking out the suckers* on the shoot on the wire *whilst quite small* and leaving only the six which have developed into the six canes seen in Fig. 96, A, these have become sufficiently strong to be used as bearers.

In this way I have covered 6 ft. of wire with six short bearers and led the last cane another 2 ft. on the wire. These bearers and the whole vine developed normally during the next summer and bore much fruit. I have not yet seen this method of covering a cordon by partial suckering described elsewhere. The necessary suckers must be removed when only 1-2 in. long—the sooner the better.

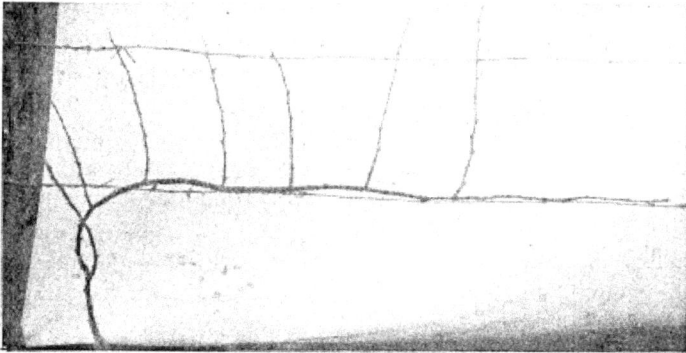

A

B

FIG. 96.—Establishing a cordon by suckering. A. Greengrape with well-developed laterals that are used as short bearers. This vine covered 6 ft. on the wire in one summer. Original. B. Sultana vine grafted over in Sept. 1922, was led for 12 ft. on the wire in one direction, suckers were left every 12 in. and the rest removed whilst small (in spring 1922). During the next winter (1923) the cordon was established for 12 ft. with twelve bearers of two eyes each, which are seen with their young shoots on Fig. 96, B, which represents the vine in Sept. 1923, precisely one year after it had been grafted. Original.

Fig. 96, B, shows a Sultana vine which I have treated in the same manner. In August 1922 a nine-year-old vine, Muscat of Alexandria grafted on Jacquez, was grafted over with Sultana. In the early summer of 1922 the undesirable suckers were removed in time, and the growing shoot was led for 12 ft. along the bottom wire.

I allowed it to keep twelve suckers which developed into strong shoots and were pruned to twelve short bearers each with one to two eyes above the stem in the winter of 1923. The photo for Fig. 96, B, was taken in September 1923, *i.e.* one year after the scion had been in¬serted into the decapitated vine. In 1924 this vine bore a splendid crop of grapes, as can be seen on Fig. 97, and it reached this development in eighteen months from being grafted. In grafting over, therefore, we lost only one crop, namely, that of 1923.

FIG. 97.—The same vine as on Fig. 96, B, in Feb. 1924. Original.

I have found that another way of soon forming a long cordon is to *remove all unnecessary eyes in winter. It is the number of eyes that matters at budding and afterwards, and not so much the distance they are apart.* This consideration led me to make the experiment of which Fig. 98 shows the result. It is a Rosaki vine which had been grafted like the previous Sultana vine in August 1922 (also on Muscat of Alexandria on Jacquez, nine years old). During the summer of 1922-23 it grew very vigorously and the leader was taken along the bottom wire for more than 15 ft. On the first 6 ft. on the wire the unnecessary suckers were removed whilst small, and thus canes were developed that were pruned as short bearers with one to two eves in the winter of 1923, when a further 9ft. of cane (leader) was kept on the wire, thus making a total length of 15 ft. covered in one season. The last 9 ft. had no laterals—the few that had developed in summer were suppressed when pruning. On it all the bottom eyes and some others were cut out, only two eyes being kept about every 15 in., so that I kept about fourteen eyes on the last 9 ft. of the leader. All these eyes budded and formed normal shoots in the following summer (1923-24). This vine bore a heavy crop of grapes in 1924, as can be seen on Fig. 98.

Fig. 98.—Establishing a cordon by suckering and by cutting out dormant buds (Rosaki vine in Feb. 1924; grafted over in Sept. 1922). Original.

This method of forming a cordon I have tested also on other vines with good results. I simply take the leader up to the next vine. Where vines are missing in a row I cover the gap by continuing the cordon, unless the gap be too great. The vine shown in Fig. 98 has grown well in the summer 1924-25, notwithstanding the fact that I had formed a cordon of 15 ft. in one year.

I do not know whether the above described methods of forming a cordon rapidly by suckering in spring and disbudding in winter are already known, as I have not found them described anywhere. The speed with which we can form the cordon is determined by the vine's vigour.   It is wrong to attempt to proceed faster than the vine's vigour will allow.

   Trellising for this Cordon according to Perold.— Whereas we sometimes find that with this cordon only one wire is put up to carry the cordon, or a second one 12-15 in. higher and vertically above the first wire to tie the growing shoots to, or two thinner wires about 6 in. apart and above the bottom wire to let all the shoots pass between them, I have worked out a system for *table grapes* which is calculated to let the bunches hang freely and prevent the bloom being rubbed off.

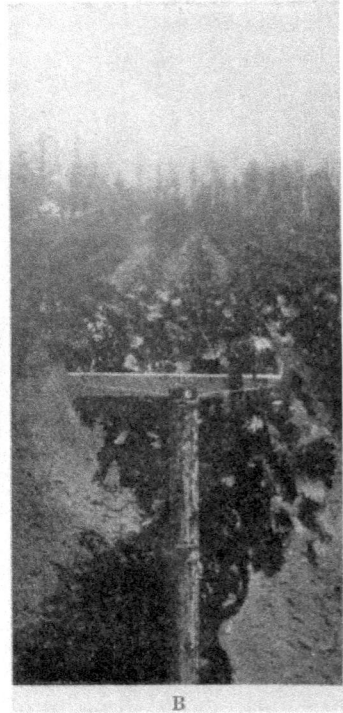

FIG. 99.—Author's system of training and trellising table grapes.
A. First form. B. Improved form. Original.

Fig. 99 shows my system of trellising table grapes. Fig. 99, A, shows the trellising done in the winter of 1923, and Fig. 99, B, that done in the winter of 1924. Here we have three wires. The bottom wire carries the cordon, and is No. 8 or No. 10 plain galvanised iron wire. The upper two wires are also plain galvanised iron wire, but thinner: No. 10 where strong winds blow in summer, otherwise No. 12. They are in a horizontal plane and are *twice as far apart as they are above the bottom wire*. The bottom wire can be 15-24 in. above the ground. The two upper wires should be about 20-24 in. apart, and must then be 10-12 in. above the bottom wire, *i.e.* the cross-piece carrying the two top wires must then be fixed to the standard 10-12 in. above the bottom wire.

As soon as the young shoots are sufficiently long they are tied alternately to the right and left top wires, and then topped at the second leaf from the wire. These shoots will now form an angle of nearly 45° with the horizontal plane, which means that their bunches will hang freely and their bloom will not be damaged easily. To be able to apply this system and still cultivate between the rows, these should not be less than 7 ft. and preferably 8 or 9 ft. apart. This system has given very satisfactory results on the Stellenbosch University Farm, where it was first tried. It is invaluable where strong winds might otherwise damage the vines and the grapes. Other advantages are that the bunches are readily accessible for pre-thinning and usually hang in the shade, thus being protected against sunburn. This latter advantage is of very great value where table grapes are grown in hot districts. The system is of no value to the wine grape-grower.

In putting up the trellis use H-section iron standards as was described for trellising Cabernet Sauvignon under the "Médoc system", but use longer cross-pieces, 20-24 in. long instead of only 9 in. long, and have the holes about ½ in. from each end. Put the standards 18 ft. apart and two droppers between every two standards. The corner posts can also have a cross-piece of iron 18 in. long, which should, however, be strong enough to resist the strain of the two wires. I would suggest angular iron or else a piece of an H-section standard. Fig. 99, A, shows that the flat iron used there as a cross-piece was unsuitable for the purpose. I am of opinion that the device adopted in Fig. 99, B, is better than a cross-piece on the corner post. Here a piece of hard wood (say Jarrah), 18" x 1½" x 1½", is tied to the corner post about 2 ft. away from it. The piece of wood gets three holes, one in the middle and one near each end. The two top wires pass through the holes near its ends and are then tied to the corner post. To prevent this piece of wood from turning, it should be fastened independently from each hole to the corner post. Through the central hole it is tied to the corner post. This device answers well.

4. **Horizontal Cordon with Short and Long Bearers.**—Of this system I wish to describe three types which differ mainly in the way the long bearers are tied. During the first couple of years they are all pruned as was described for the cordon with short bearers. We therefore commence at the stage shown in Fig. 95, C. In principle it makes no difference whether the cordon is somewhat shorter or longer and has fewer or more bearers.

(a) *Cazenave Cordon.*—In the winter following the stage shown in Fig. 95, C, we can assume that each bearer will have at least two canes as shown in Fig. 100, A. They are now all pruned as indicated by the dotted lines P. The lowest cane is used for a short bearer with two good eyes, and the best remaining cane is used as a long bearer with about six good eyes. If the cordon had six short bearers in the previous winter, it will now get six short and six long bearers. The cordon is frequently formed gradually with long and short bearers. This depends upon its vigour. The long bearers are all tied in a slanting position to the second wire, and they project slightly above this wire. They then occupy roughly the position indicated on Fig. 100, A. The next winter every arm on the cordon looks more or less like Fig. 100, B, before being pruned. When pruning we remove the old long bearer C with its canes *e* by cutting it at the dotted line P. If the old short bearer has only one good shoot, it is pruned as a short bearer, whilst the lowest good cane on the old long bearer C is used for a new long bearer. The arms on the cordon gradually become higher, hence we should use shoots arising low down the arm as short bearers, when the arm can be sawn off as was previously described under the Médoc system. This system requires a third wire to which the shoots from the long bearer can be tied.

(b) *Royat Cordon with Long Bearers.*—-Here we proceed as with the Cazenave system with this difference, that only two wires are put up, and the long bearers are bent down-awards and tied to the permanent stem or better to the bottom wire. This system of pruning I apply successfully with my system of trellising table grapes that require long pruning, *e.g.* Barlinka, Gros noir des Beni-Abès, when the upper wire with the Royat cordon is replaced by my two upper wires.

(c) *Fish-spine Cordon.*—This system I saw for the first time at the Viticultural Experiment Station of Velletri (Regia Cantina sperimentale di Velletri) near Rome during my visit in August 1909. The width between the rows was 2.30 metres or 7 ft. 6½ in. and the vines were 1.85 metres or 6 ft. 1 in. apart in the rows. The posts were 2.50 metres long, of which 1.50 metres or 59 in. was above ground.

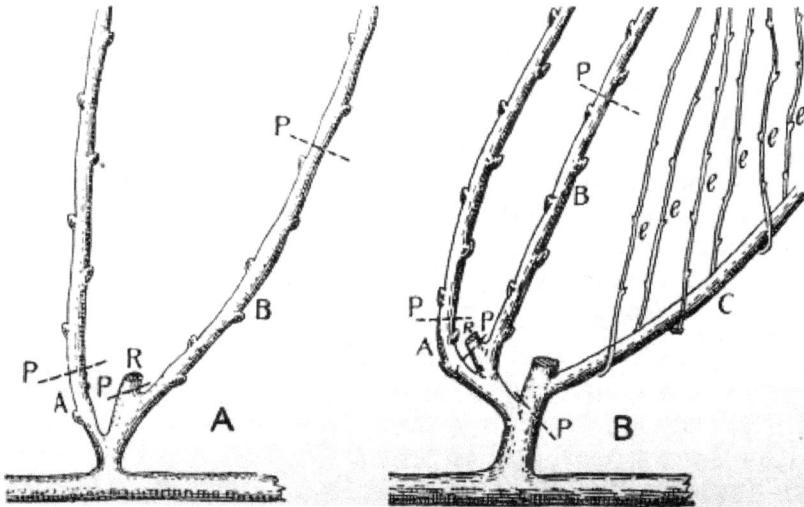

FIG. 100.—A. Arm on Cazenave cordon in fourth or fifth year. B. Arm on Cazenave cordon in fifth or sixth year. Author's illustration *(Union Agric. Journ., Jan.-Mar. 1913).*

The first wire was 50 cm. or nearly 20 in. above ground and the second wire was 80 cm. or 31½ in. higher. On each side of the bottom wire is a side wire 25 cm. or about 10 in. away and slightly higher than the bottom wire.

The permanent stem of the vine is trained on the middle bottom wire up to the next vine, and is formed into a cordon having a short bearer of two eyes and a long bearer of about seven eyes more or less for every foot of the cordon. The long bearers are tied alternately to the right and left side wire. It is this arrangement of the cordon and its long bearers that gave rise to the name fish-spine cordon. The shoots of the short bearers grow straight up and are tied to the top wire. With this system of pruning and trellising the vines can bear very heavy crops. The grapes hang in the shade and rarely become sunburnt. For producing first-class table grapes I prefer my own system of pruning and trellising.

    5. **Horizontal Cordon with Long Bearers.**—This is the *Sylvoz cordon.*   Here the horizontal stem is trained on the second wire and gets a long bearer every 10-12 in., without any short bearers. The cordon is gradually formed as the vine's vigour allows. The middle or second wire is 1.20 metres or 47 in. above the ground, the lowest wire is 40 cm. or 15¾ in. lower, and the top wire is 50 cm. or nearly 20 in. higher. This system of training the vine diminishes the danger of spring frosts, especially if the bearers are bent down and tied to the bottom wire very late in the season. This is done in moist but not too cold weather as the canes can then be easily bent without breaking. The bending of the bearers increases their productiveness and brings the grapes fairly near the ground so that they can ripen well. This system is adopted a good deal in Savoy and the Isère.

    6. **Maroger System.**—Here the bearers are taken from the head of the trunk, where in course of time some longer or shorter arms are formed which can be lowered as was indicated with other systems, and on which the short and long bearers are pruned.   Fig. 101 illustrates this. Bioletti, from whose Bulletin this illustration was taken, says that it is used a good deal for Sultana or Thompson's Seedless in California. I have successfully used it for Sultana on the Stellenbosch University Farm. The long bearers are tied to the wires in equal numbers to the right and left. The following winter the old long bearers and their canes are removed, the canes on the short bearers being used for new short and long bearers. Where it is desired to give the vine a greater development, we can use some of the lower canes on the old long bearers for new long bearers. The feature of this system is the liberal pruning, *i.e.* great length of the long bearers.

    I call this system after Monsieur E. Maroger, who has very successfully applied it for the last thirty years on his farm, as one can read in his work, *La Goutte d'eau,* to which I have previously referred. We have seen how he plants and cultivates his vines.
Here we are concerned with his method of pruning and trellising.

    Maroger places his posts 5-6 metres or about 18 ft. apart and 20 in. in the ground. His bottom wire is 40 cm. or

15¾ in. above the ground, and the other wires are each 25-30 cm. or 10-12 in. higher than the one below it. For four wires the post has to have a length of 1.65 m. *i.e.* 5 ft. 5 in. or 1.80 m. *i.e.* 5 ft. 11 in. according as the wires are 10 or 12 in. apart. For six wires the post must be 7 ft. 1 in. or 8 ft. long. As Maroger always obtained higher yields with six than with four wires, he adopted six wires as his normal trellising. He further prefers to have the wires 30 cm. or 12 in. apart. Therefore he normally uses 8 ft. posts and six wires, of which the bottom one is 15¾ in. above the ground and the others 1 ft. above each other.

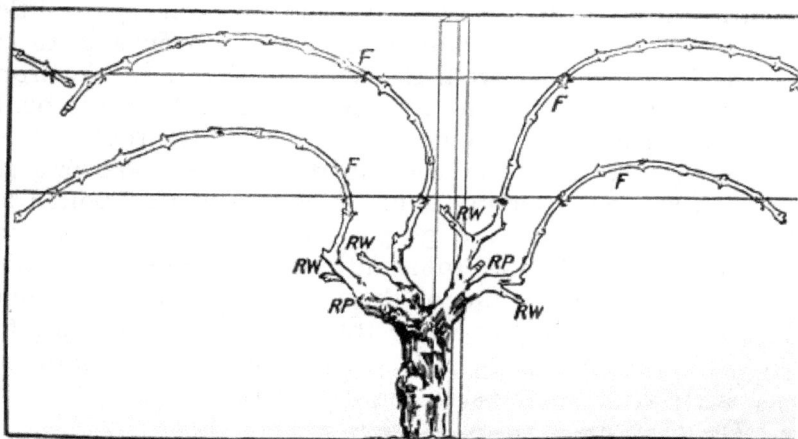

FIG. 101.—Maroger's system of pruning and trellising vines. From Bioletti, *Vine Pruning in California.* Bull. No. 246 (Oct. 1914). Univ. of California Publications

The holes in the posts are 3 ft. from the bottom end, and there is an extra hole every 1 ft. up. He uses plain galvanised iron wire No. 15 with a diameter of 2.4 mm. or $1/10$ in. I recommend our plain galvanised iron wire No. 10 with a diameter of 3 mm. Our No. 8 has a diameter of 4 mm.

For his long bearers Maroger chooses the most vigorous canes, prunes them to fourteen to eighteen eyes (see Maroger, *l.c.* p. 226), and ties them right and left to the wires.

Every vine gets at least four long bearers and some short ones of two eyes. With six wires every vine gets eight long bearers, four being tied to the right and four to the left. For tying he uses thin wire rather than twine or raffia. The long bearers of one vine should touch those of the next vine, as is seen in Fig. 101.

7. **The Thomery System**.—At Thomery, which lies near Fontainebleau on the Seine, the vines are trained against walls. Here (at Thomery) the Chasselas doré or Chasselas of Fontainebleau is thus trained and gives valuable crops. The walls are built of stone and clay, are plastered with lime and sand, and then whitewashed with grey-buff coloured lime. Each wall is covered by a little pitched roof of flat tiles which slopes downwards until it projects 10 in. over the wall. Every metre (39.37 in.) an iron standard is fixed in the wall and below the roof, and is bent downwards, projecting 50 cm. or 191 in. from the roof. They serve to support an equally wide strip of wooden boarding or watertight sail or other material which is put up about the 15th September when the grapes are ripe, and serves to protect the grapes against rain and rotting (see Perraud,[1] pp. 64-65). The great value of these grapes is due to their splendid keeping qualities when thus grown. According to Viala-Vermorel, *l.c.* vol. ii. p. 9, these grapes will sometimes keep perfectly for six to seven months from the time they are ripe.

The vines are trained on wires against the walls. The first wire is 30 or 40 cm. *(i.e.* 12 or 15¾ in.) above the ground, and the other wires each 20 or 25 cm. *(i.e.* 8 or 10 in.) higher. The walls are usually 3 m. or 9 ft. 10 in. high and 30 m. or 98 ft. 5 in. apart. Between them are rows of vines trellised on wire, 3-4 ft. high and not nearer than 7 ft. to the walls. The walls naturally run from east to west and the vines are trained along their southern side to be well exposed to the sun so that they can ripen their grapes and wood well. The number of wires depends upon the height of the wall. When the wall has a height of 3 metres, it gets ten to twelve wires. The vines are planted 40 cm. or 15¾ in. apart, and trained as is seen in Fig. 102, A and B, where the vines are shown after the leaves have dropped but before they have

---

[1] J. Perraud, *La Taille de la vigne,* 3rd ed., 1905.

been pruned. We see everywhere two canes together that have grown from a short bearer with two eyes.

When pruning in lowest cane is pruned to a short bearer with two eyes, and the other one is removed. Fig. 102, A, shows us a vertical Thomery cordon whilst Fig. 102, B, shows us a horizontal Thomery spalier. The formation of this cordon and spalier ought to present no difficulty to the reader who knows what I have already written about pruning in this chapter.

Fig. 102. — Thomery trellising. A. Vertical cordon. B. Horizontal spalier. From Chancrin, *Viticulture Moderne,* 1908. Hachette, Paris.

8. **The Almeria Trellis.**—This is a trellis in a horizontal plane which is adopted for growing the Almeria (Ohanez) grapes in the Province of Almeria in Spain. In this locality these vines are planted 20 × 20 ft. square and trained on this trellis. Strong posts, standing 8 ft. above the ground, are planted 10 × 10 ft. square. The wires run across and are fixed on the heads of these posts. Every 20 in. a wire is fixed, and similarly at right angles to these. Plain galvanised iron wire, Nos. 12 and 16, are used. On the Paarl Viticultural Experiment Station I had such a trellis put up on a small scale, of which Fig. 103 shows us a portion before the grapes were picked.

The vines reach the top of the trellis after one or at most two years, as this is a most vigorous variety. The following winter they get long and short bearers according to their vigour. Gradually the whole trellis is covered. The long bearers are pruned extra long (4-8 ft.) in order to be able to get a good crop. The stems have no bearers until the trellis is reached, when they can be allowed to divide into several branches on which we prune the short and long bearers.

Fig. 103.—Almeria grapes (Ohanez) on Almeria trellis at the Paarl Vit. Expt. Station. Original.

9. **The Tyrolese Pergolas.**—This is a type standing between systems 7 and 8, *i.e.* an inclined trellis. In Tyrol, *e.g.* at Bozen and San Michele, they are constructed exclusively of wood or of wood and wire. On the Paarl Viticultural Experiment Station I had them made exclusively of iron standards and plain galvanised iron wire No. 12. There are two types. The simple pergola seen in Figs. 104 and 105, and the double pergola. Fig. 105 shows us a vineyard at the Agricultural School of San Michele of South Tyrol. It was reproduced from a photo I took of it during my visit on 27th July in 1908. These pergolas are of the simple or single type. The vines are planted 55 or 80 or 100 cm. *(i.e.* about 22 or 32 or 40 in.) apart in the rows. For these single pergolas the rows are planted 2.50-3 m. or 8' 2½" - 9'10" apart,

FIG. 104.—Tyrolese pergolas at Paarl Vit. Expt. Station. Original.

FIG. 105.—Tyrolese pergolas on the Agric. School and Expt. Station at San Michele, Southern Tyrol (on 28th July 1908). Original.

and 5 m. or 16′ 5″ apart for the double pergolas. For the double pergola the vines are planted about 40 in. apart, hence much closer together in the rows than for the single pergola.

For the single pergolas, as seen in Fig. 105, the tops of the posts are 9-11 ft. above the ground. The cross poles are fixed 3-5 ft. above the ground on the lower side and 7-9 ft. above the ground on the higher side. The corner posts are well strutted as is seen in Fig. 105. The posts are about 3 m. or 9 ft. 10 in. apart in the rows. About seven wires are fixed, of which the first (lowest) one runs over the row of vines and at the height of the lowest end of the cross pole. The other wires are fixed at equal distances apart. The stems of the vines are taken straight up, and get on the pergola in the third year, when they get one or two long bearers which are fastened to the wires. The pergolas are gradually covered by pruning to short and long bearers. They give heavy yields. This high trellising renders the vines fairly safe against spring frosts which are otherwise to be dreaded in these parts (Tyrol).

With the double pergolas the posts in the rows of vines are about 3-5 ft. high, and those in the middle of two rows are about 7-9 ft. high. The corner posts must be strutted. The construction is similar to that of the single pergolas, but simply with two sides falling to the middle like an incompletely opened book. The pruning differs from the previous one in this respect that every vine is divided into two equal halves on reaching the pergola, unless they are trained alternately on the right and left sides of the pergola. In Tyrol it is usually wine varieties that are grown on the pergolas, and it is particularly the single pergolas that produce good wine. They are good for growing table grapes that are very susceptible to sunburn, but it is a troublesome task to pre-thin the bunches on a pergola.

V. PRACTICAL EXECUTION OF PRUNING

As was previously pointed out, this work should be done by able and experienced workmen. It is in any case necessary carefully to control the pruning. The instrument almost universally employed for pruning is the sécateur. There are those with a sharp blade and a blunt surface, and others with two sharp blades or surfaces. The first type is most generally used. Perkins[1] recommends the use of large shears for pruning old vines. At present the best sécateur is made by Rieser of Courcelles-Neuchâtel, Switzerland. It is also the most expensive, costing about 15s. to 17s. 6d. in Cape Town. In the long run it is the cheapest since it can be used intensively for many years. It is made of hard steel, which practically never bends, remains sharp for a long time, and when the blade has to be sharpened, it can easily be detached.

Clean your sécateur regularly and never put it away unless dry, so as to prevent rusting. Use a little vaseline on a rag for cleaning it. *A sharp sécateur makes a clean cut,* which is most desirable, as such wounds heal quickest and best.

An important part of the sécateur is the spring which opens it again after every cut is made. One form of spring is a worm (spiral spring), and another consists of two or more elastic metal blades fixed inside the handles, as in the Rieser sécateur. This latter type is the best when it is good. The worm is liable to get lost and is more easily broken.

When not in use, the sécateur is kept shut by means of a catch on one of the handles or by a loose strip of bent flat steel. The latter is used with the Rieser and some other sécateurs, and is the most convenient. Care should merely be taken that it is not lost.

To cut off an arm we use a pruning saw. On very vigorous vines it is sometimes used to cut off thick two-year-old wood.

If the sécateur is taken in the right hand, the sharp blade is farthest away. In pruning, the blunt surface should press against the part that is being removed, when the sharp blade will be on the side of the wood that remains on the vine. This is done, since a bruising of the cane will take place more easily against the blunt surface than on the side of the sharp blade.

Another important practical point in pruning is *the exact place where the cut is made.* In France Dezeimeris has recommended that it should pass through the node immediately above the highest eye. If this is done so carefully that the cut passes through the diaphragm so that the pith below it is not exposed, it offers the advantage of limiting the drying out of the bearer to a minimum and offering less opportunity

---

[1] A. J. Perkins, *Vine Pruning: Its Theory and Practice,* Adelaide, 1895.

for dangerous microbes to enter the wound and harm the vine. I recommend this practice for trellised vines if it is done with great care. Where long pruning is necessary, it offers the further advantage that the end of the long bearer can then be firmly tied to the wire without in any way injuring the development of any eye.

Untrellised vines with short bearers should not be pruned in this way, as it lengthens the bearers by one internode which will cause them to be more easily broken off when the vineyard is cultivated by means of machinery. Rather prune these in the internode. Most French, Italian, and many other authorities recommend the cut to begin about one inch above the highest eye, and continue in a slanting direction downwards to the opposite side of the cane. With this opinion I cannot agree in every respect. It is better to make the cut about half an inch above the highest eye, and not to make it too slanting, as thereby a great wound is made. This is the practice of our best grape growers at the Cape, and is also recommended by Prof. Bioletti in his Bulletin, No. 241, p. 42.

If my advice previously given is acted upon (namely, not to prune before the wound is visibly moist), there will be no danger of the highest eye suffering from dessiccation owing to the wound being made so near to it.

VI. PRACTICAL EXECUTION OF TRELLISING

In discussing the different systems of pruning and trellising, I have already stated how many wires each requires, and where these are to be fixed. Here I wish to discuss some other points of practical importance in connection with trellising.

Kind of Posts.—Wooden posts, especially of a hard wood like Jarrah or an almost imperishable wood like cedar, can be used, but iron standards are usually to be preferred. They usually cost somewhat more than ordinary wooden posts, but are almost imperishable, and offer less opportunity to insects for hibernating on them. We usually employ T-bulb iron standards. Where cross-pieces have to be fixed to the standards, we should use H-section standards.

The standards in the rows need not weigh more than 2 lbs. per foot. Corner posts have to be somewhat heavier, say $2^2/_3$ lbs. per foot. Thus 6 ft. 6 in. standards would weigh 13 and $17^1/_3$ lbs. respectively.

Length of Posts.—This we obtain by adding the depth the post should be in the soil to the height it should stand above the ground. In firm soil the post need not be more than 18 in. in the ground, whilst in loose soil it should be at least 2 ft. (and preferably more) in the ground. For my system of trellising table grapes the posts need not be longer than 4 ft. to 4 ft. 6 in. We have already seen that Maroger uses 6 ft. posts for four wires and 8 ft. posts for six wires.

Distance Posts are Apart.—For Maroger's system of trellising and for mine (for table grapes) the posts should be about 18 ft. apart. One or two droppers may be put between every two posts. When trellising Cabernet Sauvignon, as was recommended under the Médoc system of pruning, the posts are about 32 ft. apart, with a shorter post in between to take only the bottom wire.

Fixing of Corner Posts.—It is of the utmost importance that the corner posts should be well fixed, so that they can retain their position when the wires are strained. Sometimes the corner posts are planted leaning away from the end vine, and are anchored to the ground by means of No. 8 plain galvanised wire passing through a hole near its top and another near the middle. The anchor wire is passed through the top hole of half an iron standard showing about 3 in. above the ground and firmly fixed in a block of concrete 18" × 18" × 12" (thick) in the soil. The top of the concrete block should be about 18 in. below the surface of the ground.

A cheaper way is to use a fairly large angular stone, wind No. 8 plain galvanised iron wire around it, bury it about 2 ft. deep in the ground and 15-18 in. away from the corner post on its outside and in a line with the row of vines to be trellised. The No. 8 anchoring wire is passed through holes in the post as before, and attached to the stone anchor before it is covered with soil.

Owing to the risk of breaking the anchor wires when cultivating or ploughing the vineyard, many farmers prefer to strut the corner post at about ¼ from the top. The head of the strut is well fixed to the corner post, and a stone is put in the ground against which the strut's lower end is placed to prevent it giving when the wires are strained. To prevent the corner post from lifting it is anchored to a stone. Here the anchor is near the post and the anchoring wire hardly shows above ground, so that it will not be damaged by machinery.

The Wire, its Straining and Fixing.—Use only well galvanised, plain iron wire. The thicknesses have already been given in each case. Thin steel wire possessing the same tensile strength as much thicker galvanised iron, is not so good as the latter, and will sooner cut the shoots when strong winds blow. *Do not strain the wire too much.* If the wires are strained very much in warm weather, they might snap in cold weather or pull out the corner posts. Autumn is a good time for putting up the trellis. If this is done after the vines have been pruned, some bearers are sure to be broken off. Do not put up the fence until the vines are ready to be put on the wire, hence one and a half to two and a half years after the vines have been planted. Do not cut the wire too close when putting up the trellis, so that it can easily be loosened again if it should subsequently have to be strained again.

Use *only* well galvanised, plain iron wire as binding wire. Numbers 12-16 are suitable for this purpose. Whoever uses cheap baling wire will soon regret this slight economy. This kind of wire soon rusts and causes the wire to rust with which it is in contact.

Cost of Trellising.—This will vary very much with local prices and conditions. A good deal will depend upon the length of the rows and the distance between the rows. I recommend that the sections be made 150-200 yards long where feasible, and that the rows of vines be planted at least 8 ft. apart. If we now trellise a block of vines 200 yards broad and 150 yards long (in the direction of the trellis), we have seventy-five trellises of 150 yards long, hence a total length of 11,250 yards. On the Stellenbosch University Farm my system of trellising table grapes has cost about 6d. per yard where the rows had a length of about 500 ft. Calculated on this basis, the above mentioned block of vines will cost £281 : 5s. to trellise, or £45 : 7 : 6 per acre.

596

Under ordinary circumstances, and with other systems of trellising it might cost less or a good deal more. The benefits derived from trellising will usually more than compensate for the cost thus incurred.

   Note.—The above cost of trellising was given me by Mr. W. C Starke, Manager of the University Farm who put up this trellis and worked out the cost.[1]

---

[1] 2012 *Editor's Note.*—In 1929 Willie Starke bought Meerendal Estate in Durbanville. Professor Perold regularly visited his old friend there. With Perold's encouragement Willie Starke planted the Cape's first mother block of Shiraz in 1932 and in 1953 he was one of the first to plant Perold's creation, Pinotage. Meerendal today produce a single vineyard 'Heritage Block' Pinotage from vines planted in 1955. PFM

CHAPTER XII

SUMMER TREATMENT OF THE VINE

1. REMOVAL OF YOUNG SHOOTS

VINES usually produce some shoots we do not want. They are either weakly or in places where they are not wanted. We usually begin with this operation after the young shoots have developed sufficiently to show which of them have fruit blossoms. By this time the more advanced shoots will have reached a length of 12-18 inches.

When a vine is first put on the wire, the lower shoots are removed when still small. The wild shoots from the American stock are removed whilst still soft, when they will break off clean against the old stump. We should, in particular, remove the shoots arising out of the old wood, as these are usually unfertile and merely devitalise the vine. On the old wood (arms and trunk) we leave only such shoots as will be needed in the next winter as bearers or have fruit blossoms, and then only if the vine has insufficient fruit blossoms without these.

If a vine has too many fruit blossoms, as often happens with Muscat of Alexandria, and sometimes with young vines of other varieties, young shoots with fruit blossoms have to be removed to limit the crop. We should merely see that the vine retains enough shoots to be well supplied with leaves for maintaining a powerful carbon-assimilation. In removing young shoots, the vine's vigour should be kept in mind. With weak vines we must be particularly careful not to remove too many shoots. In any case it is better for the vine to have a limited number of strong shoots rather than a large number of weak ones.

It is sometimes said that the removal of young shoots weakens the vines, but this will not happen unless too many are removed. By removing the superfluous shoots when young, we need not do so in winter when this entails much work and takes more time. The shoots that are kept can *then* become more vigorous and will produce better grapes. The vines will then also be better ventilated, and the sunlight can reach the leaves of the remaining shoots more easily, *which will* mean a healthier vine and better bearers for the following season. The shoots on the bearers are usually not removed; if any of these are removed at all, it is done more often on young vines than on adult vines.

2. THE PINCHING AND TOPPING OF THE SHOOTS

By *pinching* is meant the removal of the tips of the young shoots by means of the thumb-nail, the shoot being between thumb and index finger, or by using a knife. This is done for the first time when the young shoots are 15-18 in. long, and is repeated once or twice until the blossoms open. *Topping* takes place later, and involves the removal of larger portions of the shoots than pinching does.

When pinching we remove merely young leaves which grow at the expense of the older part of the shoot. By removing them, we therefore help the vine for the moment, and favour especially the development of the eyes near the base of the shoot, of which some will later be on the bearer. By judicious pinching we can thus increase the next season's crop. The shoots are generally pinched at two to three leaves above the highest fruit blossom. Where this blossom is inserted very high up the shoot and strong winds are to be feared, the shoot can be pinched immediately above the highest fruit blossom—sometimes desirable with Raisin blanc. By thus checking the shoot's growth in length and giving the remaining portion of the shoot an opportunity of becoming stronger, we lessen the danger of shoots being blown off by strong winds.

At its top the pinched shoot develops a new shoot (an axillary shoot or sucker) which continues that shoot's growth in length. When pinching trellised vines for the second time, we should, as far as possible, pinch only the shoots on the long bearers and not those that are to serve as future bearers, when these will develop all the more vigorously. The pinched shoots will develop a number of axillary shoots or *suckers* which will provide shade and later protect the grapes against sunburn. In forming a cordon we can make the weaker shoots develop more vigour by pinching or topping the more vigorous ones, and thus get them to become about equally vigorous.

By pinching the young shoots a couple of times, we help them to grow more erect, thus giving young bush vines a better shape, and enabling the shoots to become stronger so as not to bend so easily under the weight of ripening grapes and directly expose the grapes to the sun's rays and cause sunburn. Vines that lack vigour should not be pinched or topped.

It is well known that pinching can help against coulure, but this will be the case only *if the vines are topped at the moment the blossoms begin to open.* If this is done about eight days earlier, so that the shoots have begun to develop axillary shoots, and these are vigorously developing when flowering begins, too much nourishment will be absorbed by these and coulure will be worse than if no pinching or topping has taken place at all. By topping when flowering begins, the parts of the shoots thus removed no longer absorb any nourishment, and the flowering blossoms are all the better supplied with it. This explains why such judicious topping can further the better setting of the berries, which largely depends upon an ample food supply just at this stage. With very vigorously growing vines this beneficial effect of topping will be most marked.

We thus see that pinching (or topping) up to the time flowering commences, is nearly always advantageous. Whether further topping is advisable will depend upon the local circumstances. Where vineyards are irrigated, and especially where brack is to be feared, topping is now usually stopped. The shoots will then gradually cover the ground and help to keep it cool. Just shortly before the vintage the shoots between the rows of vines are topped with sickles to be able easily to pass between them when picking the grapes.

Where vines are grown under the system of dryland farming, we should, after the berries have set, top at least so much that cultivation can take place without breaking off any shoots. Usually wine varieties are topped only once after the berries have set, and then not severely. With vines lacking in vigour this practice must be omitted. Vines on moist soil will continue growing later than elsewhere, and should consequently be topped more.

It is a bad practice to top vines shortly before they turn colour or the berries get soft *(véraison)*, as the vine should then not waste any energy on suckers. Hence it is a good practice to pinch the tips of growing shoots at this stage, but care should be taken not to remove any well-developed leaves. If this is correctly done it will assist in making the grapes ripen earlier and make them sweeter when ripe.

Table grape varieties can be topped more often than wine varieties, as it may help to increase the size of the berries, which with them is a great advantage. The amount of topping applied depends in any case upon the vine's vigour, and so also upon soil and climate. Where the water supply in the soil is limited, the vines will have to be topped accordingly. If they are allowed too large a development, too much water will be evaporated, and the vines might suffer from drought, when their wood and grapes are likely to ripen badly, and the grapes are more susceptible to sunburn.

By judicious topping we can thus do a lot of good, but excessive and too late topping will damage both vine and grapes. In this connection it should always be remembered that it is the leaves that form the vine's organic food and the sugar we find in the grapes. If we allow the vine to keep too few leaves, it cannot ripen its grapes properly and produce sweet grapes. Local experience should teach us when to cease topping.

3. THE TYING UP OF SHOOTS

Apart from the colder wine countries where the shoots are tied to stakes, this operation is performed almost exclusively with trellised vines. In hot wine countries the shoots of bush vines are sometimes tied together when the grapes begin to ripen. This results in more acid and less sugar being present in the fully ripe grapes. For making light wines it might be found advantageous, although it is not often done.

With trellised wine grapes this can be avoided by having two wires, 6-8 in. apart, about 11-12 in. above the bottom wire, as has been previously described for Cabernet Sauvignon, or by interlacing between two or more additional wires all in one vertical plane, as is done by M. E. Maroger. With trellised table grapes it is best to tie the shoots to a wire. As tying material we can use binder twine. It is no good using raffia or similar material that is easily broken where strong winds prevail in summer. Growing shoots should be loosely tied to allow for the growth in thickness of the shoot during the season. Some growers of table grapes tie the fruit-bearing shoots first to a second wire (in a vertical plane with the bottom wire), and later loosen them, bend them down and tie them to the bottom wire in order to prevent the bloom on the berries from being rubbed off. This causes extra labour and expense. It is really this practice that led me to work out my previously described method of tying the growing shoots to higher wires, which are so placed that the shoot, when tied, will form an angle of about 45° with the vertical; which will enable the bunch to hang free and thus prevent the bloom from being damaged. In my system the shoots are pinched at one or two leaves beyond the wire. By tying the shoots alternately to the right and left wire, as was previously suggested, the trellis is sufficiently dense to let the grapes hang in the shade and protect them against sunburn. Such grapes can be sulphured late in the season with hardly any danger of being burnt.

If a growing shoot is to be ring-barked, it must first be tied, as it is otherwise sure to be broken off by wind or in cultivating the vineyard. When tying the shoots, we have an opportunity of keeping them apart and thus preventing two bunches from touching each other.

If a growing shoot has been tied too tight and is subsequently strangled, this obstruction to the free flow of descending sap has very much the same effect as ring-barking, and grapes on such shoots may ripen a couple of weeks earlier than normal.

Where strong winds blow in summer, we simply have to tie the shoots of table varieties in order to prevent the shoots from being blown off and the grapes from being damaged.

4. RING-BARKING OR CINCTURATING OR GIRDLING

By this is meant the removal of a ring of bark. It is done to combat coulure and to make the berries and bunches grow to a greater size and to ripen earlier than they would otherwise do. This is of special value in connection with show exhibits. It can therefore also be applied in growing high-class table grapes. In growing the Zante currant it is generally applied in Australia and elsewhere.

FIG. 106. The result of ring-barking (on Bicane at the Stellenbosch University Farm, Feb. 1925). Original.

In Chapters VII. and VIII. the causes of coulure have been discussed. By removing a ring of bark below the lowest bunch, the descending stream of elaborated sap is interrupted, which now accumulates in the parts above the wound and thus favours the fructification. By ring-barking during flowering—it is safest to do it when the blossoms commence to open—we can prevent coulure. In 1924 I showed on a bunch of Madeleine Angevine that it can still be successfully done when flowering is over and the berries are setting.

The experiments I made on the Stellenbosch University Farm in 1924 with Madeleine Angevine, Bicane, Olivette blanche, Ohanez (White Almeria) and Gros noir des Beni-Abès have shown that ring-barking during flowering has *radically prevented coulure. There is, however, no proof that it has caused a single additional large berry with seeds to be formed.* Next to a few large berries with seeds the bunches were packed with small, round, seedless berries. *Ring-barking therefore promotes parthenocarpy to a wonderful extent, but has no influence upon the real fertilisation,* at least not with the varieties experimented upon.

By ring-barking after the berries have set, we can only increase the size of the berries and hasten their date of ripening. As has already been stated, it might in this way become valuable to the grower of high-class table grapes, and this will probably be its only value to him.

Prof. L. O. Bonnet[1] says in this connection: "In ringing for hastening maturity the healing over of girdles is not generally expected, and it is advisable to ring only the parts of vines that will be removed at the following pruning. Fruit cane spurs, portions of spurs, and shoots can be girdled by removing a ring of bark one-eighth of an inch wide.

"To obtain an earlier ripening of a variety, the vine must be girdled when the berries are about half grown, that is to say, six to ten weeks before the normal time of ripening. The shorter time applies to early grapes and the longer to late grapes. For mid-season grapes the time at which they should be girdled precedes the normal time of ripening by about eight weeks." It should be noticed that to hasten ripening, girdling has to take place long (fully a month and more) after flowering. If done during flowering, the girdles usually heal fairly soon, when there will be no appreciable hastening of the ripening as a result of such girdling.

By picking girdled bunches and ungirdled bunches when the former were considered ripe, and analysing their musts, Bonnet obtained the following results, which clearly show the effect of earlier ripening induced by girdling.

---

[1] L. O. Bonnet (University of California), "New Facts about Girdling," *California Fruit Grower,* vol. vi. No. 4, p. 2, April 1925, San Francisco.

Varieties.	Girdled.				Ungirdled.			
	Weight of 1 Berry per bunch.	Sugar per cent. by Balling.	Acidity per cent. as Tartaric.	Ratio S/A.	Weight of 1 Berry per bunch.	Sugar per cent. by Balling.	Acidity per cent. as Tartaric.	Ratio S/A.
Dizmar	3·68	22·3	·68	32·3	3·54	17·8	·70	25·4
Rish Baba . . .	3·94	14·3	·37	37·6	3·99	13·3	·47	27·7
Gros Sapat . . .	3·15	20·8	·84	24·7	2·57	18·3	—	—
Angulata . . .	3·44	16·3	·70	23·2	3·26	14·3	·79	18·1
Muscat Blanc . .	3·52	18·0	1·12	16·0	2·46	13·9	1·21	11·4
Danugue . . .	3·00	16·8	·78	21·2	1·90	13·7	1·16	11·8
Cornichon . . .	4·56	15·8	·93	16·8	2·68	10·3	1·77	5·8

The higher sugar content and lower acidity in the girdled as compared with the ungirdled bunches, is a clear proof that the former were riper than the latter when picked. Bonnet says, "For the Dizmar, an early grape, the advance of ripening was about two weeks, while for the Cornichon it reached about one month. This variety was picked just a little before full maturity, when girdled bunches were fully coloured and the ungirdled bunches of the same vine almost green. The influence of girdling on the size of the berry is also well marked."

Owing to the disturbance girdling causes in the downward movement of the elaborated sap to the roots and other parts of the vine below the girdles, it is dangerous to apply it to all the shoots at the same time or to the whole vine, as it will weaken the vine. We can do this only to vines growing too vigorously. Usually only the rods or long bearers are girdled, and only half of the shoots when short pruning has been adopted. A variety like the Zante Currant or Black Corinth, that has to be girdled and produces many fruit blossoms with short pruning, should be long pruned to facilitate girdling. We girdle the long bearer below the place where the lowest fruit-bearing shoot emerges from it. Sometimes the stem of the vine is girdled, but this I consider dangerous to the life of the vine, and therefore do not recommend it. According to Bonnet, girdling on fruit canes is cheaper than it is on the trunk, which constitutes another argument in its favour.

On fruit canes the girdle should be about one-eighth of an inch wide, and requires no covering. It is done during flowering Bonnet says that experiments have shown that the results are not so satisfactory when girdling is carried out a few days before or a few days after blossoming.

The tool most frequently used in girdling is a double blade incisor, of which several satisfactory types are made. By putting it around the cane to be girdled, pressing just sufficiently, and turning it to the right and left, we can rapidly make two circular cuts through the bark down to the cambium, when the strip of bark (about one-eighth of an inch wide) can easily be removed. In about eight weeks' time the wound will again be healed, if made during blossoming. It is very desirable that the girdle should be healed well before the growth of the vine has stopped for that season, in order to feed the roots, etc., well and provide them with stored food for beginning the next period of growth. See that your tools are sharp to make clean cuts that will heal well and quickly, and do not cut into the wood.

CHAPTER VII

THE PRODUCTION AND SALE OF TABLE GRAPES FOR EXPORT

THE production of table grapes for local markets is a relatively simple matter. For this purpose we can use also such varieties as cannot stand long transport well or may have too small berries, etc. Thus exceptionally early varieties like Madeleine Angevine, though too small and soft berried for export *(by export I mean transportation to distant markets!),* can sometimes be profitably grown for local markets. Black Prince and Hermitage (Cinsaut) as table grapes are mainly sold in the home markets, but they can also be exported, as is regularly done from the Cape to London where they fetch remunerative prices, mainly because they come early on the market. It should be obvious that table grape varieties grown for export, should be very carefully selected, grown, packed and transported in order to reach the distant markets in good condition and find a ready sale at remunerative prices. As transportation (including packing and packing materials) in this case swallows up a great deal of the proceeds of the sale of such grapes on the distant markets, it is obvious that the greatest care should be taken to grow the most perfect grapes possible and to choose varieties that can realise good prices on these markets. This business should be undertaken seriously or severely left alone to avoid losses.

A. THE PRODUCTION OF TABLE GRAPES FOR EXPORT

Under suitable conditions and when properly carried out, this is one of the best paying and most intensive forms of agricultural production. It is eminently suitable to the small-holder who wants to work his small plot, 2-5 acres, with the assistance of his wife and children. By concentrating all attention on his work and doing everything at the right time and in the best manner, the best results can be obtained, the more so as the factor of self-interest plays a very important role in this particular case.

I know of a gentleman at the Cape who concentrates all his energy on the production of high-class table grapes for export and does the bulk of the most important work with the aid of his family. He grows probably our most perfect table grapes and usually gets the highest prices in London. His London sales amounted to a gross return of £375-£750 per acre. These are, however, to be regarded as exceptional figures.

Under suitable soil and climatic conditions, and provided the right varieties are grown in the right way, we can count upon an *average annual gross income of £125 per acre,* although considerably higher returns will frequently be obtained. The first thing to remember is that *quality* must be aimed at throughout. The profit made will primarily depend upon the quality of the grapes exported and to a much lesser extent upon the number of boxes shipped. We shall first study the conditions to be satisfied by export grapes, and then the different factors and operations that are necessary to be able to fulfil these conditions.

I. THE CONDITIONS TO BE SATISFIED BY TABLE GRAPES FOR EXPORT

1. *They must reach the market in a perfectly sound condition.* For this the grapes must be perfectly sound when packed, the berries must have firm flesh, a tough and fairly thick skin, and adhere well to the pedicels; packing and transportation must be suitable for the purpose.

2. *They must be pretty to look at,* hence the bunches should be even in size and shape, loose, and weigh 1-2 lbs. a piece; the berries should be of *even size* and *large,* have a good colour and as much bloom as possible, and look sound and fresh.

3. *They should keep for a considerable time after the boxes have been opened.* This accounts for the good prices realised by the Almeria grape (Ohanez), which otherwise cannot in every respect be regarded as a first-class table grape.

4. *They should preferably be varieties that are well known and in demand.* This naturally refers to the market

where they are being sold. An unknown variety needs several years to become well known and establish proof of its excellence.

5. *They should have a pleasant taste.* This condition, even with luxury varieties, gains in importance as more grapes of each variety are being sold on the market, so that they are no longer used merely for decorating the tables of the rich, but are also bought for consumption by the middle classes.

Although all these conditions should be satisfied in order to obtain the highest prices, it is the appearance of the grapes that mainly decides the price. With this is coupled the good name of the grower and his trade mark, which rests on the quality of his grapes and upon his own honesty and ability, so that buyers know beforehand what they will get when buying his grapes.

6. *From the grower's point of view* the varieties grown should regularly produce good crops which ripen well and are not too much exposed to diseases and other damage.

7. *They must be passed for export by the Government Fruit Inspector.* The conditions he requires to be complied with, will be discussed under the selling of table grapes for export.

II. THE FACTORS AND OPERATIONS THAT PLAY AN IMPORTANT PART IN THE PRODUCTION OP TABLE GRAPES FOR EXPORT

1. **Climate**

This is naturally the first consideration and one of the most important limiting factors in the production of table grapes in the open air, to which I shall here limit myself. The ideal climate for this purpose is one with a good winter rainfall, little rain in early summer, and practically no rain from the time the grapes turn colour until they have been picked. We have already seen that rain in January (at the Cape) causes the berries of one of our most valuable grapes for export, *i.e.* Gros Colman, to crack, and renders them worthless.

In order to be able to grow first-class table grapes for export, the annual rainfall should be about 25-30 in. If the soil is deep and most of the rain can be absorbed and retained by the soil, a rainfall of 20 in. per annum will suffice. With a lower rainfall irrigation will usually be necessary.

If it rains frequently in early summer, and if this continues until blossoming is over, Anthracnose can cause very serious damage to certain varieties of table grapes.

Varieties like Rosaki, Gros noir des Beni-Abès, Molinera Gorda, Muscat of Alexandria are then badly damaged. Barlinka is hardly affected and Gros Colman is practically immune to this disease.

Varieties subject to coulure, *e.g.* Muscat of Alexandria, Ohanez (Almeria grape), Olivette blanche, Bicane, Gros noir des Beni-Abès, then set their berries very badly and produce few bunches suitable for export.

Rains at the time when the grapes are being picked or are nearly fit to be picked, can cause the grapes to rot on the vines and make them more or less watery, thus lowering their keeping qualities very much indeed and rendering them less fit or totally unfit for long transportation.

Strong winds can cause very great damage to table grapes grown for export. They can occur during the vine's whole period of growth from budding until the grapes have been picked, but the most critical stage is from the time the grapes turn colour until they are picked. At this stage the berries can easily be bruised, *e.g.* by grains of sand blown against them or by leaves rubbing against them, damaging their bloom as well as the skin, and thus rendering them unfit for export.

In parts where such winds are to be feared, windbreaks should be provided; my system of trellising and training table grapes should be adopted, and fairly hardy varieties like Molinera gorda, Gros Colman, Barlinka, and Ohanez should be grown, provided that in other respects they suit the local conditions. It is best not to grow export varieties where strong winds are prevalent.

Extremes of temperature can also cause great damage or make the production of table grapes in certain localities an altogether unprofitable business. Sometimes it applies only to certain varieties. Thus Flame - coloured Tokay is too susceptible to sunburn in Paarl (Cape) to be grown there with a profit, while in Constantia it is hardly ever burnt and produces splendid grapes for export.   As many of the best export varieties ripen somewhat late, their cultivation in the open air requires a fairly long and warm summer and autumn.

We have previously seen that a warm and *dry* summer and autumn promote productiveness in the vine. This is probably the main reason why Ohanez is so productive in the Hex River Valley and near Prince Alfred's Hamlet in the Warm Bokkeveld.

2. **Soil**

The soil should be deep, cool, and naturally well drained. Its moisture content during summer and autumn should be sufficient to allow the berries to develop to their maximum size. Soils of decomposed granite and shale, and broken to sandy soils are suitable for this purpose, but heavy loams and clay soils are unsuitable. The soil should not be too moist, as this would produce grapes that are too soft. If it is a fertile soil, this is an advantage, but the. most important points are its physical state and depth. It should not contain too much nitrogen as this results in rank growth and bad setting of the berries, extra susceptibility to diseases, rather soft grapes for long transport, and a lowered productiveness. For this reason the low-lying moist, dark soils rich in humus are unsuitable for the production of good export grapes.

3. **Site**

This includes both the locality and the slope or special situation of the ground. The site should be favourable from the point of view of climate and transportation to the export harbour. It is naturally an advantage to be close to the railway station and to the harbour. The farther away you are from the railway station, the better the wagon road should be and the hardier the variety of grape grown. Given good roads, most varieties could be grown up to eight miles from the nearest station, provided a satisfactory railway service is in existence. It will also depend on the state of the railway service how far the railway station can be from the port.

As regards the special site of the soil, I wish to emphasise that soil with a fair slope is usually to be preferred to level land. A slope facing the midday sun is to be preferred. Where the land is level but lies on a flat topped hill or terrace with good natural drainage, it can also produce splendid export grapes, as can be seen in different parts of Constantia. Low-lying-soils are unsuitable for this purpose.

4. **Varieties of Grapes**

At the Cape we now grow a couple of dozen varieties that are suitable for export, and from which every one can make a selection to suit his special conditions. Varieties that do well in certain localities may not answer in others. In selecting the varieties to be grown, we should *inter alia* reckon with the question of transportation and labour. Where we have long road transport or bad roads to deal with, only very hardy varieties should be grown, *e.g.* Ohanez, Molinera gorda, Rosada, Tribodo bianco and Barlinka.

Where labour is scarce we should grow varieties that do not require much pre-thinning, *e.g.* Rosaki. I wish to point out that one and the same variety, *e.g.* Muscat of Alexandria, may require very little pre-thinning in some parts and a great deal in other parts. This naturally depends to a very great extent upon the climatic conditions under which blossoming and the setting of the berries take place. It is therefore impossible to classify the export varieties with regard to the amount of thinning they will require, unless it be for a certain locality.

It is in any case wise to select the varieties in such a way that they do not all ripen at the same time and give a fairly long period for harvesting the crop. Where possible one should therefore grow fairly early, mid-season, and more or less late varieties. It is also advisable, as far as possible, to grow white, red, and black varieties.

The following varieties are some of the best (for Cape conditions at least) from which a selection can be made :

White Varieties: *(a) Fairly early and mid-season:* Rosaki (Waltham Cross), Muscat of Alexandria, Formosa, Bicane(?). *(b) Late mid-season and late:* Olivette Barthelet, Tribodo bianco, Raisin blanc, Ohanez, Olivette blanche (?).

Red Varieties: *(a) Fairly early and mid-season:* Molinera gorda, Flame-coloured Tokay, Red Hanepoot *(i.e.* Red Muscat of Alexandria), *(b) Late mid-season and late:* Rosada.

Black Varieties: *(a) Fairly early and mid-season:* Hermitage, Muscat Madresfield Court, Gros Colman, Gros Maroc, Gros noir des Beni-Abès (?). (6) *Late mid-season and late:* Henab Turki, Prune de Cazouls, Barlinka, Cornichon Violet, Lady Downe's Seedling, Barbarossa (Danugue), Bonnet de retord.

5. Stocks

The choice of the stock will depend upon soil, climate, and grape variety to be grafted upon it. It is always desirable to use more than one stock, as this is safer and will cause the grapes of the same variety to ripen at somewhat different times. Here we want stocks that will promote vigorous growth and favour the production of beautiful grapes, which should, however, ripen well. If the berries do not set very well, it will be an advantage in producing export grapes so long as enough berries set to fill the bunch well. The following stocks are useful for this purpose at the Cape: Jacquez, Riparia gloire de Montpellier, 101-14, 1202, Rup. du Lot, 420A and 333. In the chapters on Ampelography the reader will find the necessary information to enable him to choose a suitable stock. Where export grapes are not grown under irrigation, the stocks mentioned, with the exception of Riparia gloire and Rupestris du Lot in certain cases, can be successfully used on all the good soils and for practically all the export varieties.

6. Preparation of the Soil

As we are dealing here with the most intensive form of viticulture, we can spend more on the preparation of the soil than in other cases. In order to provide the vine with an ample supply of moisture once it is in full bearing, the soil should now be trenched to a depth of 30 to 36 in. Especially when dealing with soils of decomposed granite or shale, we should be prepared to incur the cost of trenching to a depth of at least 30 in. It is a good practice to give the soil a heavy dressing of phosphate, say 1000-2000 lbs. basic slag or finely-ground tricalcium phosphate per acre just before it is trenched, when the phosphate will get well mixed with the soil. Steep slopes are made into terraces. The soil is also levelled and drained where necessary.

7. **Planting and Trellising of Vines**

We have already seen that the rows of vines should run east to west. Along steep slopes this will sometimes be impossible, as this might make cultivation very difficult. In growing table grapes it is best to have the rows a good distance apart and to trellise the vines. On the whole, 8 ft. between the rows will be good. The distance the vines are planted apart in the rows will mainly depend upon the variety grown. A very vigorous grower like Ohanez or Molinera gorda should be planted 6 to 10 ft. apart in the rows, while more moderate growers like Muscat of Alexandria and Rosaki should be planted 4 to 5 ft. apart in the rows. The system of pruning also has some bearing on this point. While the distance apart in the row matters little with the cordon system, it should be about 3' 6" to 4' when the double Guyot system of pruning is adopted.

As regards trellising, my own system seems to me the best for producing high-class table grapes. For Ohanez (Almeria grape) the Almeria trellis and an ordinary trellis like that of Maroger are to be recommended, as it usually requires a fair number of very long bearers to ensure good crops, and as its grapes are very hardy and form a separate class.

8. **Manuring**

As it is our object to produce grapes with very large berries, we should give the vines a liberal dressing of nitrogen food. This we can bring about by giving the vines a heavy dressing of farmyard manure, say 10 tons per acre per annum. Where sufficient of this manure is not available, green manuring should be resorted to. I wish to remind the reader that a good supply of organic material in the soil (derived from farmyard manure and green manuring), together with a sufficient supply of moisture and air (preserved and introduced by diligent cultivation), will favour a strong development of soil bacteria that are able to bind atmospheric nitrogen and thus enrich the nitrogen content of the soil very considerably.

It is at the same time essential that the grapes should remain healthy, ripen well, and be sufficiently hardy to stand long transportation well.

Hence dressings of potash and phosphate should not be neglected in making up manurial dressings. We can, for this purpose, apply the formulae given in Chapter IX., taking the highest figures there given and doubling the dressing of farmyard manure. For the rest our own experience should teach us how far we should depart from those formulae, especially as regards the nitrogen dressing.

9. **Soil Cultivation**

This is the same as for ordinary vines, except that every operation should be performed with great care and at the right time. If the soil contains enough clay to prevent the fertilisers being washed out by the winter rains, apply the manure early in winter, and bring it fairly deep into the soil, say 8 in., by ploughing or digging. Here I would prefer digging-in, as we can in this way spread the manure evenly over the soil and get it deep into the soil without much harm to the bigger roots.

Where irrigation is necessary, it should be given during winter as far as possible. It is very desirable to discontinue irrigation as soon as possible, and during the last month before the grapes are cut for export, the vines should not be irrigated.

After the winter rains are over (or when the soil has dried off sufficiently after irrigation) *the soil should constantly be kept absolutely free from weeds and loose in the top two inches.* In this the grower should be very precise. Hence he should have the soil carefully cultivated after every summer rain as soon as the soil is fit for cultivation. In the 7-8 ft. row he must let the Planet Jr. cultivator with 5-7 points pass at least twice. Any weeds in the rows of vines should from time to time be destroyed. By careful and diligent surface cultivation we can best preserve moisture in the soil and grow large berries.

Even if no rain falls from the time the berries have set until the grapes are picked, the vineyard should occasionally be cultivated to keep the soil loose and well supplied with air. Do not do this when it is very hot and dry and there is danger of the grapes burning. The dust raised by cultivation will partly settle on the berries and sooner cause them to be sunburnt.

If there is any danger of the berries setting too badly, do not cultivate the vineyard when the vines are blossoming and setting their berries. Otherwise we can safely cultivate them even at this time.

10. **Pruning**

This is regulated by the variety and local conditions. While we do not want very heavy crops, we desire to have a good crop every year, and therefore we have to prune the vines in winter in such a way that they can produce more bunches than are required, of which we can later on remove those that are not wanted. Varieties that bear well with short pruning should get a short bearer with two good eyes about every foot on the cordon. Those that require long pruning get in addition a long bearer of eight to eighteen good eyes for every short bearer, according to the nature of the variety and the vigour of the individual vine.

Varieties that set their berries well can be pruned early (May and June at the Cape) if we want the grapes to ripen fairly early, otherwise they are pruned a month or so later. Varieties that set their berries very badly and such as grow vigorously, but are not sufficiently productive, should be pruned late. Therefore Muscat of Alexandria, Gros noir des Beni-Abès, Barlinka, Ohanez, Bicane, Olivette blanche, etc., should not be pruned at the Cape until August. All these varieties, excepting the first, should be pruned long in order to produce good crops.

Topping here is very necessary. It is done for the first time when the shoots are tied to the wires, and at one to two leaves above the wire to which they are tied. When the laterals are 12-15 in. long, they also are topped or pinched. After this we top just enough to give the vines the desired development. As our object here is very big berries, we usually top table grapes somewhat more than wine grapes.

11. **Limiting the Crop**

Where we strive after the highest quality, quantity has to be strictly limited. This is one of the greatest difficulties of the young beginner. He is usually inclined to leave too many bunches on the vine.

The first limitation of the crop takes place when superfluous shoots are removed. The next limitation takes place by *removing bunches*. This is done after the berries have set and when we can clearly see which bunches, are likely to grow out best. It is done when the bunches are being pre-thinned for the first time. If the thinners are smart enough to judge for themselves which bunches should be kept, it is best first to thin those bunches that are to be kept and then to remove the surplus bunches, as a bunch might occasionally snap while being thinned. Otherwise some capable person should remove the surplus bunches before thinning takes place, and leave two extra bunches per vine to provide for those that might be broken off in thinning.

Where strong winds or sunburn is feared, we should in any case keep a couple of bunches more than are really wanted. Retain those bunches that are already the furthest advanced, that are the largest and have the best shape, and that occupy the most favourable positions relative to sun and wind. Where the rows run from east to west, we preserve preferably the bunches on the southern side (at the Cape), and on the eastern side if the rows run from north to south. It should further be remembered that a bunch on a vigorous shoot is more likely to grow out well than one on a weak shoot. Keep as far as possible only one bunch to a cane.

How many bunches should we keep per vine? This is a very important question which every grower of export grapes has to answer, and we can know beforehand that the answer will differ considerably in different cases. The vine's vigour, the soil's water supply, the number of square feet of soil per vine, the variety of grape, and the treatment the vineyard is given, will all influence the answer to the above question. Where numbers are given below, they are meant for vineyards that are well cultivated and grow vigorously so that the grapes can reach a very high quality. In order to decide upon a number we have to take into account the number of square feet of soil per vine and the nature of the grape variety. The first point needs no further elucidation, as every one will feel that with one and the same variety the number of bunches should be proportional to the number of square feet per vine.   As regards the grape variety, we know that some varieties can produce much larger berries and bunches than others. Hence we shall retain less bunches if the bunches and berries are larger.

If the vines are planted 4' x 7' 6" rectangular, every vine has 30 square feet of soil. With varieties like Muscat of Alexandria, Gros Colman and Molinera gorda, that readily produce bunches weighing 1-1½lbs., we can reckon *one bunch per 5 to 6 square feet of soil.* This is what we count upon harvesting, hence we shall leave seven to eight bunches per vine when pre-thinning for the first time. Varieties such as Muscat Madresfield Court, Lady Downe's Seedling, and Barlinka, that usually produce bunches of ½-1 lb., will in this case get *one bunch per 4 sq. ft. of soil* or seven to eight bunches to be harvested per vine, so that we shall retain nine to ten bunches per vine at the first pre-thinning. The two reserve bunches can be removed as soon as it can be done with moderate safety.

It is impossible to lay down hard and fast rules in this connection. The numbers above quoted are merely to serve as a guide. The size of the bunches retained will greatly influence their number. The same variety, like Muscat of Alexandria, Barlinka, Rosaki, produces both big and smallish bunches. The experienced grower will almost know, when pre-thinning, to what size the bunches are likely to grow, and he will usually give preference to the large bunches. We should at all events try to produce one export box containing 10 lbs. grapes net per 60 sq. ft. of soil, *i.e.* one such box from two vines planted as above, which means 726 boxes per acre.

12. Trimming and Thinning the Bunches

Trimming is done when one is pre-thinning. Bunches that are too large, as sometimes with Flame-coloured Tokay and Barbarossa (Danugue), are now reduced in size by shortening the shoulders and cutting off the bottom part at a suitable height, so as to give the bunch a good shape. Where bunches have a side or secondary bunch arising at the fertile node, as is often seen with Gros Colman, it is now removed if the main bunch is large enough. By judicious trimming we can produce bunches of fairly even size and shape, which will subsequently facilitate the grading and packing of the grapes.

Be careful not to reduce the bunches unnecessarily in size. *Do not cut off any part that need not be removed.*

Thinning is perhaps the most important and laborious operation in the production of high-class table grapes for export. I consider it as the limiting factor in the growing of export grapes where the natural conditions are favourable. The amount of thinning depends upon the variety and how well the berries have set. Our object is to produce a fairly large loose bunch in which the berries will have reached their maximum size, be of even size, and just nicely fill the bunch without leaving any holes or bare parts. It will therefore be obvious that the bigger the berries the fewer they will be for the same-sized bunches. As the same variety, with the same treatment, will not form equally large berries in different places, it is impossible to lay down a hard and fast rule about the amount of thinning that will be required.

The time of thinning is important. Pre-thinning can be commenced when the berries have reached about the size of buck-shot or are about one-eighth of an inch in diameter, and should be completed before the berries have fully developed their bloom. In the Western Province (Cape) pre-thinning is done from about the middle of November to about the middle of January, and on the same farm it lasts about five to six weeks. We can count on having six weeks in which to do this work. The trouble is to get the work done within this time, and it is particularly in this respect that thinning becomes such an important limiting factor. The great difficulty is to get enough experienced thinners to finish the work in time. We can ease the situation somewhat by growing varieties that come on one after the other, and thus extend the time limit, and by growing varieties like Rosaki and Muscat of Alexandria (in certain places, *e.g.* Paarl) that do not require much pre-thinning.

At the *first thinning* we remove all small and bad berries and as many good berries as it may still be necessary to remove. Remember that the dark green berries will grow to a larger size than the whitish or pale green ones. Therefore keep these as far as possible. Further, we remove most of the inside berries and make the bunch a kind of hollow shell. See to it that the berries are well divided over the bunch and cover it everywhere, *i.e.* leave no gaps or bare spaces.

Some thinners, when thinning for the first time, remove all the berries on the lower side of each little branch-cluster (branch of the main cluster) when held horizontally, as well as the small and bad ones. This is a good plan, and accelerates the work very much when dealing with very well-set bunches. I do not recommend the removal of whole branch-clusters, unless with varieties branching very heavily, *e.g.* Barbarossa (Danugue), and then not the main branch-clusters, but only some of their sub-branches.

The second thinning serves to remove any diseased or bad berries and some sound ones if the bunch would otherwise not be loose enough. A third thinning usually consists in removing any bad berries, and is done after the bunches have been picked for packing. The first thinning is the most important, and takes up most time. During thinning, particularly during the second and third thinning, the berries should not be touched, in order not to spoil the bloom. Use a forked wooden branch or a piece of bent wire to hold the bunch in the desired position.

It takes a good deal of experience to be able to pre-thin well. At this stage some reserve berries should be kept to fill the places of those that might become damaged in any way. If only those berries that are desired in the ripe bunch can now be kept, they have the best chance of attaining their maximum size.

The thinning is done by means of special thinning scissors with *blunt ends* or by the fingers. Those accustomed to use their fingers at the first thinning can do the work much quicker than is possible with scissors. For subsequent thinning scissors will have to be used. These should be light, should preferably have a spring, should have blunt points so as not to puncture the berries, and be sharp. Keep them sharp by cleaning and sharpening them from time to time.

The percentage of berries removed in thinning varies considerably with the different varieties and from bunch to bunch for the reasons already given. In 1914 I determined these percentages for a number of varieties at the Paarl Viticultural Experiment Station by counting the berries removed and those kept on the bunch, from which data I calculated the percentage of the berries originally present that was removed.

Here are some of the results thus obtained. It may be stated that every bunch was thinned to form a sufficiently loose bunch when ripe.

Date.	Variety.	Percentage of Berries removed.	
Dec. 1	Prune de Cazouls . .	23-57 (on 13 bunches)	Average 43
,, 1	Henab Turki . . .	40, 42, 46, 47, 50, 52, 56	,, 48
,, 9	Trifère du Japon = Henab Turki . . .	41, 41, 43, 53	,, 44
,, 1	Gros Colman . .	42, 44, 45, 46, 52	,, 46
,, 8	Gros Maroc . . .	40, 42, 48, 53, 56	,, 48
,, 8	Barlinka . . .	38, 40, 41, 44, 47	,, 42
,, 9	Bonnet de retord . .	50, 51, 52, 55, 58	,, 53
,, 9	Hermitage . . .	42, 45, 47, 51, 53	,, 48
,, 9	Formosa . . .	50, 51, 52, 55, 60	,, 54
,, 9	Tribodo bianco . .	39, 44, 48, 50, 50	,, 46
,, 8	Bailey (like Raisin blanc)	48, 50, 53, 54	,, 51
,, 8	Kirsten . . .	64, 65, 66, 69	,, 66
,, 1	Rosaki . . .	Only the few tiny berries removed	

From these figures we notice that all the bunches do not require an equal amount of thinning, and that some varieties on the whole require a good deal more thinning than others. The average percentage of berries removed varied from 0 to 66. Roughly we can say that one-third to two-thirds of the berries should be removed, though every bunch should always be treated upon its own merits. In my tests I removed too few rather than too many berries.

13. Removal of Leaves and Tendrils

Certain leaves should be removed to prevent their rubbing against the berries when strong winds blow and to give free access to air and light to the grapes. Berries of which the bloom has been rubbed off by leaves look bad and are usually further damaged, so that they rot easily and therefore lose their greatest value. We shall naturally remove only such leaves as may cause damage. This is done some time after the first thinning, and when one can easily see which leaves ought to be removed. Where no strong winds blow during the last two months the grapes are on the vines, the removal of leaves for this purpose will hardly be necessary.

If table grapes are grown according to my system, they will usually be well protected against sunburn.

621

Some growers remove enough leaves shortly after the berries have set to expose the grapes early to the sun and thus to harden them against sunburn, and have obtained good success through this method. It can be assumed that the results will not be equally successful everywhere.

Bunches that hang free and are fairly exposed to the air and diffused daylight are less subject to disease and colour up better than those that are covered with a dense foliage. In 1921 I noticed on the Stellenbosch University Farm that bunches of Barlinka that were completely shut in by leaves and shoots and hung practically in the dark, had merely a pale reddish colour when they were fully ripe. Three days after I had removed enough leaves to expose them to the sun's rays, these grapes had a fine black colour. This shows that grapes do not colour up well unless they are fully exposed to the air and light, although it be only diffused daylight, as is well known from other instances (pergolas, etc.).

If it rains for several days in succession once the grapes are ripe, the removal of leaves and direct exposure to the sun's rays will help very materially to prevent the grapes from rotting.

When thinning for the first time, the bunches should be freed of all tendrils, as it is usually impossible to do this later without damaging the bunches. The sooner these tendrils are removed the better. It is better to do this before the first thinning, when they can be removed by pinching.

14. **Control of Diseases**

Here we apply the control measures described and recommended in the chapter on diseases. The grower of table grapes has merely to carry out these measures with extra care and diligence. Varieties that are sensitive to anthracnose had better be treated with 4 per cent sulphuric acid every winter in localities where this disease is to be feared. Varieties that are somewhat susceptible to oidium or powdery mildew—this applies to most export varieties—we should sulphur very thoroughly and a couple of times more than wine grapes. If need be, we can even fight the fungoid diseases and mealy bug by spraying with hot water at 75° C. or 167° F. on a hot day.

Insects can sometimes be caught by hand *(e.g.* caterpillars, mealy bug, calandra), and against the calandra we can put "tanglefoot" or some other sticky material on the stems of the vines and the standards and droppers to prevent them getting on to the vines.

B. THE SELLING OF TABLE GRAPES FOR EXPORT

1. **Picking the Grapes**

The *time to pick* is the first point to settle. One should guard against the extremes of picking "too green" and "too ripe", although the former is a more serious fault than the latter. Grapes that are too green spoil the market, and grapes that are too ripe do not keep very well and no longer look fresh and pretty. Commence picking when the grapes are just sweet enough to be eaten with pleasure. Have only such bunches picked as are sufficiently ripe, and remember that the grapes do not get riper in the box. It is difficult to lay down numerical rules for determining ripeness. Different varieties have different sugar and acid contents when ripe. With the same variety these vary considerably from one year to the next, from place to place, with the size of the crop and the system of trellising adopted (high or low).

It is too high acidity rather than too low sugar content that makes grapes uneatable. Grapes with a low acidity can be eaten when their sugar content is still fairly low. This explains why White French (Palomino) with a fairly low sugar content can be eaten when Muscadel or Gros Colman with the same sugar content will still be uneatable. If we therefore desire to fix the ripeness or eatableness of table grapes by means of numbers, the total acidity of their sap is a safer guide than the sugar content.

In the *California Grape Grower,* vol. v. No. 5, p. 2, May 1924, the new "United States Grades for Table Grapes" are given with the following definitions about ripeness:

" *'Well matured',* in grades for table grapes, means that none of the grapes shall show a sugar test of less than 18 per cent solids in juice [evidently degrees Balling; A. I. P.], excepting Emperor, Gros Colman, Pierce Isabella, and Cornichon, which shall show a sugar test of 17 per cent.

" 'Mature', in grades for table purposes, means that none of the grapes shall show a sugar test of less than 17 per cent soluble solids in juice, excepting Emperor, Gros Colman, Pierce Isabella, and Cornichon, which shall show a sugar test of 16 per cent."

Here the acidity is clearly not taken into consideration, which I consider a pity.

We now come to the *best time of the day* for picking the grapes. Hardy, firm flesh varieties can be picked as soon as the dew is off. If no dew has fallen during the night, we can begin picking the grapes when work is commenced in the morning and continue during the rest of the day. It seems to me best not to pick between 12 noon and 3 P.M. on a hot day, as the berries will then lack moisture and be somewhat flabby. Grapes that have very big, juicy berries had better be cut early in the afternoon, when they will be less juicy and stand a better chance of arriving on the market in a sound condition.

The manner of picking the grapes and transporting them to the packing-house is also very important. Take hold of the bunch by the peduncle and *cut it near the cane,* then put it into a shallow box or on a tray. The boxes should be lined, and the tray covered with a layer of soft wood-wool to prevent the grapes from being damaged. Put the bunches side by side in a single layer. If boxes of 2' × 3' × 6" (deep) are used, they can be stacked on the vehicle transporting the grapes to the packing-house. This vehicle must be well sprung and moved slowly, as we are dealing with a valuable product that must not be bruised.

2. **Final Trimming of Bunches**

When the grapes arrive at the packing-house experienced hands take every bunch by the peduncle, examine it carefully, and remove any bad berries or such as are too small. Use only blunt-pointed scissors for this work to avoid any berries being punctured, which would certainly rot and spoil the whole bunch, if not the entire box. The final inspection and cleaning up should be done most carefully. Hence it should take place in a well-lighted place.

Grapes that have been carefully grown (and pre-thinned) under favourable conditions will at this stage require little attention. Bunches that have been dealt with are put aside on a tray with a layer of soft wood-wool. It is best if the bunches do not touch each other.

3. **The Grading of the Grapes**

Grapes exported from the Union of South Africa are graded in the following four grades, which are described as follows in Government Regulation No. 1866 of 6th November 1923, and subsequently somewhat modified in the Government Gazette of 31st October 1924.

"The grades shall be:

"(1) *Extra Selected.*—In this grade the bunches shall be wrapped singly *(i.e.* one bunch in each wrapper) and shall be properly trimmed and thinned and of uniform size, and the berries large and of uniform size and colour.

"(2) *Selected.*—In this grade the bunches shall be wrapped singly *(i.e.* one bunch in each wrapper). The berries shall be of good and uniform size and the bunches trimmed.

"(3) *Choice.*—In this grade the bunches shall be trimmed and the berries of good size, but not falling within the preceding two grades.

"(4) *Graded.*—In this grade shall fall all grapes not included in the preceding grades, and also all varieties not recommended for export.

"No unripe grapes shall be passed by the inspector. The grading of new varieties shall be left to the discretion of the inspector."

The grading of grapes for export is a very difficult task to the beginner. To get a good idea of what the different grades represent, one should go down to the Cape Town docks, where these grapes are inspected. One cannot grade properly by merely reading the printed definitions. *The grader should know what size the berries of the particular variety can reach under favourable conditions, for it is about the maximum size that is required for "Extra selected".* He should thus also know that the berries of Gros Colman, Prune de Cazouls, Rosaki, and Muscat of Alexandria should be considerably larger than those of Molinera gorda,

Barbarossa (Danugue), Hermitage, and Barlinka for the first grade.

Besides size of berry, colour and ripeness also count in grading. If a bunch would be fit for the first grade, except that its colour is poor, then it goes to the second grade. In the first grade only the very best grapes come, and they should be as near perfect as possible, since a very high standard of excellence is insisted upon by the inspectors.

In actual practice the ablest person, preferably the grower himself, should do the grading. He first picks out those bunches that are fit to go into the first grade, and puts them down in two groups according to size of bunch on a tray or table covered with a layer of wood-wool over which some calico has been fixed. When all the bunches of the first grade have been removed, he removes those of the other grades in succession. These are kept apart to avoid confusion. If grapes of different grades are packed in the same box, those of the lowest grade determine the grade of the whole box. When necessary, the inspector degrades the grower's grade to what he considers to be the correct grade.

4. **Packing the Grapes**

In the first place we require a cool, well-lighted *packinghouse.* Though any suitable room can be used for this purpose, it will usually pay the large grower to build a special packing-house for this purpose. It is a great advantage to have this house shaded by trees, as this helps to keep it cool. A thatched roof and thick walls will help to keep the building cool. If the roof is covered with galvanised iron, the floor of the loft should be covered with a layer of clay at least an inch thick, or otherwise should have ceiling boards nailed on the underside of the beams, whereby stationary air will be enclosed between the flooring boards on top and the ceiling boards below. In this way we can have a cool packing-house with a galvanised iron roof. Provide enough windows to have ample light inside.

The box wood can be stored in the loft, where the boxes can be nailed together. Have an inclined shoot by means of which these boxes can slide down into the packing-room. Arrange the staircase to the loft on the outside and save room inside.

Give the packing-room a cement floor. The furniture includes suitable packing tables, a cupboard for the rubber stamps, wrapping paper, labels, etc., a scale weighing up to say 25 lbs., and one weighing up to 500 or 1000 lbs. to check the weight of individual boxes and to get the total weight of every consignment, and a fairly large box about 4 ft. deep to keep the loosened wood-wool clean in.

The *packing material* includes box-wood, wood-wool, wrapping paper and nails. Use first class white Swedish pine or similar wood with as few knots as possible. This wood should be *thoroughly dry.* Moist wood warps, and easily causes the grapes to rot. The boxes should measure 12″ × 18″ externally, and may have any depth. Usually their depth is 5″ or 5½″ or 6″. For a complete box of 12″ × 18″ × 5½″ we require: sides, two pieces each 18″ × 5″ × 3/16″, two top and two bottom pieces each 18″ × 5½″ × 3/16″ two end pieces each 11½″ × 5½″ × 5/8″, and two cleats each 11″ × ¾″ × ½″.

The *wood-wool* should be soft and white, and free from dust. Coarse wood-wool is quite unsuitable, as it can easily break the paper wrapper around the bunches and damage the grapes. Use only the best wood-wool and have it thoroughly loosened before it is used. Store the wood-wool where it will not get damp or dirty, and keep the loosened wood-wool in a clean box. It is best not to colour the wood-wool. The cleats should be coloured green on grape boxes.

The paper for wrapping the bunches in should be a good, soft, white paper. It is generally used in sizes 12″ × 18″. The *nails* are ordinary wire nails 1¼″ long, and fairly thin (diameter about 0.08 in.). They are bought in kegs of 100 or 112 lbs. Too thick nails are uneconomic, and may cause the wood to burst.

The *packing of the grapes* is very important. There are numbers of different ways of packing, all of which I shall not describe here. To begin with, the box is lined with thoroughly loosened wood-wool. The bunches of grapes that have been graded are carefully wrapped one by one in paper (with the two lower grades two bunches can be wrapped together) and handed to the packer. Sometimes the bunch is wrapped in such a way as to keep the stalk (peduncle) and top of the bunch open; sometimes the whole bunch is covered so that only the stalk is uncovered. Let the top of the packing table slope towards the packer. Put some extra wood-wool between the bottom and the lower end of the end nearest the packer, who puts the first bunch in the bottom left-hand corner. Now he puts a thin layer of wood-wool around the first bunch and puts the second bunch firmly against it, and continues in this way until the first row of bunches (three or four) is packed parallel with the bottom end. Now pack the second row in the same manner and continue until the box is full. We usually pack four rows of bunches in such a box. The size of the individual bunches determines their number. The box should be packed so tightly that it can be turned upside down when packed without the grapes dropping out. Loosely-packed grapes can easily be bruised during transport and arrive in a wasty condition on the distant market.

Bunches that were wrapped so as to leave their tops open are usually packed in an erect position with wood-wool between them; a sheet of white paper is put over them, and then the top layer of wood-wool. When the bunches have been completely wrapped in paper, the packed box gets a top layer of wood-wool. The two top pieces are put in position and a cleat is put across them and nailed down to the box in two or three places. The same is done at the other end. The outer edges of the top pieces are put in a line with the ends of the two end pieces, thus leaving an opening of about half an inch between them.

Any projecting wood-wool is neatly clipped by means of sheep shears. On the one end of the box stencil the initials of your agent and the city the fruit is sent to; *e.g.* T. J. P. and below that L., which stands for "T. J. Poupart, London". On the other end paste your label which gives your brand, name and address, description of the contents, *e.g.* Barlinka Grapes, Selected, 10 lbs. net, and often in addition the name and address of the agent or salesman to whom the grapes are consigned. See that you supply the full weight of grapes. A box marked "10 lbs. Net" will take about 10½ lbs. fresh grapes when packed, and the box when ready for shipment should weigh 14¾ lbs. Before nailing down the tops one can put them together with cleats on the scale to see if the weight is 14¾ lbs. If less, repack to get more grapes into the box, until the weight is right, or mark down the weight to its true value.

After some practice one can easily get the right weight of grapes into the boxes. For varieties like Gros Colman that do not get very sweet until they are very ripe, get boxes 6 in. deep to take 10 lbs. grapes net. As the grapes become riper and sweeter it becomes easier to pack full weight into the boxes. Boxes that are under weight are marked down by the inspector. It is better to have too much rather than too little weight of grapes in the boxes. Whilst the boxes should be packed full, they must not be packed so full as to have any of the berries cracked when nailing down the tops.

5. **Transportation from the Farm to the Market**

Here we distinguish the following five stages: transportation from the farm to the station, railway transport to the docks, inspection and keeping in cold store till shipped, transportation in ship's cold store to harbour of disembarkation, transportation from this harbour to the market.

(a) *Transportation from Farm to Railway Station.*—Use only well-sprung vehicles and keep the roads in a good state of repair. It is a good policy to load a large number of boxes on the wagon and have them moved slowly to the station to avoid damage. If the roads are good and the station is about six to ten miles away, a motor lorry may be very useful, as it can bring several loads daily to the station.

(b) *Railway Transport to the Docks.*—The trucks into which the grapes are loaded should be under a shed which should preferably be wide enough to make unloading in the shade possible. The trucks with the export fruit should be ventilated but closed on top, and they should be taken to the docks during the night, where they should be unloaded and inspected in the morning, when they are placed in the cold store as soon as possible, where they remain until they are taken into the ship's cold store.

Boxes containing grapes for export should be taken from the wagon and packed immediately into the railway trucks. They must be carefully handled and not thrown like bricks. By careless handling all the farmer's expenditure of labour and care can be nullified very easily. Boxes containing grapes should be handled as carefully as if they contained eggs. Where the railway transport takes the better part of twenty-four hours, the trucks should be iced. A very important point is that the *trucks must be packed well* to prevent any boxes tumbling about as the train stops or starts.

(c) Inspection and Storage in Cold Chambers.—Five per cent of the boxes are selected at random for the inspection of every variety and grade of grapes sent in by every grower. To pass this Government inspection, the grapes must be packed in accordance with the Government regulations, and they must in every case be *quite sound* and *sufficiently ripe*. The strict inspection of our South African fruit has given it a good name on markets overseas and has helped to educate our growers to the high level they have now reached. Boxes that have been opened for inspection are stamped to mention this fact. The Government Fruit Inspector can refuse or reject fruit for export, lower it from one grade to another, and mark the net weight down where necessary. Every box of a consignment of fruit for export passed by the inspector gets his stamp to mention this fact.

The grapes passed for export are taken into a cold store as soon as possible and kept there until they can be taken into the ship's cold store. All fruit for export must have been pre-cooled in a cold store for at least thirty-six hours, and must be taken from there straight into the ship's cold store. It is best to have the cold stores on land as close as possible to the spot where the ships will be loading in the fruit. The ideal site is alongside the ship on the wharf. The cold stores should not be over filled so as to be able to cool down the fruit rapidly. Unless this happens some of the grapes might begin to rot before reaching the ship's cold store. Cold stores should regularly be disinfected.

(d) Transportation by Boat.—With the exception of Almeria (Ohanez) grapes which can be carried in the ventilated hold, and can be packed in other than the standard box, the boxes with grapes for export are carried in the ship's cold stores. It is of the utmost importance to disinfect these cold stores regularly, to bring down their temperature quickly to the desired degree (about 36° F. or ½° C.) once they have been filled, and to keep this temperature constant everywhere in the cold chamber during the whole of the voyage. The fruit continually forms water-vapour and carbon-dioxide by respiration, and therefore the air in the cold chambers becomes foul after a while.

To get an even temperature right throughout the cold chamber strong electric fans are used. The foul air should from time to time be replaced by fresh, dry, cold air. Hence the best system of cold storing fruit is that where the air is cooled and dried in one room and then pumped through the chambers to keep them at the desired low temperature and to remove the foul air. In the older system the cold brine pipes run through the chambers which remain closed during the voyage and retain their foul air. We have had the experience with this out of date installation, that fruit next to the brine pipes gets frozen and becomes valueless, or fruit in the middle of the chamber gets too high in temperature when that near the pipes has the right temperature. This evil can be overcome to some extent by installing electric fans in such chambers. The system with circulating cold dry air is absolutely the best for carrying soft fruits over a long distance in cold chambers.

(e) *Transportation from Ship to Market.*—Our grapes fortunately arrive in the northern hemisphere during the latter part of winter and early spring, when it is still cold and in any case certainly not warm. If the temperature in the unloading harbour is several degrees below the freezing point of water, the grape boxes must be packed into a room or truck where they will not be liable to freeze; they must still be handled very carefully until they are actually on the market, which they should reach as soon as possible. Fruit for the continent should be landed in one of the great continental harbours like Antwerp, Rotterdam, Hamburg, etc. Fruit for Great Britain now goes mainly to London, although a considerable quantity is shipped directly to the markets in Southampton, Birmingham, Manchester, Liverpool, Hull, etc.

6. The Sale of our Export Grapes Overseas

The grapes are usually consigned to some commission agent or wholesale dealer in fruit. Most of our grapes go to London and are sold chiefly in Covent Garden Market either by public auction or by private treaty. The agent can freely dispose of the fruit and hold it over for some time if he considers this advisable, or have it sold on another market. The grower gets a statement about the prices realised but buyers' names are not supplied. The grower is thus completely dependent upon the honesty of his agent, and should therefore try to get an honest and able agent or dealer to consign his grapes to. Most London agents now charge 7½ per cent commission on the prices realised. After deducting further costs the net proceeds are paid out by the South African representative to the grower.

7. The Economics of the Growing of Table Grapes for Export

The cost of growing table grapes at the Cape for export has not yet been determined with any accuracy. I shall therefore only be able to give certain data in order to calculate it approximately. Next to the cost of production comes the cost of marketing. The gross amount realised on the market less the cost of marketing gives us the net proceeds. We get our profit by deducting the cost of production from the net proceeds.

Cost of Production. — This will naturally vary a great deal in individual cases. The higher cost of production caused by more expensive vineyard soil will usually be made good by a higher quality or a more regular yield. For this purpose I shall assume the table grapes to be grown in Paarl, Stellenbosch, or Constantia on four acres of land which is very suitable for this purpose and has a suitable site, so that good and regular crops can be relied upon. We shall suppose that the soil has been trenched to a depth of at least 30 in., and that the vines are planted 4 ft. in the row and 8 ft. 0 in. between the rows. We shall also assume the vines to reach the age of thirty-three years before they need be taken out, so that we can count upon thirty average crops, and that we annually get one 10 lb. box of export grapes from three vines. We shall also assume the land to be so treated during the thirty-three years that it will have the same value at the end of this period as it had at its beginning. The value of the trellis will not amount to much at the end of this period. Any residual value it may then still have we shall offset against the cost of repairs during this period.

The *initial cost* we can calculate as follows:

Cost of trenching 4 acres	about £100
Cost of 5445 grafted vines for 4 acres	60
Cost of trellising 4 acres	190
Total initial cost . .	£350

This amount with compound interest for thirty-three years is wiped out by making thirty annual payments of £30, 5s. 8d. during the thirty years of production, if interest is calculated at 6 per cent per annum.

Supposing the bare soil originally cost £400, *i.e.* £100 per acre, then compound interest on it for thirty-three years at 6 per cent p.a. will be covered by an annual payment of £29, 11s. during the thirty years of production. We can now calculate the *cost of production* as follows:

Interest and sinking fund on initial outlay		£30	5	8	p.a
„ on purchase of ground		29	11	0	„
Cost of cultivating and ploughing the 4 acres	about	15	0	0	„
Cost of manuring the 4 acres	„	15	0	0	„
Winter pruning	„	3	0	0	„
Summer treatment including pre-thinning	„	50	0	0	„
Controlling diseases	„	8	0	0	„
Depreciation of implements, etc. .	„	5	0	0	„
Total about		£155	16	8	„

This does not include any remuneration for the owner or manager. If we make an allowance for this, the total cost of production will rise to £300. This will be for an estimated crop of about 1800 boxes of 10 lbs. each. Hence *the cost of producing one 10 lb. box of export grapes will be about 3s. 4d.* It should generally be less as the production might well be one 10 lb. box per two vines, when the cost of production will be 2s. 3d. per 10 lb. box.

The *cost of marketing* we can calculate as follows:

Picking and packing of grapes about	4d. per box
Packing material	7d. per box
Transportation charges until on the market (this includes inspection fees, levy by S.A. Fruit Growers Exchange, cold storage charges in Cape Town, insurance, etc.) about	1s.10d. per box
Total about	2s. 9d. per box

To this must be added the commission of about 7½ per cent taken by the overseas agent or dealer. If the box of export grapes is sold for 10s. on the market, the commission will be 9d., and the total calculated *cost of marketing* will be 3s. 6d. per box. In this case the net proceeds will be 6s. 6d. per box, and the *net profit* will be 6s. 6d. less 3s. 4d. or 3s. 2d. per box. The *net return per acre* will be 450 × 6s. 6d. or £146, 5s. per annum, and the *net profit per acre per annum* will be 450 × 3s. 2d. or £71, 5s. or £285 for the 4 acres. If the owner himself is the manager, this will give him an annual net income of £285 + £144, 3s. 4d. = £429, 3s. 4d., which is excellent for such a small holding.

If the grapes fetched only 6s. 8d. per box on the London market, the commission would have been 6d. per box and the total cost of marketing 3s. 3d. per box. Here the net profit will be only 6s. 8d. - (3s. 4d. + 3s. 3d.) = 1d. per box or £7,10s. for the 4 acres. *Hence 6s. 8d. per box on the London market is very near the price level where it no longer pays to export grapes grown at considerable cost.* Whoever labours under favourable conditions and performs all operations with care, will easily get an average of 10s. gross per box, and frequently considerably more.

The following figures give the prices obtained by the South African export grapes per 10 lb. box on the Covent Garden market during 1924. Although this list does not include all the prices, it suffices to give us some idea of the prices realised by the different varieties.

The table was drawn up by my student, Mr. C. J. Theron, B.Sc.Agric. It is very instructive, and I wish to draw the reader's attention to the following points:

(1)   That Black Prince figures as qualitatively the poorest of the varieties exported, but pays on account of its earliness.

(2)   That Raisin blanc is qualitatively little better than Black Prince and pays best when it comes on the market very late and is of good quality.

(3)   That Hermitage pays well at first when it is not competing with the better varieties, but later in the season hardly pays to be exported.

(4)   That White and Red Hanepoot (R. & W. Muscat of Alexandria) realise throughout good and fairly constant prices, so long as they arrive in a good condition.

Left Cape Town	Black Prince	Hermitage	White Hanepoot	Red Hanepoot	Gros Colman	Waltham Cross	Rosaki	Barbarossa	Prune de (Cazouls)	Henab Turki	Barlinka	Raisin Blanc	Flaming Tokay	Almeria (Ohanez)
Jan. 4.	8/													
,, 11.	8/	12/												
,, 18.	8/	12/												
,, 25.		6-8/					15/							
Feb. 1.		6-8/	8-12/	10-12/	15-20/	10-14/	10-15/	12/						
,, 8.		5-6/	6-12/	10-12/	15-25/	10-12/	8-12/	12/	12/				16-20/	
,, 15.		5-6/	6-12/	7-12/	12-25/	10-12/	10-12/	12/	15/	..				
,, 22.		6-8/	10-16/	10-16/	15-25/	12/	12/	12-15/	:	..				
,, 29.			12-16/	12-16/	15-25/	..	11-14/	10-12/	:	10-15/		..		
Mar. 7.			8-15/	8-15/	12-30/	12-15/	..	14-15/	:		..	8/		
,, 14.			8-10/	8-10/	10-18/	8-10/	14-15/	12-15/	:	14/	..	8-10/		
,, 21.			4-15/	4-15/	:	8-10/	:	:	:		25/	5-10/		
,, 28.			10-15/	10-15/	:	10-15/	:	:		:	16/	10-12/	12-15/	
Apr. 4.			5-10/	5-10/	:	15/	:			:	..	5-12/	..	15/
,, 11.			8-10/	10-14/	:	8-14/	:				..	..	..	12-14/
,, 18.										10-12/		8-12/	14/	17/6-20/

The low prices were due to the wasty condition in which the respective consignments arrived. The red variety sometimes realises somewhat higher prices than the white variety.

(5) That Gros Colman realises the highest average price, and is closely followed by Barlinka as second best.

(6) That the remaining varieties possess more or less equal value as export varieties, and, according to the figures quoted, realised fully 13s. 3d. per box, which can be considered as very satisfactory.

In conclusion I wish to add the following statistics. In 1923 and 1924 were *exported* respectively 225,214 and 234,909 *boxes of grapes* (according to Bull. No. 74 of July 11, 1924, of the S.A. Fruitgrowers' Exchange).

The *grades of the grapes exported in 1924* were as follows (see Bull. No. 77 of Dec. 5, 1924, of the S.A. Fruitgrowers' Exchange):

Extra selected.	Selected.	Choice.	Graded	Rejected.	Degraded
1.5%	55.9%	41.0%	1.6%	2.3%	6.4%

This should be regarded as somewhat abnormal. In a normal year the Extra selected grade will show a higher percentage and the Selected grade proportionately less.

CHAPTER XIV

PRODUCTS OF THE VINE

BY products of the vine I really mean products of its grapes. The very name "vine" (Latin, *vinum* = wine) indicates wine as the main product of this crop plant. Wine is the product of the alcoholic fermentation of grape juice that has subsequently received proper cellar treatment. Wine-making is an intricate business that is described in special text-books dealing with this subject. Strictly speaking, it does not form part of *Viticulture,* hence I shall not say more about it here.

Out of wine we can distil brandy, which contains the wine's alcohol together with certain secondary products in a concentrated form. If properly distilled from good wine, the product will be good brandy which, however, develops its good qualities only after being matured in oak hogsheads for at least three to five years. Brandy is a strong liquor that should be used with care. It can often be a valuable medicine.

By distilling wine in so-called patent stills (*i.e.* stills with rectifying columns), we can extract practically only the pure ethyl alcohol with a very slight admixture of water from it. Hence the product of distillation is known as **spirits of wine**, not brandy. While brandy is distilled at a strength of 50-70 vol. per cent alcohol (it is sold for consumption at a strength of 43-45 vol. per cent alcohol), spirits of wine is distilled at a strength of 85-96 vol. per cent alcohol. In order to obtain a so-called *silent* spirit it must be distilled at a strength of at least 94 vol. per cent or 65° O.P. Such spirits are used to fortify wine, make sweet wine, make a compounded brandy, make spirit vinegar, as a solvent for essences and certain medicines, fuel, etc.

If a light wine is exposed to air when it is not very cold, the wine usually turns sour fairly soon.

It then becomes vinegar, *i.e.* sour wine. This product will be discussed later. In the wine casks **argol** collects and constitutes a valuable bye-product, out of which *cream of tartar* or potassium bi-tartrate and *tartaric acid* are prepared.

Out of the grape-seeds the **grape - seed oil** can be extracted as was shown in a previous chapter. Fresh grapes can be turned into raisins, grape syrup, unfermented grape juice, etc. In this chapter only the three last-named products and vinegar will be discussed.

VINEGAR

The essential ingredient of vinegar is acetic acid. This acid can be obtained by fermentation or by the destructive distillation of wood. In the latter process the pure acid is prepared chemically as a colourless, strongly acid liquid with a pungent smell, which becomes "acid vinegar" when sufficiently diluted with water.

When alcohol *(i.e.* ethyl alcohol) is present in dilute solution as in light wine or beer, it can be oxidised to acetic acid by acetic bacteria in the presence of air. The change is expressed by the following chemical equation:

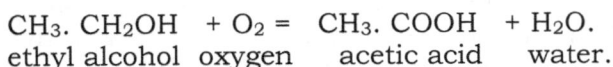

$$CH_3 . CH_2OH + O_2 = CH_3 . COOH + H_2O.$$
ethyl alcohol oxygen   acetic acid   water.

This equation shows that oxygen or air (which contains oxygen) is necessary for the acetic fermentation. As in all processes of oxidation, a certain amount of heat is liberated during this fermentation. This becomes an important consideration in the quick vinegar process where the generators become overheated if the fermentation proceeds too rapidly.

When an alcoholic liquid is exposed to the air and turns sour, we notice that a mycoderma or surface growth, called *mycoderma aceti,* has been formed at its surface. This mycoderma consists of the acetic bacteria which bring about the acetic fermentation, by means of which the ethyl alcohol is changed into acetic acid. These bacteria are most minute organisms which propagate themselves by scissiparity or splitting in two. This usually takes place most rapidly at a temperature of about 28-30° C. or 82.4-

86° F., which indeed is the most favourable temperature at which undergoing an acetic fermentation. The mycoderma can be composed of various races of acetic bacteria. In my "Unter-suchungen über Weinessigbakterien"[1] I have found that, when a dry wine turns sour naturally, the mycoderma formed generally consists of only one kind or race of acetic bacterium, and that on different wines different kinds or races of acetic bacteria frequently develop. The origin and composition of the wine have a decisive influence in this connection.

The different races differ particularly in the rate of acetification they can cause, in the maximum concentrations of alcohol and acetic acid they can stand, hence also in the highest concentration of acetic acid they can form in a wine. Some kinds of acetic bacteria, *e.g.* *Bacterium xylinum,* possess the property of further oxidising the acetic acid formed to carbon dioxide and water. Such a vinegar then naturally loses in strength again. This particular bacterium forms a tough felt-like skin on the surface of the liquid, and produces a vinegar with a characteristic flavour, which soon begins to lose in strength. It can, therefore, cause great damage in a vinegar factory. It is most desirable to operate in such a factory with bacteria that do not destroy the acetic acid once formed, that can stand a high concentration of acetic acid, and that produce vinegar with a pleasant flavour.

In the *quick vinegar process* a wooden vat is filled with wood (usually beechwood) shavings, provided with a false bottom and a perforated lid, and with several holes in the sides to admit air. Through some holes thermometers are inserted to indicate the temperature inside the vat. The vat is started with a good culture of acetic bacteria— preferably a pure culture. The alcoholic liquid is periodically poured on the top of the cask. In a modern plant this feeding is automatic. When the liquid has run through the cask a couple of times, nearly all the alcohol will have been changed into acetic acid. In this cask the alcoholic liquid is spread over a very large surface and brought into contact with much air, which allows the bacteria to multiply very fast and conduct a very rapid acetification. The air in the room in which the casks are

---

[1] Dr A I. Perold, "Untersuchungen über Weinessigbakterien," *Centralblatt für Bakt.* Abt. II. Band 24 (1909), pp. 13-55.

placed is kept at a temperature of about 27° C. or 80.6° F. As soon as the temperature inside the cask goes too high (say above 30° C. or 86° F.), some of the air-holes are closed to diminish the rate of acetification and thus allow the temperature inside the cask to sink. The alcoholic liquid used in this process should contain little extract. Hence wines to be used in this process are first fortified by means of spirits of wine or brandy to a strength of 20 vol. per cent alcohol or 35.2 per cent proof spirit and then diluted with pure water to a strength of 6-10 vol. per cent alcohol or 10.7-17.7 per cent proof spirit, *i.e.* they are diluted two or three times.

The finest vinegar is made from a *good dry wine.* Sweet-sour wines, *i.e.* wines suffering from a mannitic fermentation, cannot produce good vinegar. Wines that have turned sour because an acetic fermentation has been set up in them, but have otherwise nothing wrong with them, are quite suitable for the manufacture of vinegar. They will be quite or fairly bright, and will have a mycoderma on the surface. It is safest and best to begin with a good, sound dry wine. In all cases the wine should first be clarified by filtration or fining, unless it is bright to begin with, and should then be mixed with 10 per cent vinegar which will render it unfit for consumption and protect it against diseases.

The Orléans Process of Making Wine Vinegar

The finest flavoured and best vinegar is a wine vinegar made according to the Orléans process. In 1908 I visited such a vinegar factory in Orléans itself. The wine to be turned into vinegar lies in wooden casks each holding 55 gallons, in a room kept at 28-30° C. or 82.4-86° F. Each cask gets 50 gallons wine. One inch above the level of the wine in the cask, a round hole is made in its front end, large enough to let an egg pass through it. Just above and slightly to the right or left of the first hole, a second hole of equal size is made in the same end of the cask. These two holes are kept open to admit air into the cask, but the bunghole is closed to keep out dirt.

How to start a New Cask.—Steam the cask until the condensed water runs off colourless. Then pour 22 gallons very good, bright vinegar into it and add half a gallon of wine. After a week add two-thirds of a gallon of wine; after further week add a gallon of wine, and continue in this way until the cask contains 50 gals. liquid, when a commencement is made with the drawing off of vinegar. These casks lie in rows above each other in a framework.

Drawing off the Vinegar.—In the factory I visited in Orléans this takes place as follows: On 1st September (say) 4½ gals. vinegar are drawn off and after that 2¼ gals. wine are added to the cask; on 8th September a further 2¼ gals. wine are added to the cask; on 15th September a further 4½ gals. vinegar are drawn off and 2¼ gals. wine afterwards added to the cask, and so on. Every cask thus produces 4½ gals. vinegar in a fortnight. [By adding wine to the cask every week, the vinegar drawn off always contains some alcohol and has the pleasant aroma of acetic ether or ethyl acetate, which gives wine vinegar its excellent flavour. Out of ethyl alcohol and acetic acid are formed ethyl acetate and water by chemical reaction:

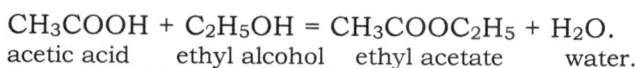

$$CH_3COOH + C_2H_5OH = CH_3COOC_2H_5 + H_2O.$$
acetic acid   ethyl alcohol   ethyl acetate   water.

When all the alcohol has been oxidised to acetic acid, the acetic bacteria continue oxidising the alcohol formed through the inter-action of ethyl acetate and water (the above reaction being reversible), until no ethyl acetate is left over. This explains why vinegar made from light wine loses its pleasant flavour if the acetic fermentation is allowed to proceed until all the alcohol has been oxidised to acetic acid.] The vinegar is syphoned out by means of a piece of rubber tubing through one of the holes in the cask's front end, and the wine is poured in through the lower hole by means of a tinned beaker with a long, bent spout. During both operations the mycoderma layer should be disturbed as little as possible. Only wooden taps should be used for vinegar. The vinegar drawn off contains 6-8 per cent acetic acid. It is put into stukvats (large casks) which are completely filled and well closed to prevent any further fermentation. After one month it is fined with isinglass. The bright vinegar is pumped into a cask which is kept full and closed until the vinegar is sold. The product is excellent.

Vinegar for Domestic Use

The first essential here is a good *vinegar cask,* which can be a hogshead or smaller cask with a wooden tap for drawing off the vinegar. Insert a loose plug of cotton-wool or place a little sand-bag on the bung-hole to keep out dirt and still allow the air free access. The cask is filled up to three-quarters of its capacity. A somewhat more complicated though still fairly simple arrangement is shown in Fig. 107.

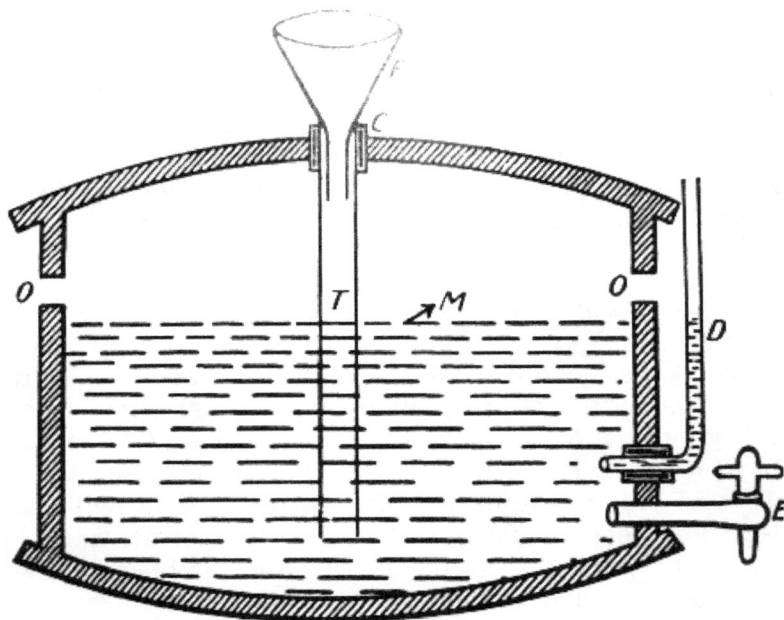

Fig. 107.—Cask for domestic production of vinegar. After Ruilopez, *La fabricatión doméstica del vinagre,* 1908. Imp. de la Sue. de M. Minuesa de los Rios, Madrid, Somewhat modified by the author.

It is taken from A. P. Ruilopez's *La fabricatión doméstica del vinagre,* Madrid, 1908. O, O are air-holes in the ends of the cask, with a diameter of about 1¼ in. and at such a height that the cask will contain 45-50 gallons when filled as in Fig. 107-the cask has a capacity of 50-55 gallons. C is a cork in the bung-hole through which a glass tube T passes till well below the surface of the liquid. F is a glass funnel through which the wine is poured into the cask. With this arrangement the mycoderma M, on the surface of the liquid in the cask is not disturbed when wine is

642

poured in. D is a bent glass tube indicating the level of the liquid in the cask, and E is a wooden tap for drawing off vinegar. The holes, O, O, and the open ends of T and D are closed with loose cotton wool plugs for keeping out dirt and undesirable microbes while allowing free access to the air.

The cask is washed till clean and then sterilised by steaming or washing with one tablespoonful of caustic soda in a gallon of water, and is then washed at least three times in succession with clean water to remove the caustic soda. Now pour half a gallon of clear, boiling vinegar into the cask and shake it thoroughly to allow the vinegar to penetrate well into the wood all over the inside of the cask. Keep it like this for a few days; then fill it with wine until just below the holes O, O, as is shown in Fig. 107. Now add some good, unboiled vinegar or a pure culture of good acetic bacteria, when the acetic fermentation will automatically proceed. If the cask is placed in a room where the temperature is kept at 28-30° C. or 82.4-86° F., one-third of its content can be drawn off after the expiration of one month, when an equal quantity of wine is immediately added slowly through the glass funnel T. This operation can be repeated every twenty days provided the room's temperature is kept at the temperature previously stated. This cask will then produce about 15 gals. or 90 bottles of vinegar in twenty days. At a lower temperature the acetification will take place more slowly. Keep the funnel absolutely clean, use it only for this purpose and wash it well with clean water every time it has been used.

The Relation between Alcohol and Acetic Acid.— According to the equation previously given for the formation of acetic acid out of alcohol, 46 grams of ethyl alcohol can produce 60 grams acetic acid. On this assumption, Ruilopez, *l.c.* p. 50, drew up the following table, to which I have added the wine's alcoholic strength expressed in per cent proof spirit according to Thorpe. [1]

---

[1] Edward Thorpe, *Alcoholometric Tables,* London, 1915.

Grams Alcohol per 100c.c. Wine.	Vol. per cent Alcohol in Wine	Per cent Proof Spirit in Wine.	Grams acetic Acid per 100c.c. Vinegar
1	1.26	2.2	1.304
2	2.51	4.6	2.608
3	3.76	6.8	3.912
4	5.00	8.9	5.217
5	6.24	11.0	6.521
6	7.48	13.2	7.825
7	8.72	15.3	9.130
8	9.95	17.7	10.434
9	11.17	19.8	11.738
10	12.40	21.8	13.043
11	13.62	24.1	14.347
12	14.84	26.1	15.651
13	16.05	28.2	16.955
14	17.26	30.3	18.260
15	18.48	32.4	19.590

Where vinegar is manufactured on a large scale, we should assume a loss of about 10 per cent of the alcohol originally present in the wine. If we wish to retain about 1 vol. per cent alcohol in the finished vinegar in order to keep its full flavour, we must deduct this and the previous amount from the wine's alcoholic content when calculating the quantity of vinegar that can be produced from it. If the wine originally had 21.8 per cent proof spirit or 12.40 vol. per cent alcohol, we deduct from this amount 10 per cent for loss in manufacture, hence 12.40-1.24 = 11.16, from which we deduct the 1 per cent to be retained in the vinegar, and thus get 10.16 vol. per cent alcohol that can yield acetic acid according to the above table. In this case the vinegar would contain 10.651 grams acetic acid per 100 c.c. As vinegar must contain at least 4 per cent (grams per 100 c.c.) acetic acid when sold in the Union of South Africa, 100 gals. wine of 21.8 per cent proof spirit will yield about 266 gals. vinegar of legal strength.

Influence of Alcohol and Acetic Acid on the Acetic Fermentation.—Wines with more than 12 vol. per cent alcohol or 21.2 per cent proof spirit do not acetify rapidly enough, and wines with 16 vol. per cent alcohol do not turn sour. In my previously quoted publication on acetic bacteria in wines, I recorded my experience that these bacteria grow in wine containing 15.5 vol. per cent or 27 per cent proof spirit, but cannot turn such wine into vinegar.

The vinegar manufacturer should, when necessary, dilute his wine or other alcoholic liquid to a strength of not more than 10-12 vol. per cent alcohol. In practice this will mean that out of the 12 - ($^{1}/_{10}$ × 12 + 1) =9.8 vol. per cent alcohol theoretically 10.27 per cent acetic acid can be produced. In my own laboratory experiments 8.97 was the highest percentage of acetic acid produced, but this does not prove that more could not have been formed, as all the alcohol had then been oxidised to acetic acid. The wine used in my experiments originally contained only 8.5 vol. per cent alcohol. I consider it to be in the best interest of the vinegar manufacturer to dilute his wine to about 10 vol. per cent alcohol or even somewhat further, as the acetic fermentation will then proceed more rapidly and go far enough.

The Clarification of Vinegar.—In order to get vinegar perfectly bright it is fined or filtered. Very turbid vinegar can often be clarified much better by fining than by filtering. For this purpose isinglass or skimmed milk or Spanish earth can be used. *Freshly skimmed milk* is used at the rate of 1-3 gallons per leaguer (127 gallons) of vinegar according to the degree of turbidity of the vinegar. *Spanish earth* is used at the rate of 1-1¼ lbs. per leaguer of vinegar. *Isinglass* is used at the rate of 3-5 grams per hectolitre or ¾-1 oz. per leaguer of vinegar. It is cut in small pieces, left in clean water over night, taken out the next morning and well pressed. Take 1¼ gals. cold water in which one ounce of tartaric acid has been dissolved for every ounce of isinglass to be used. Throw the isinglass into it and leave standing for a while. The isinglass will gradually turn into a gelatinous mass which is diluted to ten times its volume with vinegar, well stirred, and then poured into the vinegar that is to be fined and very well mixed with it by stirring or pumping—as is necessary in all cases of fining. Leave the cask standing closed for three to five weeks, when the perfectly bright vinegar is pumped into a clean, sulphured cask which must be kept full and closed.

The simplest *filter* consists of a cask filled with wood-shavings through which the vinegar has to run repeatedly until it is bright. It is no easy matter to construct a good vinegar filter, as the vinegar should not come in contact with metal parts. In his excellent work on *Fermentation Vinegar,* Paul Hassack,[1] *l.c.* pp. 259-263, describes his own vinegar filter, to which I refer the interested reader. The materials used consist of sulphite cellulose, calico, and clean sea-sand.

Colouring of Vinegar.—With the exception of spirit vinegar or distilled vinegar, which in the Union of South Africa must be sold colourless, vinegar is usually coloured light golden yellow for sale. For this purpose the vinegar is treated with some colouring material such as caramel, which can be prepared as follows: Put 10 lbs. white sugar in a copper pot of about 4 gals. capacity; melt it slowly over an open fire, stirring all the time. As the mass of sugar becomes hotter its colour becomes darker, forms bubbles, and emits acrid vapours. Continue stirring to prevent the mass burning at the bottom; as soon as it is sufficiently thick (this is decided by examining a sample from time to time), the pot is taken off the fire and ¾-1 gallon of boiling water is poured into it. Now warm slowly over the fire until the whole mass is dissolved. Pour the solution into a 25-50 gallon cask and fill it with good vinegar. Mix the content well and use it for colouring vinegar.

Vinegar Diseases.—When vinegar is exposed to air in a bottle, a layer of mycoderma aceti might easily form on its surface and render it turbid. In order to prevent this the vinegar should be pasteurised before being bottled. *Vinegar-eels* are another source of trouble which the vinegar manufacturer has to guard against. If vinegar in a little bottle shows something like moving silken threads in it when held to a good light, it contains vinegar-eels. They live on acetic bacteria and damage the mycoderma while at the same time consuming oxygen. Vinegar with 10 per cent acetic acid is not troubled by them. Once they get into a generator in a vinegar factory, that cask with its content has to be pasteurised. They are all killed when kept at a temperature of 45° C. or 113° F. for one minute. From the finished vinegar they can be removed by filtration through a good cellulose filter. By pasteurising the vinegar for one minute at 60-70° C. (or 140-158° F.) all eels, bacteria, etc., are killed.

---

[1] Paul Hassack, *Gärungsessig,* 1904.

Use of Vinegar.—Vinegar is used a great deal in everyday life. In connection with food it is used in making salad or pickled fish, for keeping venison in for a couple of days, and for preserving vegetables, *e.g.* onions, tomatoes, artichokes, chillies, cucumbers, green beans, maize, etc.

Vinegar is also medicinally used in making up embrocation, buchu-vinegar, which is excellent for treating a bruised or sprained ankle or other part of the body, and for other medicines.

It would be best if vinegar were sold at a strength of 6 to 8 per cent acetic acid—the price can be fixed accordingly—as it keeps better than the 4 per cent vinegar.

GRAPE SYRUP

This term includes any syrup obtained from the grape by concentrating its juice. The practice of concentrating must or grape juice by boiling it in an open pot over a fire is very old indeed. In this way grape syrup has been made for the last two centuries at the Cape under the name "Moskonfyt". When concentrated in this way, the syrup has a brown colour and cooked taste and flavour due to caramelisation of some of the sugar and overheating of some of the organic acids in the must.

When the must is concentrated by letting live steam into it or by means of a steam coil placed in the must, no overheating will take place, and the syrup will have a lighter colour than when boiled over an open fire. When boiled in an open pot it will still have a light brown colour and a cooked taste, which are characteristic of "Moskonfyt".

If the must is concentrated in vacuum pans under diminished pressure at a temperature of about 30-65° C. or 86-149° F., or under ordinary air-pressure, but at a very low temperature, an excellent grape syrup can be obtained. The lower the temperature at which concentration took place, the more of the fruity flavour and taste of the grape juice will be retained in the syrup.

In connection with the manufacture of grape syrup I wish to discuss: (1) the preparation of the juice; (2) preservation of the juice; (3) the necessary concentration; (4) concentrating the juice; and (5) treatment of the syrup.

1. **Preparation of the Grape Juice or Must**

(a) *Extraction of the Juice.*—Take grapes that are fully ripe, crush them in a crusher and stemmer ("fouloir-egrappoir"), and press out immediately the juice that did not run off by itself freely. Do not apply a very high pressure in order to prevent the syrup from becoming too astringent in flavour.

For a *red syrup* we use black grapes. After these have been crushed and stemmed as in wine-making, the skins and pulp are heated to extract the colour. By heating the skins their cells are killed, when the pigment is readily dissolved. Cruess[1] says on p. 402: "The colour will dissolve slowly at 105° F. to 120° F. and almost instantly at 160° F. to 170° F. The higher temperatures have proved most satisfactory from the practical standpoint, although a better flavoured product is obtained at the lower temperatures." Avoid all contact between the hot juice and iron, which it would dissolve fairly rapidly, thus injuring the flavour and colour of the juice and syrup. Cruess, *l.c.* p. 403, says in this connection: "Copper dissolves to a slight extent; it is doubtful whether injurious amounts of this will dissolve during heating of the crushed grapes. Aluminium, tin, 'Monel' metal, or silver-plated copper may be used safely. Tin-lined copper, glass-lined steel, or plain aluminium steam-jacketed kettles have all been successfully used."

Keep stirring the crushed grapes during heating to prevent local overheating. When the mass of crushed grapes has reached a temperature of 160-170° F. it should at once be transferred to the press. Cruess says that "a juice of better flavour and of as deep colour can be obtained by heating the crushed grapes to only 110° to 120° F., transferring to wooden vats, and allowing to stand overnight. This lower temperature is less injurious to the flavour, but is more troublesome to apply."

If the heated grapes cannot be pressed at once, the temperature should not be allowed to go much beyond 120° F., as too much tannin would otherwise be dissolved out of the skins and seeds and get into the juice.

---

[1] W. V. Cruess, *Commercial Production of Grape Syrup*, Bull. No. 321, May 1920. University of California Press, Berkeley, 1920.

By thus heating the crushed grapes a pretty dark red juice is obtained.

(b) *Clarifying the Juice.*—Before the juice is concentrated it should be clarified. Turbid juice gives a turbid syrup. Perfectly bright juice gives a bright syrup of fine quality. A partial clarification can be brought about by straining the freshly pressed juice through screens, and by allowing it to settle for about twelve hours in fairly shallow tanks, as used for fermentation in wine-making. Then the somewhat clarified juice can be filtered through a fairly coarse wood-pulp filter under pressure.

Instead of filtration we can clarify by means of fining. As such, Cruess, *l.c.* p. 405, recommends "the addition of about four to five gallons of a solution of casein containing three ounces of odourless commercial casein per gallon. This solution is made by dissolving a weighed amount of the casein in dilute ammonia water, boiling off the excess ammonia, and diluting with water so that each gallon contains three ounces of casein." Cruess does not state to what quantity of juice he adds the above solution of casein, but I presume it is per 100 gallons of juice. The casein solution is thoroughly mixed with juice, which is then allowed to stand for about twelve hours. The clarified juice can then easily be made bright by filtration. Professor Ventre of Montpellier recommends fining with 2-3 ounces of gelatine per leaguer (127 gallons) of juice.

(c) *De-acidification of the Juice.*—Do not reduce the acidity of the grape juice if the grape syrup is to be consumed as a beverage after diluting it with water or soda-water. If, however, it is to be consumed like golden syrup as a sort of jam, the acidity of the must should be greatly reduced before being boiled into syrup. If the syrup is to be used for sweetening wines, the de-acidification is not absolutely necessary, though desirable.

De-acidification is best done by using a pure, high-grade, finely ground marble or limestone. My own investigations (see Perold,[1] *l.c.* p. 8) have shown that 3½-4 lbs. calcium carbonate per 100 gallons must give the desired result with musts having a total acidity of 5-6‰ (here always as tartaric acid). If the limestone contains $x\%$

---

[1] Prof. Dr. A. I. Perold, "Ondersoekings omtrent Moskonfyt," *Annale v. d. Uniwersiteit van Stellenbosch,* Jaarg. I., Reeks A., Afl. 1 (April 1923).

CaCO$_3$ and the must y per mil $(= y‰)$ total acidity, we shall require $\frac{100}{x} \times 3\frac{1}{2} = \frac{70y}{x}$ lbs. ground limestone. If the must has a total acidity of 6‰ and the limestone contains 70% CaCO$_3$, we shall require $\frac{70 \times 6}{70} = 6$ lbs. of this ground limestone per 100 gals. must.

In one of my experiments I boiled 20 gals. must with 5.43‰ total acidity and 23.48° Balling in an open copper pot with 1 lb. ground limestone containing 70.7% CaCO$_3$ (according to my formula 1.07 lbs. limestone was necessary for the 20 gals. must). The syrup was concentrated to 69.7° Brix (Balling) and then had 5.52‰ total acidity. It was a fine "moskonfyt". If all the acid in the must is neutralised the syrup has not such a good flavour. If slaked lime is used for de-acidification too much might be used. If such alkaline must is concentrated, the syrup will be almost black and very bitter, hence worthless. Therefore I recommend ground limestone, where an excess can do no harm beyond leaving too little acid in the syrup. If a syrup with practically no acid is required, double the quantity of ground limestone recommended above should be used.

Where ground limestone is used it should be added to the pot in which the must is boiled to syrup, as it neutralises the acidity better during boiling than in the cold must, where it would lie at the bottom of the container and act much more slowly. Prof. J. Ventre,[1] l.c. p. 26, recommends the neutralisation of all the acid in the must by means of calcium carbonate after having previously been treated with animal charcoal to decolorise it and then fined with 10-15 gm. gelatine or 100-150 c.c. fresh blood per hectolitre must (i.e. about 1⅔-2½ ozs. gelatine or 0.1-0.15 gal. fresh blood per 100 gals. must). The neutralisation of the must will proceed during boiling. He further recommends using as many grams calcium carbonate as the must contains total acid calculated as sulphuric acid, since 98 gr. sulphuric acid are neutralised by 100 gm. calcium carbonate. He recommends this for producing a grape syrup that can compete with "Golden Syrup".

---

[1] Prof. J. Ventre, *Les Utilisations possibles de la vendange en dehors de la production proprement dite du vin,* Montpellier, 1921.

2. **Preservation of Juice**

It is best to prepare the juice and concentrate whilst fresh. Whatever cannot be concentrated at once must be preserved for some time. Cruess, *l.c.* pp. 413-414, suggests one of the following three ways of doing this, *i.e.* by *pasteurisation* at 175-185° F. (or 80-85° C.) which is costly and troublesome; by *freezing storage*—Cruess says that grape juice will ferment at 32° F. *(i.e.* 0° C), but will keep perfectly at 15-20° F. *(i.e.* -9.4° to -6.7° C.)—which is practicable where a large cold storage plant is in existence for other purposes; and finally by means of *sulphur dioxide.*

The last method is the oldest, cheapest, and easiest, but causes some difficulty in boiling such juice, since sulphurous acid attacks most metals. Therefore Cruess suggests that it be removed before the actual concentration begins. He says in this connection (p. 414): "The passing of steam through the boiling juice in the open air removes the sulphurous acid more rapidly than does a current of air; steam passed through the juice in a vacuum pan removes the sulphurous acid the most rapidly of any method tested, but the method offers mechanical difficulties when applied on a large scale. Dry steam must be used, and the juice must be kept at the boiling-point during the passage of the steam to prevent a great increase in volume by condensation." Concentration in a vacuum pan will remove about 90 per cent of the sulphur dioxide, but then the pan should be lined with glass or made of Monel metal.

The quantity of sulphur dioxide required to preserve the juice for an indefinite time varies with its sugar content and its temperature. Little sugar in the must and a low temperature require less sulphur dioxide than very sweet must at a fairly high temperature. Laborde,[1] *l.c.* p. 107, gives 500-1000 mg. SO_2 per lit. must as the dose required to prevent fermentation. In one of my own experiments, (Perold,[2] *l.c.* pp. 5-6) 1362 mg. SO_2 per lit. must of 23.17° Brix prevented fermentation for sixteen days, when fermentation set in and the must fermented dry. According to the results of my investigations, I recommended (see

---

[1] Prof. J. Laborde, *Cours d'Oenologie,* Bordeaux, 1908.
[2] Prof. Dr. A. I. Perold, *Ondersoekings omtrent Moskonfyt,* 1923.

Perold,[1] *l.c.* p. 6) the use of 1600 mg. SO_2 per lit. must or 3-15 lbs. potassium meta-bisulphite per 100 gals. must for South African conditions. I subsequently saw that Cruess, *l.c.* p. 414, recommended 1200-1500 mg. SO_2 per lit. must for hot localities and 750-1000 mg. SO_2 per lit. for cooler localities. As I have already stated, the sugar content of the must influences the quantity of sulphur dioxide to be used, as some of the sulphurous acid enters into chemical combination with the sugar, when it loses its preservative action.

We can introduce the sulphur dioxide into the must by burning pure sulphur and pumping the sulphur dioxide thus formed into the must until the necessary quantity has been taken up by the must. This point we determine by rapidly titrating the sulphured must with a $\frac{N}{50}$ iodine solution, using some starch solution as indicator. Otherwise we use the 6 per cent solution of sulphur dioxide in water or liquid sulphur dioxide, of which 1 c.c. weighs about 1.42 gr. at 20-25° C, or potassium meta-bisulphite and assume it to contain 50 per cent sulphur dioxide.

3. **The Desired Concentration and how it is Determined**

Syrup that has not been brought to a sufficiently high concentration is liable to ferment later on. This spoils the syrup for direct eating and drinking purposes, and worst of all, such syrup might burst its containers. It is therefore of the utmost importance to concentrate the syrup so highly that it can no longer ferment. Cruess, *l.c.* p. 416, says on this point: "Syrup should be concentrated to 68° to 70° Balling if it is to be kept without sterilisation". Ottavi,[2] *l.c.* p. 268, says the following concerning the grape syrup that has been boiled by Fratelli Favara since 1888 in vacuum pans at a temperature below 40° C. or 104° F. : "Il concentrato del Favara presenta una densitá media di 1,35 [= about 70° Brix; A. I. P.] e offre questa costituzione:

---

[1] Prof. Dr. A. I. Perold, *Ondersoekings omtrent Moskonfyt,* 1923.
[2] O. Ottavi, *Enologia teorico-pratica, 6ᵉ* ed., 1906.

	Per ettolitro.	Per quintale [=100 kg.; [A.I.P]
Glucosio (sugar)	Kg. 90.000	Kg. 66.66
Acidi (acids)	Kg. 2.400	Kg. 1.77."

Hence Favara's syrup is concentrated to about 70° Brix (Balling). Such a syrup contains 94.5 gr. extract per 100 c.c. In one of my experiments a syrup of 69.75° Brix and specific gravity $1.3494\left(\frac{17.5°C}{17.5°C}\right)$ began to ferment spontaneously. Hence 70° Brix might be enough, but in the light of my experience just quoted I consider it safest to concentrate the syrup to 71° Brix or Balling. Such a syrup has a specific gravity of $1.3572\left(\frac{17.5°C}{17.5°C}\right)$. A delicious moskonfyt made by a farmer (my father), Mr. I. S. Perold of Prince Alfred's Hamlet, in 1922, had precisely this concentration with a total acidity of 4.25‰. Syrup of 71° Brix weighs 13 lbs. 9 ozs. per gallon at 17.5° C. or 63.5° F. At 100° C. or 212° F. its specific gravity is 1.3065. Syrup showing under 70° Brix must be pasteurised or sterilised if it is to keep for an indefinite period.

How is the concentration of the boiling syrup determined?

(*a*) Directly by means of a Brix or Balling *saccharometer.* For syrup of 60-70° Brix Cruess gives the following corrections:

Temperature in		Degrees Balling or Brix to be added to Hydrometer Reading.	Temperature in		Degrees Balling or Brix to be added to Hydrometer Reading.
Degrees F.	Degrees C. [calculated; A. I. P.].		Degrees F.	Degrees C. [calculated: A. I. P.],	
64	17.8	0.0	126	52.2	2.9
72	22.2	0.3	130	54.4	3.1
75	23.9	0.4	135	57.2	3.4
82	27.8	0.7	140	60	3.7
86	30	0.8	149	65	4.2
90	32.2	1.1	158	70	4.3 (4.7? A.I.P.)
97	36.1	1.4	167	75	5.2
100	37.8	1.6	176	80	5.8
108	42.2	1.9	185	85	6.3
110	43.3	2.0	194	90	6.9
15	46.1	2.3	203	95	7.5
121	49.4	2.6	212	100	8.2

If the Balling (Brix) saccharometer reads 61.8° at 100° C, the true concentration is 61.8 + 8.2 =70° Balling.

(*b*) By means of a *pycnometer*. This is a narrow-necked glass bottle with a mark on the neck. Weigh the bottle when empty and dry, then again when filled with distilled water at 100° C. The difference between the two weights gives us the weight of water just filling the bottle at 100° C. Supposing the empty bottle weighed 20 gr., and 65 gr. when filled with water at 100° C, then the water weighed 45 gr. Since syrup of 71° Brix has a specific gravity of 1.3065 at 100° C, the bottle filled with syrup of 71° Brix at 100° C. will weigh 20 + 45 × 1.3065 = 78.8 grams. Take a balance weighing accurately to one-tenth gram, put 78.8 gr. on the side where the weights are put, fill the bottle from time to time with syrup up to the mark on the neck, and put it on the balance. As soon as the balance shows the bottle to have the weight of 78.8 gr. the syrup is at 71° Brix. Though the temperature of the syrup in the boiler is above 100° C, it will be near 100° C. when filled in the bottle.

(*c*) By means of a *thermometer* suspended in the syrup. As the syrup becomes more concentrated its boiling point rises. Syrup boiling under ordinary atmospheric pressure will have reached the concentration of 71° Brix when the boiling point has gone up to 107-108° C. or about 225° F. At 108° C. or 226.4° F. the syrup may already be somewhat too thick. The lower temperature will be applicable to large quantities of syrup boiled at a time, and the higher temperature where small quantities are boiled in a pot. These temperatures I found in boiling syrup in an open pot and in porcelain dishes in the laboratory. The pressure of the air naturally influences the boiling point. I worked at an altitude of about 370 ft.

4. Concentrating the Juice

If ordinary, more or less turbid juice is boiled in an open vessel, a thick layer of froth collects at the surface, which should be removed. Bright juice does not form froth. If boiled in an open vessel over direct fire, a fairly light coloured syrup can be obtained by boiling very rapidly. By steam heating, either direct or through a steam coil, we get a lighter-coloured syrup than when boiling over an open fire.

654

Where grape syrup is produced on a large scale it is generally boiled in a vacuum pan where the high vacuum results in a low boiling point. The higher the vacuum, *i.e.* the lower the pressure, the lower the boiling point of the liquid. In *Bevande analcooliche* the publishers, Fratelli Marescalchi of Casalmonferrato, Italy, gave a resume of some of the work done by Cruess and Monti in this connection. From this little work I took the following table, to which I have added the columns giving the degrees Fahrenheit and the vacuum in inches.

Boiling Points of Water under Diminished Pressure.							
Boiling Point.		Vacuum.		Boiling Point.		Vacuum.	
° C.	° F.	mm.	inch.	° C.	° F.	mm.	inch.
98·1	208·6	50	1·97	61·6	142·9	600	23·62
96·1	205·0	100	3·94	53·6	128·5	650	25·59
91·7	197·0	200	7·87	41·7	107·1	700	27·56
86·5	187·7	300	11·81	34·2	93·6	720	28·35
80·4	176·7	400	15·75	22·4	72·3	740	29·13
72·5	162·5	500	19·68	11·8	53·2	750	29·53

Cruess, *l.c.* p. 406, says that 24-26 in. vacuum is the degree of vacuum ordinarily used in commercial practice, when water boils at 140-125° F. and syrup about 10° F. higher, *i.e.* at 150-135° F. or 65.6-57.2° C.   He adds, however, that in order to obtain the best results the vacuum should be at least 28 in., when the syrup will boil at about 108° F. or 42.2° C.   He recommends a "dry vacuum pump" in combination with a barometric condenser system, and says further that, in addition to a good vacuum pump, a large supply of water for condensing purposes is necessary to maintain a high vacuum.   "At 28 inches vacuum (according to an estimate furnished by the Ingersoll-Rand Company) and use of a barometric condensing system, approximately 5.5 gallons[1] of water at 75° F. will be needed to condense each pound of water vapour; or over forty gallons of water for each gallon of water evaporated from the juice."   The higher the vacuum the more water is required to condense the water vapour formed in concentrating the juice.   Some vacuum pans are continuous, *i.e.* juice continually flows in and syrup out;

---

[1] One American gallon = 3.785 litre = 8⅓ lbs., hence 5.5 American gallons = 45⅚ or nearly 46lbs.

others are charged with juice which is concentrated to the desired point, when the syrup is run off and a new charge of syrup is put in.

A totally different method of concentration is applied by Gore in America and Monti in Italy. They concentrate the juice *by freezing*. The juice is frozen to a solid mass, then broken in an ice-crushing machine, and then centrifuged to separate the syrup from the ice. Gore freezes the juice at 10-15° F., and again freezes the syrup at 0-10° F. to a mushy mass of ice crystals and syrup, which is centrifuged, and then gives a syrup of 50-60° Balling (according to Cruess, *l.c.* p. 412). About 1 per cent of sugar is lost in the ice. Gore's work was done upon apple juice. Cruess repeated Gore's work with Muscat grape juice and obtained a syrup of 55° Balling. Cruess, *l.c.* p. 413, says that a 60° Balling syrup is the sweetest syrup recorded as having been produced by this process. The syrup so produced is of the highest quality, but must either be concentrated further in a vacuum pan or be blended with an extra thick syrup produced in a vacuum pan to get a finished product of at least 70° Balling. The thinner syrup will not keep unless it is pasteurised.

5. **Treatment of the Syrup**

When the syrup is run off, it must *soon be cooled* to a temperature of 100° F. or 37.8° C. or less in order to avoid caramelisation of the flavour and a darkening of the colour. According to Cruess, *l.c.* p. 410, Mr. O. S. Newman (Manager of Woodbridge Vineyard Association, Woodbridge, California) "uses a long, shallow copper pan, around the sides and bottom of which is circulated cold water. The syrup is cooled as it flows over the surface of the pan." A vertical cooler as used in breweries can also be used for cooling the syrup.

Leave the syrup for some time in a cool place to allow tartrates to crystallise out before putting it into cans or bottles. Syrup that is boiled only to 66° or 68° Balling must be sealed hot in the cans or bottles, as it otherwise is liable to ferment. If the juice had not been bright before concentration began, the syrup should be stored for some time to clarify before being bottled or canned. The great difficulty with syrup boiled thick enough (70-71° Balling) is not the crystallising out of tartrates only, but sugaring.

The sugaring of grape syrup is a great trouble which, as far as I know, occurs in winter with all grape syrups that have been concentrated sufficiently (70-71° Balling). I have not found any mention made of it anywhere. My own investigations on it are still proceeding. The results obtained thus far would seem to indicate that sugaring might be prevented by heating the syrup, after it had been stored for some months, to a temperature between 90° and 100° C, probably 100° C. or 212° F. will be the best, and then allowing it to cool slowly in the containers.

6. **Determination of the Degree of Concentration of the Syrup**

This is usually done directly by means of a *pycnometer* or *specific gravity bottle.* Weigh the bottle empty and then again when filled up to the mark with distilled water at 15° C. or at 17.5° C. Deduct the former from the latter weight, when you will get the weight of the water filling the bottle up to the mark at 15° C. or at 17.5° C. The dry, empty bottle is now filled and weighed with syrup up to the mark and at the same temperature at which it was previously filled with water. (To get the liquid in the bottle at the right temperature, leave it for 15 min. in a trough with a considerable quantity of water which has the right temperature.) By deducting the weight of the empty bottle we get the weight of the syrup it contained, and on dividing this by the weight of the water previously obtained we get the specific gravity of the syrup at 15° or at 17.5° C. From extract or sugar tables we can then immediately read off the strength of the syrup in degrees Balling or Brix. It should be remembered that *every degree Balling or Brix is one per cent by weight.* Thus 70° Balling means that such a syrup contains 70 lbs. total extract (sugar, tartrates, etc.) in 100 lbs. syrup, not 70 grams per 100 c.c.

This explains why one cannot dilute the syrup say three times, put a Balling saccharometer into it, take the reading, multiply it by three, and then imagine the result to give the correct strength of the syrup in degrees Balling.

Supposing a syrup of 70° Balling is diluted with water to three times its original volume at 15° C. Then the Balling saccharometer placed in the diluted liquid will read 28.1°, but 3 × 28.1 is not 70. A syrup of 70° Balling contains 94.5 grs. extract per 100 c.c. at 15° C.; after being diluted three times with pure water, it contains 31.5 g. extract per 100 c.c. at 15° C, and according to Windisch this corresponds with 28.1° Balling.

The *dilution method* can, however, be quite well employed as follows. Take 100 c.c. syrup at 15° C. and dilute it accurately to 300 c.c. with distilled water or rain water at the same temperature. Place the Balling saccharometer in the diluted syrup and take the reading. Let it be 28.1 (work at the temperature at which the instrument is graduated). From extract tables we find (by interpolation) that 28.1° Balling corresponds with 31.5 g. extract per 100 c.c. at 15° C. Now multiply 31.5 by 3 and we get 94.5 g. extract per 100 c.c. at 15° C. as the original amount present in our syrup, which according to the tables corresponds with 70° Balling, *i.e.* the true strength.

The following table can be used in determining the concentration of grape syrup.

UNFERMENTED GRAPE JUICE

The United States of America is the country where this kind of non-alcoholic beverage is produced and consumed on the largest scale. One is therefore not surprised to find good American publications on this subject. Of these I wish to mention the following: Hartman and Tolman, *Concord Grape Juice: Manufacture and Chemical Composition;* Charles Dearing, *Unfermented Grape Juice: how to make it in the Home;* and Cruess, *Unfermented Fruit Juices,* to which I refer the interested reader for further information.

Since yeast cells occur on ripe fruits and will set up an alcoholic fermentation in their juice once they have been crushed, every one will understand that the suppression or prevention of all fermentation is a primary essential in the manufacture of unfermented fruit juices. Where such grape juice is manufactured on a large scale it usually contains a little alcohol, which is formed in crushed berries while the grapes are being transported from the vineyard to the factory or are waiting to be treated in the factory. According to Hartman and Tolman, *l.c.* p. 4, the juice of greatly damaged

PARTS OF EXTRACT TABLE ACCORDING TO K. WINDISCH
(taken from Röttger, *Nahrungsmittel-Chemie*, 3te AufL, 1907).

Spec. Grav. at 15° C. $d\left(\frac{15°C}{15°C}\right)$	Weight per cent Sugar = degrees Balling.	Grams Sugar or Extract per 100 c.c.	Spec. Grav. At 15° C $d\left(\frac{15°C}{15°C}\right)$	Weight per cent Sugar = degrees Balling.	Grams Sugar or Extract per 100 c.c.
1.111	26.07	28.94	1.337	67.77	90.53
1.112	26.28	29.20	1.338	67.93	90.81
1.113	26.50	29.47	1.339	68.09	91.09
1.114	26.71	29.73	1.340	68.25	91.38
1.115	26.92	29.99	1.341	68.41	91.66
1.116	27.13	30.26	1.342	68.57	91.94
1.117	27.35	30.52	1.343	68.73	92.23
1.118	27.56	30.79	1.344	68.89	92.51
1.119	27.77	31.05	1.345	69.05	92.79
1.120	27.98	31.31	1.346	69.21	93.08
1.121	28.19	31.58	1.347	69.37	93.36
1.122	28.40	31.84	1.348	69.53	93.65
1.123	28.61	32.11	1.349	69.69	93.94
1.124	28.82	32.37	1.350	69.85	94.21
1.125	29.03	32.64	1.351	70.01	94.50
1.126	29.24	32.90	1.352	70.16	94.79
1:127	29.45	33.17	1.353	70.32	95.07
...	...	...	1.354	70.48	95.35
1.320	65.01	85.74	1.355	70.64	95.64
1.321	65.17	86.02	1.356	70.80	95.93
1.322	65.34	86.30	1.357	70.96	96.21
1.323	65.50	86.58	1.358	71.12	96.49
1.324	65.66	86.86	1.359	71.27	96.78
1.325	65.82	87.14	1.360	71.43	97.07
1.326	65.99	87.43	1.361	71.59	97.35
1.327	66.15	87.71	1.361	71.75	97.64
1.328	66.31	87.99	1.363	71.90	97.92
1.329	66.48	88.27	1.364	72.06	98.21
1.330	66.65	88.55	1.365	72.22	98.50
1.331	66.81	88.84	1.366	72.38	98.78
1.332	66.96	89.12	1.367	72.53	99.07
1.333	67.12	89.40	1.368	72.69	99.35
1.334	67.29	89.69	1.369	72.85	99.64
1.335	67.45	89.97	1.370	73.00	99.92
1.336	67.61	90.25			

grapes can contain as much as 0.30 grams alcohol per 100 c.c., *i.e.* 0.-4 vol. per cent alcohol or 0-7 per cent proof spirit.

In connection with the manufacture of unfermented grape juice, I wish briefly to discuss the following: the grapes, the extraction of the juice, the clarification of the juice, the first pasteurisation of the juice, the storing of the juice, the bottling and repasteurisation of the juice.

1. **The Grapes**

For a long time Concord has been the standard grape used for this purpose in North America, and probably is still the principal one used, although other varieties are also used. Of the European varieties, Muscat of Alexandria and other Muscats, Traminer, etc., can be used on account of their flavour, and blended with other varieties.   This can be done with a view to obtaining a higher total acidity (*e.g.* with Folle blanche, Burger, West's Prolific) or a darker colour (*e.g.* with Alicante Bouschet, Cabernet Sauvignon, Hermitage, Barbera). If a pretty dark red colour is wanted, black varieties will be used almost exclusively.

According to Cruess, *l.c.* pp. 12-13, it is desirable that the juice of the varieties of Eastern America (*e.g.* Concord), when harvested for this purpose, should show 17-18° Balling, whilst Muscat of Alexandria, Sémillon, Traminer, and other aromatic varieties should have reached 22-23° Balling. Varieties blended with these on account of their high acidity should be picked at about 17° Balling. Cruess states that the finished product should have a total acidity of 9-11‰ (calculated as tartaric acid).

According to Hartman and Tolman, *l.c.* p. 24, who have analysed 104 commercial juices made from Concord grapes, the average minimum total acidity calculated as tartaric acid was 8.1‰, the average maximum was 12.8‰, and the average total acidity of the 104 samples was 10.1‰, which agrees fairly well with Cruess's figures.

If the juice has too low an acidity, it can be raised by blending with juice richer in acid or by adding citric acid of which 1 gr. per lit. or 1 lb. per 100 gallons juice will raise the total acidity by fully 1‰. If citric acid is used for this purpose, the diminution in the acidity during the subsequent storage of the juice, after the first pasteurisation, will be less than if tartaric acid is used. If the juice contains too little sugar, it can be sweetened by adding cane-sugar, but then it must be stated on the label that citric acid and cane-sugar have been added.

Hartman and Tolman, *l.c.* p. 5, write as follows about Concord: "Careful air-ripening, however, mellows the fruit, develops flavour, and admits of a better condition for the process of juice manufacture than can be attained by using freshly picked fruit". They point out, however, that only sound and well-matured grapes should be used; these should be gathered in wooden crates of about 25 lbs. capacity which must not be overfilled, and continue: "For air-ripening it is of the utmost importance that the boxes of fruit should be stacked in a cool place in such a manner that air may circulate freely throughout the stack. . . . With proper care grapes may be kept for several days without particular harm to the fruit." Grapes that have to be transported a considerable distance to reach the factory, and grapes with a rather low acidity, should be crushed immediately they reach the factory in order to prevent an alcoholic fermentation and loss of acid by respiration. Dirty grapes must first be washed in clean water, which will then be allowed to drain off well, while diseased grapes should be discarded.

2. The Extraction of the Juice

The grapes are crushed in a grape-mill or crusher. With white grapes the skins and stalks remain with the juice, as the presence of the stalks makes it easier to press out the juice. With black grapes a crusher and stemmer is used to leave only the skins and juice together, as these have subsequently to be heated for extracting the colour, when the presence of the stalks would make the juice too astringent. In working at home on a small scale, the grapes can be crushed by hand or in a small hand-crusher. The stalks can be removed by hand unless the berries have been stripped off the stalks before being crushed, which will give the best results. If necessary heat in an aluminium pot.

In the factory the crushed white grapes are immediately pressed. The black grapes are pressed only after being *heated to extract colour.* This is usually done by heating the skins and juice in a steam-jacketed aluminium kettle with a stirrer to 160° F. or 71.1° C. and keeping it at

661

this temperature for a few minutes, and then immediately pressing it. During heating the mass should be well stirred to prevent local overheating.. Do not bring the heated juice in contact with iron, copper, tin, etc., but use aluminium as stated above.

The colour can also be extracted by putting the skins and juice in a press, extracting ½-⅔ of the juice by applying a slight pressure, heating the juice in a pasteuriser to 140° F., and then running it on to the pomace (skins with some juice) in a clean wooden vat where it is thoroughly mixed and repeatedly stirred. After four to eight hours the juice will have extracted sufficient colour and we proceed to press. According to Cruess this second method gives better results, as overheating is avoided.

Hartmann and Tolman, *l.c.* p. 7, state that the juice and skins should be heated to between 135 and 150° F. or 58 and 65° C. During heating the juice extracts not only colour from the skins, but also other ingredients. Hence the heated juice is richer in extract than the cold juice. The following analytical data bearing on this point I have taken from Hartmann and Tolman, *l.c.* p. 7.

CHEMICAL COMPOSITION OF HOT AND COLD PRESSED CONCORD GRAPE JUICE

Experiment.	Juice pressed before or after Heating.	Solids, per 100 c.c.	Sugar as Invert before Inversion per 100 c.c.	Non-sugar Solids, per 100 c.c.	Total Acid as Tartaric Acid per 100 c.c.	Cream of Tartar per 100 c.c.	Tannin and Colouring Matter per 100 c.c.
		gms.	gms.	gms.	gms.	gms.	gms.
1	Before	17.20	14.36	2.84	0.78	0.56	0.08
	After.	17.83	14.58	3.25	1.12	1.05	.24
2	Before	16.33	13.88	2.45	.74	.62	.07
	After.	17.25	13.62	3.63	1.01	.99	.19
3	Before	16.10	13.74	2.36	.84	.42	.06
	After.	17.17	13.74	3.43	1.16	.93	.20

From these data we see that the dry extract, total acidity, cream of tartar, and tannin and colouring matter in the juice were much increased by heating, while the sugar content remained practically unaltered.

The juice should not come in contact with iron in the press. Hartmann and Tolman, *l.c.* p. 9, write as follows about the *pressing of the heated fruit:* "The hot pulp is enveloped in strong, coarse-meshed cloths to form layers which are stacked on top of one another. These stacks or 'cheeses' as they are called, are then subjected to pressure.... As a rule, ten layers (48 × 48 in.) constitute a cheese... Before the pressure is applied the cheese should be allowed to settle. This settling period is of importance, as it allows the coarse particles of pulp gradually to shift with the flow of juice to the sides of the cloths, followed up by finer material; thus a very efficient filter for the juice is provided." Fifteen minutes is quite sufficient for settling.

3. **Clarification of the Juice**

Allow the juice to stand in a cool (below 50° F. or 10° C.) place for twelve to forty-eight hours to settle. Pour the relatively clear juice through a double thickness of cheesecloth and then through a flannel jelly-bag. Pour it two to three times through the latter until it runs off clear. Otherwise the juice from the press can be filtered at once through several thicknesses of coarse linen. Filtration under pressure can be applied as when clarifying must that is to be made into grape syrup, if a very clear juice is wanted.

4. **First Pasteurisation of the Juice**

In order to avoid a turbidity being formed later in the bottles, the juice should be heated to coagulate the albumenoids. If we wish to prevent tartrates crystallising out later in the bottles, the juice should be stored at a low temperature for several months, and in order to prevent fermentation during this period of time, it should first be pasteurised. This can be done in several ways. The juice can be heated in the same steam-jacketed aluminium kettles as before (turn on steam only after the kettles have been charged with juice!) at 176-190° F. or 80-87.8° C. for four to two minutes according to the temperature. The glass demijohns of 5-20 gals. capacity into which the pasteurised juice is poured whilst hot, are packed beforehand into a box and sterilised directly with live steam. They must be still very hot when the hot juice is poured into them, as they might otherwise crack very easily. They are filled to within a few inches from the top, and immediately closed with good corks that have been dipped in molten paraffin wax to close up their pores and to sterilise them. Pour a little molten paraffin wax over the corks when in position and allow the wax to solidify before the demijohns are removed.

Another method is to pass the juice through a pasteuriser and collect it aseptically in sterilised demijohns or cans or casks, and close these up as previously described.

5. The Storage of the Juice

Store the demijohns for a couple of months in a room which is kept at 32° F. or 0° C. as nearly as possible. Now tartrates, etc. separate out and the suspended material settles so that the clear juice can later be siphoned off. For this purpose use a rubber tube (½ in. diameter) which is attached to a bent aluminium tube (³/₈ in. diameter) that goes into the demijohn to near the sediment. Siphon off as much of the clear liquid as possible without disturbing the sediment. The residue in the demijohns is poured through a double cloth, when the juice that passed through is collected and again stored to settle and clarify. The clear liquid (that of Concord is somewhat brownish-red, more or less turbid and has a pleasant taste and aroma) is then immediately treated further. In order to accelerate the crystallisation of the tartrates the demijohns are shaken after having been in the cold store for a couple of days. To facilitate the subsequent siphoning, the demijohns are at first placed at a suitable height above the floor.

6. The Bottling and Repasteurisation of the Juice for the Trade

The siphoned juice can be poured through a double coarse linen cloth or it can be filtered bright, which is quickly done if the juice has been siphoned off carefully. Juices that settle badly can be fined in the cold store. The clear liquid is now filled into quarts or smaller bottles which are filled up to 1½ in. from the top. This space is necessary to allow for the expansion of the juice during repasteurisation. Before being filled the bottles must be sterilised by means of live steam or boiling water, but they must have cooled down when they are being filled. Also the metal capsules or the corks used to close them must first be sterilised by means of steam or boiling water for one minute just before they are used,

The closed bottles are now immediately pasteurised. Pack the bottles horizontally on the false bottom of the pasteuriser to have the corks or capsules also covered by the hot juice and sterilised. Fill the vat or kettle with water until all the bottles are covered by the water and now heat by means of steam or over an open fire until the temperature of the water has gone up to 170-172° F. or 76.7-77.8° C, and keep it at this temperature for thirty minutes. Experiments have shown that the temperature inside the bottles will differ by two degrees from that of the water. Run off the water and allow the bottles to cool inside the pasteuriser or out of it in a room free from draught.

Now the whole process is complete and the bottles can be labelled and capsuled before being sent to the trade to be sold. Before this happens, however, they must be kept at the temperature of a warm room for four weeks to see whether they remain clear and do not develop mould.

RAISINS

By raisins we mean *dried grapes*. Nowadays the term "dried grapes" is applied to the dried grapes of varieties that have in the past not been used for making raisins, and are really dried only to be turned into syrup, wine, or brandy. It is principally wine grapes that are thus dried, and this is one of the results of the introduction of total prohibition in the United States of North America. Here I shall limit myself to the true raisins of which the Muscatels, Sultanas, and Currants are the three well-known types.

A. **Muscatels.**

Under this type I wish to include, for the purposes of this discussion, not only the raisins made from the Muscat of Alexandria or Hanepoot grape, but also those made from Rosaki and similar varieties. Here we can again distinguish between two main types according as the grapes have been dried directly or only after being dipped in a lye.

(*a*) **Malaga Raisins.**

This is the world-famed type of Muscat of Alexandria raisin which is dried directly in the sun without

665

being previously dipped in a lye. Because they were produced first and on the largest scale in the neighbourhood of the important South Spanish coast town Málaga, they are known as Málagas or Málaga stalk raisins since the dried bunches are sold intact and not as loose raisins.

In passing I wish to observe that in California, Australia, and elsewhere special dehydrators or evaporators have been constructed to dry grapes and other fruits and vegetables. Drying in such a dehydrator naturally takes place in the shade, and is usually completed in eighteen to twenty-four hours. Such raisins are therefore lighter in colour than when sun-dried.

The annual export of Málaga raisins from the Málaga district amounts to about 3500 tons in boxes containing at most 22 lbs. raisins. Also in California and elsewhere malagas are now produced on a fairly large scale. At the Cape Mr. P. J. Cillie, C son ("Piet California") was the first man to produce any quantity of malagas. This was after he returned from his visit to California, and after reading *The Raisin Industry,* by Gustav Eisen in 1890, as he informed me in a letter in 1919. In 1918 and 1919 Mr. Cillie produced 12 and 10 tons respectively of first-class malagas.

Picking the Grapes.—When the grapes are very well matured, their juice should contain at least 25 per cent sugar for producing a first-class raisin. Take hold of the bunch by the stalk and cut the latter near the cane. Put the bunches on a tray without damaging the bloom. Have all bad and small berries removed by little scissors (the same as used for pre-thinning grapes) and group the bunches according to size, putting those of approximately the same size on the same tray. This is done since a small bunch will dry out quicker than a large bunch.

Since all the bunches are not ripe at the same time, begin as soon as there are sufficient well-matured bunches. The vineyard should be gone through at least three times for cutting the ripe bunches. Grapes that are not well matured do not produce a first-class raisin. Hence the bunches should be very carefully selected when picking grapes. Ripeness can be determined by testing the pressed out juice of several ripe bunches with a saccharometer, which should show 25° Balling when the grapes are well matured. This is particularly valuable when beginning to pick for the first time. Further ripeness is judged by the eye and by tasting. In some cases the grapes will be fully ripe before reaching a sweetness of 25° Balling. Then they have to be picked when fully ripe, *i.e.* when the sugar content no longer increases without the berries shrivelling. *Pick the grapes in the morning as soon as the dew is off the grapes.*

FIG. 108.—Earth floors for drying Malaga stalk raisins at Campanillas near Malaga. Original (photo 1909).

The Drying of the Grapes.—In Málaga this is done on ground floors lying side by side in rows along southerly slopes, as is seen in Fig. 108. Every floor is about 10 ft. wide and 40 ft. long, has a little brick wall at each end with a depression in the middle at the highest point in which a pole is placed towards sunset. This pole rests on the two walls and on three iron supports in between. Along the borders of each floor is a row of bricks, and one foot away on the outside a row of Spanish reeds is fixed. Towards sunset a suitable cover (sails) is put over each pole acting as ridge and tied to the reeds on each side to prevent dew falling on the drying grapes during the night. About an hour after sunrise the sails are again removed. When rain threatens the drying floors are covered in this way. On Fig. 108 the drying floors are still covered with grass as the grapes were still young during my visit on the 18th June 1909. Grapes for malagas are picked at Campanulas from about 15th August. The floors lie in a plane forming 20-30° with the horizontal. They are cleaned and hardened before the grapes are placed on them. After two to two and a half weeks the bunches are turned carefully without touching the berries with the hand. After three to four weeks the raisins are sufficiently dry to be taken up. Between the drying floors are pathways a couple of feet wide.

Generally speaking, trays are preferable to ground floors for drying fruit. They can more easily be kept clean than ground floors, are easy to stack against dew or rain, can be placed everywhere, and can be removed to the packing house with the raisins on them, thus avoiding unnecessary handling. Also the bunches can be turned by placing an empty tray on the full one and turning them through 180°. Two people can easily do this if the trays are small. Eisen speaks of trays 2' x 3' which were used in California in 1890 and can be placed between the rows of vines. Each of them takes about 18-20 lbs. of grapes. They are handy and, I believe, are still used.

Eisen rightly recommends that the bunches should be packed close together. They soon dry out so much as no longer to lie close together. On pp. 138-139 of his book Eisen says: "A tray 2 by 3 ft. may be made to comfortably hold from eighteen to twenty pounds of grapes. The first crop should be placed pretty close on the trays, not allowing any part of the tray to be visible, as the reflected heat will be too great and may injure the raisins. ... The warmer it is the closer should the bunches be packed on the trays, and on the contrary when later on in the season, or when the drying weather is unfavorable, plenty of space should be given the grapes. . . . The heat necessary and favorable for drying the grapes is different in different localities. At certain temperatures the raisins will get cooked and spoil, assume a red color, lose their sweetness, become sour and hard, and covered with large, sharply defined corrugations—signs of a very inferior or even entirely worthless raisin. I would think that from 90 to 103 degrees (Fahrenheit, A. I. P.) in the shade would be the best temperature for drying perfectly ripe and sweet Muscat grapes. When the grapes are very ripe, a much higher temperature will not injure them, while unripe and sour grapes, especially of the second crop, will burn or cook at a lower temperature than would be the proper one for ripe grapes."

In South Africa, *trays* of 3' x 9' are most commonly used. They have a drying surface of 2' 9" × 8' = 22 square ft. = nearly 2 square metres.

Tribolet[1] *l.c.* p. 21, gives the following "Plan of fruit-drying tray and list of wood for making same: Wood, 2 pieces for sides, 9' × 2" × 1½"; 2 pieces for ends, 2' 9" × 2" × 1½"; 33 boards 3' x 3" x ⅜" to cover framework of tray less 4 in. at each end left for grips or handles for working tray; 22' 4"× 1" × ½" beading to nail round top of tray to prevent grapes or prunes from rolling off; 2 pieces 8' × 1" × ½" to nail lengthwise at equal distances apart on underneath side of tray to give it stability."

The trays must be stacked towards sunset and covered on top with an empty tray to keep off dew. Soon after sunrise they are again spread out over the drying ground.

The Packing of Malagas.—Pack the dried bunches straight from the trays into the boxes or cartons in which they are to be sold. In Malaga they use ¼, ½ and whole boxes, respectively about 1, 2, 4 in. deep, × 10" × 22". About four bunches are carefully laid upon each other, each little bundle is then pressed flat by hand without damaging the blue bloom. This is done for the London market. Paris and other continental markets do not want these raisins pressed flat. Some raisins are packed in cartons. The boxes are beautifully lined with two kinds of paper and, with the finest grades, a picture of a pretty girl is put on top for an advertisement. In the box the bundles of bunches are tied separately to prevent shaking and damage. In packing, the bunches are classified into four grades according to colour and size.

Where trays are used it is a good practice to stack them about 5 ft. high when the raisins are nearly dry enough, and place an inverted empty tray over the top tray of drying grapes. In this way the raisins will dry out more evenly, and will be somewhat equalised in this respect. It

---

[1] I Tribolet, *Raisin-making in South Africa,* Union of S.A., Dept. of Agric., Bulletin No. 1 of 1916.

is of the utmost importance to know *when the raisins are sufficiently dry.* The drier should learn to judge this by experience. When the raisin is rolled between the fingers no water should come out of it, but the flesh should move easily inside the skin, hence it should still be sufficiently soft. Eisen points out that the raisin must not be too hard or too soft. Concerning raisins that have been taken up too wet, he says: "Such raisins will 'sugar' in course of time and not keep a year". Concerning raisins that have been taken up too dry, he says: "Such over-dried raisins will not again become first-class raisins; their skin will always be tough, and their colour will be somewhat inferior. If but slightly over-dried, they may be brought out by equalising. To know when the raisins are in a proper condition to be taken up is most important to every raisin-man, and he should never neglect to watch his trays early and late. Upon his good judgment and watchfulness depend the quality of his crop."

The Málaga stalk-raisin is the most expensive raisin. It has, if carefully made, a bluish colour, due to the bluish bloom covering the dark brown skin of the raisin. It has a pleasant taste and flavour differing from that of the lye raisin, and is sold principally in the Christmas trade.

(b) **Muscat of Alexandria Lye Raisins.**

Picking the Grapes.—For this purpose the grapes should also be well matured and sweet. In Denia (Spain) it is considered that the grapes should have 260-300 grams sugar per litre must for the production of first-class raisins. Do not let the grapes get over-ripe. The skins of shrivelled (overripe) berries crack with great difficulty, and they produce dark, ugly raisins. *Pick the grapes early in the morning.* Here the dew does not matter, as the grapes are steeped in lye. Early in the morning the berries are full of juice and hard, and will crack more easily and evenly when dipped in the lye than if picked later in the day, when they will have lost some of their turgor.

See that only ripe bunches are picked, and have all bad berries removed from the bunches as they are being picked. The vineyard should be picked over a number of times, as all the bunches will not be fit at the same time. When producing loose raisins it is not necessary to remove the small berries, as these are easily removed by the raisin-grader in the packing-house. It is now no longer necessary to avoid touching the bunches or to cut the stalks very long. The grapes can be cut in bushel boxes or baskets, but care should be taken not to break any berries, as this would tend to produce sticky raisins. Hence the grapes should be transported from the vineyard to the lye pots with care.

The Lye.—This is a very important matter to the producer of lye raisins. We can distinguish between lye made from plant-ash and lye made from caustic soda or potash, usually with certain additions.

The best and most exhaustive publication I have found on raisins is *Les Raisins secs en Tunisie,* by N. Minangoin and F. Couston, next to Eisen's *Raisin Industry.* According to Minangoin and Couston, Columella in his *De re rustica,* published in the first century of the Christian era, recommended the use of the ash of vine canes together with a little olive oil for making a lye for the production of raisins. According to Minangoin and Couston, *l.c.* pp. 47-48, the lye is made as follows on the Island Pantellaria: "Pour 200 litres water in a large pot and heat nearly to the boil. Now put into it 30 kg. sifted *vine cane ash,* hence 15 kg. per 100 lit. [or 15 lbs. per 10 gallons; A. I. P.]; heat until the water boils, meanwhile stirring with a stick. After a few moments the contents of the pot is poured into a wooden tub to settle. The next day the clear lye is poured or siphoned off. The sediment is thrown away and the clear liquid is used as a lye for dipping the grapes in. Before raisin-making begins, sufficient of this lye is made for the season and stored in casks." M. and C. strongly recommend this lye.

My own experiment with the ash of vine canes.—I boiled 100 g. ash with 500 c.c. water (same ratio as 20 lbs. ash per 10 gals. water) for 15 min., allowed to settle and cool, then filtered. I obtained 236 c.c. clear filtrate with an alkalinity equal to 19.2 g. Na_2CO_3 = 25.0 g. K_2CO_3 per litre, which is just about the right strength.

Lye from "Ganna" Ash.—At the Cape the "ganna" bush, *Salsola aphylla,* is burnt in the little Karroo (Robertson, Montagu, etc.), and its ash used for making lye. Usually about 1½-2 lbs. of this ash is used per gallon of water for making the lye. It is made as was described under lye made from the ash of vine canes on the Island Pantellaria.

My own investigation on "Ganna" ash from Robertson.—100 g. ash was boiled for 15 min. with 1 litre water (same ratio as 1 lb. ash per gallon of water), stirred and crushed fine, allowed to settle and cool. Then it was decanted when 840 c.c. lye A was obtained. The residue was treated with 250 c.c. water, boiled for 15 min., allowed to settle and cool, then decanted, giving 200 c.c. lye B. On analysis lye A showed an alkalinity equal to 34.49 g. Na_2CO_3 = 44.90 g. K_2CO_3 per litre, and its specific gravity at 28° C. was 1.038. Lye B showed an alkalinity equal to 13.2 g. Na_2CO_3 = 17.2 g. K_2CO_3 per litre.

On the Stellenbosch University Farm lye from "ganna" ash was made by putting 50 lbs. ash in a wooden tub, adding 25 gallons boiling water to it, stirring well and allowing to stand for two days. The clear lye was excellent, and more than sufficiently strong. Its alkalinity was equal to 69.54 g. Na_2CO_3 = 90.53 g. K_2CO_3 per litre. Here 2 lbs. ash was used per gallon of water, and the alkalinity was twice as strong as where 1 lb. per gallon was used on a small scale in the laboratory. The ash had a total alkalinity equal to 90.9 g. Na_2CO_3 = 118.34 g. K_2CO_3 per 200 g. ash, which is the quantity used for 1 lit. water when 2 lbs. ash are used to 1 gallon of water in making the lye. With the method of preparation adopted, $\frac{69.54}{90.9} \times 100$ = 76.5 per cent of the ash's alkali was thus dissolved.

According to Minangoin and Couston, *l.c.* p. 46, the following lye is used at Smyrna for dipping Rosaki and Sultana grapes:

Wood ashes (from oak or vine canes)	20 lbs.
Olive oil	1 lb.
Water	100 lbs. or 10 gals.

The olive oil is put in the clean lye in the pot. It is supposed to give the raisins a better gloss and to improve their value. In Spain and elsewhere it is not much used.

Caustic soda and *caustic potash* can also be used, but it is better to use them together with their chlorides and sulphates. Therefore we find that the *commercial lyes* usually contain only about 95 per cent caustic soda (or caustic potash) and about 5 per cent sodium sulphate and sodium chloride. Of such lyes ¾-1 lb. is used per 20 gallons water. As they are very strong, these lyes should be carefully weighed or else bought in 1 lb. tins, and the water should be accurately measured. The less caustic *potassium carbonate* can be used with greater safety. According to M. H. Leon (quoted from M. and C, *l.c.* p. 49), the following lye is used for Rosaki and Sultana grapes at Smyrna:

Potassium carbonate	6	lbs. (6 kg.)
Water	10	gals. (100 lit.)
Olive oil	⅕	gal. (2 lit.).

Sometimes the lye is made aromatic by putting into it some aromatic plant, *e.g.* lavender, rosemary. Sometimes plants are added to the lye to give the raisins a pretty yellowish, light colour. Thus in Robertson the dry, yellowish Kraal-bush is sometimes used, and elsewhere the dry branches of *Artemisia herba alba*.

Dipping the Grapes in the Lye.—About 20 lbs. of grapes are put in a basket with long handles or in a wire basket or metallic vessel with numerous holes. The grapes should be dipped for a second in clean boiling water in order to wash them before being dipped in the lye, as the lye would otherwise soon become dirty. The clean water should boil to prevent the wet grapes from unnecessarily cooling the boiling lye. After the water has drained off, the grapes are dipped in the boiling lye for at most two seconds, when they are taken out and held over the lye pot to allow most of the lye to flow back into the pot. Then they can be immediately put out on the trays, or only after being dipped into cold, running water to wash off the lye. In Denia and in many other places where lye raisins are made, this last washing is omitted, as it is considered unnecessary, and it is believed that the little lye remaining on the grapes helps to keep them sound. It is in any case too little to harm the people who eat the raisins or use them in other ways.

The grapes must not be kept too long in the lye, as they would thereby be cooked and lose their fine taste. If the berries have big and deep cracks or cuts the lye must have been too strong, and also the grapes have possibly been immersed in the lye too long. If they show no cracks at all, the lye is usually too weak. Over-ripe berries do not crack easily, whilst under-ripe berries crack very easily. The lye should simply remove the bloom from the berries and develop only very *fine cracks* or *checks* in the skins to allow the grapes to dry rapidly. Lye raisins have an amber colour, whereas the Málagas, which have retained their bloom, have a bluish colour.

If the berries burst or crack too deep, some of the juice will flow out and make the berries sticky, thus making the raisins also sticky. If everything is in order, the cracks should be very fine and very numerous. They should rather be too fine than too deep. If the lye is too strong, add water; if it is too weak, add some strong lye. After some time throw away the used lye and replace it with fresh lye of the proper strength.

The Drying-ground.—It should be so situated that no dust will blow on the drying grapes, and that the sun will shine on it during the greater part of the day. It should also be close to the lye pots and as near as possible to the vineyard. The drying-ground should have a hard surface. The trays are placed on it in regular rows. Where the grapes are dried directly on earth floors, the surface should be rendered particularly hard and clean. It is, however, much better to dry the grapes on trays, especially when making lye raisins.

The Drying of the Grapes.—After two to three days the grapes have usually dried out sufficiently to be turned. After five days they are usually dry enough to be taken up. It is of the greatest importance to take up the raisins when they are just dry enough. What was previously said with regard to this point under Málaga raisins applies also here. By stacking the trays one to two days before the raisins will be sufficiently dry, we get lighter-coloured and more evenly dried raisins than when they are allowed to dry in the sun until they are fit to be taken up.

Taking up the Raisins.—Remove bad bunches and berries, especially badly-coloured berries, from the trays before taking up the raisins. No machine can grade raisins according to colour; it does so only according to size of berry. The underripe berries produce reddish raisins and the over-ripe ones dark-coloured raisins. Both kinds should be removed by hand and kept separate from the rest. The reddish raisins are better not sold, as they have a fairly sour taste and constitute a bad commercial article.

674

The raisins should not be put in ordinary bags, as the loose fibres of the latter stick to the raisins and are difficult to remove later. It is much better to pack the raisins in clean boxes and deliver them as soon as possible to the nearest co-operative dried fruit company or factory. Every raisin producer of any importance ought to be a member of some co-operative association.

Packing and selling Raisins.—Raisins are treated and packed in special packing-houses. What is not sold as stalk raisins is put through the stemmer and grader, which removes the stalks and grades the raisins according to size. If necessary, these raisins are further put through machines that wash and clean them. The raisins are then "placed in 'sweat-boxes', holding about 60 lbs., or in heaps on a clean floor, for ten days or so, to undergo a sweating or evening-up process. During this sweating time those on the floor are turned over a few times with a wooden shovel, and those in the sweat-boxes are changed into other boxes two or three times. This tends to give them uniformity of dryness and texture. After this they may be packed according to grade, in parcels of size and get-up most suitable for the trade" (Tribolet, *l.c.* p. 23).

To be exported from the Union of South Africa raisins must be packed in boxes holding 25 lbs. raisins. These must not contain more than 15 per cent moisture, and all the raisins in the same box must have been made from one and the same variety of grape, which can be Muscat of Alexandria (Hanepoot), Waltham Cross, Rosaki (which is the same as Waltham Cross), Sultana, Thompson's Seedless (same as Sultana), or any other variety placed on this list from time to time by the Minister of Agriculture.

The small seedless berries of Muscat of Alexandria are separated from the rest by the grader and sold as seedless raisins. The seeds from the ordinary Muscat or other raisins can be extracted by special machines known as seeders. It would be a great step forward if some one were to succeed in producing a seedless, large-berried Muscat variety of grape, both, for table use and for the production of seedless raisins.

675

B. **Sultanas**

What has previously been said about ripeness, picking of grapes, lye, etc., under Muscat lye raisins applies here also. The dipped Sultana grapes *must,* however, be dipped in clean running water after coming out of the lye, as any remaining lye will hinder the action of the sulphur dioxide in the sulphuring chamber. The grapes are now put on trays, which are stacked in a special room or enclosure *to be sulphured.* The trays should be stacked in such a way as to allow free passage of air between them. Burn 1 lb. of sulphur for every 100-150 cubic feet of air in the sulphuring chamber. The best plan is to lay a trolley line into the chamber, put a low trolley on the line and stack the trays carefully on the trolley outside the chamber. Push in the trolley with the trays when loaded to a height of 5-6 ft. Now put a light to the sulphur and close the door at once. Construct the chamber of concrete (roof, walls and floor), say 10 ft. long by 4(-5) ft. wide and 8 ft. high, and the door of wood, which should be air-tight and should give a tight fit. Leave the grapes for a couple of hours in the sulphuring chamber, then take out the trolley, and run or take the trays to the drying-ground, where they are put on the ground as was previously described.

Turn the bunches on the second day and once more after that. By stacking the trays before the sultanas are sufficiently dry we get a lighter-coloured sultana than if dried in the sun to a finish. The colour is still lighter if they are dried in the shade after being exposed to the sun for a day or two. Other qualities being about equal, the light-coloured sultana fetches a higher price than the dark-coloured sultana.

By sulphuring the grapes they are bleached and a lighter-coloured sultana is obtained, but it thereby loses a good deal of its flavour and has a more acid and less sweet taste. It would be ideal to dry sultanas in a dehydrator, or at all events only in the shade and without previous sulphuring. This would certainly improve their flavour and taste, while still giving a fairly light-coloured sultana. Meanwhile it is still the fashion to demand a light-coloured sultana, which explains why most Sultana grapes are sulphured before they are dried.

Like Muscat raisins the sultanas are passed through stemmers and graders to be freed from stalks and graded.

C. **Currants**

At the Cape most currants have so far been made from the Cape Currant grape which has been described in Chap. V. as a small-berried, seedless Red Muscadel. It has a distinct Muscat flavour not possessed by the Grecian or Zante Currant grape. The world's supply of currants is almost exclusively obtained from the Zante Currant grape.

As soon as the grapes are fully matured the bunches are cut and put immediately on trays to dry in the sun. They should not be allowed to dry on the vine, as the dry currants easily drop and get lost. The grapes can be dried on sails, fine netting, or trays. It would appear that the best quality is obtained by drying in the shade, but then the weather must be hot and dry, and remain so. For this purpose the trays can be stacked after two days, strips of wood or something else being put between the trays to allow the air to circulate freely between them. Turn the bunches a couple of times if they lie on a sail or on trays in order to let them dry more rapidly and more evenly.

As soon as the currants have dried out sufficiently they must be taken up. The currants can be rubbed off the stems and cleaned in the wind, or better still by means of proper machinery.

Currants are used principally in making cakes, but can also be used for making syrup, wine, vinegar, brandy, etc. When the phylloxera had destroyed the greater part of the French vineyards and there was an acute shortage of wine in France, very large quantities of Grecian currants were imported into France to be manufactured into wine. This, however, was stopped years ago. In the *California Grape Grower,* vol. vi., March 1925, J. K. Kromidakis, M.S., published an article on "History of Greece's Currant Industry", and stated that France imported 70,000 tons of currants from Greece in 1889-1890, but that this has stopped completely since 1897.

At the present time, according to Kromidakis, there are about 150,000 acres in Greece under currant vines, and the annual crop amounts to from 120,000 to 135,000 tons of dried currants. Greece is still the world's main producer of currants.

BIBLIOGRAPHY

(LIST OF PUBLICATIONS QUOTED AND REFERRED TO IN THIS BOOK)

1. RAVAZ, L., Recherches sur la culture de la vigne, 1909. Coulet et fils, Montpellier.
2. BABO u. Mach, Handbuch des Weinbaues, Erster Halbband, 4. Aufl., 1923. Paul Parey, Berlin.
3. GULLON, J.-M., Étude générale de la vigne, 1905. Masson et Cie, Paris.
4. OTTAVIO-MARESCALCHI, I principii della viticoltura, 1909. Tip. Lit. C. Cassone. Casale Monferrato.
5. MŰLLER-THURGAU, Die Rebenknospe. Schweizer. Zeitschrift f. Obst- u. Weinbau, 1892, p. 11; also in Weinbau u. Weinhandel, 1892, p. 63.
6. BEHRENS, J., Aufbau u. Wachstum des Rebensprosses. Weinbau u. Weinhandel, 1897, p. 437.
7. HEDRICK, U. P., The Grapes of New York, 1908. Albany, J. B. Lyon, State Printers.
8. MAS et PULLIAT, Le Vignoble, 3 vols., 1874-1879. G. Masson, Paris.
9. PACOTTET, P., Vinification, 1904. J.-B. Baillière et fils, Paris.
10. OUDEMANS EN DE VRIES, Leerboek der Plantenkunde, 4e druk, 1906. Haarlem, H. D. Tjeenk Willink en Zoon.
11. STRASBURGER'S Text-book of Botany, re-written by Fitting, Schenck, Jost, Karsten. 5th English edition by Lang, 1921. Macmillan and Co., Ltd., London.
12. RATHAY, Die Geschlechsverhältnisse der Reben und ihre Bedeutung für den Weinbau. Vienna. Part I., 1888 ; Part II., 1889.
13. MACH, Die Geschlechsverhältnisse der Reben und ihre Bedeutung für den Weinbau. Tiroler Landw. Blatter, 1888, Nr. 13 and 14.
14. PORTELE, Studien über die Entwicklung der Trauben-beere. Mitt. a. d. Laboratorium der landw. Landesanstalt in San Michele, 1883.
15. HAMBÖCK, C, Annalen der Oenologie, Bd. VIII.
16. TAMARO, Prof. Dott. D., Uve da tavola. 4th ed., 1915. Ulrico Hoepli, Milano.
17. F. HOUDAILLE et J.-M. GUILLON, Contribution à l'etude des pleurs de la vigne.
18. HALES, La Statique des végétaux.
19. WIELER, Das Pluten der Pflanzen. Cohns Beiträge zur Biologie der Pflanzen, 6, 1892, p. 1.
20. F. HOUDAILLE et J.-M. GUILLON, Recherches sur l'action exercée par l'absorption de différents liquides sur la végétation de la vigne.
21. CANSTEIN, Über das Tränen oder Bluten der Weinstöcke im Frühjahre. Annal. d. Oenologie, IV. p. 517.

22. NEUBAUER, Untersuchung des im Frühjahr aus den frisch geschnittenen Reben ausfliessenden Saftes, der sogenannten Rebtränen. Annal. d. Oenol. IV. p. 499.
23. MILLARDET, A., Essai sur l'hybridation de la vigne.
24. MACAGNO, Recherches sur les fonctions des feuilles de la vigne. Comptes rendus LXXXV., 1877, p. 810.
25. J. LEWIS, M.A., The Development of the Grape. Agric. Journ. Dept. of Agric, Cape of Good Hope, Nov. 1910.
26. GRÜNHUT, Dr. L., Die Chemie des Weines. In Sammlung chem. u. chem.-techn. Vorträge, herausgegeben von Prof. Ahrens, Bd. II., Stuttgart, 1897. Ferdinand Enke.
27. VIALA-VERMOREL, Traité générale de viticulture. Ampélographie, 7 vols. 1901-1910. Masson et Cie, Paris.
28. MŰNTZ, Les Vignes, 1895. Berger-Levrault et Cie, Paris et Nancy.
29. CZAPECK, F., Atmung der Pflanzen. In Handwörterbuch der Naturwissenschaften in 10 Bde., 1912-1915. Vol. I. p. 716, 1912. Gustav Fischer, Jena.
30. GIRARD et LINDET, Recherches sur la composition des raisins des principaux cépages de la France. Revue de Viticulture, IV. pp. 317, 341; VI. pp. 173, 201, 225, 249.
31. MACH, Die Gärung und Technologie des Weines. Wien, 1884.
32. JACQUEMIN, E., Développement des principes aromatiques par fermentation en présence de certaines feuilles. Comptes rendus CXXV., 1897, p. 114; and Nouvelles observations sur le développement des principes aromatiques par fermentation alcoolique en présence de certaines feuilles. Comptes rendus CXXVIII., 1899, p. 369.
33. WINDISCH, K., Die chemischen Vorgange beim Werden des Weines, 1905. Friedrich Find. Plieningen.
34. OTTAVI-MARESCALCHI, I residui della vinificazione, 1901. Tip. Lit. C. Cassone. Casale Monferrato.
35. MŰLLER-THURGAU, H., Die Edelfäule der Trauben. Landw. Jahrbücher, 1888.
36. LABORDE, J., Cours d'oenologie, 1908. Feret et fils, Bordeaux.
37. KÖVESSI, M. F., Recherches biologiques sur l'aoûtement des sarments de la vigne, 1901. Imprimerie Le Bigot Frères. Lille.
38. RAVAZ, L., Influence des opérations culturales sur la végétation et la production de la vigne, 1909. Coulet et fils, Montpellier.
39. MOLON, Prof. G., Ampelografia, 2 vols., 1906. Ulrico Hoepli, Milano.
40. FOËX, G., Cours complet de viticulture. 4th ed., 1895. C. Coulet, Montpellier; G. Masson, Paris.
41. RAVAZ, L., Les Vignes américaines, Porte-greffes et Producteurs-directs, 1902. Coulet et fils, Montpellier; Masson et Cie, Paris.
42. HELBLING, S., Beschreibung der in der Wiener Gegend gemeinen Weintrauben. Prague, 1777.
43. PULLIAT, V., Mille variétiés de vignes. 3rd ed., 1888. C. Coulet, Montpellier; Delahaye et Lecrosnier, Paris.
44. GOETHE, H., Ampelographisches Wörterbuch, 1876. Fösy und Frick, Wien.
45. PEROLD, Dr. A. I., De Vernieuwing van Wingerden in Kalkgronden op geschikte Amerikaanse Stokken. Unie van Z.A. Landbouw Dept., No. 48, van 1913.
46. PEROLD, Dr. A. I., and CRAWFORD, D. C, Some Preliminary Investigations into the Chemical Composition of certain Vineyard Soils in the Montagu and Robertson districts. S.A. Journal of Science, June 1915
47. CHANORIN, E., Viticulture moderne, 1908. Librairie Hachette, Paris.

48. TELEKI, Andor, Die Rekonstruktion der Weingärten. Wien und Leipzig, 1907. Hartlebens Verlag.
49. PEROLD, Dr. A. I., and TRIBOLET, L, American Stocks for Cape Vineyards. Report of an inquiry into the suitability of the American stocks on which the Vineyards in the Cape Province have thus far been reconstituted. Agric. Journ. for Union of S.A., July and August, 1912.
50. CORNU, Max, Études sur le *Phylloxera vastatrix*. Memoires présentés par divers savants à l'Académie des Sciences de l'Institut National de France, tome XXVI., No. 1, pp. 1-357. MDCCCLXXVIII. Imprimerie nationale, Paris.
51. MILLARDET, A., in Comptes rendus de l'Academie des Sciences, 1878; and Pourridié et Phylloxéra (1882); and Altérations phylloxériques, in Revue de Viticulture, 1898, Part II. pp. 692-698, 717-722, 753-58.
52. RAVAZ, L., Étude sur la résistance phylloxérique. Rev. de Vitic. VII. (1897), pp. 109-114, 137-142, 193-199, and VIII. (1897) 688-694.
53. PETRI, Dr. L., Nodositätenbildung auf den Rebenwurzeln durch die Reblaus in sterilisiertem Mittel. Centralblatt für Bakt., Abt. II., Bd. 24 (1909), pp. 146-154.
54. PETRI, Dr. L., L'Acidité des sucs et la résistance phylloxérique. Rev. de Vitic. XXXV., pp. 487, 505, 544.
55. COMES, Prof., Del fagiuolo comune (Phaseolus vulgaris). Napoli, 1909, and Giornale di Viticoltura ed Enologia. Avellino, XVII. (1909).
56. AVERNA SACCA, Dr., L' acidita dei succhi nelle viti americane in rapport alia loro resistenza alia filossera. Giorn. di Vitic. ed Enol. Avellino XVII. (1909), p. 350.
57. DANIEL, L., La Question phylloxérique, le greffage et la crise viticole, 1908-1911 (3rd fascicule appeared only in 1919). G. Gounouilhou, Bordeaux.
58. P. VIALA and L. RAVAZ, American Vines: Adaptation, etc. Translated by Dubois and Wilkinson. Melbourne, 1901, from "Les Vignes américaines," Montpellier, 1892.
59. PLANCHON, Les Vignes américaines. Montpellier, 1875.
60. P. J. CILLIE, A. I. PEROLD, S. W. VAN NIEKERK, Report of Commission on Grafted Vines. Union of S.A. Agric. Journal, Sept. 1920.
61. LEROUX, S., La Vigne et le vin en Algérie et en Tunisie, 2 vols., 1894. Imprimerie admin, et comm. Heintz, Alger.
62. MARÉS, Henri, Description des cépages principaux de la région méditerranéenne de la France, 1891. Coulet et fils, Montpellier.
63. STRUCCHI, A., I migliori vini d' Italia, 1908. Ulrico Hoepli, Milano.
64. VILLA MAIOR, Douro illustrado, 1876.
65. HOGG, Dr. Robert, The Fruit Manual. 5th ed., 1884. Journal of Horticulture Office, London.
66. CLAASSEN, HAZELOOP EN SPRENGER, Leerboek voor de Fruitteelt, 2ᵉ druk, 1917. W. E. J. Tjeenk Willink, Zwolle.
67. RAVAZ, L., Recherches sur le bouturage de la vigne. In Comptes rendus de l'Acad. des Sc., Sept. 15, 1890.
68. BALTET, CH., L'Art de greffer, 8th ed., 1907. Masson et Cie, Paris.
69. DANIEL, L., La Théorie des capacites fonctionnelles et ses conséquences en agriculture, 1902. Imprimerie Fr. Simon, Succʳ. d'A. le Roy, Rennes.
70. FONZES DIACON, Revue de vitic., LIX. p. 231 (1923).
71. RAVAZ, L., Traité général de viticulture IIIᵐᵉ. partie, tome III. Le Mildiou, 1914. Coulet et fils, Montpellier; Masson et Cie, Paris.

72. MOLISCH, Prof. HANS, Pflanzenphysiologie als Theorie der Gärtnerei, 5th ed., 1922. G. Fischer, Jena.
73. SORAUER, P., Handbuch der Pflanzenkrankheiten, 3rd ed. 3 vols., 1909. Paul Parey, Berlin.
74. BABO u. MACH, Handbuch des Weinbaues. Zweiter Halbband, 4th ed., 1924. Paul Parey, Berlin.
75. PORTES et RUYSSEN, Traité de la vigne et de ses produits, 3 vols., 1886-1889. Octave Doin, Paris.
76. MOLZ, Dr. EMIL, Untersuchungen über die Chlorose der Reben. Centralblatt fur Bakt., Abt. II., Bd. XIX., 1907.
77. MÜLLER-THURGAU, H., Kernlose Traubenbeeren und Obstfrüchte. Landw. Jahrb. d. Schweiz, 1908.
78. RAVAZ, L., La Brunissure de la vigne, 1904. Coulet et fils, Montpellier; Masson et Cie, Paris.
79. VIALA, P., Les Maladies de la vigne, 1893. C. Coulet, Montpellier; G. Masson, Paris.
80. PANTANELLI, E., Roncet, in La Viticoltura moderna, anno XVII. (1911), nos. 10 and 11.
81. PIERCE, N. B., The California Vine Disease. U.S. Dept. of Agric. Division of Vegetable Pathology, Bull. No. 2, 1892.
82. RAVAZ, L., Influence de la surproduction sur la végétation de la vigne, 1906. Coulet et fils, Montpellier.
83. MARÈS, H., La Maladie de la vigne, 1856.
84. MACH, E., Zur Frage über die Art und Weise in welcher Weise der zur Bekämpfung des Oidiums angewendete Schwefel wirkt. Weinlaube, 1884, and Tiroler landw. Blätter, 1879, and May 14, 1884.
85. HEDRICK, U. P., Manual of American Grape-growing, 1919. The Macmillan Company, New York,
86. ISTVANFFI, G. de, Études sur le rot livide de la vigne *(Coniothyrium diplodiella)*. Annales de l'Instit. centr. ampélog. royal hongrois. Tome II., 1902.
87. MÜLLER-THURGAU, H., Der rote Brenner des Weinstocks. Centralblatt für Bakt., Abt. II., Bd. X. (1903), pp. 1-38.
88. BARON CARL VON BABO, Reports on Viticulture in the Cape Colony, 1885 and 1886.
89. BÖRNER, C, Experimenteller Nachweis einer biologischen Rassendifferenz zwischen Rebläusen aus Lothringen und Südfrankreich. *Peritymbia (Phylloxera) vitifolii pervastatrix,* 1910. Zeitschr. f. angew. Entomologie, 1914, Bd. I., p. 59.
90. OTTAVI, O., , Viticoltura teorico-pratica, 4th ed., 1922.Fratelli Ottavi. Casale Monferrato.
91. LOUNSBURY, C. P., The Calandra of the Vine *(Phlyctinus callosus,* Bohem). Agric. Journ. C. G. H., vol. 37 (Oct. 1910), pp. 448-450.
92. MÜLLER-THURGAU, H., Die Milbenkrankheit der Reben (Verzwergung, Courtnoué, Kräuselkrankheit, etc.). Centralblatt f. Bakt., Abt. II., Bd. XV.
(1905), pp. 623-629.
93. SCHNEIDEWIND, Prof. Dr. W., Die Ernährung der landwirtschaflchen Kulturpflanzen, 5th ed., 1922. Paul Parey, Berlin.
94. Handwörterbuch der Naturwissenschaften, 10 vols., 1912-1915. Gustav Fischer, Jena.
95. MÜNTZ, Recherches sur les exigences de la vigne. Paris, Comptes rendus des séances de l'Acad. d. Sciences, 1895, seance du 4 mars

96. MICHAUT ET VERMOREL, Les Engrais de la vigne, 1905. Coulet et fils, Montpellier; Masson et Cie, Paris.
97. PEROLD, Dr. A. I., Manuring of Vines. Lectures given at Paarl on May 16 and 30, 1911. Agric. Journ. of Union of S.A., July and August 1911.
98. LYON, FIPPIN, BUCKMAN, Soils, their Properties and Management. 1915. The Macmillan Company, New York.
99. RUSSELL, Dr. E. J., Soil Conditions and Plant Growth. 3rd ed., 1917. Longmans, Green and Co., London.
100. ZACHAREWICZ, ED., Expériences sur les engrais appliqués à la culture de la vigne, 1900. Coulet et fils, Montpellier; Masson et Cie, Paris.
101. PEROLD, DR. A. I., The Establishment and Cultivation of a Vineyard. Agric. Journ. of Union of S.A., Jan.-March 1913. Reprint No. 5 of 1913.
102. MAROGER, E., La Goutte d'eau, 1922. Sociéte générate d'imprimerie et d'édition, Paris.
103. D'ARMAILHACQ, A., De la culture des vignes dans le Médoc. 3rd ed., 1867. P. Chaumas, Bordeaux.
104. RAVAZ, L., Taille hâtive ou taille tardive? 1912. Coulet et fils, Montpellier.
105. BIOLETTI, F. T., Vine Pruning in California. Part I., Bull. No. 241; Part IL, Bull. No. 246 (Oct. 1914). University of California Publications. College of Agriculture, Agric. Expt. Stat., Berkeley, California.
106. PERRAUD, J., La Taille de la vigne. 3rd ed., 1905. Coulet et fils, Mont pellier; Masson et Cie, Paris.
107. PEROLD, Prof. Dr. A. I, "Wynbou" in Populair Wetenskaplike Leesboek, Vol. 5, 1920. Nasionale Pers Bpk., Kaapstad.
108. WAGNER, Dr. PAUL, Forschungen auf dem Gebiete der Weinbergdüngung. Arbeiten der Deut. Landw. Gesellschaft, Heft 124, Feb. 1907.
109. QUINN, GEO., Fruit Tree and Grape Vine Pruning. 5th ed., 1915. Govt. Printer, Adelaide.
110. California Grape Grower, Vol. V., No. 5, p. 2, May 1924,
111. PÉE-LABY, E., La Vigne nouvelle. 2nd ed., 1923. J.-B. Bailltere et fils, Paris.
112. PEROLD, Dr. A. I., Untersuchungen über Weinessigbakterien. Centralblatt für Bakt., Abt. II., Bd. 24 (1909), pp. 13-55.
113. RUILOPEZ, A. P., La fabricación doméstica del vinagre. Madrid, 1908.
114. THORPE, SIR EDWARD, Alcoholometric Tables, 1915. Longmans, Green and Co., Ltd., London.
115. HASSAK, Paul, Gärungsessig. 1904.
116. PEROLD, Prof. Dr. A. I., Ondersoekings omtrent Moskonfyt. Annals of Univ. of Stellenbosch, April 1923.
117. CRUESS, W. V., Commercial Production of Grape Syrup. Bull. No. 321, May 1920. Univ. of California Publications. College of Agriculture, Agric. Expt. Stat., Berkeley, California.
118. VENTRE, J., Les Utilisations possibles de la vendange en dehors de la production proprement dite du vin, 1921. Coulet et fils, Montpellier.
119. OTTAVI, O., Enologia teorico-pratica. 6th ed., 1906. Tip. Lit. C. Cassone. Casale Monferrato.
120. CRUESS, W. V., and MONTI, E., Bevande analcooliche. Edit. Fratelli Marescalchi. Casale Monferrato.
121. HARTMANN, B. G., and TOLMAN, L. M., Concord Grape Juice: Manufacture and Chemical Composition. United States Dept. of Agrio., Bull. No. 656.May 8, 1918.

122. DEARING, CHARLES, Unfermented Grape Juice: how to make it in the Home. Farmers' Bull. 1075, U.S. Dept. of Agric., Oct. 1919; reprinted Sept. 1920.

123. CRUESS, W. V., Unfermented Fruit Juices. Univ. of California, College of Agriculture, Agric. Expt. Station, Circular No. 220, July 1920.

124. EISEN, GUSTAV, The Raisin Industry, 1890. H. S. Crocker and Co., San Francisco.

125. TRIBOLET, I., Raisin Making in South Africa, Union of S.A., Dept. of Agric. No. 1, 1916. p

126. MINANGOIN, N., et COUSTON, F., Les Raisins sees en Tunisie, 1907. Imprimerie moderne (J. Orliac). Tunis.

127. J. K. KROMIDAKJS, M.S., History of Greece's Currant Industry. California Grape Grower, Vol. VI. No. 3, March 1925. San Francisco.

128. BONNET, L. O. (Univ. of California), New Facts about Girdling. California Grape Grower, Vol. VI. No. 4, p. 2, April 1925. San Francisco.

129. BASSERMANN-JORDAN, Dr. FRIEDRICH, Geschichte des Weinbaues, 3 Bde., 1907. Heinrich Keller, Frankfurt-am-Main.

130. GODÉE MOLSBERGEN, Dr E. C, De Stichter van Hollands Zuid-Afrika Jan van Riebeeck, 1618-1677, 1912. S. L. van Looy, Amsterdam.

131. TAMARO, D., Uve da tavola, 1915. Ulrico Hoepli, Milano.

132. CANDOLLE, A. de, L'Origine des plantes cultivées. 3rd ed., 1886. Félix Alcan, Paris.

ALPHABETICAL LIST OF GRAPE SPECIES AND VARIETIES

A. AMERICAN STOCKS AND GRAPE VARIETIES

B. EUROPEAN GRAPE VARIETIES (VITIS VINIFERA)

690

GENERAL INDEX

Grading of table grapes, requirements for, 513
Graft-hybrids, 395, 417
Grafting by approach, 378
   „   on the spot, 387
Grafting wax, 377
Granite, 20
Grape berry, 53, 54
Grape juice, composition of, 137
Grape-seed oil, 139
Grape seeds, composition of, 138
Grape syrup, 647
Graufäule, 479
Green-grafting, 381
Green manuring, 532
   „   „ some quantitative data on, 533
Green of Montpellier, 475
Grey rot, 467, 479
   „   „ control of, 482
Growing period of the vine, 147
Guard cells, 72
Guignardia Bidwellii, 477
Guyot system of pruning, 573
Gynaecium, 51

Hail, 431
Hail cannons, 431
Hair roots, 29
Hard bast, 91
Healthy, definition of, 423
Heliotropism, 31
Heterodera radicicola, 495
Heu und Sauerwurm, 504
Hilum, 59
Hybrides greffons, 353
Hybrides producteurs directs, 353
Hydrotropism, 31

Icterus, 434
Integument, 51
Intercellular spaces, 68, 69
Intercutis, 84
Intermediate graft, 401
Internode, 32, 40
Intine, 101

Krompokkel, 507

Lamella, primary or middle, 67
„   secondary or thickening, 67
Lamellae, 66
Large green caterpillar, 507
Lateral, 35
Leaf, asymmetry of, 169
   „ bullate ("bullee"), 174
   „ cordate, 169
   „ crinkled ("gaufrée"), 173
   „ downy ("duveteuse"), 174
   „ entire, 171
   „ felt-like ("cotonneuse"), 174
   „ five-lobed, 171

Leaf, folded ("tourmentée") 173
   „ glabrous, 174
   „ "pubescente" 174
   „ reniform, 169
   „ round or orbicular, 169
   „ trilobed, 171
   „ truncated, 169
   „ undulating ("onduée"), 174
   „ wedge-shaped or cuneiform, 169
   „ woolly or tomentose, 174
Leaf-blade, 45
Leaf-skeleton, 45
Leaf-stalk, 45
Lederbeerenkrankheit, 465
Lenticel, 95
Leucoplasts, 65
Life of the grafted vine, 399
Lime, 535
Lime chlorosis, 435
Limestone, 24
Longevity of the grafted vine, 404
Lye for raisin-making, 671
Lye raisins, 670
Lyes, commercial, 672

Macroconidia, 472
Magarodes capensis, 508
Malaga raisins, 665
Malagas, packing of, 669
Mally's Fruit Fly Bait, 502
Manuring, a system of, 528
Manuring of vines, 521
Maroger system of pruning & trellising, 586
Mealy Bug, 496
Mediterranean Fruit Fly, 501
Medoc system of pruning, 574
Medullary rays, 88
   „   „   primary, 88
Meristem, 65
Meristematic cells, 65
Mesophyll, 45, 97
Micropyle, 51
Mildew downy, 465
   „   powdery, 454
Mildiou, 465
Millerandage, 122, 439
Mixed grafting, 403
Moskonfyt, 647
Mycelium, 457
Mycoderma aceti, 638

Nectaries, 50
Nematodes, 495
Nerves, 97, 98
Noble rot, 141, 479
Nodes, 32, 40
Non-setting, 121
Nucleolus, 63
Nucleus, 63

Oenophthira pilleriana, 502

694

695

THE END

A Year in Paarl with A I Perold: *Vine and Wine Experiments 1916*
by **Peter F May** and **A I Perold**

Contains the complete unabridged text of *"Some Viticultural and Oenological Experiments conducted at the Paarl Viticultural Experiment Station during 1915-1916"* by Dr A I Perold, issued as a pamphlet by the Government Printer in Pretoria in 1916. It is Dr Perold's annual report as Government Viticulturist for the Union of South Africa on his work at the viticultural research station at Paarl, near Cape Town, South Africa.

May puts the work into context with an introduction, biography of Dr Perold, explanatory notes, photographs, glossary and index.

Paperback: 88 pages
ISBN-13: 978-0956152312
Publisher: Inform & Enlighten Ltd, 2011

*

PINOTAGE: *Behind the Legends of South Africa's Own Wine*
by **Peter F May**

Peter May investigates various legends and myths about the origins and parentage of Pinotage and travels to four continents to interview Pinotage winemakers and winery owners. The book details how Pinotage is grown, made and marketed, profiles its creator and early pioneers, covers its history and finds the oldest vineyards.

As well as Pinotage in South Africa, May provides a comprehensive review of Pinotage in other countries. This is the first book on Pinotage.

Paperback: 248 pages
ISBN-13: 978-0956152305
Publisher: Inform and Enlighten Ltd, 2009

Also available as an eBook from
Amazon for Kindle, Apple for the iPad, Barnes & Noble for the Nook
*
"May is an oenologist of some distinction" – Satisfaction magazine

www.ingramcontent.com/pod-product-compliance
Lightning Source LLC
Chambersburg PA
CBHW021804270326
41932CB00007B/48